Studies in Systems, Decision and Control

Volume 271

The series "Studies in Systems, Decision and Control" (SSDC) covers both new developments and advances, as well as the state of the art, in the various areas of broadly perceived systems, decision making and control–quickly, up to date and with a high quality. The intent is to cover the theory, applications, and perspectives on the state of the art and future developments relevant to systems, decision making, control, complex processes and related areas, as embedded in the fields of engineering, computer science, physics, economics, social and life sciences, as well as the paradigms and methodologies behind them. The series contains monographs, textbooks, lecture notes and edited volumes in systems, decision making and control spanning the areas of Cyber-Physical Systems, Autonomous Systems, Sensor Networks, Control Systems, Energy Systems, Automotive Systems, Biological Systems, Vehicular Networking and Connected Vehicles, Aerospace Systems, Automation, Manufacturing, Smart Grids, Nonlinear Systems, Power Systems, Robotics, Social Systems, Economic Systems and other. Of particular value to both the contributors and the readership are the short publication timeframe and the world-wide distribution and exposure which enable both a wide and rapid dissemination of research output.

** Indexing: The books of this series are submitted to ISI, SCOPUS, DBLP, Ulrichs, MathSciNet, Current Mathematical Publications, Mathematical Reviews, Zentralblatt Math: MetaPress and Springerlink.

More information about this series at http://www.springer.com/series/13304

Martin Steinberger · Martin Horn ·
Leonid Fridman
Editors

Variable-Structure Systems and Sliding-Mode Control

From Theory to Practice

Springer

Editors
Martin Steinberger
Institute of Automation and Control
Graz University of Technology
Graz, Austria

Martin Horn
Institute of Automation and Control
Graz University of Technology
Graz, Austria

Leonid Fridman
Control Engineering and Robotics
Department
National Autonomous University of Mexico
Mexico City, Distrito Federal, Mexico

ISSN 2198-4182 ISSN 2198-4190 (electronic)
Studies in Systems, Decision and Control
ISBN 978-3-030-36620-9 ISBN 978-3-030-36621-6 (eBook)
https://doi.org/10.1007/978-3-030-36621-6

This Springer imprint is published by the registered company Springer Nature Switzerland AG
The registered company address is: Gewerbestrasse 11, 6330 Cham, Switzerland

Dedicated to
Stefanie, Georg and Klemens
Astrid, Ilona and Lorenz
and Millie

Preface

Variable-structure systems (VSS) and sliding-mode control (SMC) and observation are known to be one of the most efficient robust control and observer design techniques. Especially the ability to completely reject matched perturbations, i.e., perturbations that act in the input channels, is a stand-alone feature of these approaches compared to other robust methods.

In the last years, research was pushed further leading to new algorithms and important extensions of existing methods and more and more applications exploit the nice properties of sliding-mode techniques. This was also evident at the 15th International Workshop on Variable Structure Systems and Sliding Mode Control (VSS 2018) which was held at Graz University of Technology, Austria. The conference featured five sessions focusing on sliding-mode theory and four sessions on applications based on submissions from 30 countries.

The present book covers theoretical and practical aspects related to VSS and SMC. It is divided into four parts comprising 14 self-contained chapters that allow separate reading in any preferred order.

The first part introduces *New HOSM Algorithms*. New homogeneous controllers obtained by means of a recursive procedure for any relative degree of the system are proposed. The controllers are accompanied by new filtering sliding-mode-based differentiators. In addition, discontinuous integral control for systems with arbitrary relative degree capable of tracking smooth but unknown reference signals under the presence of Lipschitz continuous perturbations is presented.

Part II addresses *Properties of Continuous Sliding-Mode Algorithms* such as the computation and estimation of the reaching time of the super-twisting algorithm and two chapters dealing with the issue of constrained sliding-mode control, which is inevitable for real-world implementations. Also, an analysis of the orbital stability of self-excited periodic motions in a Lure system as well as a comparison of the chattering using continuous and discontinuous sliding-mode controllers is treated.

Part III covers the *Usage of VSS Controllers for Solving Other Control Problems*. Sliding-mode stabilization of SISO bilinear systems with delays is considered where Volterra operator theory is exploited to perform stability and robustness analysis. After that, a comparison of classical results and recent methods using integral and HOSM is carried out in the next chapter in order to investigate their ability to compensate for unmatched disturbances.

The last part of the book is dedicated to *Applications of VSS*. Three chapters related to power electronics show the capability of sliding-mode techniques in this field. First, a grid-connected shunt active LCL control via continuous SMC and HOSM control techniques is presented. After that, the robust distributed secondary control of islanded inverter-based microgrids as well as local and wide-area sliding-mode observers in power systems is investigated.

The last two chapters deal with the application of sliding-mode-based methods for vehicle platooning, i.e., the task to form tight vehicle formations on one lane on a highway, and an application to a single-loop integrated guidance and control intercept strategy that makes use of HOSM control.

Enjoy reading!

Graz, Austria — Martin Steinberger
Graz, Austria — Martin Horn
Mexico City, Mexico — Leonid Fridman
July 2019

Acknowledgements

We gratefully acknowledge the financial support of (i) the Christian Doppler Research Association, the Austrian Federal Ministry for Digital and Economic Affairs and the National Foundation for Research, Technology and Development; (ii) CONACYT (Consejo Nacional de Ciencia y Tecnologia) grant 282013; PAPIIT-UNAM (Programa de Apoyo a Proyectos de Investigación e Innovación Tecnológica) grant IN 115419; and (iii) the European Union's Horizon 2020 research and innovation programme under the Marie Skłodowska-Curie grant agreement No 734832.

Graz, Austria — Martin Steinberger
Graz, Austria — Martin Horn
Mexico City, Mexico — Leonid Fridman

Contents

Nomenclature

$\mathbb{N}$	The natural numbers
$\mathbb{R}$	The field of real numbers
$\mathbb{C}$	The field of complex numbers
$\mathrm{Re}[z]$	The real part of the complex number z
$\mathrm{Im}[z]$	The imaginary part of the complex number z
$\mathbb{R}_+$	The set of strictly positive real numbers
$\mathbb{R}^{n\times m}$	The set of real matrices with n rows and m columns
$\|a\|$	The absolute values of the real (or complex) number a
$\mathrm{sign}\,(\cdot)$	The signum function
A^T	The transpose of the matrix A
$det(A)$	The determinant of the square matrix A
A^{-1}	The inverse of the square matrix A
A^+	The (left) pseudo-inverse of the matrix A
$\mathrm{rank}\,(A)$	The rank of the matrix A
$\lambda(A)$	The spectrum of the square matrix A, i.e., the set of eigenvalues
$\lambda_{max}(A)$	The largest eigenvalue of the square matrix A
$\lambda_{min}(A)$	The smallest eigenvalue of the square matrix A
$\mathscr{R}(A)$	The range space of the matrix A (viewed as a linear operator)
$\mathscr{N}(A)$	The null space of the matrix A (viewed as a linear operator)
I_n	The $n \times n$ identity matrix
$A > 0$	The square matrix A is symmetric positive definite
$A > B$	The square matrix $A - B$ is symmetric positive definite
$\\|\cdot\\|$	Euclidean norm for vectors and the spectral norm for matrices
$x^{(n)}$	n-th derivative of the variable x with respect to time
$\equiv$	Equivalent to
$\times$	Cartesian product
$\perp$	Orthogonal complement
$:=$	Equal to by definition

Part I
New HOSM Algorithms

Chapter 1
New Homogeneous Controllers and Differentiators

Avi Hanan, Adam Jbara and Arie Levant

Abstract Unlimited number of new homogeneous output regulators are produced by means of a recursive procedure in the form of control templates. The templates are valid for any relative degree of the system, any homogeneity degree and include a number of parameters to be found recursively by simulation. In particular, infinitely many new sliding-mode (SM) controllers are produced. The controllers are accompanied by new filtering SM-based differentiators which are exact in the absence of noises, robust, and asymptotically optimal in the presence of bounded noises, and filter out *unbounded* noises of small average values.

1.1 Introduction

Sliding-mode (SM) control (SMC) [24, 27, 70, 71, 78] remains one of the main approaches to control uncertain systems. The idea is to suppress the uncertainty by exactly keeping a proper output σ (the sliding variable) at zero.

Due to the uncertainty of the system, one does not know the exact control which would accomplish the task. The solution is to keep $\sigma \equiv 0$ by applying a sufficient control effort each time a deviation of σ from zero is detected. It results in the infinite-frequency switching of control. The corresponding motion is called SM.

The high-frequency control switching generates undesirable system vibrations called chattering effect [8, 12, 16, 33, 78]. Another restriction of the classical SMC [27, 78] is the requirement that the control explicitly appear already in $\dot{\sigma}$.

In the following, we restrict ourselves to scalar sliding variables $\sigma \in \mathbb{R}$. Introduce the notation

A. Hanan · A. Jbara · A. Levant (✉)
Applied Mathematics Department, Tel-Aviv University, 6997801 Tel Aviv, Israel
e-mail: levant@tauex.tau.ac.il

A. Hanan
e-mail: Avi.Hanan@elbitsystems.com

A. Jbara
e-mail: adam_jbara_94@hotmail.com

M. Steinberger et al. (eds.), *Variable-Structure Systems and Sliding-Mode Control*,
Studies in Systems, Decision and Control 271,
https://doi.org/10.1007/978-3-030-36621-6_1

$$\vec{\sigma}_k = (\sigma, \dot{\sigma}, \ldots, \sigma^{(k)})^T, \ k \in \mathbb{N}.$$

The appearance of the control in $\dot{\sigma}$ means that the relative degree r [37] of the sliding variable σ is 1, and the SMC keeps $\vec{\sigma}_{r-1} = (\sigma)$ at zero.

Both described restrictions are connected. Indeed, the discontinuity of $\dot{\sigma}$ due to the control switching exaggerates the chattering of σ [48] and diminishes the accuracy of the SM [43]. In the case when $r > 1$, the discontinuous control appears in $\sigma^{(r)}$ and one usually introduces the auxiliary variable

$$\Sigma = \beta^T \vec{\sigma}_{r-1}, \ \beta = (\beta_0, \ldots, \beta_{r-2}, 1)^T \in \mathbb{R}^r$$

of the relative degree 1. Unfortunately, keeping $\Sigma \equiv 0$ by discontinuous control results in the same vibration magnitude level [75] of all the components of $\vec{\sigma}_{r-1}$ and prevents the establishment of $\sigma \equiv 0$ in finite time (FT). It still diminishes the vibration energy (i.e., the chattering [48]) of lower derivatives.

High-order SMs (HOSMs) were introduced to deal with the restrictions of the relative degree and the chattering, and to improve the accuracy of keeping $\sigma = 0$. The corresponding rth-order SM (r-SM) controllers establish the equality $\vec{\sigma}_{r-1} \equiv 0$ in FT by means of discontinuous control. It potentially improves the accuracy of keeping $\sigma = 0$ [43] and can significantly diminish the chattering by moving the highest chattering level to $\sigma^{(r-1)}$ [9, 48]. The HOSM approach has become popular in the last two decades [3, 10, 18, 20, 22, 25, 28, 32, 35, 41, 42, 60, 62, 65–69, 72, 74, 81] to mention just a few publications.

Establishing the r-SM $\vec{\sigma}_{r-1} \equiv 0$ does not automatically improve the SM accuracy [43, 75]. In particular, such improvement is attained by the homogeneity approach [45, 46] and its modifications, but other methods are also possible [1, 77].

This paper deals with the homogeneity-based stabilization of σ. For that end one intentionally increases system uncertainty reducing the output dynamics to an autonomous differential inclusion (DI) of the form $\sigma^{(r)} \in H(\vec{\sigma}_{r-1}) + G(\vec{\sigma}_{r-1})u$. The homogeneity theory [7, 13–15, 38, 46, 63] is further applied to make σ vanish in FT or asymptotically.

We present a simple, effective method of general homogeneous feedback stabilization of such DIs. The method does not require validation by complicated Lyapunov-stability analysis. The generated controllers contain a number of parameters to usually be found by experiment, and therefore we call them *control templates*. We demonstrate the technique by designing *a completely new family of controllers* valid for any relative degree r, and any possible homogeneity degree including positive, negative, and zero values.

The approach naturally requires estimating the derivatives $\vec{\sigma}_{r-1}$ in real time. Moreover, the exact nth-order differentiation problem can itself be reformulated as FT stabilizing the output $\sigma = z - f(t)$ of the auxiliary control system $z^{(n+1)} = u$. The differentiation problem lies in the core of the observation methods [6, 31]. SM observation and differentiation always were a significant part of SMC research [24, 34, 71, 76, 79, 82].

HOSM-based differentiators are quite well known [4, 5, 19, 29, 39, 45, 46, 53, 64, 73]. In particular, in the absence of input noises, nth-order differentiators [45] are exact on the signals with the bounded $(n+1)$th derivative. They also provide for the asymptotically optimal accuracy in the presence of bounded noises [56].

The new *filtering* nth-order differentiators [50, 56, 58] are capable of filtering out *unbounded noises* featuring small average values, while still remaining exact and asymptotically optimally accurate on the input signals with the known Lipschitz constant of the nth derivative. That property can be considered as the realization of the approximability principle introduced by Prof. Bartolini and his collaborators [11, 12].

These robustness features are analogue to the well-known features of linear filters. Linear filters are practically insensitive to large high-frequency input components. It results in filtering out many noises, but also prevents exact differentiation. We show here that the nonlinear filters are capable to preserve the exactness while still filtering out certain classes of large noises including high-frequency ones.

Notation. A binary operation $\diamond$ of two sets is defined as $A \diamond B = \{a \diamond b |\ a \in A, b \in B\}$. In that context, we define $a \diamond B = A \diamond B$ for $A = \{a\}$. A function of a set is the set of function values on this set. The zeroth derivative of a function is the function itself. The norm $||x||$ stays for the standard Euclidean norm of x, $B_\varepsilon = \{x \mid ||x|| \le \varepsilon\}$; $||x||_h$ is a homogeneous norm; $\lfloor a \rceil^b = |a|^b \operatorname{sign} a$, $\lfloor a \rceil^0 = \operatorname{sign} a$, $a \in \mathbb{R}$, is a power function extension.

1.2 Preliminaries: Weighted Homogeneity

Denote the tangent space to $\mathbb{R}^{n_x}$ at the point x by $T_x\mathbb{R}^{n_x}$. Recall that a solution of a differential inclusion (DI)

$$\dot{x} \in F(x),\ x \in \mathbb{R}^{n_x},\ F(x) \subset T_x\mathbb{R}^{n_x} \tag{1.1}$$

is any locally absolutely continuous function $x(t)$, satisfying the DI for almost all t.

DI (1.1) is called *Filippov DI*, if $F(x)$ is nonempty, compact, and convex for any x, and F is an upper-semicontinuous set function [30, 46]. The latter means that the maximal distance from the vectors of $F(x)$ to the set $F(y)$ vanishes as $x \to y$. Filippov DIs feature existence and extendability of solutions, but not the solution uniqueness [30].

Introduce the *weights* $m_1, m_2, \ldots, m_{n_x} > 0$ of the coordinates $x_1, x_2, \ldots, x_{n_x}$ in $\mathbb{R}^{n_x}$. Define [7] the *dilation* $d_\kappa(x) = (\kappa^{m_1}x_1, \kappa^{m_2}x_2, \ldots, \kappa^{m_{n_x}}x_{n_x})$ for $\kappa \ge 0$.

A function $f : \mathbb{R}^{n_x} \to \mathbb{R}^m$ is said to have the homogeneity degree (weight) $q \in \mathbb{R}$, $\deg f = q$, if the identity $f(x) = \kappa^{-q} f(d_\kappa x)$ holds for any x and $\kappa > 0$. We do not distinguish between the weight of the coordinate x_i and the homogeneity degree of the coordinate function $c_{x_i} : x \mapsto x_i$: $\deg x_i = \deg c_{x_i} = m_i$.

A vector-set field $F(x) \subset T_x\mathbb{R}^{n_x}$ (DI (1.1)) is called *homogeneous of the degree* $q \in \mathbb{R}$, if the identity $F(x) = \kappa^{-q} d_\kappa^{-1} F(d_\kappa x)$ holds for any x and $\kappa > 0$ [46].

Thus, the homogeneity of the vector-set field $F(x) \subset T_x\mathbb{R}^{n_x}$ implies the invariance of DI (1.1) with respect to the combined time-coordinate transformation

$$(t, x) \mapsto (\kappa^{-q}t, d_\kappa x), \quad \kappa > 0, \tag{1.2}$$

where $-q$ can naturally be interpreted as the weight of t, $\deg t = -q$. Hence, (1.2) establishes a one-to-one correspondence between solutions of (1.1) for any $\kappa > 0$.

The standard definition [7] of homogeneous differential equations (DEs) is a particular case here. Note the difference between the homogeneity degree of a vector function taking values in $\mathbb{R}^m$ and of a vector (or vector set) field which takes the values in the tangent space $T\mathbb{R}^{n_x}$.

A homogeneous function $\varphi : \mathbb{R}^{n_x} \to \mathbb{R}$ is called *a homogeneous norm*, if it is continuous, positive definite and $\deg \varphi = 1$. Note that it is not a real norm. Any two homogeneous norms $||\cdot||_h$ and $||\cdot||_{h*}$ are equivalent in the sense that there exist $\gamma, \gamma_* > 0$ such that $\forall x : \ ||x||_h \le \gamma ||x||_{h*}, \ ||x||_{h*} \le \gamma_* ||x||_h$.

If DI (1.1) is homogeneous with the homogeneity degrees $q, m_1, \ldots, m_{n_x}$ then the homogeneity degrees $\gamma q, \gamma m_1, \ldots, \gamma m_{n_x}$ are also valid for any $\gamma > 0$. Such rescaling of weights certainly does not preserve the homogeneous norms.

A function is called quasi-continuous (QC) [47], if it is continuous everywhere except the origin. In particular, any continuous function is QC.

We say that the homogeneous Filippov DI (1.1) is *asymptotically stable (AS)*, if $x(t) \equiv 0$ is its Lyapunov-stable solution, all solutions are extended till infinity in time and $\lim_{t \to 0} x(t) = 0$ holds for each solution $x(t)$. The same DI (1.1) is called *finite-time (FT) stable*, if for each solution $x(t)$, $t \ge 0$, there exists $T > 0$ such that $\forall t \ge T : \ x(t) = 0$. It is called *fixed-time (FxT) stable*, if such T can be chosen the same for all solutions.

We call a ball B_ε *FxT attracting* if there exists $T > 0$ such that for each solution $x(t)$, $t \ge 0$, independently of $x(0)$ the inequality $t \ge T$ implies $x(t) \in B_\varepsilon$.

The *contractivity* [46] of the homogeneous Filippov DI (1.1) is equivalent to the existence of three positive numbers $T, R, r > 0$, $R > r$, such that for all solutions $||x(0)|| \le R$ implies $||x(T)|| \le r$.

A Filippov DI $\dot{x} \in \tilde{F}(x)$ is called a *small homogeneous perturbation* of (1.1) if it has the same degree and weights, and $F(x) \subset \tilde{F}(x) + B_\varepsilon$, $\tilde{F}(x) \subset F(x) + B_\varepsilon$ hold over the unit ball B_1 for some small $\varepsilon \ge 0$.

Theorem 1.1 ([51, 53]) *Let the Filippov DI* (1.1) *be homogeneous,* $\deg F = q$. *Then asymptotic stability and contractivity features are equivalent, and are robust with respect to small homogeneous perturbations.*

- *If $q < 0$ the asymptotic stability implies the FT stability, and the maximal (minimal) stabilization time is a well-defined upper (lower) semicontinuous function of the initial conditions [53]. Moreover, the FT stability of* (1.1) *implies that $q < 0$.*
- *If $q = 0$ the asymptotic stability is exponential.*
- *If $q > 0$ the asymptotic stability implies FxT attractivity of any ball B_ε, $\varepsilon > 0$. The convergence to 0 is slower than exponential.*

Example 1.1 Any linear time-invariant system $\dot{x} = Ax$, $x \in \mathbb{R}^{n_x}$, $A \in \mathbb{R}^{n_x \times n_x}$, is equivalent to the homogeneous Filippov DI $\dot{x} \in \{Ax\}$ of the homogeneity degree $q = 0$ with $\deg x_i = 1$, $i = 1, \ldots, n_x$. For any $\varepsilon > 0$ any smooth dynamic system $\dot{x} = f(x)$, $f(0) = 0$, $f'(0) = A$, implies $\dot{x} \in \{Ax + \varepsilon||x||B_1\}$ in some small vicinity of 0. Thus, in the case $q = 0$, Theorem 1.1 validates the standard linearization method of the stability verification. □

The rigorous formulation of the following result [45] requires some natural initial conditions [26, 53] omitted here for the clarity of presentation. Consider the retarded "noisy" AS DI of the negative homogeneity degree $q < 0$ [46]

$$\dot{x} \in F(x(t - [0, \tau]) + B_{h\varepsilon}),$$

where $\tau, \varepsilon \geq 0$, $B_{h\varepsilon} = \{x \in \mathbb{R}^{n_x}, \mid ||x||_h \leq \varepsilon\}$. Then, the accuracy,

$$x \in \mu B_{h\rho}, \ \rho = \max[\varepsilon, \tau^{-1/q}], \tag{1.3}$$

is established in FT for some $\mu > 0$.

In the case $q = 0$, that accuracy is established for $\rho = \varepsilon$ and sufficiently small τ [26]. In the case $q > 0$, the accuracy is also established for $\rho = \varepsilon$, but the initial values $x(0)$ are to be uniformly bounded, and τ, ε are to be sufficiently small (it is the most delay, noise-sensitive case [26]).

1.3 Homogeneity Approach to Output Regulation Under Uncertainty

Consider a smooth dynamic system

$$\dot{x} = a(t, x) + b(t, x)u, \quad \sigma = \sigma(t, x), \tag{1.4}$$

where $x \in \mathbb{R}^{n_x}$, $u \in \mathbb{R}$ is the control, the output $\sigma : \mathbb{R}^{n_x+1} \to \mathbb{R}^n$, and a, b are unknown smooth functions. The dimension n_x is nowhere used in the sequel. Here and further, differential equations are understood in the Filippov sense [30]. Solutions of (1.4) are assumed infinitely extendible in time (forward complete) for any Lebesgue-measurable bounded control $u(t)$.

The informal *control task* is to keep the real-time measured output-regulation (tracking) error σ as small as possible.

The system relative degree is known to be r which means [37] that

$$\sigma^{(r)} = h(t, x) + g(t, x)u, \quad \forall t, x : \ g(t, x) \neq 0, \tag{1.5}$$

where the smooth functions g, h are unknown.

The main idea is to choose the feedback control $u = U(\vec{\sigma}_{r-1})$ so that the uncertain Eq. (1.5) be replaced with a certain AS homogeneous Filippov DI,

$$\sigma^{(r)} \in H(\vec{\sigma}_{r-1}) + G(\vec{\sigma}_{r-1})u(\vec{\sigma}_{r-1}), \tag{1.6}$$

on $\vec{\sigma}_{r-1} = (\sigma, \dot{\sigma}, \ldots, \sigma^{(r-1)})^T$. The system uncertainty and the time dependence are to be absorbed by the sets H, G. Realization of this approach requires some proper assumptions as well as the real-time estimation or availability of $\vec{\sigma}_{r-1}$.

Here and further, in order to get a Filippov DI, the Lebesgue-measurable control $u(\vec{\sigma}_{r-1})$ is replaced in (1.6) with the result $[u]_F$ of the Filippov procedure [30]

$$[u]_F(\vec{\sigma}_{r-1}) = \bigcap_{\delta>0} \bigcap_{\mu_L N=0} \overline{\text{co}}\, u((\vec{\sigma}_{r-1} + B_\delta) \setminus N). \tag{1.7}$$

Here, $\mu_L N$ is the Lebesgue measure of N and $\overline{\text{co}}M$ denotes the convex closure of M.

Let the weight of σ be 1, $\deg \sigma = 1$, and the system homogeneity degree be $q \in \mathbb{R}$, i.e., $\deg t = -q$. Then $\deg \sigma^{(k)} = 1 + kq$, $k = 0, 1, \ldots$.

In order for (1.6) to be a Filippov DI, the inequality $\deg \sigma^{(r)} = 1 + rq \geq 0$ is necessary [53]. Thus, $q \geq -1/r$ is required.

The set functions H, G and the control $u = U(\vec{\sigma}_{r-1})$ are to be homogeneous, $\deg H = \deg G + \deg u = 1 + rq$, they also are to be compact and convex. Without losing the generality assume that $\deg G = 0$, $\deg u = 1 + rq$. Indeed, one can redefine $u = ||\vec{\sigma}_{r-1}||_h^{-\deg G} U(\vec{\sigma}_{r-1})$ for any homogeneous norm $|| \cdot ||_h$.

Fix an arbitrary homogeneous norm $|| \cdot ||_h$. Then, maximally enlarging the sets H, G obtain the assumptions for (1.5)

$$|h(t, x)| \leq C||\vec{\sigma}_{r-1}||_h^{1+rq}, \; 0 < K_m \leq g(t, x) \leq K_M \tag{1.8}$$

and the DI of the form [49]

$$\begin{array}{l} \sigma^{(r)} \in [-C, C]||\vec{\sigma}_{r-1}||_h^{1+rq} + [K_m, K_M]u(\vec{\sigma}_{r-1}), \\ C \geq 0, \; 0 < K_m \leq K_M, \;\; 1 + rq \geq 0, \; \deg u = 1 + rq. \end{array} \tag{1.9}$$

As follows from the equivalence of the homogeneous norms, choosing another homogeneous norm just implies proportionally changing the parameter C.

In the particular case, when $q > -1/r$ conditions (1.8) are natural if $r = n_x$. Then $\vec{\sigma}_{r-1}$ are simply new coordinates in $\mathbb{R}^{n_x}$. Stabilization of power-integrator chains [36] is reducible to that problem statement [23].

The important case $q = -1/r$ corresponds to the standard high-order SMC approach (HOSMC) [45, 46]. Thus, $\deg \sigma^{(r)} = 0$, and (1.6) gets the well-known form

$$\sigma^{(r)} \in [-C, C] + [K_m, K_M]u(\vec{\sigma}_{r-1}), \;\; \deg u = 0. \tag{1.10}$$

The corresponding assumptions $|h| \leq C$ and $g \in [K_m, K_M]$ are always true at least locally for some C, K_m, K_M.

Due to Theorem 1.1 DI (1.10) is to become FT stable and the control feedback function $u(\vec{\sigma}_{r-1})$ is necessarily discontinuous at $\vec{\sigma} = 0$. The motion on the set $\vec{\sigma} = 0$ is said to be in the rth-order SM (r-SM), and the control is called r-SMC [45, 46].

There are a number of homogeneous SM controllers solving the problem in the case $q = -1/r < 0$, $\deg u = 0$. The recently established powerful method [17] exploits the knowledge of a concrete homogeneous control Lyapunov function for the system $\sigma^{(r)} = u$ in order to generate an r-SM controller. Constructing a new control Lyapunov function becomes the initial nontrivial design step.

The alternative approach [21, 57] starts from the knowledge of a homogeneous AS DE. It removes any control differentiability conditions and, respectively, yields more controllers. The following general method in a simple way produces new homogeneous AS DEs and controllers for any r and q.

1.4 Homogeneous Control Templates

We call two scalar functions $\omega, \varpi : \Omega \to \mathbb{R}$, $\Omega \subset \mathbb{R}^{n_\omega}$, *sign-equivalent* in Ω, if $\operatorname{sign} \omega(s) \equiv \operatorname{sign} \varpi(s)$ whenever $s \in \Omega$, and $\omega(s) \neq 0$ or $\varpi(s) \neq 0$.

Consider an $(r-1)$th-order AS homogeneous DE

$$\sigma^{(r-1)} + \tilde{\varphi}(\vec{\sigma}_{r-2}) = 0, \tag{1.11}$$

where $\tilde{\varphi}$ is a continuous function, $\deg \tilde{\varphi} = 1 + (r-1)q, q \geq -1/r$.

Theorem 1.2 ([49]) *Let $q \geq -1/r$. Choose any homogeneous norm $N(\vec{\sigma}_{r-1})$, and let $\varphi(\vec{\sigma}_{r-1})$ be any homogeneous QC scalar function, $\deg \varphi \geq 0$, which is sign-equivalent to $\sigma^{(r-1)} + \tilde{\varphi}(\vec{\sigma}_{r-2})$ for $\vec{\sigma}_{r-1} \neq 0$. Consider the homogeneous controls of the form*

$$u = \alpha U(\vec{\sigma}_{r-1}), \tag{1.12}$$

where U is defined by one of the formulas

$$U(\vec{\sigma}_{r-1}) = -[N(\vec{\sigma}_{r-1})]^{1+qr-\deg \varphi} \varphi(\vec{\sigma}_{r-1}), \tag{1.13}$$

$$U(\vec{\sigma}_{r-1}) = -[N(\vec{\sigma}_{r-1})]^{1+qr} \operatorname{sign} \varphi(\vec{\sigma}_{r-1}). \tag{1.14}$$

Then, for any sufficiently large $\alpha > 0$, these controllers asymptotically stabilize DI (1.9). *In particular, the homogeneous DE*

$$\sigma^{(r)} + \alpha U(\vec{\sigma}_{r-1}) = 0 \tag{1.15}$$

is AS for any sufficiently large α.

The function (1.13) *is QC (i.e., discontinuous only at $\vec{\sigma}_{r-1} = 0$) for $q = -1/r$. It is continuous for $q > -1/r$, provided $U(0) = 0$ is assigned. DI* (1.9) *(in particular* (1.15)*) is FT stable for $q < 0$, and exponentially stable for $q = 0$. If $q > 0$ any ball B_ε attracts solutions in FxT.*

Multiplication of controllers by any locally bounded Lebesgue-measurable function $k(t, x) \geq 1$ does not interfere with the convergence.

Note that the chattering of the QC r-SM controller (1.12), (1.13), obtained in the case $q = -1/r$, is significantly lower than that of (1.12), (1.14) [47]. Also, although controller (1.12), (1.14) looks as a classical SM controller, in general, it cannot keep the SM $\varphi(\vec{\sigma}_{r-1}) = 0$ due to $\varphi(\vec{\sigma}_{r-1})$ having infinite total derivatives.

1.4.1 Recursion in r

Theorem 1.2 actually establishes a recursion from the $(r-1)$th-order AS DE (1.11) to the new rth-order AS DE (1.15). The recursion is only possible for $q \geq -1/r$.

It is easy to see that $A + B$ and $\lfloor A \rceil^{\gamma} + \lfloor B \rceil^{\gamma}$ are sign-equivalent for any $A, B \in \mathbb{R}$ and $\gamma > 0$.

The initial step. Let $r = 2$, $q \geq -1$. The AS DE (1.11) of the order $r - 1 = 1$ can always be chosen as

$$\dot{\sigma} + \beta_0 \lfloor \sigma \rceil^{1+q} = 0, \ \ \beta_0 > 0. \tag{1.16}$$

Now one can perform a recursive step producing a homogeneous stabilizer for $r = 2$, provided $q \geq -1/2$.

The recursive step. Let an $(r-1)$th-order AS DE (1.11) be given, $q \geq -1/r$. Choose two arbitrary homogeneous norms $||\cdot||_h$, $||\cdot||_{h*}$ and some $m > 0$. Let $\phi(s)$, $\phi : \mathbb{R} \to \mathbb{R}$, be any continuous function sign-equivalent to s. Then the following are some possible simple options for choosing $\varphi(\vec{\sigma}_{r-1})$:

$$\begin{array}{ll} 1.\ \varphi(\vec{\sigma}_{r-1}) = \lfloor \sigma^{(r-1)} + \tilde{\varphi} \rceil^m, & \deg\varphi = m > 0, \\ 2.\ \varphi(\vec{\sigma}_{r-1}) = \lfloor \sigma^{(r-1)} \rceil^m + \lfloor \tilde{\varphi} \rceil^m, & \deg\varphi = m > 0, \\ 3.\ \varphi(\vec{\sigma}_{r-1}) = \phi\left(\frac{\lfloor \sigma^{(r-1)} + \tilde{\varphi} \rceil^m}{||\vec{\sigma}_{r-1}||_h^{m(1+(r-1)q)}} \right), & \deg\varphi = 0. \end{array} \tag{1.17}$$

More options are available in (1.17), e.g., algebraic combinations like

$$\varphi(\vec{\sigma}_{r-1}) = |\sigma^{(r-1)} + \tilde{\varphi}|^{m_1} \left| \lfloor \sigma^{(r-1)} \rceil^{m_2} + \lfloor \tilde{\varphi} \rceil^{m_2} \right| \phi \left(\frac{\lfloor \sigma^{(r-1)} + \tilde{\varphi} \rceil^{m_3}}{||\vec{\sigma}_{r-1}||_h^{m_3(1+(r-1)q)}} \right)$$

with $\deg\varphi = m_1 + m_2$ are possible for any $m_2, m_3 > 0$, $m_1 + m_2 \geq 0$, etc. Obviously, the number of such constructions is infinite for $r \geq 2$.

Now, due to Theorem 1.2, obtain the new homogeneous controls (1.12) of the order r,

$$\begin{aligned} u_r &= -\alpha ||\vec{\sigma}_{r-1}||_{h*}^{1+qr-\deg\varphi} \varphi(\vec{\sigma}_{r-1}), \\ u_r &= -\alpha ||\vec{\sigma}_{r-1}||_{h*}^{1+qr} \operatorname{sign} \varphi(\vec{\sigma}_{r-1}), \end{aligned} \tag{1.18}$$

and the rth-order AS DE (1.15)

$$\sigma^{(r)} + \alpha||\vec{\sigma}_{r-1}||_{h*}^{1+qr-\deg\varphi}\varphi(\vec{\sigma}_{r-1}) = 0. \tag{1.19}$$

The new equation contains uncertain parameters and functions of the recursion step (1.17), as well as the uncertain parameter α. It is natural to call (1.18) *a controller template*. If $q \geq -1/(r+1)$ one can now perform one more recursive step, etc.

In general, one needs $r-1$ recursive steps to come to a controller of the order r, provided $q \geq -1/r$. It is reasonable to immediately assign proper values to additional design parameters which appear at each recursion step. Usually, it is done by simulation.

1.4.2 Examples of Recursive Homogeneous Control Design

Start with (1.16). Provided $q \geq -0.5$, Theorem 1.2, (1.12), (1.14) and option 2 of (1.17) produce

$$u_2 = -\alpha||\vec{\sigma}_1||_h^{1+2q}\,\mathrm{sign}(\lfloor\dot{\sigma}\rceil^{\frac{1}{1+q}} + \beta_0^{\frac{1}{1+q}}\sigma),\ q \geq -\tfrac{1}{2}.$$

Taking $q = -0.5$ obtain the homogeneous non-singular terminal 2-SM control (NSTSMC) [28]. Then, (1.12), (1.13) produce a quasi-continuous 2-SM controller [47]. The standard version of NSTSMC [28] is obtained for $q > -0.5$ by replacing the multiplier $||\vec{\sigma}_1||_h^{1+2q}$ with a proper nonvanishing function.

Some other possible continuous controller templates for $q > -0.5$ and $r = 2$ are of the form

$$\begin{aligned}
u_2 &= -\alpha\left\lfloor \lfloor\dot{\sigma}\rceil^{\frac{1}{1+q}} + \beta_0\sigma\right\rceil^{1+2q},\\
u_2 &= -\alpha(|\sigma| + |\dot{\sigma}|^{\frac{1}{1+q}})^q(\dot{\sigma} + \beta_0\lfloor\sigma\rceil^{1+q}),\\
u_2 &= -\alpha(\lfloor\dot{\sigma}\rceil^{\frac{1+2q}{1+q}} + \beta_0\lfloor\sigma\rceil^{1+2q}),\\
u_2 &= -\alpha\left\lfloor \lfloor\dot{\sigma}\rceil^{\frac{1+3q}{1+q}} + \beta_0\lfloor\sigma\rceil^{1+3q}\right\rceil^{\frac{1+2q}{1+3q}} \text{ for } q > -\tfrac{1}{3},
\end{aligned} \tag{1.20}$$

where $\beta_0 > 0$ is any number and $\alpha > 0$ is sufficiently large. The last controller of (1.20) allows the next recursive step, etc.

In such a way, one easily proves that for each r and $q > -1/r$, there exist the controller template and the corresponding AS DE

$$u_r = -\alpha\left(\lfloor\sigma^{(r-1)}\rceil^{\frac{1+rq}{1+(r-1)q}} + \beta_{r-2}\lfloor\sigma^{(r-2)}\rceil^{\frac{1+rq}{1+(r-2)q}} + \cdots + \beta_0\lfloor\sigma\rceil^{1+rq}\right), \tag{1.21}$$

$$\sigma^{(r)} + \tilde{\beta}_{r-1}\lfloor\sigma^{(r-1)}\rceil^{\frac{1+rq}{1+(r-1)q}} + \tilde{\beta}_{r-2}\lfloor\sigma^{(r-2)}\rceil^{\frac{1+rq}{1+(r-2)q}} + \cdots + \tilde{\beta}_0\lfloor\sigma\rceil^{1+rq} = 0, \tag{1.22}$$

where $\tilde{\beta}_{r-1} = \alpha$, $\tilde{\beta}_j = \alpha\beta_j$, $j = 0, 1, \ldots, r-2$. Asymptotic stabilizers of the form (1.21) (DE (1.22)) are well known from [15] where they are only developed for $q < 0$, provided $|q|$ is sufficiently small.

Similarly, the controller

$$u_r = -\alpha \left\lfloor \lfloor \sigma^{(r-1)} \rceil^{\frac{\omega}{1+(r-1)q}} + \beta_{r-2} \lfloor \sigma^{(r-2)} \rceil^{\frac{\omega}{1+(r-2)q}} + \cdots + \beta_0 \lfloor \sigma \rceil^{\omega} \right\rceil^{\frac{1+rq}{\omega}} \quad (1.23)$$

is shown to be valid for any $\omega > 0, q \geq -1/r$. The following is its quasi-continuous version for $q = -1/r$:

$$u_r = -\alpha \frac{\lfloor \sigma^{(r-1)} \rceil^{\frac{\omega}{1+(r-1)q}} + \beta_{r-2} \lfloor \sigma^{(r-2)} \rceil^{\frac{\omega}{1+(r-2)q}} + \cdots + \beta_0 \lfloor \sigma \rceil^{\omega}}{|\sigma^{(r-1)}|^{\frac{\omega}{1+(r-1)q}} + \beta_{r-2} |\sigma^{(r-2)}|^{\frac{\omega}{1+(r-2)q}} + \cdots + \beta_0 |\sigma|^{\omega}}. \quad (1.24)$$

Similar results are obtained in [17, 21] for $q = -1/r$. In particular, for $r = 5, q = -0.2$, $\omega = 0.2$ obtain the following quasi-continuous 5-SM controller (1.24) [21]:

$$u_5 = -\alpha \frac{\sigma^{(4)} + 6\lfloor \dddot{\sigma} \rceil^{\frac{1}{2}} + 5\ddot{\sigma}^{\frac{1}{3}} + 3\lfloor \dot{\sigma} \rceil^{\frac{1}{4}} + \sigma^{\frac{1}{5}}}{|\sigma^{(4)}| + 6|\dddot{\sigma}|^{\frac{1}{2}} + 5|\ddot{\sigma}|^{\frac{1}{3}} + 3|\dot{\sigma}|^{\frac{1}{4}} + |\sigma|^{\frac{1}{5}}}.$$

The coefficients have been found one by one by simulation according to the design recursion steps. No Lyapunov function is known for that controller. Once one valid parametric set $\{\beta_j\}$ is found, the convergence rate is easily regulated in a standard way by changing β_j [21].

It seems that all known continuous/QC homogeneous asymptotic stabilizers can be developed in such a way. One cannot obtain the nested SM controllers [45], since the corresponding recursive procedure [45] involves superposition of discontinuous functions.

The accuracy (1.3) of the controllers for any q in the presence of noises and discrete sampling is calculated in the papers [46, 51, 53] (Sect. 1.2). In particular, let $\epsilon_i \geq 0$ be the sampling noise magnitude of $\sigma^{(i)}$, and $\tau \geq 0$ be the maximal sampling time step, then the accuracy

$$\begin{aligned} &|\sigma^{(i)}| \leq \mu_i \rho^{1+iq}, && i = 0, 1, \ldots, r-1, \\ &\rho = \max[\|\vec{\epsilon}_{r-1}\|_h, \tau^{1/|q|}] && \text{for } q < 0, \\ &\rho = \|\vec{\epsilon}_{r-1}\|_h && \text{for } q \geq 0 \end{aligned} \quad (1.25)$$

is established for some $\{\mu_i\} > 0$. It is established in FT if $q < 0$ or $\|\vec{\epsilon}_{r-1}\|_h > 0$. In the case $q = 0$, the maximal sampling interval τ is to be small enough. In the case $q > 0$, the initial conditions are to be bounded and both $\vec{\epsilon}_{r-1}$, τ are to be small enough.

1.5 Design of a Completely New Control Family

In the following, we do not pretend to develop high-quality controllers, but only demonstrate the simplicity of creating new controller families for any relative degree r and any homogeneity degree $q \geq -1/r$.

Let $\phi(s)$, $\phi : \mathbb{R} \to \mathbb{R}$, be any continuous function sign-equivalent to s. Denote

$$||\vec{\sigma}_k||_{h1} = |\sigma| + |\dot{\sigma}|^{\frac{1}{1+q}} + \cdots + |\sigma^{(k)}|^{\frac{1}{1+kq}}. \tag{1.26}$$

Fix any $\omega > 0$ and $\beta_0 > 0$.

Then the following recursion procedure successively uses option 3 of (1.17) and, therefore, introduces a new family of valid control templates:

$$\begin{array}{ll} \psi_0 = \lfloor\sigma\rceil^{1+q}, & i = 0 \\ \psi_i = ||\vec{\sigma}_i||_{h1}{}^{1+(i+1)q}\, \phi\left(\frac{\lfloor\sigma^{(i)}\rceil^{\omega} + \lfloor\beta_{i-1}\psi_{i-1}(\vec{\sigma}_{i-1})\rceil^{\omega}}{||\vec{\sigma}_i||_{h1}{}^{(1+iq)\omega}}\right), & i = 1, \ldots, r-1, \\ u_r = -\alpha\psi_{r-1}(\vec{\sigma}_{r-1}), \alpha > 0. & \end{array} \tag{1.27}$$

Here $\beta_1, \ldots, \beta_{r-2} > 0$ are chosen sufficiently large in the list order and α defines the magnitude of the produced control u_r. The corresponding AS DE is

$$\sigma^{(r)} + \beta_{r-1}\psi_{r-1}(\vec{\sigma}_{r-1}) = 0$$

for sufficiently large $\beta_{r-1} > 0$. The procedure can be continued, provided $q \geq -1/(r+1)$. In particular, it can be continued indefinitely if $q \geq 0$.

1.5.1 Simulation

Consider the discontinuous dynamic system

$$\sigma^{(4)} = [8\{2.1\cos(15.73t)\}_f - 4]|\sigma|^{1+4q} + (2 + 2\{\sin(9t)\}_f)\, u. \tag{1.28}$$

Here $\{x\}_f$ denotes the fractional part of x, i.e., $\{1\}_f = 0$, $\{1.3\}_f = 0.3$, $\{-1.3\}_f = 0.7$. Solutions of (1.28) obviously satisfy the DI

$$\sigma^{(4)} \in [-4, 4] \cdot ||\vec{\sigma}_3||_{h1}{}^{1+4q} + [2, 4]u(\vec{\sigma}_3). \tag{1.29}$$

Inclusion (1.29) becomes homogeneous of the homogeneity degree q, provided the proper choice of the feedback control $U(\vec{\sigma}_3)$. Choose $\omega = 1$, $\phi(s) = \lfloor s\rceil^{4/5}$, $(\beta_0, \beta_1, \beta_2) = (1, 5, 5)$, $\alpha = 15$.

The fourth-order controller of the family (1.27) takes the form

$$\begin{aligned} u &= -15 \cdot ||\vec{\sigma}_3||_{h1}{}^{1+4q} \cdot \left(\frac{\dddot{\sigma} + 5\psi_2(\vec{\sigma}_2)}{||\vec{\sigma}_3||_{h1}{}^{1+3q}}\right)^{\frac{4}{5}}, \\ \psi_2 &= ||\vec{\sigma}_2||_{h1}{}^{1+3q} \cdot \left(\frac{\ddot{\sigma} + 5\psi_1(\vec{\sigma}_1)}{||\vec{\sigma}_2||_{h1}{}^{1+2q}}\right)^{\frac{4}{5}}, \\ \psi_1 &= ||\vec{\sigma}_1||_{h1}{}^{1+2q} \cdot \left(\frac{\dot{\sigma} + \lfloor\sigma\rceil^{1+q}}{||\vec{\sigma}_1||_{h1}{}^{1+q}}\right)^{\frac{4}{5}}. \end{aligned} \tag{1.30}$$

In general, one needs specific $\beta_0, \beta_1, \ldots, \beta_{r-2}$ for each r and q, then one has to choose specific α for each concrete C, K_m, K_M. In our case, the same

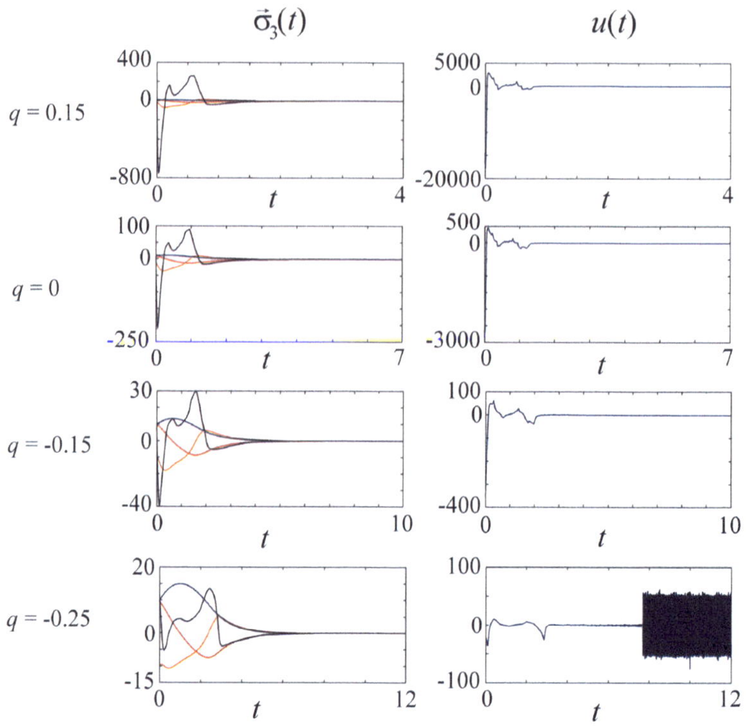

Fig. 1.1 Performance of the developed controller (1.30) for $r = 4$ and $q = 0.15, 0, -0.15, -0.25$

values have been found appropriate for the system (1.29) for the values $q = 0.15, 0, -0.15, -0.25$. It certainly means that in each of these cases the parameters can be adjusted to get better performance.

Choose the Euler integration method with the integration step $\tau = 10^{-4}$ and the initial conditions $\vec{\sigma}_3(0) = (10, 10, -10, 10)$. The performance of the controllers is demonstrated in Fig. 1.1.

One can see from Fig. 1.1 that the control u is continuous for any $q > -0.25$. The asymptotic convergence for $q = 0.15$ is very slow near the origin, but very fast at large distances. The accuracy $|\sigma| \leq 2 \cdot 10^{-5}$, $|\dot{\sigma}| \leq 8 \cdot 10^{-5}$, $|\ddot{\sigma}| \leq 2.5 \cdot 10^{-6}$, and $|\dddot{\sigma}| \leq 7.8 \cdot 10^{-7}$ is maintained for $t > 25$. In the case $q = 0$, the convergence is exponential, and the accuracy $|\sigma| \leq 8 \cdot 10^{-12}$, $|\dot{\sigma}| \leq 2.3 \cdot 10^{-11}$, $|\ddot{\sigma}| \leq 4.7 \cdot 10^{-11}$, $|\dddot{\sigma}| \leq 5.3 \cdot 10^{-11}$ is established for $t > 23$.

The cases $q < 0$ correspond to the FT convergence. The accuracy $|\sigma| \leq 1.7 \cdot 10^{-22}$, $|\dot{\sigma}| \leq 7.3 \cdot 10^{-19}$, $|\ddot{\sigma}| \leq 8.4 \cdot 10^{-15}$, and $|\dddot{\sigma}| \leq 5.3 \cdot 10^{-10}$ is kept for $q = -0.15$ and $t \geq 14$.

In the case $q = -0.25$, the 4-SM $\sigma \equiv 0$ is established by the QC controller (1.30). The corresponding steady-state 4-SM accuracy is $|\sigma| \leq 9.1 \cdot 10^{-12}$, $|\dot{\sigma}| \leq$

$7.3 \cdot 10^{-9}$, $|\ddot{\sigma}| \leq 4.5 \cdot 10^{-5}$, and $|\dddot{\sigma}| \leq 8.9 \cdot 10^{-2}$ for $t > 10$. The control remains continuous till the very entrance into the SM at about $t = 7.7$.

1.6 Differentiation

Differentiation problem naturally accompanies the output regulation. Indeed, in order to realize control (1.12), one needs to estimate $\vec{\sigma}_{r-1}$ in real time. Since the output feedback capabilities of the SM-based differentiators are well known [2, 19, 29, 45, 46, 53, 64] in the following, we only present some new differentiators and their features.

1.6.1 Asymptotically Optimal Differentiation

The general exact differentiation problem is obviously ill-posed, since even small noises generate arbitrarily large derivative deviations. The issue is resolved by filtering. A smooth component of the signal is extracted by a filter and differentiated, while the difference is considered a noise and is neglected.

Popular linear observers/filters/differentiators [6, 31, 61] are based on the frequency-domain analysis considering high-frequency signals as noises. Correspondingly, all high-frequency signal components are rejected, making exact differentiation practically impossible. Indeed, even a smooth slowly changing signal, in general, contains high harmonics in its Fourier representation.

Nonlinear filtering can avoid that trade-off phenomenon. The following assumption [54] defines a large class of functions which allow exact differentiation in the absence of noises.

Let $\mathrm{Lip}_n(L)$ be the set of all scalar functions defined on $\mathbb{R}_+ = [0, \infty)$, and featuring the Lipschitz constant $L > 0$ of their nth derivative.

Assumption 1.1 ([44, 45]) The input signal $f(t)$, $f(t) = f_0(t) + \eta(t)$, consists of a bounded Lebesgue-measurable noise $\eta(t)$ and an unknown basic signal $f_0(t)$, $f_0 \in \mathrm{Lip}_n(L)$. The noise η is bounded, $|\eta| \leq \varepsilon$. The number $\varepsilon \geq 0$ is unknown. □

Differentiation problem. The problem is to evaluate the derivatives $f_0^{(j)}(t)$ in real time. The estimation is to be exact in the absence of noises and robust with respect to small noises.

Theorem 1.3 ([56]) *For any $t_0 > 0$, there exists such $\varepsilon_0 > 0$ that for any ε, $0 < \varepsilon \leq \varepsilon_0$, and any $f_0, f_1 \in \mathrm{Lip}_n(L)$ the inequality $\sup_{t \geq 0} |f_1(t) - f_0(t)| \leq \varepsilon$ implies the inequality*

$$\sup_{t \geq t_0} |f_1^{(i)}(t) - f_0^{(i)}(t)| \leq K_{i,n} (2L)^{\frac{i}{n+1}} \varepsilon^{\frac{n+1-i}{n+1}}, \quad i = 0, 1, \ldots, n, \tag{1.31}$$

where $K_{i,n} \in [1, \pi/2]$ are the Kolmogorov constants. The inequality is sharp, i.e., for each ε, i it becomes equality for certain functions.

The Kolmogorov constants appear in the celebrated Landau–Kolmogorov sharp inequalities [40]

$$M_i(\phi) \le K_{i,k} M_{n+1}^{\frac{i}{n+1}}(\phi) M_0^{\frac{n+1-i}{n+1}}(\phi),\ i = 0, \dots, k, \tag{1.32}$$

which hold for the functions $\phi : \mathbb{R} \to \mathbb{R}$ and $M_k \phi = \sup_{\mathbb{R}} |\phi^{(k)}|$. The constants are exactly calculated and satisfy $K_{i,n} \in [1, \pi/2]$, e.g., $K_{1,1} = \sqrt{2}$.

Let the differentiator to be constructed produce the outputs $z_0(t), z_1(t), \dots, z_n(t)$ as the real-time estimations of the derivatives $f_0(t), \dot{f}_0(t), \dots, f_0^{(n)}(t)$. Let the differentiator be exact on each function of $\mathrm{Lip}_n(L)$ after a finite-time transient. Then, in particular, it has to be exact on the functions f_1, f_0 appearing in (1.31) with f_1 and $\eta = f_1 - f_0$, respectively, considered as the input and the noise. Thus, we obtain that the best possible accuracy guaranteed for *any* input satisfies $\inf \sup |z_i - f_0^{(i)}| \ge K_{i,n} (2L)^{\frac{i}{n+1}} \varepsilon^{\frac{n+1-i}{n+1}}$.

Theorem 1.3 justifies the following definition. We say that a differentiator is **asymptotically optimal** if under Assumption 1.1 there exist such constants $\mu_i > 0$ that for each $f_0 \in \mathrm{Lip}_n(L)$, the accuracy,

$$|z_i(t) - f_0^{(i)}(t)| \le \mu_i L^{\frac{i}{n+1}} \varepsilon^{\frac{n+1-i}{n+1}},\ i = 0, 1, \dots, n, \tag{1.33}$$

holds after some finite-time transient.

An asymptotically optimal differentiator is known since [45]. The recursive form of the differentiator is

$$\begin{aligned}
\dot{z}_0 &= -\lambda_n L^{\frac{1}{n+1}} \lfloor z_0 - f(t) \rceil^{\frac{n}{n+1}} + z_1, \\
\dot{z}_1 &= -\lambda_{n-1} L^{\frac{1}{n}} \lfloor z_1 - \dot{z}_0 \rceil^{\frac{n-1}{n}} + z_2, \\
&\dots \\
\dot{z}_{n-1} &= -\lambda_1 L^{\frac{1}{2}} \lfloor z_{n-1} - \dot{z}_{n-2} \rceil^{\frac{1}{2}} + z_n, \\
\dot{z}_n &= -\lambda_0 L \ \mathrm{sign}(z_n - \dot{z}_{n-1}).
\end{aligned} \tag{1.34}$$

An infinite sequence of parameters λ_i can be built, valid for all natural n. In particular, $\{\lambda_0, \lambda_1, \dots\} = \{1.1, 1.5, 2, 3, 5, 7, 10, 12, \dots\}$ suffice for $n \le 7$ [54, 56]. In the absence of noises, the differentiator provides for the exact estimations in finite time. Equations (1.34) can be rewritten in the usual non-recursive form

$$\begin{aligned}
\dot{z}_0 &= -\tilde{\lambda}_n L^{\frac{1}{n+1}} \lfloor z_0 - f(t) \rceil^{\frac{n}{n+1}} + z_1, \\
\dot{z}_1 &= -\tilde{\lambda}_{n-1} L^{\frac{2}{n+1}} \lfloor z_0 - f(t) \rceil^{\frac{n-1}{n+1}} + z_2, \\
&\dots \\
\dot{z}_{n-1} &= -\tilde{\lambda}_1 L^{\frac{n}{n+1}} \lfloor z_0 - f(t) \rceil^{\frac{1}{n+1}} + z_n, \\
\dot{z}_n &= -\tilde{\lambda}_0 L \ \mathrm{sign}(z_0 - f(t)).
\end{aligned} \tag{1.35}$$

Table 1.1 Parameters $\tilde{\lambda}_0, \tilde{\lambda}_1, \ldots, \tilde{\lambda}_n$ for the differentiation orders $n = 0, 1, \ldots, 7$

0	1.1							
1	1.1	1.5						
2	1.1	2.12	2					
3	1.1	3.06	4.16	3				
4	1.1	4.57	9.30	10.03	5			
5	1.1	6.75	20.26	32.24	23.72	7		
6	1.1	9.91	43.65	101.96	110.08	47.69	10	
7	1.1	14.13	88.78	295.74	455.40	281.37	84.14	12

It is easy to see that $\tilde{\lambda}_0 = \lambda_0$, $\tilde{\lambda}_n = \lambda_n$, and $\tilde{\lambda}_j = \lambda_j \tilde{\lambda}_{j+1}^{j/(j+1)}$, $j = n-1, n-2, \ldots, 1$. The corresponding parameters are listed in Table 1.1.

In the presence of discrete measurements, the differentiator is applied with the difference $z_0(t_j) - f(t_j)$ fixed between the sampling moments t_j, t_{j+1}. In that way, the differentiation of a piece-wise constant signal is avoided, since it would invoke renewed transient of all z_i to 0 after each measurement.

Provided the sampling time intervals are uniformly bounded, $t_{j+1} - t_j \leq \tau$, after a FT transient differentiator (1.35) has the steady-state accuracy

$$\begin{aligned} &|z_i(t) - f_0^{(i)}(t)| \leq \mu_i L \rho^{n+1-i}, \; i = 0, 1, \ldots, n, \\ &\rho = \max[(\varepsilon/L)^{1/(n+1)}, \tau] \end{aligned} \tag{1.36}$$

for some constants $\mu_i > 0$ [45]. Hence, it is asymptotically optimal in the case $\tau = 0$ of continuous sampling. The same accuracy (with different constants μ_i) is maintained by properly discretized differentiator [54, 59] (see further).

1.6.2 Filtering Differentiators

Introduce the number $n_f \geq 0$ which is further called *the filtering order*. The new differentiator is formally defined as follows:

$$\begin{aligned}
\dot{w}_1 &= -\tilde{\lambda}_{n+n_f} L^{\frac{1}{n+n_f+1}} \lfloor w_1 \rceil^{\frac{n+n_f}{n+n_f+1}} + w_2, \\
&\dots \\
\dot{w}_{n_f-1} &= -\tilde{\lambda}_{n+2} L^{\frac{n_f-1}{n+n_f+1}} \lfloor w_1 \rceil^{\frac{n+2}{n+n_f+1}} + w_{n_f}, \\
\dot{w}_{n_f} &= -\tilde{\lambda}_{n+1} L^{\frac{n_f}{n+n_f+1}} \lfloor w_1 \rceil^{\frac{n+1}{n+n_f+1}} + z_0 - f(t), \\
\dot{z}_0 &= -\tilde{\lambda}_{n} L^{\frac{n_f+1}{n+n_f+1}} \lfloor w_1 \rceil^{\frac{n}{n+n_f+1}} + z_1, \\
&\dots \\
\dot{z}_{n-1} &= -\tilde{\lambda}_{1} L^{\frac{n+n_f}{n+n_f+1}} \lfloor w_1 \rceil^{\frac{1}{n+n_f+1}} + z_n, \\
\dot{z}_n &= -\tilde{\lambda}_0 L \operatorname{sign}(w_1).
\end{aligned} \tag{1.37}$$

In the case $n_f = 0$, the first n_f equations disappear, and $w_1 = z_0 - f(t)$ is formally substituted for w_1, which results in the standard differentiator (1.35). Show that differentiator (1.37) demonstrates significant filtering properties.

A (noise) function $\nu(t), \nu : [0, \infty) \to \mathbb{R}$, is called a *signal of the (global) filtering order* $k \geq 0$ if ν is a locally integrable Lebesgue-measurable function, and there exists a uniformly bounded solution $\xi(t)$ of the equation $\xi^{(k)} = \nu$. Any number exceeding $\sup |\xi(t)|$ is called the kth-*order integral magnitude of* ν.

The bounded noise described in Assumption 1.1 is of the zeroth filtering order and the zeroth-order integral magnitude ε.

Assumption 1.2 In addition to Assumption 1.1 assume that the input signal contains additional possibly unbounded noises, $f(t) = f_0(t) + \eta(t) + \nu_1(t) + \cdots + \nu_{n_f}(t)$, where each ν_k, $k = 1, \dots, n_f$, is a signal of the filtering order k and the kth-order integral magnitude $\delta_k \geq 0$. □

The following assumption is the natural extension of Assumption 1.2 to the discrete sampling case.

Assumption 1.3 Let the signal $f(t)$ be sampled at times $t_j, 0 < t_{j+1} - t_j < \tau, t_j \to \infty$. The discrete sampling at times t_j creates piece-wise-constant noises, $\tilde{\nu}_k(t) = \nu_k(t_j)$, $t \in [t_j, t_{j+1})$, of the global filtering order k and the integral magnitude δ_k, $f(t_j) = f_0(t_j) + \eta(t_j) + \nu_1(t_j) + \cdots + \nu_{n_f}(t_j)$. □

The latter assumption is definitely restrictive. In particular, it holds in the SMC, and allows to directly differentiate the chattering SMC, producing the equivalent control and its derivatives [58].

As follows from (1.36), the standard differentiator (1.35) is robust with respect to the noises $\eta(t)$ of the filtering order 0 [45]. The following theorem shows that differentiators (1.37) of the filtering order n_f are robust with respect to sums of the bounded noise $\eta(t)$ and of *possibly unbounded* noises of the filtering orders $1, \dots, n_f$.

Theorem 1.4 *Under Assumptions 1.1, 1.2, 1.3 let the input be sampled with the sampling step not exceeding* $\tau \geq 0$*, whereas the case* $\tau = 0$ *formally corresponds to the continuous (not discrete) sampling. Then differentiator* (1.37) *in FT provides for the accuracy*

$$
\begin{aligned}
&|z_i(t) - f_0^{(i)}(t)| \le \mu_i L \rho^{n+1-i}, \ i = 0, 1, \dots, n, \\
&|w_1(t)| \le \mu_{w1} L \rho^{n+n_f+1}, \\
&\rho = \max[(\tfrac{\varepsilon}{L})^{\frac{1}{n+1}}, (\tfrac{\delta_1}{L})^{\frac{1}{n+2}}, \dots, (\tfrac{\delta_{n_f}}{L})^{\frac{1}{n+n_f+1}}, \tau],
\end{aligned}
\tag{1.38}
$$

for some $\{\mu_i\}$, μ_{w1} *only depending on the choice of* $\{\lambda_l\}$, $l = 0, \dots, n + n_f$.

Thus, the filtering differentiator with $n_f > 0$ is also asymptotically optimal, i.e., is robust and exact, but it also possesses strong filtering abilities. Assumptions 1.2, 1.3 can be difficult to check, but the differentiator will "itself detect" whether they hold.

Actually, though we keep the same notation δ_k for $\tau = 0$ and $\tau > 0$ preserving the same accuracy formula (1.38), the discrete and continuous sampling cases are quite different in the filtering context. The filtering order, as well as the integral magnitude of a continuous-time noise, can change or disappear in discrete sampling. The phenomenon is well known in the traditional linear filtering and is directly treated in Sect. 1.7.

Proof According to the filtering-order definition, introduce the functions $\xi_k(t)$, $|\xi_k| \le \delta_k$, $\xi_k^{(k)}(t) = \nu_k(t)$ for $\tau = 0$ and $\xi_k^{(k)}(t) = \tilde{\nu}_k(t)$ for $\tau > 0$. Let

$$
\begin{aligned}
&\omega_1 = w_1 + \xi_{n_f}, \omega_2 = w_2 + \dot{\xi}_{n_f} + \xi_{n_f-1}, \dots, \\
&\omega_{n_f} = w_{n_f} + \xi_k^{(n_f-1)} + \dots + \dot{\xi}_2 + \xi_1; \\
&\sigma_i = z_i - f_0^i, \ i = 0, \dots, n.
\end{aligned}
\tag{1.39}
$$

Then $f = f_0 + \eta + \dot{\xi}_1 + \dots + \xi_{n_f}^{(n_f)}$, and one can rewrite (1.37) in the form

$$
\begin{aligned}
\dot{\omega}_1 &= -\tilde{\lambda}_{n+n_f} L^{\frac{1}{n+n_f+1}} \lfloor \omega_1 - \xi_{n_f} \rceil^{\frac{n+n_f}{n+n_f+1}} + \omega_2 - \xi_{n_f-1}, \\
\dot{\omega}_2 &= -\tilde{\lambda}_{n+n_f-1} L^{\frac{2}{n+n_f+1}} \lfloor \omega_1 - \xi_{n_f} \rceil^{\frac{n+n_f-1}{n+n_f+1}} + \omega_3 - \xi_{n_f-2}, \\
&\dots \\
\dot{\omega}_{n_f-1} &= -\tilde{\lambda}_{n+2} L^{\frac{n_f-1}{n+n_f+1}} \lfloor \omega_1 - \xi_{n_f} \rceil^{\frac{n+2}{n+n_f+1}} + \omega_{n_f} - \xi_1, \\
\dot{\omega}_{n_f} &= -\tilde{\lambda}_{n+1} L^{\frac{n_f}{n+n_f+1}} \lfloor \omega_1 - \xi_{n_f} \rceil^{\frac{n+1}{n+n_f+1}} + \sigma_0 + \eta, \\
\dot{\sigma}_0 &= -\tilde{\lambda}_n L^{\frac{n_f+1}{n+n_f+1}} \lfloor \omega_1 - \xi_{n_f} \rceil^{\frac{n}{n+n_f+1}} + z_1, \\
&\dots \\
\dot{\sigma}_{n-1} &= -\tilde{\lambda}_1 L^{\frac{n+n_f}{n+n_f+1}} \lfloor \omega_1 - \xi_{n_f} \rceil^{\frac{1}{n+n_f+1}} + \sigma_n, \\
\dot{\sigma}_n &\in -\tilde{\lambda}_0 L \ \mathrm{sign}(\omega_1 - \xi_{n_f}) + [-L, L],
\end{aligned}
\tag{1.40}
$$

which is a small perturbation of the FT stable homogeneous error dynamics of the standard $(n + n_f)$th-order differentiator (1.35). Obviously, $\deg \omega_k = n + n_f + 2 - k$, $\deg z_i = n + 1 - i$, $\deg t = -q = 1$.

It follows from (1.3) that the accuracy $\sup |\sigma_i| = O(\rho^{n+1-i})$, $\sup |\omega_k| = O(\rho^{n+n_f+2-k})$ [53] is established in FT. Now the accuracy of z_i is directly obtained from these relations. Taking into account that $\sup|\dot{\omega}_k| = O(\rho^{n+n_f+1-k})$ obtain the accuracy of w_k from (1.39). □

It is not always easy to check the global filtering order of the noise. The following notion allows to circumvent that condition.

A locally (essentially) bounded Lebesgue-measurable function $\nu(t), \nu : [0, \infty) \to \mathbb{R}$ is said to have the *local filtering order* $k > 0$ if there exist numbers $T > 0$, $a_0, a_1, \dots, a_{k-1} \geq 0$, such that for any $t_1 \geq 0$ there exists a solution $\xi(t)$, $t \in [t_1, t_1 + T]$, of the equation $\xi^{(k)}(t) = \nu(t)$ which satisfies $|\xi^{(l)}(t)| \leq a_l$ for $l = 0, 1, \dots, k-1$. Each number a_l is called the *local* $(k-l)$*-th-order integral magnitude of* ν. Signals of local sampling filtering order 0 by definition are uniformly bounded signals of the magnitude a_0.

Introduce the homogeneous norm $||a||_{h\infty} = \max_{i=0,\dots,k-1} |a_i|^{1/(k-i)}$ for $k > 0$.

For example, the signal $\cos \omega t$, $\omega > 0$, is of any global filtering order. It is also of any local filtering order $k \geq 0$, with $||a||_{h\infty} = 2/\omega$ for $k > 0$.

The following lemma is a modified and corrected version of the similar lemma from [50].

Lemma 1.1 *Any signal* $\nu(t)$ *of the* ***local*** *filtering order* $k \geq 0$ *can be represented as* $\nu = \eta_0 + \eta_1 + \eta_k$, *where* η_0, η_1, η_k *are signals of the global filtering orders* $0, 1, k$.

Fix any number $a_* > 0$. *Then, provided* $||a||_{h\infty} \leq a_*$, *the integral magnitudes of the signals* η_0, η_1, η_k *are calculated as* $\gamma_0 ||a||_{h\infty}/T$, $\gamma_1 ||a||_{h\infty}$, $\gamma_k ||a||_{h\infty}^k$, *respectively, where the constants* $\gamma_0, \gamma_1, \gamma_k > 0$ *only depend on k and* a_*.

In the important particular case $k = 1$ *get* $||a||_{h\infty} = a_0$, $\nu = \eta_0 + \eta_1$, *and independently of* a_* *get* $\gamma_0 = 1$, $\gamma_1 = 2$, *i.e.,* $|\eta_0| \leq a_0/T$, *and the first-order integral magnitude of* η_1 *is* $2a_0$.

In the case $k = 1$, the proof is very simple. Indeed, it is enough to take

$$\nu_0(t) = \frac{1}{T}\int_{mT}^{(m+1)T} \nu(s)ds, \;\; \nu_1 = \nu - \nu_0, \;\; t \in [mT, (m+1)T),$$
$$m = 0, 1, \dots$$

Obviously, the first-local-filtering-order condition trivially implies the kth-local-order condition for any $k > 1$. Thus, according to the lemma, any signal of the first local filtering order can be represented as the sum of a signal of any predefined filtering order, a signal of the first global filtering order and a bounded signal. Differentiators of higher filtering orders are capable of treating more representations of such kind. Hence, provided the differentiation order is fixed, in the presence of noise one can expect higher accuracy from the differentiators of higher filtering order *for sufficiently small sampling steps.*

1.7 Discrete Differentiators

Contrary to the feedback controllers of Sect. 1.5, differentiators have their own dynamics, which require proper realization on computer-based devices.

Once more consider the sampling at the times $t_0, t_1, \ldots, t_0 = 0, t_{j+1} - t_j = \tau_j \leq \tau$, $\lim t_j = \infty$. In Sect. 1.6.2, the differentiator was described by differential equations over the sampling intervals. Any practical application involves discrete sampling and some kind of *numeric integration* of the differentiator dynamics between the sampling instants.

It follows from the Nyquist–Shannon sampling rate principle that not all sampling times' sequences properly represent a noise, since noises small in average under one sampling sequence can be large under another. Suppose that sampling time sequences are available for any $\tau > 0$. In particular, the number of such sampling sequences is infinite.

Notation. Denote $\delta_j \phi = \phi(t_{j+1}) - \phi(t_j)$ for any sampled signal $\phi(t)$.

A discretely sampled signal $\nu : [0, \infty) \to \mathbb{R}$ is said to have the global *sampling filtering order* $k \geq 0$ and the global kth-order *integral sampling magnitude* $a \geq 0$ if for each available sampling sequence t_j there exists a discrete vector signal $\xi(t_j) = (\xi_0(t_j), \ldots, \xi_k(t_j))^T \in \mathbb{R}^{k+1}$, $j = 0, 1, \ldots$, which satisfies the relations $\delta_j \xi_i = \xi_{i+1}(t_j)\tau_j$, $i = 0, 1, \ldots, k-1$, $\xi_k(t_j) = \nu(t_j)$, $|\xi_0(t_j)| \leq a$.

Assumption 1.4 The sampled noise signal is comprised of $n_f + 1$ components, $f(t_j) = f_0(t_j) + \eta(t_j) + \nu_1(t_j) + \cdots + \nu_{n_f}(t_j)$. Each discretely sampled signal $\nu_l(t_j)$, $l = 1, 2, \ldots, n_f$, has the global sampling filtering order l and integral magnitude δ_l and possibly is unbounded. □

Denote the (n_f, n)-differentiator (1.37) by $(\dot{w}, \dot{z})^T = D_{n_f,n}(w, z, L)$. The simplest discrete differentiator is based on the Euler integration and gets the form $(w(t_{j+1}), z(t_{j+1}))^T = (w(t_j), z(t_j))^T + D_{n_f,n}(w(t_j), z(t_j), L)\tau_j$. Unfortunately, in that case the errors $z_i - f_0^{(i)}$, $i = 1, \ldots, n$, become proportional to τ [59] which is much worse than the continuous-time accuracy (1.40).

The following discretization utilizes the ideas [59] to restore the accuracy (1.38):

$$
\begin{aligned}
(w(t_{j+1}), z(t_{j+1}))^T = (w(t_j), z(t_j))^T + D_{n_f,n}(w(t_j), z(t_j), L)\tau_j \\
+ T_{n_f,n}(z(t_j), \tau_j),
\end{aligned}
$$

$$
\begin{pmatrix} T_0 \\ \ldots \\ T_{n_f-1} \\ T_{n_f} \\ \ldots \\ T_{n_f+i} \\ \ldots \\ T_{n_f+n-2} \\ T_{n_f+n-1} \\ T_{n_f+n} \end{pmatrix} = \begin{pmatrix} 0 \\ \ldots \\ 0 \\ \frac{1}{2!} z_2(t_j)\tau_j^2 + \cdots + \frac{1}{n!} z_n(t_j)\tau_j^n \\ \ldots \\ \sum_{s=i+2}^{n} \frac{1}{(s-i)!} z_s(t_j)\tau_j^{s-i} \\ \ldots \\ \frac{1}{2!} z_n(t_j)\tau_j^2 \\ 0 \\ 0 \end{pmatrix}. \tag{1.41}
$$

Here $T_{n_f,n} \in \mathbb{R}^{n_f+n+1}$, also see example (1.42). In particular, $T_{n_f,0}(w, z, \tau) = 0 \in \mathbb{R}^{n_f+1}$, $T_{n_f,1}(w, z, \tau) = 0 \in \mathbb{R}^{n_f+2}$.

Theorem 1.5 *Under Assumptions 1.1, 1.4 discrete differentiator* (1.41) *provides for the same accuracy asymptotics* (1.38)*, as its continuous-time analogue* (1.37) *(the asymptotics coefficients can be different).*

The proof is similar to [58].

Similarly to the definition of the continuous-time local filtering order, one defines the local sampling filtering order using first-order divided finite differences instead of derivatives. The discrete analogue of Lemma 1.1 is also true for the discrete sampling, i.e., any signal of the local sampling filtering order k can be presented as the sum of the sampled noises of the global sampling filtering orders 0, 1, and k. The case $k = 1$ is especially simple as well.

There is an analogue of the Nyquist sampling rate principle. Let the continuous-time noise signal $\nu(t)$ be of the global/local filtering order k. Then its sampled version can have no sampling filtering order at all. But under the additional condition that it has a Lipschitz constant, though possibly very large (the model example is $\cos \omega t$), for sufficiently small τ the sampled signal $\nu(t_j)$ is of the *local sampling* filtering order k. Then the accuracy of differentiator (1.41) can be directly estimated.

1.7.1 Numeric Differentiation

Consider the simple input signal

$$f(t) = 0.8 \cos t - \sin(0.2t) + \nu(t),$$

where ν is a noise. Obviously, $f_0(t) = 0.8 \cos t - \sin(0.2t)$, $|f_0^{(i)}| \le 1$ for $i = 1, 2, \ldots$. Let the sampling step be constant, $\tau_j = \tau$. In most experiments, $\tau = 10^{-6}$ is taken to reveal the filtering properties.

The main problem of the numeric differentiation is the trade-off between the differentiation accuracy in the absence of noises and its robustness to small sampling noises. The simplest method is based on the standard MATLAB divided differences. Indeed, it has no transient and works quite well in the absence of noises. The estimation $\hat{f}_0^{(4)}$ of $f_0^{(4)}$ has the accuracy of about 0.01 for $\tau = 10^{-3}$ (Fig. 1.2a). Unfortunately, in spite of the absence of noises that estimation explodes already for $\tau = 10^{-4}$ due to the digital roundup errors (Fig. 1.2b). The error is already of the order of $6 \cdot 10^5$ for $\tau = 10^{-5}$.

Another option is the classical high-gain observer (HGO) [6] with the characteristic polynomial $(p + 1000)^5$ and the sampling period $\tau = 10^{-6}$. In the absence of noises, it provides for very high accuracy (Fig. 1.2c). Its best accuracy (see further for the filtering differentiator) is obtained for $\tau = 10^{-5}$, $\sup |z_i - f_0^{(i)}| \le 2.2 \cdot 10^{-15}$, $8.5 \cdot 10^{-12}$, $1.2 \cdot 10^{-8}$, $9.8 \cdot 10^{-6}$, $4.3 \cdot 10^{-3}$ for $i = 0, 1, 2, 3, 4$, respectively. It remains practically the same for smaller τ.

Unfortunately, in the presence of a small Gaussian noise with the distribution $N(0, 0.001^2)$, the accuracy deteriorates to $\sup |z_i - f_0^{(i)}| \le 3.3 \cdot 10^{-4}$, 0.58, $5.7 \cdot$

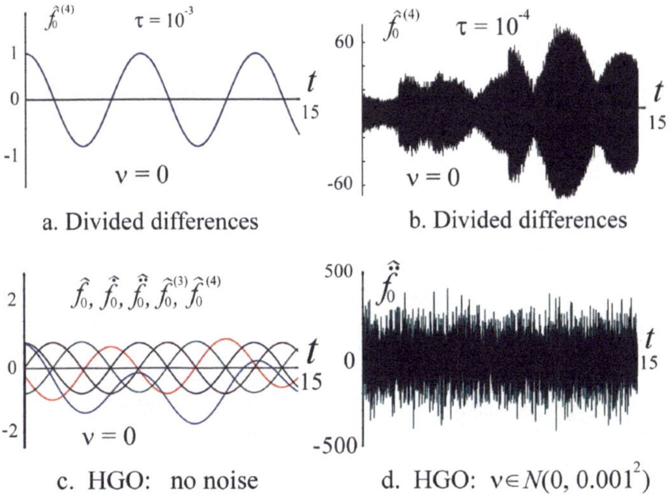

Fig. 1.2 Difficulty of numeric differentiation. **a** The divided differences' estimation of $f_0^{(4)}$ in the absence of noises, $\nu = 0$, for $\tau = 10^{-3}$; **b** the same estimation for $\tau = 10^{-4}$. Differentiation by the HGO with the multiple eigenvalue -1000 for $\tau = 10^{-6}$. The graphs are cut from above and from below to remove the high transient values (up to 10^{11}). **c** in the absence of noises the accuracy is excellent; **d** estimation $\hat{\ddot{f}}_0(t)$ of $\ddot{f}_0(t)$ in the presence of the Gaussian noise $\nu \in N(0, 0.001^2)$

$10^2,\ 2.8 \cdot 10^5,\ 5.6 \cdot 10^7$ for $i = 0, 1, 2, 3, 4$, respectively, for $\tau = 10^{-6}$ (Fig. 1.2d). Note [80] that increasing the small parameter one gets accuracies similar to those of the SM-based differentiators in the presence of noises not exceeding ± 0.002, but this deliberately sacrifices the differentiator accuracy in the absence of noises.

Now consider the discrete filtering differentiator (1.41) of the differentiation order $n = 4$ and the filtering order $n_f = 3$,

$$\begin{aligned}
\delta_j w_1 &= [-12L^{1/8} \lfloor w_1(t_j) \rceil^{7/8} + w_2(t_j)]\tau_j, \\
\delta_j w_2 &= [-84.14L^{2/8} \lfloor w_1(t_j) \rceil^{6/8} + w_3(t_j)]\tau_j, \\
\delta_j w_3 &= [-281.37L^{3/8} \lfloor w_1(t_j) \rceil^{5/8} + z_0(t_j) - f(t_j)]\tau_j, \\
\delta_j z_0 &= [-455.40L^{4/8} \lfloor w_1(t_j) \rceil^{4/8} + z_1(t_j)]\tau_j + z_2(t_j)\tfrac{\tau_j^2}{2} + z_3(t_j)\tfrac{\tau_j^3}{6} + z_4(t_j)\tfrac{\tau_j^4}{24}, \\
\delta_j z_1 &= [-295.74L^{5/8} \lfloor w_1(t_j) \rceil^{3/8} + z_2(t_j)]\tau_j + z_3(t_j)\tfrac{\tau_j^2}{2} + z_4(t_j)\tfrac{\tau_j^3}{6}, \\
\delta_j z_2 &= [-88.78L^{6/8} \lfloor w_1(t_j) \rceil^{2/8} + z_3(t_j)]\tau_j + z_4(t_j)\tfrac{\tau_j^2}{2}, \\
\delta_j z_3 &= [-14.13L^{7/8} \lfloor w_1(t_j) \rceil^{1/8} + z_4(t_j)]\tau_j, \\
\delta_j z_4 &= [-1.1L\ \mathrm{sign}(w_1(t_j))]\tau_j.
\end{aligned} \tag{1.42}$$

Parameters $\tilde{\lambda}_i$ are taken from the row number $n + n_f = 4 + 3 = 7$ of Table 1.1. Let $L = 1$. Choose the initial values $z = 0 \in \mathbb{R}^5$, $w = 0 \in \mathbb{R}^3$ at $t = 0$ to make the transient evident.

First, consider the case $\nu = 0$ to show that the filtering differentiator is exact. Convergence of estimations $z_0, \ldots, z_4$ to the derivatives $f_0, \ldots, f_0^{(4)}$ is shown in Fig. 1.3a. The resulting accuracies are

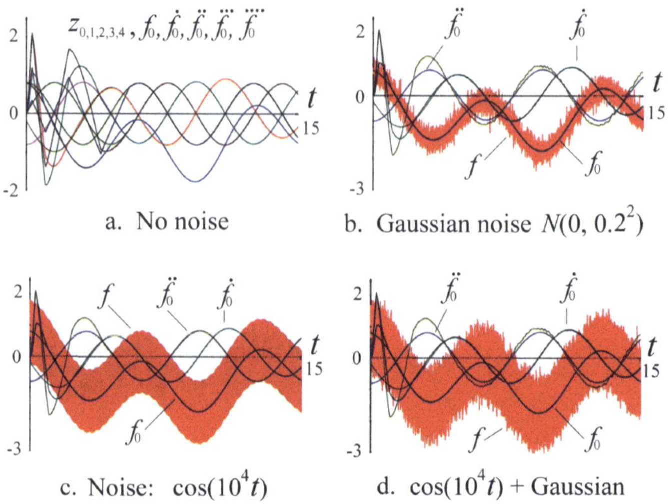

Fig. 1.3 Performance of the discrete filtering differentiator (1.42) for $L = 1$ and $\tau = 10^{-6}$. **a** in the absence of noise; **b** for the Gaussian noise $\nu = \nu_G \in N(0, 0.2^2)$; **c** for the noise $\nu = \cos 10^4 t$; **d** for the noise $\nu = \cos 10^4 t + \nu_G$. The sampled signal f is shown in red

$\sup|z_i - f_0^{(i)}| \leq 1.5 \cdot 10^{-14},\ 3.4 \cdot 10^{-11},\ 3.7 \cdot 10^{-8},\ 2.0 \cdot 10^{-5},\ 6.0 \cdot 10^{-3}$, respectively, for $i = 0, 1, 2, 3, 4$ and $\tau = 10^{-4}$, and $\sup|z_i - f_0^{(i)}| \leq 2.4 \cdot 10^{-15},\ 6.5 \cdot 10^{-12},\ 8.3 \cdot 10^{-9},\ 7.2 \cdot 10^{-6},\ 3.8 \cdot 10^{-3}$, respectively, for $i = 0, 1, 2, 3, 4$ and $\tau = 10^{-5}$. These are already the best possible accuracies to be obtained by standard software due to the digital roundup computer errors [56, 59]. They practically coincide with the above best accuracy of the HGO.

Estimations of the derivatives 0, 1, 2 are shown for the noises $\nu = \nu_G \in N(0, 0.2^2)$, $\nu = \cos 10^4 t$, and $\nu = \cos 10^4 t + \nu_G$ in Fig. 1.3b and Fig. 1.3d, respectively. Noise ν_G is normally distributed with the standard deviation 0.2. Note that the differentiator almost removes the large high-frequency noise $\nu = \cos 10^4 t$, providing for the accuracies $\sup|z_i - f_0^{(i)}| \leq 2.4 \cdot 7.3^{-6},\ 3.6 \cdot 10^{-4},\ 7.3 \cdot 10^{-3},\ 6.8 \cdot 10^{-2},\ 0.33$, respectively, for $i = 0, 1, 2, 3, 4$ and $\tau = 10^{-6}$, and demonstrates the accuracy $\sup|z_i - f_0^{(i)}| \leq 3.7 \cdot 10^{-3},\ 3.7 \cdot 10^{-2},\ 0.2,\ 0.60,\ 0.97$, respectively, for $i = 0, 1, 2, 3, 4$ for the combined noise $\nu = \cos 10^4 t + \nu_G$.

1.8 Conclusion

New controllers and differentiators have been demonstrated.

The presented method of control template design easily produces infinite number of new controllers stabilizing the output σ for any relative degree r and any homogeneity degree $q \geq -1/r$, $\deg \sigma = 1$. Instead of choosing a concrete already published controller family valid for all r and then adjust the coefficients, one can now directly develop a new controller which fits a concrete problem and its concrete relative degree.

The developed controllers are of the simple feedback form and do not have internal dynamics. Therefore, their discrete realization is not relevant to this chapter. The situation changes if the output feedback control is considered that involves differentiator (1.35) in the feedback, or an integrator is inserted in order to diminish the chattering. Corresponding discretization issues are considered in detail in [52, 55]. Filtering differentiators have still not been studied in that aspect.

The main qualitative properties of the new controllers are directly determined by their relative and homogeneity degrees. Nevertheless, the controllers are very different in their sensitivity to noises and time delays, convergence rate, and calculation complexity.

One can expect some further optimization for each concrete control template choice. For example, for each relative degree, all linear controllers constitute one template featuring the homogeneity degree 0. There are infinitely many more control templates even for the fixed zero homogeneity degree.

To demonstrate the technique, we have presented a completely new family of stabilizers valid for any r, $q \geq -1/r$. The controllers are classified as QC r-SM controllers for $q = -1/r$, i.e., the corresponding control assures finite-time stabilization and remains continuous until the very establishment of $\sigma \equiv 0$. The generated control is continuous for the homogeneous degrees $q > -1/r$. In particular, in the case of the homogeneity degree $q = 0$, new nonlinear *continuous* controllers exponentially stabilize linear disturbed systems with discontinuous matched disturbances. Controllers corresponding to $q > 0$ feature fixed-time stabilization to any vicinity of the origin.

New differentiators are called filtering differentiators. They demonstrate high filtering abilities in the presence of unbounded noises of small average values. At the same time, they remain exact in the absence of noises and feature the asymptotically optimal accuracy in the presence of bounded noises.

The evaluated accuracy estimation depends on the expansion of the noise into a sum of noises of different filtering orders. There are infinitely many such representations, which means that the resulting accuracy automatically corresponds to the representation providing for the minimal estimated differentiation errors.

That simple logic often leads to surprisingly good differentiation accuracies in the presence of large chaotic noises. The authors believe that the new differentiators are to replace the standard SM-based differentiators [45] in numerous applications.

References

1. Acary, V., Brogliato, B.: Implicit Euler numerical scheme and chattering-free implementation of sliding mode systems. Syst. Control Lett. **59**(5), 284–293 (2010)
2. Angulo, M., Fridman, L., Levant, A.: Output-feedback finite-time stabilization of disturbed LTI systems. Automatica **48**(4), 606–611 (2012)

3. Angulo, M.T., Fridman, L., Moreno, J.A.: Output-feedback finite-time stabilization of disturbed feedback linearizable nonlinear systems. Automatica **49**(9), 2767–2773 (2013)
4. Angulo, M.T., Moreno, J.A., Fridman, L.M.: Robust exact uniformly convergent arbitrary order differentiator. Automatica **49**(8), 2489–2495 (2013)
5. Apaza-Perez, W.A., Fridman, L., Moreno, J.A.: Higher order sliding-mode observers with scaled dissipative stabilisers. Int. J. Control **91**(11), 2511–2523 (2018)
6. Atassi, A.N., Khalil, H.K.: Separation results for the stabilization of nonlinear systems using different high-gain observer designs. Syst. Control Lett. **39**(3), 183–191 (2000)
7. Bacciotti, A., Rosier, L.: Liapunov Functions and Stability in Control Theory. Springer, London (2005)
8. Bartolini, G.: Chattering phenomena in discontinuous control systems. Int. J. Syst. Sci. **20**, 2471–2481 (1989)
9. Bartolini, G., Ferrara, A., Usai, E.: Chattering avoidance by second-order sliding mode control. IEEE Trans. Autom. Control **43**(2), 241–246 (1998)
10. Bartolini, G., Pisano, A., Punta, E., Usai, E.: A survey of applications of second-order sliding mode control to mechanical systems. Int. J. Control **76**(9/10), 875–892 (2003)
11. Bartolini, G., Punta, E., Zolezzi, T.: Approximability properties for second-order sliding mode control systems. IEEE Trans. Autom. Control **52**(10), 1813–1825 (2007)
12. Bartolini, G., Zolezzi, T.: Variable structure systems nonlinear in the control law. IEEE Trans. Autom. Control **30**(7), 681–684 (1985)
13. Bernuau, E., Efimov, D., Perruquetti, W., Polyakov, A.: On homogeneity and its application in sliding mode control. J. Frankl. Inst. **351**(4), 1866–1901 (2014)
14. Bernuau, E., Polyakov, A., Efimov, D., Perruquetti, W.: Verification of ISS, iISS and IOSS properties applying weighted homogeneity. Syst. Control Lett. **62**(12), 1159–1167 (2013)
15. Bhat, S.P., Bernstein, D.S.: Geometric homogeneity with applications to finite-time stability. Math. Control, Signals Syst. **17**(2), 101–127 (2007)
16. Boiko, I., Fridman, L., Pisano, A., Usai, E.: Analysis of chattering in systems with second-order sliding modes. IEEE Trans. Autom. Control **52**(11), 2085–2102 (2007)
17. Cruz-Zavala, E., Moreno, J.A.: Lyapunov approach to higher-order sliding mode design. Recent Trends in Sliding Mode Control, pp. 3–28. Institution of Engineering and Technology IET, London (2016)
18. Cruz-Zavala, E., Moreno, J.A.: Homogeneous high order sliding mode design: a Lyapunov approach. Automatica **80**, 232–238 (2017)
19. Davila, J., Fridman, L., Pisano, A., Usai, E.: Finite-time state observation for non-linear uncertain systems via higher-order sliding modes. Int. J. Control **82**(8), 1564–1574 (2009)
20. Defoort, M., Floquet, T., Kokosy, A., Perruquetti, W.: A novel higher order sliding mode control scheme. Syst. Control Lett. **58**(2), 102–108 (2009)
21. Ding, S.H., Levant, A., Li, S.H.: Simple homogeneous sliding-mode controller. Automatica **67**(5), 22–32 (2016)
22. Dinuzzo, F., Ferrara, A.: Higher order sliding mode controllers with optimal reaching. IEEE Trans. Autom. Control **54**(9), 2126–2136 (2009)
23. Dvir, Y., Efimov, D., Levant, A., Polyakov, A., Perruquetti, W.: Acceleration of finite-time stable homogeneous systems. Int. J. Robust Nonlinear Control, submitted (2016)
24. Edwards, C., Spurgeon, S.K.: Sliding Mode Control: Theory and Applications. Taylor & Francis, London (1998)
25. Edwards, E., Shtessel, Y.B.: Adaptive continuous higher order sliding mode control. Automatica **65**, 183–190 (2016)
26. Efimov, D., Levant, A., Polyakov, A., Perruquetti, W.: Discretization of asymptotically stable homogeneous systems by explicit and implicit Euler methods. In: 55th IEEE Conference on Decision and Control, CDC'2016, Las-Vegas, 12–14 Dec 2016
27. Emelyanov, S.V., Utkin, V.I., Taran, V.A., Kostyleva, N.E.: Theory of Variable Structure Systems (in Russian). Nauka, Moscow (1970)
28. Feng, Y., Yu, X., Man, Z.: Non-singular terminal sliding mode control of rigid manipulators. Automatica **38**(12), 2159–2167 (2002)

29. Ferreira, A., Bejarano, F.J., Fridman, L.M.: Robust control with exact uncertainties compensation: with or without chattering? IEEE Trans. Control Syst. Technol. **19**(5), 969–975 (2011)
30. Filippov, A.F.: Differential Equations with Discontinuous Right-Hand Sides. Kluwer Academic Publishers, Dordrecht (1988)
31. Fliess, M., Join, C., Sira-Ramírez, H.: Non-linear estimation is easy. Int. J. Model., Identif. Control **4**(1), 12–27 (2008)
32. Floquet, T., Barbot, J.P., Perruquetti, W.: Higher-order sliding mode stabilization for a class of nonholonomic perturbed systems. Automatica **39**(6), 1077–1083 (2003)
33. Fridman, L.: Chattering analysis in sliding mode systems with inertial sensors. Int. J. Control **76**(9/10), 906–912 (2003)
34. Golembo, B.Z., Emelyanov, S.V., Utkin, V.I., Shubladze, A.M.: Application of piecewise-continuous dynamic-systems to filtering problems. Autom. Remote Control **37**(3), 369–377 (1976)
35. Harmouche, M., Laghrouche, S., Chitour, Y., Hamerlain, M.: Stabilisation of perturbed chains of integrators using Lyapunov-based homogeneous controllers. Int. J. Control **90**(12), 2631–2640 (2017)
36. Hong, Y.G.: Finite-time stabilization and stabilizability of a class of controllable systems. Syst. Control Lett. **46**(4), 231–236 (2002)
37. Isidori, A.: Nonlinear Control Systems I. Springer, New York (1995)
38. Kawski, M.: Homogeneous stabilizing feedback laws. Control Theory Adv. Technol. **6**, 497–516 (1990)
39. Koch, S., Reichhartinger, M.: Discrete-time equivalent homogeneous differentiators. In: 15th International Workshop on Variable Structure Systems (VSS), pp. 354–359 (2018)
40. Kolmogoroff, A.N.: On inequalities between upper bounds of consecutive derivatives of an arbitrary function defined on an infinite interval. Am. Math. Soc. Transl., Ser. **1**(2), 233–242 (1962)
41. Laghrouche, S., Harmouche, M., Ahmed, F.S., Chitour, Y.: Control of PEMFC air-feed system using Lyapunov-based robust and adaptive higher order sliding mode control. IEEE Trans. Control Syst. Technol. **23**(4), 1594–1601 (2015)
42. Laghrouche, S., Plestan, F., Glumineau, A.: Higher order sliding mode control based on integral sliding mode. Automatica **43**(3), 531–537 (2007)
43. Levant, A.: Sliding order and sliding accuracy in sliding mode control. Int. J. Control **58**(6), 1247–1263 (1993)
44. Levant, A.: Robust exact differentiation via sliding mode technique. Automatica **34**(3), 379–384 (1998)
45. Levant, A.: Higher order sliding modes, differentiation and output-feedback control. Int. J. Control **76**(9/10), 924–941 (2003)
46. Levant, A.: Homogeneity approach to high-order sliding mode design. Automatica **41**(5), 823–830 (2005)
47. Levant, A.: Quasi-continuous high-order sliding-mode controllers. IEEE Trans. Autom. Control **50**(11), 1812–1816 (2005)
48. Levant, A.: Chattering analysis. IEEE Trans. Autom. Control **55**(6), 1380–1389 (2010)
49. Levant, A.: Non-Lyapunov homogeneous SISO control design. In: 56th Annual IEEE Conference on Decision and Control (CDC), pp. 6652–6657 (2017)
50. Levant, A.: Filtering differentiators and observers. In: 15th International Workshop on Variable Structure Systems (VSS), pp. 174–179 (2018)
51. Levant, A., Efimov, D., Polyakov, A., Perruquetti, W.: Stability and robustness of homogeneous differential inclusions. In: Proceedings of the 55th IEEE Conference on Decision and Control, Las-Vegas, 12–14 Dec 2016
52. Levant, A., Livne, M.: Uncertain disturbances' attenuation by homogeneous MIMO sliding mode control and its discretization. IET Control Theory Appl. **9**(4), 515–525 (2015)
53. Levant, A., Livne, M.: Weighted homogeneity and robustness of sliding mode control. Automatica **72**(10), 186–193 (2016)

54. Levant, A., Livne, M.: Globally convergent differentiators with variable gains. Int. J. Control **91**(9), 1994–2008 (2018)
55. Levant, A., Livne, M., Lunz, D.: On discretization of high order sliding modes. In: Fridman, L., Barbot, J.P., Plestan, F. (eds.) Recent Trends in Sliding Mode Control, pp. 177–202. IET, London (2016)
56. Levant, A., Livne, M., Yu, X.: Sliding-mode-based differentiation and its application. In: Proceedings of the 20th IFAC World Congress, Toulouse, 9–14 July, France (2017)
57. Levant, A., Pavlov, Y.: Generalized homogeneous quasi-continuous controllers. Int. J. Robust Nonlinear Control **18**(4–5), 385–398 (2008)
58. Levant, A., Yu, X.: Sliding-mode-based differentiation and filtering. IEEE Trans. Autom. Control **63**(9), 3061–3067 (2018)
59. Livne, M., Levant, A.: Proper discretization of homogeneous differentiators. Automatica **50**, 2007–2014 (2014)
60. Man, Z., Paplinski, A.P., Wu, H.: A robust MIMO terminal sliding mode control scheme for rigid robotic manipulators. IEEE Trans. Autom. Control **39**(12), 2464–2469 (1994)
61. Mboup, M., Join, C., Fliess, M.: Numerical differentiation with annihilators in noisy environment. Numer. Algorithms **50**(4), 439–467 (2009)
62. Moreno, J.A.: On strict Lyapunov functions for some non-homogeneous super-twisting algorithms. J. Frankl. Inst. **351**(4), 1902–1919 (2014)
63. Orlov, Y.: Finite time stability of switched systems. SIAM J. Control Optim. **43**(4), 1253–1271 (2005)
64. Pisano, A., Davila, J., Fridman, L., Usai, E.: Cascade control of pm dc drives via second-order sliding-mode technique. IEEE Trans. Ind. Electron. **55**(11), 3846–3854 (2008)
65. Pisano, A., Usai, E.: Sliding mode control: a survey with applications in math. Math. Comput. Simul. **81**(5), 954–979 (2011)
66. Plestan, F., Glumineau, A., Laghrouche, S.: A new algorithm for high-order sliding mode control. Int. J. Robust Nonlinear Control **18**(4/5), 441–453 (2008)
67. Polyakov, A., Efimov, D., Perruquetti, W.: Finite-time and fixed-time stabilization: implicit Lyapunov function approach. Automatica **51**(1), 332–340 (2015)
68. Polyakov, A., Efimov, D., Perruquetti, W.: Robust stabilization of MIMO systems in finite/fixed time. Int. J. Robust Nonlinear Control **26**(1), 69–90 (2016)
69. Polyakov, A., Poznyak, A.: Unified Lyapunov function for a finite-time stability analysis of relay second-order sliding mode control systems. IMA J. Math. Control Inf. **29**(4), 529–550 (2012)
70. Sabanovic, A.: Variable structure systems with sliding modes in motion control-a survey. IEEE Trans. Ind. Inform. **7**(2), 212–223 (2011)
71. Shtessel, Y., Edwards, C., Fridman, L., Levant, A.: Sliding Mode Control and Observation. Birkhauser, Basel (2014)
72. Shtessel, Y., Taleb, M., Plestan, F.: A novel adaptive-gain supertwisting sliding mode controller: methodology and application. Automatica **48**(5), 759–769 (2012)
73. Shtessel, Y.B., Shkolnikov, I.A., Levant, A.: Smooth second-order sliding modes: missile guidance application. Automatica **43**(8), 1470–1476 (2007)
74. Sira-Ramírez, H.: Dynamic second-order sliding mode control of the hovercraft vessel. IEEE Trans. Control Syst. Technol. **10**(6), 860–865 (2002)
75. Slotine, J.-J.E., Li, W.: Applied Nonlinear Control. Prentice Hall Int, New Jersey (1991)
76. Spurgeon, S.K.: Sliding mode observers: a survey. Int. J. Syst. Sci. **39**(8), 751–764 (2008)
77. Su, W.-C., Drakunov, S.V., Ozguner, U.: An O(T2) boundary layer in sliding mode for sampled-data systems. Int. J. Syst. Sci. **45**(3), 482–485 (2000)
78. Utkin, V.I.: Sliding Modes in Control and Optimization. Springer, Berlin (1992)
79. Utkin, V.I.: Sliding mode control design principles and applications to electric drives. IEEE Trans. Ind. Electron. **40**(1), 23–36 (1993)
80. Vasiljevic, L.K., Khalil, H.K.: Error bounds in differentiation of noisy signals by high-gain observers. Syst. Control Lett. **57**(10), 856–862 (2008)
81. Yan, Y., Galias, Z., Yu, X., Sun, C.: Euler's discretization effect on a twisting algorithm based sliding mode control. Automatica **68**(6), 203–208 (2016)
82. Yu, X., Xu, J.X.: Nonlinear derivative estimator. Electron. Lett. **32**(16), 1445–1447 (1996)

Chapter 2
Discontinuous Integral Control for Systems with Arbitrary Relative Degree

Jaime A. Moreno, Emmanuel Cruz-Zavala and Ángel Mercado-Uribe

Abstract For systems with arbitrary relative degree, we propose a homogeneous controller capable of tracking a smooth but unknown reference signal, despite a Lipschitz continuous perturbation, and by means of a continuous control signal. The proposed control scheme consists of two terms: (i) a continuous and homogeneous state feedback and (ii) a discontinuous integral term. The state feedback term aims at stabilizing (in finite time) the closed-loop while the (discontinuous) integral term estimates the perturbation and the unknown reference signal in finite time and provides for perfect compensation in closed loop. By adding an exact and robust differentiator, we complete an output feedback scheme, when only the output is available for measurement. The global finite-time stability of the closed-loop system and its insensitivity with respect to the Lipschitz continuous perturbations are proved in detail using several (smooth) homogeneous Lyapunov functions for different versions of the algorithm.

Keywords Sliding modes · Variable-structure control · Lyapunov methods · Discontinuous observers · High-order sliding modes · Integral control · Continuous higher-order sliding modes

J. A. Moreno (✉) · Á. Mercado-Uribe
Instituto de Ingeniería, Universidad Nacional Autónoma de México (UNAM), Coyoacán, 04510 Ciudad de México, Mexico
e-mail: JMorenoP@ii.unam.mx

Á. Mercado-Uribe
e-mail: AMercadoU@ii.unam.mx

E. Cruz-Zavala
Department of Computer Science, University of Guadalajara, 44430 Guadalajara, Jalisco, Mexico
e-mail: emitacz@yahoo.com.mx

M. Steinberger et al. (eds.), *Variable-Structure Systems and Sliding-Mode Control*, Studies in Systems, Decision and Control 271,
https://doi.org/10.1007/978-3-030-36621-6_2

2.1 Introduction

Consider a single-input-single-output (SISO) smooth, uncertain system affine in the input

$$\dot{z} = f(t, z) + g(t, z)u, \ y = h(t, z), \tag{2.1}$$

where $z \in \mathbb{R}^m$ is the state vector, $u \in \mathbb{R}$ is the control input, and $h(t, z) : \mathbb{R} \times \mathbb{R}^m \to \mathbb{R}$ is a smooth output function. The smooth vector fields $f(t, z)$ and $g(t, z)$ on $\mathbb{R}^m$ are uncertain and the dimension m can also be unknown. A standard control problem is the output tracking [22], consisting in forcing the output y to track a (time-varying) reference signal $r(t)$. Usually, this problem has associated a (robust) disturbance decoupling or attenuation property [22, 23]. This problem can be solved using Higher-Order Sliding Modes (HOSM) [30, 32, 57]. In this case, a discontinuous controller is designed such that the tracking error signal $\sigma = y - r$ will become zero after a finite time, i.e., $\sigma(t) \equiv 0$, $\forall t \geq T$, despite uncertainties and perturbations. When the relative degree n with respect to (w.r.t.) σ is known, well defined, and constant, this is equivalent to designing a controller for the perturbed differential equation given by

$$\Sigma_T : \begin{cases} \dot{x}_i = x_{i+1}, & i = 1, \ldots, n-1, \\ \dot{x}_n = w(t, z) + b(t, z)u, & \end{cases} \tag{2.2}$$

where $x = (x_1, \ldots, x_n)^T = (\sigma, \dot{\sigma}, \ldots, \sigma^{(n-1)})^T$ and $\sigma^{(i)} = \frac{d^i}{dt^i} h(z, t)$. The scalar functions $w(t, z)$ and $b(t, z)$ may be uncertain but they are assumed to be uniformly bounded for all $z \in \mathbb{R}^m$ and all $t \geq 0$ as $0 < K_m \leq b(t, z) \leq K_M$ and $|w(t, z)| \leq C_w$, for some known positive constants C_w, K_m, and K_M. The remaining dynamics when $m > n$ (which is related to the zero dynamics [27]) is assumed to be well behaved and it will not be considered further in this work. The problem is solved if, e.g., a static feedback control law $u = \phi\left(\sigma, \ldots, \sigma^{(n-1)}\right)$ is found such that after a finite time $x(t) = \left(\sigma(t), \dot{\sigma}(t), \ldots, \sigma^{(n-1)}(t)\right)^T \equiv 0$. When $C_w \neq 0$ a continuous function $\phi(x)$ cannot achieve the objective, and HOSM provide for control functions $\phi(x)$, which are discontinuous at least on the set $\sigma = \dot{\sigma} = \cdots = \sigma^{(n-1)} = 0$. The motion on the set $\sigma = \cdots = \sigma^{(n-1)} = 0$, which consists of Filippov trajectories [15], is called *nth-order sliding mode*. Homogeneous HOSM [32] is characterized by the use of bounded memoryless feedback $\mathbf{r}_s$-homogeneous control laws of degree 0 (called also n-sliding homogeneous) with $\mathbf{r}_s = (n, n-1, \ldots, 1)$, i.e., $\phi\left(\varepsilon^n \sigma, \ldots, \varepsilon \sigma^{(n-1)}\right) = \phi\left(\sigma, \ldots, \sigma^{(n-1)}\right)$ for all $\varepsilon > 0$, discontinuous at the n-sliding set $\sigma = \cdots = \sigma^{(n-1)} = 0$, and that render this point finite-time stable for (2.2).

Discontinuity of ϕ leads to the *chattering* phenomenon, which consists of undesired oscillations caused by the high-frequency switching, that substantially degrades the closed-loop system's performance and represents the major drawback of classical HOSM control. A classical strategy in HOSM to obtain a continuous control signal consists of taking a further derivative in (2.2) and using $v = \dot{u}$ as a new control variable, what increases artificially the relative degree from n to $n+1$. The discontinuous controller on $v = \dot{u} = \tilde{\phi}\left(\sigma, \ldots, \sigma^{(n-1)}, \sigma^{(n)}\right)$ solves the problem and produces a continuous signal in u, thus possibly reducing the chattering effect [29, 32]. The main disadvantage of this approach is the necessity of estimating $\sigma^{(n)}(t)$, which involves taking the derivative of the perturbation and the uncertainties. Alternative output feedback controllers, with arbitrary smoothness level, are presented in [14, 37].

Chattering reduction using the super-twisting. For relative degree $n = 1$, a classical alternative to obtain a continuous control signal is the use of the super-twisting (ST) controller [16, 29, 35],

$$u = -k_1 \sqrt{|\sigma|}\, \mathrm{sign}(\sigma) - k_2 \int_0^t \mathrm{sign}(\sigma(\tau))\, d\tau\,,$$

that performs appealing features as: (i) it is able to compensate matched Lipschitz uncertainties/perturbations; (ii) it enforces the signals σ and $\dot{\sigma}$ to be $\sigma = \dot{\sigma} \equiv 0$ simultaneously after a finite time; (iii) it only needs the information of σ; (iv) it generates a continuous control signal; and (v) it provides better sliding accuracy than a first-order Sliding-Mode (SM) control. Due to its robustness and easy implementation properties, the ST has found ample use in applications [32, 57].

Discontinuous Integral Controllers: an extension of the ST to arbitrary order. For $n > 1$, a natural extension of the super-twisting idea is to consider the following controller with integral action:

$$u = \vartheta_1(x) + \int_0^t \vartheta_2(x)\, d\tau, \tag{2.3}$$

where $\vartheta_1(x)$ is a continuous state feedback control, while $\vartheta_2(x)$ is a bounded *discontinuous* function of the plant's state x, corresponding to the integral action. The objective of this chapter is to show that it is possible to design (homogeneous) functions $\vartheta_1(x)$ and $\vartheta_2(x)$ such that exact tracking is attained in finite time, i.e., $x(t) = (\sigma(t), \dot{\sigma}(t), \ldots, \sigma^{(n-1)}(t))^T \equiv 0$, $\forall t \geq T$. The nice properties of the ST are, therefore, extended to an arbitrary order:

(i) Lipschitz uncertainties/perturbations can be completely compensated.
(ii) All signals $\sigma = \dot{\sigma} = \cdots = \sigma^{(n-1)} = \sigma^{(n)} = 0$ are forced to zero in finite time.
(iii) For the implementation only the variable $x(t) = (\sigma(t), \dot{\sigma}(t), \ldots, \sigma^{(n-1)}(t))^T$ is required. This is an advantage compared to the classical approach in HOSM to reduce the chattering, since for that $\sigma^{(n)}$ is also required. These signals, in principle, can be calculated in real time by means of a robust and exact differentiator [10, 31, 32, 57].
(iv) The control signal $u(t)$ is continuous, thus possibly achieving chattering reduction.
(v) The sliding accuracy is higher than using HOSM control.

We use a Lyapunov approach for the design of discontinuous integral controllers (2.3). We propose different explicit (smooth) and strong Lyapunov functions, and we obtain different variations of the discontinuous integral controllers, which possess diverse specific properties. The main structural distinguishing characteristic for the different alternatives is the dependence of ϑ_2 on the state: either it depends on the full state x or it depends only partially on the state, i.e., it depends alone on $x_1 = \sigma$ or some more components of x. This induces large differences in the dynamical behavior of the controller.

Moreover, we also propose the use of families of continuous integral controllers (2.3), i.e., ϑ_2 is continuous, in order to approximate the discontinuous algorithms, and to further increase the smoothness of the control signal $u(t)$. This approximation can be arbitrarily near to the discontinuous algorithm and we use also a Lyapunov approach to study both the discontinuous and the continuous versions of the algorithms. For relative degree $n = 1$, the discontinuous integral controller reduces to the super-twisting controller, so that all versions can be seen as generalizations of the ST to arbitrary order.

State of the art. The idea of using a discontinuous integral controller is not novel. For $n = 2$, in [17, 24, 25, 60], a full state discontinuous integral controller is derived using Lyapunov functions. References [17, 24] extend this idea to arbitrary n, but lacking a proof. For arbitrary order n, [28] presents for the first time a full state integral controller, with a proof based on a non-smooth homogeneous Lyapunov function. The approach is a nice generalization of the one proposed in [48] for the super-twisting ($n = 2$), and thus leads to a non-smooth (and not Lipschitz) Lyapunov function. Moreover, the low homogeneity degree of the Lyapunov function forces the integral term ϑ_2 to be of rational form.

For $n = 2$, a partial state discontinuous integral controller is derived in [45, 46] using smooth Lyapunov functions. Moreover, an output feedback scheme is also developed using a continuous rather than a discontinuous observer to estimate the unmeasured derivatives of the output σ. This is partially extended to $n = 3, 4$ in [43, 44]. In [56, 58, 59], Lyapunov functions in the class of generalized forms [56] are obtained numerically using sum of squares methods for $n = 2$, and in [41] for $n = 3$.

Another related, but quite different, solution to the robust regulation problem is the use of the classical integral action, as, for example, in the PID control [27], which is very successful in the case of (almost) constant references/perturbations. For nonlinear systems, the integral controller introduced in [26] is able to attain robust output regulation for SISO systems to (asymptotically) constant references, while rejecting perturbations in the same class. The basic structure consists in adding an integrator at the output, and then constructing a first-order sliding-mode controller to stabilize the closed-loop system. To avoid the chattering effect of the discontinuous controller, the sign function is replaced by a saturation function and the states are estimated using a high-gain observer. The controller, in this case, has the following (basic) structure:

$$u = -k \operatorname{sign}\left(Kx + k_I \int_0^t x_1(\tau)\, d\tau\right). \tag{2.4}$$

A crucial difference of this controller (2.4) with the structure proposed here (2.3) is that the signal to be integrated is a *continuous* function of the state x (in fact, it is just the output x_1) while in (2.3) the function $\vartheta_2(x)$ is *discontinuous*. And this changes completely the convergence, steady-state, and robustness properties of the controller algorithm. This observation is also valid for the more general version of the integral controller for multivariable nonlinear systems presented in [2], which is a generalization of [26].

In this chapter, we extend the work initiated by [25, 28, 45, 46] for arbitrary order $n \geq 1$. Some of the results presented here are submitted for publication in [9, 42]. The rest of the chapter is organized as follows. Section 2.2 reviews some basic concepts on differential inclusions and homogeneity, which are used in the rest of the work. The main results are stated in Sect. 2.3. In Sect. 2.4, a simulation example illustrates the performance of the controllers. Sections 2.5, 2.6, and 2.7 are devoted to the proofs of the main results, for different versions of the algorithms. Finally, some conclusions are drawn. The appendix contains some technical lemmas used in the proofs.

2.2 Preliminaries: Differential Inclusions and Homogeneity

We recall some important concepts about Differential Inclusions (DI), homogeneity, and homogeneous DIs [1, 3–8, 13, 15, 21, 32, 39], which are used in the chapter.

Uncertain or discontinuous systems are more appropriately described by differential inclusions (DI) $\dot{x} \in F(t, x)$ than by differential equations (DE). A solution of this DI is any function $x(t)$, defined in some interval $I \subseteq [0, \infty)$, which is absolutely continuous on each compact subinterval of I and such that $\dot{x}(t) \in F(t, x(t))$ almost everywhere on I. Thus, for a discontinuous DE $\dot{x} = f(t, x)$, the function $x(t)$ is said to be a generalized solution of the DE if and only if it is a solution of the associated DI $\dot{x} \in F(t, x)$. We will consider the DI $\dot{x} \in F(t, x)$ associated to $\dot{x} = f(t, x)$ using Filippov's regularization procedure [3, 15, Sect. 1.2]. So, we refer to such DI as Filippov DI and to its solutions as Filippov solutions.

The multivalued map $F(t, x)$ satisfies the *standard assumptions* if: (H1) $F(t, x)$ is a nonempty, compact, convex subset of $\mathbb{R}^n$, for each $t \geq 0$ and each $x \in \mathbb{R}^n$; (H2) $F(t, x)$ as a set-valued map of x is upper semicontinuous for each $t \geq 0$; (H3) $F(t, x)$ as a set-valued map of t is Lebesgue measurable for each $x \in \mathbb{R}^n$; and (H4) $F(t, x)$ is locally bounded. Recall that a set-valued map $G : \mathbb{R}^{n_1} \rightrightarrows \mathbb{R}^{n_2}$ with compact values is *upper semicontinuous* if for each x_0 and for each $\varepsilon > 0$ there exists $\delta > 0$ such that $G(x) \subseteq G(x_0) + B_\varepsilon$, provided that $x \in B_\delta(x_0)$. It is well known that, see [15] or [3, Theorem 1.4], if the multivalued map $F(t, x)$ satisfies the standard assumptions then for each pair $(t_0, x_0) \in [0, \infty) \times \mathbb{R}^n$ there is an interval I and at least a solution $x(t) : I \to \mathbb{R}^n$ such that $t_0 \in I$ and $x(t_0) = x_0$. A DI $\dot{x} \in F(x)$ (a DE $\dot{x} = f(x)$) is called *globally uniformly finite-time stable* (GUFTS) at 0, if $x(t) = 0$ is a Lyapunov-stable solution and for any $R > 0$ there exists $T > 0$ such that the trajectory starting within the ball $\|x\| < R$ reaches zero in the time T.

Assume that the origin $x = 0$ is an equilibrium position of the DI $\dot{x} \in F(t, x)$, this means that $0 \in F(t, 0)$ for almost every (a.e.) $t \geq 0$. It is possible to characterize the asymptotic stability of the origin by means of a (strict) Lyapunov function (LF), in the same spirit as the classical second Lyapunov theorem for smooth systems (see, e.g., [3, Theorem 4.1]). In particular, if the Lyapunov function $V(t, x)$ is of class $\mathscr{C}^1$ (once continuously differentiable), and it satisfies

$$a(\|x\|) \leq V(t, x) \leq b(\|x\|)$$
$$\frac{\partial V(t, x)}{\partial t} + \frac{\partial V(t, x)}{\partial x} \nu \leq -c(\|x\|)$$

for a.e. $t \geq 0$, for all $x \in \mathbb{R}^n$ and all $\nu \in F(t, x)$, and for some functions $a, b, c \in \mathscr{K}_\infty$, then the origin is uniformly globally asymptotically stable (UGAS)[1] for the DI.

Continuous and discontinuous homogeneous functions and systems have a long history [1, 3–6, 8, 18, 20, 32, 39, 50–53, 61]. We recall briefly this important property. For a given vector $x = [x_1, \ldots, x_n]^\top \in \mathbb{R}^n$ and for every $\varepsilon > 0$, the dilation operator is defined as $\Delta_\varepsilon^r x := [\varepsilon^{r_1} x_1, \ldots, \varepsilon^{r_n} x_n]^\top$, where $r_i > 0$ are the weights of the coordinates, and let $\mathbf{r} = [r_1, \ldots, r_n]^\top$ be the vector of weights. A function $V : \mathbb{R}^n \to \mathbb{R}$ (respectively, a vector field $f : \mathbb{R}^n \to \mathbb{R}^n$ or a vector-set field $F(x) \subset \mathbb{R}^n$) is called $\mathbf{r}$-homogeneous of degree $l \in \mathbb{R}$ (or $(\mathbf{r}, l)$-homogeneous for short) if the identity $V(\Delta_\varepsilon^r x) = \varepsilon^l V(x)$ holds for every $\varepsilon > 0$ (resp., $f(\Delta_\varepsilon^r x) = \varepsilon^l \Delta_\varepsilon^r f(x)$ or $F(\Delta_\varepsilon^r x) = \varepsilon^l \Delta_\varepsilon^r F(x)$). Along this paper, we refer to this property as **r**-*homogeneity* or simply *homogeneity*. A system is called homogeneous if its vector field (or vector-set field) is **r**-homogeneous of some degree.

Given a vector $\mathbf{r}$ and a dilation $\Delta_\varepsilon^r x$, the homogeneous norm is defined by $\|x\|_{\mathbf{r},p} := \left(\sum_{i=1}^n |x_i|^{\frac{p}{r_i}}\right)^{\frac{1}{p}}$, $\forall x \in \mathbb{R}^n$, for any $p \geq 1$, and it is an $\mathbf{r}$-homogeneous function of degree 1. The set $S = \{x \in \mathbb{R}^n : \|x\|_{\mathbf{r},p} = 1\}$ is the corresponding homogeneous unit sphere. The following lemma provides some important properties of homogeneous functions and vector fields (some others are recalled in the appendix).

Lemma 2.1 ([3, 8, 21]) *For a given family of dilations $\Delta_\varepsilon^r x$, and continuous real-valued functions V_1, V_2 on $\mathbb{R}^n$ (resp., a vector field f) which are* **r**-*homogeneous of degrees $m_1 > 0$ and $m_2 > 0$ (resp., $l \in \mathbb{R}$), we have: (i) $V_1 V_2$ is homogeneous of degree $m_1 + m_2$. (ii) For every $x \in \mathbb{R}^n$ and each positive definite function V_1, we have $c_1 V_1^{\frac{m_2}{m_1}}(x) \leq V_2(x) \leq c_2 V_1^{\frac{m_2}{m_1}}(x)$, where $c_1 \triangleq \min_{\{z: V_1(z)=1\}} V_2(z)$ and $c_2 \triangleq \max_{\{z: V_1(z)=1\}} V_2(z)$. Moreover, if V_2 is positive definite, there exists $c_1 > 0$. (iii) $\partial V_1(x) / \partial x_i$ is homogeneous of degree $m_1 - r_i$, with r_i being the weight of x_i. (iv) The Lie's derivative of $V_1(x)$ along $f(x)$, $L_f V_1(x) := \frac{\partial V_1(x)}{\partial x} \cdot f(x)$, is homogeneous of degree $m_1 + l$.*

---

[1] Since uniqueness of solutions is, in general, not assumed, this means that *all* trajectories starting at any initial point (t_0, x_0) are uniformly stable and uniformly attractive. This concept is sometimes referred to as "strong stability", in contrast to the "weak stability" which is valid only for *some* trajectory for every initial point.

It is worth recalling that for homogeneous systems the local stability implies global stability and if the homogeneous degree is negative, asymptotic stability implies finite-time stability [3, 5, 32, 39]. Asymptotic stability of homogeneous systems and homogeneous DIs can be studied by means of homogeneous Lyapunov Functions (HLFs), see, for example, [3–6, 8, 18, 20, 32, 39, 50, 54, 61]: Assume that the origin of a homogeneous Filippov DI $\dot{x} \in F(x)$ is strongly globally AS. Then, there exists a $\mathscr{C}^\infty$ homogeneous strong LF.

The following robustness result of asymptotically stable homogeneous Filippov differential inclusions is of paramount importance for the assertion of the accuracy properties of HOSM algorithms in presence of measurements or discretization noise or also delay and external perturbations. They have been established by Levant [31, 32, 39] (see also [5]).

Proposition 2.1 *Let $\dot{x} \in F(x)$ be a globally uniformly finite-time stable homogeneous Filippov inclusion with homogeneity weights $\mathbf{r} = (r_1, \ldots, r_n)$ and degree $l < 0$, and let $\tau > 0$. Suppose that a continuous function $x(t)$ is defined for any $t \geq -\tau^l$ and satisfies some initial conditions $x(t) = \xi(t)$ for $t \in [-\tau^l, 0]$. Then if $x(t)$ is a solution of the perturbed differential inclusion*

$$\dot{x}(t) \in F_\tau\left(x\left(t + \left[-\tau^l, 0\right]\right)\right), \; 0 < t < \infty,$$

then the inequalities $|x_i| < \gamma_i \tau^{r_i}$ are established in finite time with some positive constants γ_i independent of τ and ξ.

Along this paper, we use the following notation. For a real variable $z \in \mathbb{R}$ and a real number $p \in \mathbb{R}$, the symbol $\lceil z \rfloor^p = |z|^p sign(z)$ is the sign-preserving power p of z. According to this $\lceil z \rfloor^0 = \text{sign}(z)$, $\frac{d}{dz}\lceil z \rfloor^p = p\,|z|^{p-1}$, and $\frac{d}{dz}|z|^p = p\lceil z \rfloor^{p-1}$ almost everywhere for z. Note that $\lceil z \rfloor^2 = |z|^2\text{sign}(z) \neq z^2$, and if p is an odd number then $\lceil z \rfloor^p = z^p$ and $|z|^p = z^p$ for any even integer p. Moreover, $\lceil z \rfloor^p \lceil z \rfloor^q = |z|^{p+q}$, $\lceil z \rfloor^p \lceil z \rfloor^0 = |z|^p$, and $\lceil z \rfloor^0 |z|^p = \lceil z \rfloor^p$. Note that $\lceil z \rfloor^0 = \text{sign}(z)$, which is a multivalued function, i.e.,

$$\lceil z \rfloor^0 = \begin{cases} +1 & \text{if } z > 0 \\ [-1, +1] & \text{if } z = 0 \\ -1 & \text{if } z < 0 \end{cases}.$$

For any $p \in \mathbb{N}$, $\mathscr{C}^p$ defines the set of continuous functions which are continuously differentiable up to the order p.

2.3 Integral Controller: Main Results

Consider the uncertain plant's (partial) representation (2.2) and applying the integral controller (I-controller for short) in (2.3) that we can also write as (see Fig. 2.1)

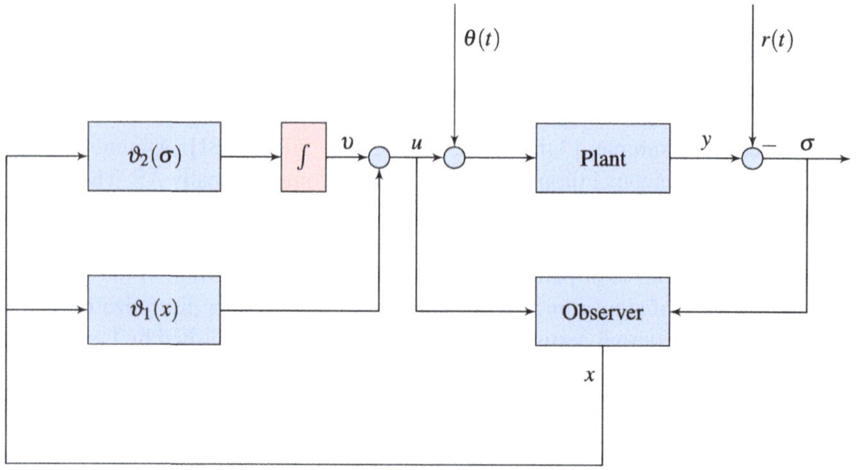

Fig. 2.1 Block diagram of the integral controller

$$\begin{aligned} u &= \vartheta_1(x) + \upsilon\,, \\ \dot{\upsilon} &= \vartheta_2(x)\,, \end{aligned} \tag{2.5}$$

we obtain

$$\Sigma_T : \begin{cases} \dot{x}_i = x_{i+1}, & i = 1, \ldots, n-1, \\ \dot{x}_n = b(t, z)(\vartheta_1(x) + x_{n+1}), \\ \dot{x}_{n+1} = \vartheta_2(x) + \frac{d}{dt}\theta(t, z)\,, \end{cases} \tag{2.6}$$

where $\theta(t, z) = w(t, z)/b(t, z)$ corresponds to the uncertain/perturbation term, which we assume to be a Lipschitz continuous-time function, and $x_{n+1} = \int_0^t \vartheta_2(x)d\tau + \theta(t, z)$ is a new state variable, comprising the integral part of the controller plus the perturbation term. We assume that the scalar functions $\frac{d}{dt}\theta(t, z)$ and $b(t, z)$ are unknown and that the only information about them are their bounds, that is, there exist known positive constants C, K_m, and K_M such that for all $z \in \mathbb{R}^m$ and all $t \geq 0$,

$$0 < K_m \leq b(t, z) \leq K_M\,, \; \left|\tfrac{d}{dt}\theta(t, z)\right| \leq C\,. \tag{2.7}$$

The closed-loop system can be represented by the Filippov DI

$$\Sigma_{DI} : \begin{cases} \dot{x}_i = x_{i+1}, \; i = 1, \ldots, n-1, \\ \dot{x}_n \in [K_m, K_M](\vartheta_1(x) + x_{n+1}) \\ \dot{x}_{n+1} \in \vartheta_2(x) + [-C, C]\,. \end{cases} \tag{2.8}$$

In Σ_{DI}, the dependencies of the original systems' dynamics disappear and the DI only "remembers" the constants n, C, K_m, and K_M. In practice, conditions (2.7) are satisfied at least in compact operating regions. Solutions of DI (2.8) are understood in the sense of Filippov [15].

Let us assign to the partial vectors $\overline{x}_i = (x_1, \ldots, x_i)^T$, for $i = 1, \ldots, n+1$, the weights $\mathbf{r}_i = (r_1, \ldots, r_i)$, that is, each variable x_i has weight $r_i = 1 - (n+1-i)\,l$ with $l \in [-1, 0]$. Note that $r_1 > \cdots > r_i$ when $l < 0$. For convenience, we define the value of $r_{n+2} = 1 + l$. We will keep the symbol x for the plant's state (2.6) without controller, i.e., $x = \overline{x}_n$, and the symbol $\overline{x}$ for the extended state vector, i.e., $\overline{x} = \overline{x}_{n+1}$. We assume that functions $\vartheta_1(x)$ and $\vartheta_2(x)$ are $(\mathbf{r}_n, 1)$-homogeneous and $(\mathbf{r}_n, 1+l)$-homogeneous, respectively, so that the DI (2.8) is $(\mathbf{r}_n, l)$-homogeneous.

When $l = -1$, the right-hand side (r.h.s.) of (2.8) is $(\mathbf{r}_{n+1}, -1)$-homogeneous so that the weights $\mathbf{r}_{n+1}$ become $\mathbf{r}_{n+1} = (n+1, n, \ldots, 1)$, $\vartheta_1(x)$ is an $(\mathbf{r}_n, 1)$-homogeneous continuous function, while $\vartheta_2(x)$ is a locally bounded and $(\mathbf{r}_n, 0)$-homogeneous, and therefore discontinuous function. In this case, $\vartheta_2(x)$ is also globally bounded [32].

In the sequel, we design the time-invariant functions $\vartheta_1(x)$ and $\vartheta_2(x)$ of the feedback controller (2.3) such that the origin $\bar{x} = 0$ of the augmented system (2.8) is finite-time stable.

We present different kinds of controllers solving the posed problem. We distinguish them basically by two properties of the integral function $\vartheta_2(x)$: (i) if it is continuous or discontinuous and (ii) if it depends on the full state x or only on part of the state, e.g., if it depends only on the output x_1. Full state I-controllers are also distinguished as (a) passivity-based or (b) nonpassive controllers according to their design strategy.

Discontinuous I-controllers provide for exact convergence despite the bounded uncertainties while continuous I-controllers only achieve input-to-state stability (ISS) with respect to the derivative of the perturbation $\frac{d}{dt}\theta(t, z)$, and therefore they solve the posed problem only approximately. The different families of controllers differ also in the Lyapunov functions used for the stability proof.

In the following subsections, we present the different families of I-controllers and discuss their properties, while proofs are given in Sects. 2.5–2.7. For this purpose, given the homogeneity degree $l \in [-1, 0]$, we choose a nondecreasing sequence of positive real numbers α_i, such that $\alpha_n \geq \cdots \geq \alpha_1 \geq r_1 = 1 - nl$, and define the sequence of $(\mathbf{r}_i, \alpha_i)$-homogeneous functions

$$\begin{aligned}\sigma_1(x_1) &= \lceil x_1 \rfloor^{\frac{\alpha_1}{r_1}}, \\ \sigma_i(\overline{x}_i) &= \lceil x_i \rfloor^{\frac{\alpha_i}{r_i}} + k_{i-1}^{\frac{\alpha_i}{r_i}} \lceil \sigma_{i-1}(\overline{x}_{i-1}) \rfloor^{\frac{\alpha_i}{\alpha_{i-1}}}, \quad i = 2, \ldots, n,\end{aligned} \tag{2.9}$$

where k_i are some positive constant gains (to be designed, see Sect. 2.5.3.2).

Furthermore, selecting $m \geq r_1 + \alpha_1$, we also define the sequence of $(\mathbf{r}_i, m)$-homogeneous, positive definite, $\mathscr{C}^1$ functions

$$\begin{aligned}V_1(x_1) &= \tfrac{r_1}{m} |x_1|^{\frac{m}{r_1}}, \\ V_i(\overline{x}_i) &= \tfrac{\alpha_i + r_i}{m} W_i^{\frac{m}{\alpha_i + r_i}}(\overline{x}_i) + \gamma_{i-1} V_{i-1}(\overline{x}_{i-1}), \; i = 2, \ldots, n,\end{aligned} \tag{2.10}$$

where, for all $i = 2, \ldots, n-1$, $\gamma_i > 0$ are arbitrary positive constants and the $(\mathbf{r}_i,\ \alpha_i + r_i)$-homogeneous, positive semidefinite, $\mathscr{C}^1$ functions $W_i\left(\overline{x}_i\right)$ are given by

$$W_i\left(\overline{x}_i\right) = \frac{r_i}{\alpha_i + r_i}\left|x_i\right|^{\frac{\alpha_i + r_i}{r_i}} + k_{i-1}^{\frac{\alpha_i}{r_i}}\left\lceil\sigma_{i-1}\left(\overline{x}_{i-1}\right)\right\rfloor^{\frac{\alpha_i}{\alpha_{i-1}}} x_i + \frac{\alpha_i}{\alpha_i + r_i} k_{i-1}^{\frac{\alpha_i + r_i}{r_i}}\left|\sigma_{i-1}\left(\overline{x}_{i-1}\right)\right|^{\frac{\alpha_i + r_i}{\alpha_{i-1}}}. \tag{2.11}$$

Besides, let $D(x)$ be any continuous, $(\mathbf{r}_n,\ 1)$-homogeneous, positive definite function. Possible values of the function $D(x)$ are, e.g., $D(x) = \|x\|_{\mathbf{r},p}$ or $D(x) = \left(\sum_{j=1}^{n}\beta_j\left|x_j\right|^{\alpha_n/r_j}\right)^{\frac{1}{\alpha_n}}$, with arbitrary $\beta_j > 0$.

2.3.1 HOSM Controllers with Discontinuous Integral Action

For all these controllers, we assume that the uncertainties/perturbations fulfill the conditions (2.7) and we fix $l = -1$.

Passivity-Based Full State Discontinuous I-Controllers

We will consider here that the control coefficient $b(t,\ z) = b$ is constant but (possibly) unknown, i.e., $0 < K_m \leq b \leq K_M$. This is done partially to simplify the presentation and in part to clearly see the passivity interpretation. Consider (2.3) with

$$\begin{aligned}\vartheta_1(x) &= -k_n\left\lceil\sigma_n(x)\right\rfloor^{\frac{1}{\alpha_n}} \\ \vartheta_2(x) &= -k_I V_n^{\frac{2-m}{m}}(x)\, W_n^{\frac{m}{\alpha_n + r_n} - 1}(x)\,\sigma_n(x)\,,\end{aligned} \tag{2.12}$$

where V_n and W_n are defined in (2.10) and (2.11), respectively.

Nonpassive Full State Discontinuous I-Controllers

We distinguish it into two types as given below:
Discontinuous controllers:

$$\vartheta_1(x) = -k_n\left\lceil\sigma_n(x)\right\rfloor^{\frac{1}{\alpha_n}},\ \vartheta_{2d}(x) = -k_I\left\lceil\sigma_n(x)\right\rfloor^{0}. \tag{2.13}$$

Quasi-continuous (QC) controllers:

$$\vartheta_1(x) = -k_n\sigma_n(x)\,,\ \vartheta_{2cq}(x) = -k_I\frac{\sigma_n(x)}{D^{\alpha_n}(x)}. \tag{2.14}$$

Partial State Discontinuous I-Controllers

$$\vartheta_1(x) = -k_n \chi(\bar{x}_n) = -k_n \left[\lceil x_n \rfloor^{\frac{r_1}{r_n}} + \sum_{j=1}^{n-1} \kappa_j \lceil x_j \rfloor^{\frac{r_1}{r_j}} \right]^{\frac{r_{n+1}}{r_1}}, \tag{2.15}$$

$$\vartheta_2(x) = -k_I \psi(\bar{x}_n) = -k_I \frac{\left\lceil \sum_{j=2}^{n} k_{I,j} \lceil x_j \rfloor^{\frac{r_1}{r_j}} + x_1 \right\rfloor^{\frac{a+r_{n+2}}{r_1}}}{\left(\sum_{j=2}^{n} \beta_j |x_j|^{\frac{r_1}{r_j}} + |x_1| \right)^{\frac{a}{r_1}}}, \tag{2.16}$$

where we select $a \geq 0$, $k_{I,j} \in \mathbb{R}$, $\eta_j > 0$ arbitrary, and set $\beta_j = \eta_j \left| k_{I,j} \right|$. Note that $x_1 \psi(x_1, 0, \ldots, 0) = x_1 \lceil x_1 \rfloor^0 > 0$ for $x_1 \neq 0$. For $a = 0$, function ψ is of "polynomial" type, while for $a > 0$, it is a "rational" form. Note that ψ depends only on the states x_j for which $k_{I,j} \neq 0$. This gives the name to the controllers.

The main result of this chapter is the following.

Theorem 2.1 *Assume that conditions* (2.7) *are satisfied, and for the passive controller* (2.12) *that b is constant. For the I-Controllers* (2.12), (2.13), (2.14), *and (2.15–2.16), there exist gains $k_i > 0$, $\kappa_i > 0$, $i = 1, \ldots, n$, and $k_I > C$, such that the origin $\bar{x} = 0$ of the closed-loop* (2.8) *is globally finite-time stable.*

The proof of this theorem is presented in Sects. 2.5–2.7 for each type of controllers, using explicit Lyapunov functions for each one. After the proof, a procedure to calculate appropriate gains is presented, making use of the corresponding Lyapunov function.

All these controllers produce an absolutely continuous control signal $u(t)$, thus potentially reducing the chattering effect, in contrast to the classical HOSM control [32], which produces a discontinuous one.

All functions $\vartheta_2(x)$ are discontinuous, but they differ in the discontinuity set. For controllers (2.13) and (2.16) with $a = 0$, the discontinuity is a manifold given by $\{\sigma_n(x) = 0\}$ and $\left\{ \sum_{j=2}^{n} k_{I,j} \lceil x_j \rfloor^{\frac{r_1}{r_j}} + x_1 = 0 \right\}$, respectively. In contrast, for controllers (2.12), (2.14), and (2.16) with $a > 0$, the discontinuity is only at the point $x = 0$. The latter type are named "quasi-continuous" by Levant [33].

Since for all controllers after finite time $x_{n+1}(t) \equiv 0$ then $\int_0^t \vartheta_2(x) d\tau = -\theta(t, z)$, meaning that the integral term of the controller estimates exactly and in finite time the perturbation, and thus is able to fully compensate it.

An important difference of the controllers lies in the fact that $\vartheta_2(x)$ depends on the full state x for controllers (2.12), (2.13), and (2.14), while for controller (2.16), the coefficients $k_{I,j} \in \mathbb{R}$ can be selected arbitrarily, also, e.g., $k_{I,j} = 0$, and thus the integral action depends only partially on the state. Note that the classical PID control (see, e.g., (2.4)) depends only on the tracking error $\sigma = x_1$, and it falls, therefore, under this category of controllers. An interesting difference between the passivity-based full state and the partial state I-controllers (what is illustrated in the example below (Sect. 2.4) is that when the integral gain k_I is increased (without changing the other gains) in the full state controllers the stability of the closed loop is maintained, while for partial state controller this leads to instability of the controlled system. This is a well-known phenomenon in classical PID control, and it is interesting to find it also here.

In contrast to the usual HOSM technique, the chattering alleviation is achieved without requiring the feedback of all $n+1$ state variables, since the proposed control schemes require only to feedback the n plant's variables. Under the extra assumption that the perturbation/uncertainty $\theta(t, z)$ is bounded, a robust and exact differentiator [10, 31, 32, 57] of $(n-1)$-th order is able to calculate exactly (in the absence of noise) in real time the required derivatives of σ up to $\sigma^{(n-1)}$ to implement the control scheme (2.3).

2.3.2 *Homogeneous and Continuous Approximations of the I-Controllers*

The discontinuous integral term $\vartheta_2(x)$ (2.13) causes undesirable abrupt changes in the derivative of the control variable $\dot{u}$. This can be avoided by replacing the discontinuous sign function in (2.13) by a saturation function, as it is usually done for sliding-mode controllers. Instead of this, we propose here a controller (2.3) with a continuous integral term $\vartheta_2(x)$, i.e., with $l > -1$ that provides for "practical" robust asymptotic stability of the origin of (2.8). When the perturbation is constant, i.e., $C = 0$, asymptotic stability of $\bar{x} = 0$ is recovered.

For all these controllers, we select $l \in (-1, 0]$ and assume that the perturbations satisfy conditions (2.7).

Passivity-Based Full State I-Controllers

$$\begin{aligned} \vartheta_1(x) &= -k_n \lceil \sigma_n(x) \rfloor^{\frac{r_{n+1}}{\alpha_n}} \\ \vartheta_2(x) &= -k_I V_n^{\frac{2-m}{m}}(\bar{x}_n)\, W_n^{\frac{m}{\alpha_n + r_n} - 1}(\bar{x}_n)\, \sigma_n(\bar{x}_n) \ . \end{aligned} \tag{2.17}$$

Nonpassive Full State I-Controllers

$$\vartheta_1(x) = -k_n \lceil \sigma_n(x) \rfloor^{\frac{1}{\alpha_n}}, \ \vartheta_2(x) = -k_I \lceil \sigma_n(x) \rfloor^{\frac{1+l}{\alpha_n}} . \tag{2.18}$$

In (2.18), it is also possible to use rational-type controllers, i.e.,

$$\vartheta_1(x) = -k_n \sigma_n(x), \ \vartheta_{2R}(x) = -k_I \frac{\sigma_n(x)}{D^{\alpha_n - 1 - l}(x)} . \tag{2.19}$$

Partial State I-Controllers

These controllers are given by expressions (2.15), (2.16) for values of $l \in (-1,\ 0]$.

Theorem 2.2 *Assume that conditions* (2.7) *are satisfied, and for the passive controller* (2.12) *that* b *is constant. For all previous continuous controllers with* $l \in (-1,\ 0]$, *there exist gains* $k_i > 0$, $\kappa_i > 0$, $i = 1, \ldots, n$, *and* $k_I > 0$, *such that*

(i) if $C = 0$ *the origin* $\bar{x} = 0$ *of the closed-loop* (2.8) *is GAS. For* $l = 0$, *the convergence is exponential, while for* $l \in (-1,\ 0)$, *the convergence is in finite time.*

(ii) if $C \neq 0$ *system* (2.6) *is input-to-state stable (ISS) with respect to the input* $\frac{d}{dt}\theta\,(t,\ z)$.

The proof of this theorem is presented in Sects. 2.5–2.7 for each type of controllers, together with the discontinuous controllers, using different explicit Lyapunov functions for each controller type. Continuous controllers can be used as approximations of the discontinuous ones, in particular, when the homogeneity degree l is near to $l = -1$. Moreover, one expects that the closer l is to -1 the smaller is the ultimate bound of the trajectories.[2] Note that the expressions for all these continuous controllers become the discontinuous ones for $l = -1$ (see, e.g., the expressions of (2.18) and (2.13)). When $l = 0$ and $C = 0$, the continuous approximation controller provides exponential stability. In fact, in the case when $l \in (-1, 0)$ and $C \neq 0$, the effect of the perturbation term can be arbitrarily reduced by making k_n large enough for a fixed l or by making l close to -1 for a suitable k_n, and therefore the steady-state error can be minimized.

Some expressions for the function $\sigma_n\,(x)$ defining the controllers (2.13), (2.14), and (2.18) and derived from Theorems 2.1 and 2.2 are listed below for $n = 1, \ldots, 4$ and $l \in [-1,\ 0]$:

$$\textbf{n=1: } \sigma_1\,(x_1) = \lceil x_1 \rfloor^{\frac{\alpha_1}{1-l}}\ ,$$

$$\textbf{n=2: } \sigma_2\,(\overline{x}_2) = \lceil x_2 \rfloor^{\frac{\alpha_2}{1-l}} + k_1^{\frac{\alpha_2}{1-l}}\,\lceil x_1 \rfloor^{\frac{\alpha_2}{1-2l}}\ ,$$

$$\textbf{n=3: } \sigma_3\,(\overline{x}_3) = \lceil x_3 \rfloor^{\frac{\alpha_3}{1-l}} + k_2^{\frac{\alpha_3}{1-l}} \left\lceil \lceil x_2 \rfloor^{\frac{\alpha_2}{1-2l}} + k_1^{\frac{\alpha_2}{1-2l}}\,\lceil x_1 \rfloor^{\frac{\alpha_2}{1-3l}} \right\rfloor^{\frac{\alpha_3}{\alpha_2}}\ ,$$

$$\textbf{n=4: } \sigma_4\,(\overline{x}_4) = \lceil x_4 \rfloor^{\frac{\alpha_4}{1-l}} + k_3^{\frac{\alpha_4}{1-l}} \left\lceil \lceil x_3 \rfloor^{\frac{\alpha_3}{1-2l}} + k_2^{\frac{\alpha_3}{1-2l}} \left\lceil \lceil x_2 \rfloor^{\frac{\alpha_2}{1-3l}} + k_1^{\frac{\alpha_2}{1-3l}}\,\lceil x_1 \rfloor^{\frac{\alpha_2}{1-4l}} \right\rfloor^{\frac{\alpha_3}{\alpha_2}} \right\rfloor^{\frac{\alpha_4}{\alpha_3}}\ .$$

By selecting different values for the parameters $\alpha_n \geq \cdots \geq \alpha_1 \geq 1 - nl$ and l, we obtain different types of controllers. For example, if $\alpha_n = \cdots = \alpha_1 \geq r_1$, a polynomial-type controller is obtained. Besides, if $l = 0$ and $\alpha_n = \cdots = \alpha_1 = 1$, we obtain a linear integral controller.

[2] A sketch of a proof of this fact is presented in Sect. 2.5.3.1 (see Remark 2.1).

2.3.3 Output Feedback Integral Controller: The Use of a Differentiator

The implementation of the I-controller (2.5) requires the values of the plant's states x. However, since these states x correspond to the successive derivatives of the error tracking signal σ, it is not measurable in general. They can be estimated in finite time by a robust and exact differentiator [31, 32] (see also [10, 11, 55] for a Lyapunov approach), if we assume further that $b(t, z)$ and $w(t, z)$ are bounded.

Constructing an observer to estimate the states for the plant (2.2), when the I-controller (2.5) is implemented using the estimated values, i.e., $u = \vartheta_1(\hat{x}) + \upsilon$, we obtain

$$\Omega : \begin{cases} \dot{\hat{x}}_i = -l_i \lceil e_1 \rfloor^{\frac{n-i}{n}} + \hat{x}_{i+1}, & i = 1, \ldots, n-1, \\ \dot{\hat{x}}_n = -l_n \lceil e_1 \rfloor^0 + \bar{b}\left(\vartheta_1\left(\hat{x}\right) + \upsilon\right), \end{cases}$$

where $e_1 = \hat{x}_1 - x_1$ and $\bar{b}$ is a nominal value for the coefficient. The estimation error $(e_i = \hat{x}_i - x_i)$ is given by

$$\Xi : \begin{cases} \dot{e}_i = -l_i \lceil e_1 \rfloor^{\frac{n-i}{n}} + e_{i+1}, & i = 1, \ldots, n-1, \\ \dot{e}_n = -l_n \lceil e_1 \rfloor^0 + \left(\bar{b} - b(t, z)\right)\left(\vartheta_1\left(\hat{x}\right) + \upsilon\right) - w(t, z) . \end{cases}$$

The estimation errors converge to zero in finite time if the signal

$$v(t) \triangleq \left(\bar{b} - b(t, z)\right)\left(\vartheta_1\left(\hat{x}\right) + \upsilon\right) - w(t, z)$$

is bounded by a known constant. This is, in general, difficult to satisfy globally. In particular, if the coefficient $b(t)$ is known and we use $\bar{b} = b(t)$, then the resulting signal $v(t) = -w(t, z)$ is bounded if $w(t, z)$ is bounded. In this particular case, the estimation in finite time is possible. In the general case, since $u = \vartheta_1(\hat{x}) + \upsilon$ is known, and assuming that $b(t, z)$ and $w(t, z)$ are bounded, it is possible to design an observer Ω with time-varying gains $l_i(t)$, whose size depends on the size of u and the bounds of $b(t, z)$ and $w(t, z)$ (see, e.g., [34, 36, 38, 40, 47]).

2.3.4 Discussion and Properties of the Controllers

There are two important distinguishing characteristics between the different families of integral controllers presented before:

1. When $\vartheta_2(x)$ is discontinuous, the controller is able to compensate exactly the perturbation while if $\vartheta_2(x)$ is continuous convergence to zero is only possible in the absence of uncertainties/perturbations. These two results are the content of Theorems 2.1 and 2.2, respectively.

2. If $\vartheta_2(x)$ depends on all state variables then, in general, the convergence is faster, more damped, and the integral gain may increase without losing stability. This latter property is attained by the passivity-based I-controller. In contrast, when $\vartheta_2(x)$ depends partially on the states, and, in particular, when it depends only on the output x_1, the convergence is less damped and by increasing the integral gain, the stability of the closed loop is lost.

A particularly interesting property of the discontinuous I-controller is its ability to track exactly and in finite time any arbitrary differentiable reference signal $r(t)$, with bounded nth derivative. This contrasts with the continuous I-controllers, also the classical I-control in, e.g., [26, 27], for which exact tracking is only achieved for constant references r. Recall that according to the internal model principle, continuous controllers are able to track exactly only the class of signals related to an exosystem.

A main feature of all (discontinuous) I-controllers is that the control signal $u(t)$ is (absolutely) continuous, in contrast to the discontinuous signal produced by the (static) HOSM controllers. Note that this difference in the properties of the control signal is reflected in the different class of uncertainties/perturbations that can be compensated: for HOSM $\theta(t, z)$ is required to be bounded while for the I-controller it has to be Lipschitz. The I-controllers are able to compensate exactly the effect of the perturbation, since the I-term estimates exactly and in finite time its value, a feature well known for classical I-controllers.

As it is well known, homogeneity together with the existence of (smooth) Lyapunov functions implies some robustness properties of the algorithms. For example, in the presence of external inputs or perturbations, there is ISS (or i-ISS) with respect to the inputs [7, 32]. This is, in particular, true if we implement any of the algorithms subjected to measurement noise: since there is ISS with respect to the measurement noise, the trajectories of the closed-loop system will not converge to the origin but will remain in a neighborhood of it, whose size is a monotonic growing function of the size of the noise. Moreover, if there are some delays and perturbations in the implementation of the control algorithm (coming, e.g., from the discretization) Proposition 2.1 shows that in steady state, the states will remain in a neighborhood of size

$$|x_i| < \gamma_i \tau^{r_i} = \gamma_i \tau^{1-(n+1-i)l},$$

where γ_i are some constants independent of τ and of the initial conditions, but depending on the parameters of the controller and system, and r_i are the weights of homogeneity.

This behavior is usually called the *accuracy* of the algorithm in the sliding-mode literature. This issue is discussed in more detail in [5, 31, 32, 39]. Note that for a linear controller (or any smooth controller that can be approximated by a linear one in the neighborhood of the origin) $l = 0$ and the accuracy is $|x_i| < \bar{\gamma}_i \tau$. However, when $l < 0$ accuracy is $|x_i| < \gamma_i \tau^{r_i}$, with $r_i > 1$, except for $i = n + 1$ for which we have $|x_{n+1}| < \gamma_{n+1} \tau$. This shows that no matter what are the values of γ_i or $\bar{\gamma}_i$, for small $\tau < 1$ the accuracy of the algorithm is better than the one for any smooth controller.

This is particularly true for the discontinuous case since $r_i = 1 + (n + 1 - i)$. The output variable has accuracy $|\sigma| = |x_1| < \gamma_1 \tau^{1+n}$.

Since for the closed-loop system we are able to provide an explicit **r**-homogeneous Lyapunov function $V(\bar{x})$ of (some) degree m, which satisfies the following inequality:

$$\dot{V}(\bar{x}) \leq -\kappa V^{\frac{m+l}{m}}(\bar{x}),$$

it is possible to assure not only finite-time convergence (for $l < 0$) but also an estimate of the convergence time can be calculated from the expression

$$T(\bar{x}_0) \leq \frac{m}{-\kappa l} V^{\frac{-l}{m}}(\bar{x}_0), \tag{2.20}$$

for $l = -1$ and $C \neq 0$, and for $l \in (-1, 0)$ and $C = 0$. This estimated convergence time is usually rather conservative and it can be very far from the true convergence time.

Another advantage of the use of explicit Lyapunov functions for the convergence proofs is that they allow the calculation of gains rendering stable the origin of the closed-loop system (2.8). These gains are calculated by maximizing some functions (obtained in the Lyapunov analysis), as it is detailed below in Sects. 2.5.3.2 and 2.7.3 for the full state and the partial state I-controllers, respectively. Unfortunately, this maximization approach to calculate the gains has two main drawbacks: (D1) it leads usually to high values of the gains since the calculation is rather conservative. And so for applications, it is, in general, better to find appropriate gains with the help of some simulations. (D2) The approach is basically of numerical nature, so that it is not possible to get analytical expressions for stabilizing gains, and we have to run a maximization procedure to find each set of gains.

To counteract this second item (D2), we can again take advantage of the homogeneity of the algorithm that allows to scale a set of stabilizing gain values to another set of stabilizing gain values. For example, for the partial state I-controller, it is possible to show that for any constant $L > 0$ the following scaling preserves the stability for any $l \in [-1, 0]$:

$$\begin{aligned} &k_n \to L^{\frac{r_n - r_{n+1}}{r_n}} k_n, \quad K_I \to L^{\frac{r_1 - r_{n+2}}{r_1}} K_I, \\ &C \to LC, \qquad \kappa_j \to L^{\left(\frac{r_1}{r_n} - \frac{r_1}{r_j}\right)} \kappa_j, \quad j = 1, \ldots, n-1, \\ &k_{I,\iota} \to L^{1 - \frac{r_1}{r_j}} k_{I,\iota}, \quad \beta_\iota \to L^{1 - \frac{r_1}{r_j}} \beta_\iota, \qquad \iota = 2, \ldots, n. \end{aligned} \tag{2.21}$$

This scaling property generates a family of stabilizing gains (parametrized by L) starting from one set of gains.

2.4 Example

To illustrate the performance of the proposed controllers, we give an elementary example. Consider a third-order system

$$\dot{x}_1 = x_2,\ \dot{x}_2 = x_3,\ \dot{x}_3 = w(t) + b(t)u,$$

where $w(t) = (2 + \cos(3t))$ y $b(t) = 1 + \cos^2(t)$. Note that $|w(t)| \le 3$, $1 \le b(t) \le 2$, and for the perturbation term $\theta(t) = \frac{w(t)}{b(t)} \le 3$ we have that $|\theta(t)| \le 3$ and $\left|\dot{\theta}(t)\right| \le 12$. For our simulation study, we design a HOSM controller using the results of [12], a discontinuous full state I-controller as given by (2.13) and a discontinuous partial state I-controller as in (2.15). Note that for this later controller the integral action depends only on the output x_1. Moreover, we also design continuous approximations of the full state I-controller for two homogeneity degrees, $l = -0.8$ and $l = 0$ (i.e., a linear controller), and we compare their behaviors. Selecting $r_1 = 1 - nl$, $r_2 = 1 - (n-1)l$, $r_3 = 1 - (n-2)l$, and we have taken $\alpha_3 = \alpha_2 = \alpha_1 = 1 - 3l$, $K_m = 1$ and $C = 12$, these controllers are given by

1. **HOSMC:** $u = -k_3\lceil x_3 + 1.5\lambda\lceil\lceil x_2\rfloor^{3/2} + \lambda x_1\rfloor^{1/3}\rfloor^0$, with $\lambda = 0.5$ and $k_3 = 5.1$.
2. **DFSIC:** $u = -k_3\lceil\sigma_3(\overline{x}_3)\rfloor^{\frac{1}{\alpha_3}} - \int_0^t k_I\lceil\sigma_3(\overline{x}_3)\rfloor^0$, with $l = -1$ and gains $k_1 = 0.5C^{1/4}$, $k_2 = 0.8C^{1/3}$, $k_3 = 4C^{1/2}$ and $k_I = 2C$.

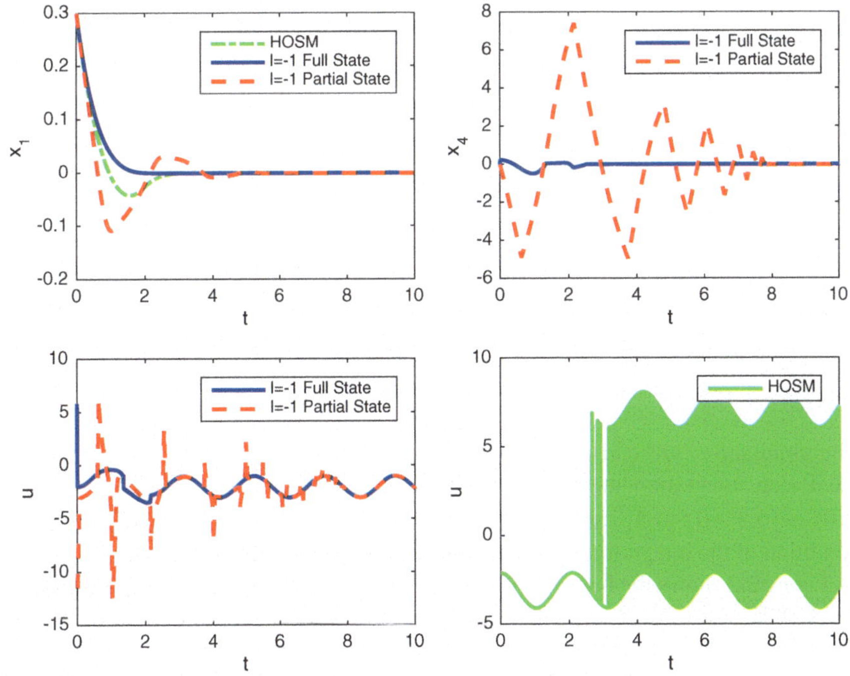

Fig. 2.2 Simulation results for the discontinuous controllers: the HOSM controller, a full state I-controller, and a partial state I-controller

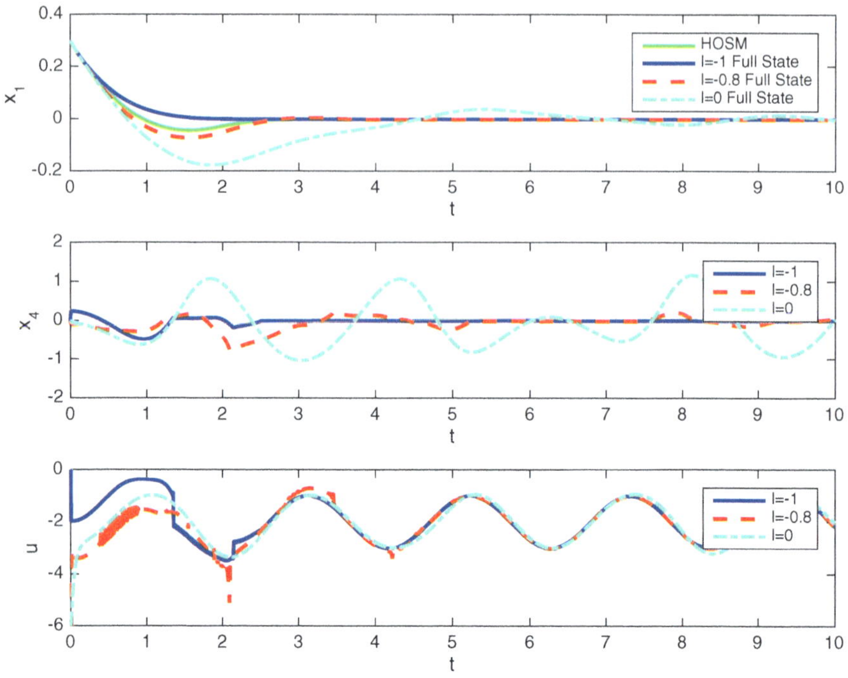

Fig. 2.3 Simulation results for the discontinuous controllers, HOSM and full state I-controller, and the continuous approximations of the I-controllers with degrees $l = -0.8$ and $l = 0$ (a linear controller)

3. **DPSIC:** $u = -k_3 \left[\lceil x_3 \rfloor^{\frac{r_1}{r_3}} + \kappa_2 \lceil x_2 \rfloor^{\frac{r_1}{r_2}} + \kappa_1 x_1 \right]^{\frac{1}{r_1}} - k_I \int_0^t \lceil x_1 \rfloor^0 d\tau$, with $l = -1$ and gains $\kappa_2 = k_2^{\frac{1-3l}{1-l}}$, $\kappa_1 = k_2^{\frac{1-3l}{1-l}} k_1^{\frac{1-3l}{1-2l}}$, where $k_1 = 0.5C^{1/4}$, $k_2 = 0.8C^{1/3}$, $k_3 = 4C^{1/2}$ and $k_I = 0.6C$.
4. **CFSIC1:** $u = -\lceil \sigma_3 (\overline{x}_3) \rfloor^{\frac{1}{\alpha_3}} - \int_0^t k_I \lceil \sigma_3 (\overline{x}_3) \rfloor^{\frac{1+l}{\alpha_3}} d\tau$, with $l = -0.8$ and gains $k_2 = k_1^{r_1/r_2}$, $k_3 = 12 k_1^{r_1/r_3}$, $k_I = 6 k_1^{2r_1/r_3}$, and $k_1 = 1$.
5. **CFSIC2:** $u = -\lceil \sigma_3 (\overline{x}_3) \rfloor^{\frac{1}{\alpha_3}} - \int_0^t k_I \lceil \sigma_3 (\overline{x}_3) \rfloor^{\frac{1+l}{\alpha_3}} d\tau$, with $l = 0$ and gains $k_2 = k_1$, $k_3 = 12k_1$, $k_I = 6k_1^2$, and $k_1 = 1$.

The simulations are carried out with sampling time $t_s = 0.0001$ [s] and Euler's fixed step integration method. As initial conditions in all simulations, we take $x(0) = \left[0.3\ -0.5\ 0.5\right]^T$ and for $x_4(0) = \upsilon(0) + \theta(0) = 0$, where $\upsilon(0)$ is the initial condition of the integrator and $\theta(0)$ is the initial value of the perturbation.

Figure 2.2 shows the behavior of the discontinuous controllers HOSMC, DFSIC, and DPSIC. All of them are able to bring the output x_1 to zero in finite time despite the perturbation $w(t)$ and the uncertain time-varying change in the control coefficient $b(t)$ (see Fig. 2.2 upper left). However, the control signal $u(t)$ for the HOSM controller is strongly discontinuous (Fig. 2.2 lower right), what leads to the unde-

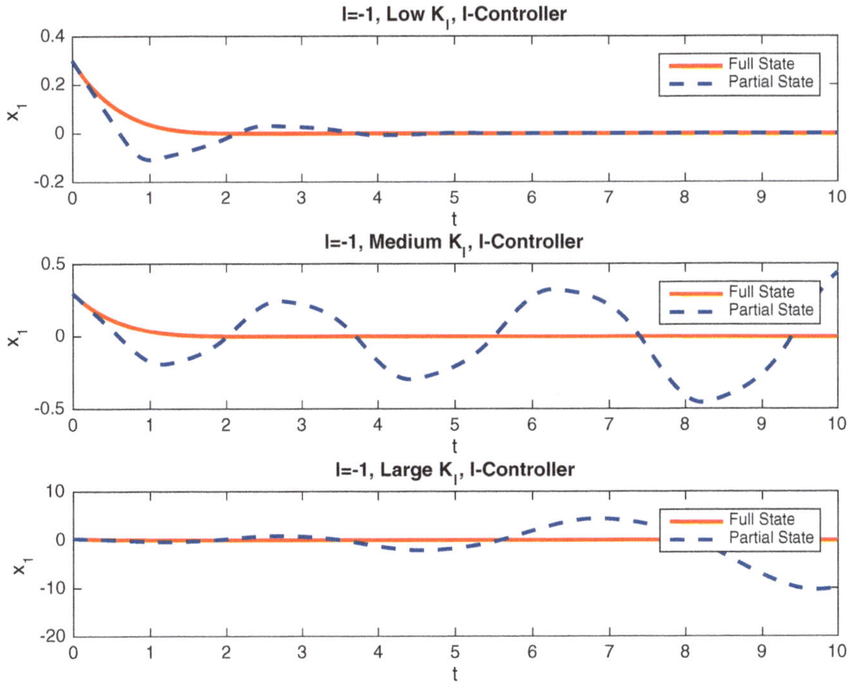

Fig. 2.4 Simulation results comparing the behavior of a full state and a partial state I-controller by increasing integral gain k_I

sirable chattering effect. In contrast, the control signal $u(t)$ of the I-controllers is continuous functions (Fig. 2.2 lower left). For these later controllers, the integral action estimates in finite time the perturbation and compensate its effect on the system, what is reflected in the fact that the variable x_4 converges to zero in finite time (Fig. 2.2 upper right). Note that the I-controller depending only on the output x_1 shows a less damped behavior, compared to the full state I-controller. This is the usual behavior of the I action in classical control, since it depends on the output alone: it is well known that the I action is "destabilizing" and its effect should be damped using the proportional action.

Instead of a discontinuous integral action, it is possible to apply a continuous homogenous integral action to approximate it. Figure 2.3 shows the behavior of the discontinuous full state I-controller DFSIC and its two approximations CFSIC1 and CFSIC2. The upper graph in Fig. 2.3 presents the behavior of the output x_1 for the three controllers (we also include the controller HOSM). Note that only the discontinuous controllers (HOSMC and DFISC) are able to bring x_1 to zero in finite time, while the continuous controllers (CFSIC1 and CFSIC2) cannot achieve exact tracking. This is also reflected in the integral signal x_4 in the middle graph of Fig. 2.3, which is responsible for compensating the perturbation, and that reaches zero for the discontinuous I-controller but not for the continuous ones. Note that the control signals of the three I-controllers are continuous and they are very similar in steady state (see lower graph of Fig. 2.3).

Figure 2.4 presents the simulation results of the discontinuous full state and partial state I-controllers DFSIC and DPSIC when the integral gain k_I is increased (all the rest is equal). The upper graph in Fig. 2.4 corresponds to a low gain, the middle graph to a middle gain while the lower graph to a large gain. We see that full state I-controller remains stable and well damped, but the partial state I-controller becomes unstable for large integral action.

2.5 Convergence Proofs of the Full State Nonpassivity-Based I-Controllers

We provide a unified proof for both Theorems 2.1 and 2.2. Given two scalar (real) variables z_1 and z_2, and three positive real numbers ρ_1, ρ_2, δ, let us define the function

$$Z(z_1, z_2) = \frac{\rho_1}{\delta}|z_1|^{\frac{\delta}{\rho_1}} + z_1 \lceil z_2 \rfloor^{\frac{\delta-\rho_1}{\rho_2}} + \left(\frac{\delta-\rho_1}{\delta}\right)|z_2|^{\frac{\delta}{\rho_2}}, \tag{2.22}$$

and the two associated functions

$$\begin{aligned} \sigma(z_1, z_2) &\triangleq \lceil z_1 \rfloor^{\frac{\delta-\rho_1}{\rho_1}} + \lceil z_2 \rfloor^{\frac{\delta-\rho_1}{\rho_2}} \\ s(z_1, z_2) &\triangleq z_1 + \lceil z_2 \rfloor^{\frac{\rho_1}{\rho_2}}. \end{aligned} \tag{2.23}$$

Lemma 2.2 *Functions Z, σ, and s have the following properties:*

1. *$Z(z_1, z_2)$ is homogeneous of degree δ and with weights (ρ_1, ρ_2), it is continuous for every $\rho_1, \rho_2, \delta > 0$ and it is continuously differentiable if $\delta \geq \rho_1 + \rho_2$.*
2. *$Z(z_1, z_2)$ is nonnegative, i.e., $Z(z_1, z_2) \geq 0$ for every $\rho_1, \rho_2, \delta > 0$ and every $(z_1, z_2) \in \mathbb{R}^2$. $Z(z_1, z_2) = 0$ if and only if $\lceil z_1 \rfloor^{\frac{\delta}{\rho_1}} = -\lceil z_2 \rfloor^{\frac{\delta}{\rho_2}}$.*
3. *If Z is differentiable, its partial derivatives are continuous and are given by $\partial_{z_1} Z(z_1, z_2) = \sigma(z_1, z_2)$ and $\partial_{z_2} Z(z_1, z_2) = \frac{\delta-\rho_1}{\rho_2}|z_2|^{\frac{\delta-\rho_1-\rho_2}{\rho_2}} s(z_1, z_2)$. Furthermore, under the same conditions, functions σ and s are continuous. $\sigma(z_1, z_2)$ is differentiable if $\delta \geq \max\{2\rho_1, \rho_1 + \rho_2\}$, while $s(z_1, z_2)$ is differentiable if $\rho_1 \geq \rho_2$.*
4. *Function $s(z_1, z_2)$ vanishes on the same set where $\sigma(z_1, z_2)$ vanishes, i.e., $s(\mathcal{N}_\sigma) = 0$, where $\mathcal{N}_\sigma = \left\{(z_1, z_2) \in \mathbb{R}^2 | \sigma(z_1, z_2) = 0\right\}$. $Z(z_1, z_2)$ vanishes only on the set $\mathcal{N}_\sigma$. Moreover, $\sigma(z_1, z_2)\, s(z_1, z_2) \geq 0$ for every $(z_1, z_2) \in \mathbb{R}^2$.*

Proof Item 2 comes from the Young's inequality, and therefore $z_1 \lceil z_2 \rfloor^{\frac{\delta-\rho_1}{\rho_2}} \leq \frac{\rho_1}{\delta}|z_1|^{\frac{\delta}{\rho_1}} + \left(\frac{\delta-\rho_1}{\delta}\right)|z_2|^{\frac{\delta}{\rho_2}}$. From this and (2.22), it follows that $Z(z_1, z_2) \geq 0$. The proofs for the other items are simple. □

2.5.1 The Lyapunov Function Candidate

Having selected $m \geq r_1 + \alpha_1$, consider the homogeneous and continuously differentiable ($\mathscr{C}^1$) Lyapunov function candidate for system (2.8), when the controllers (2.18) or (2.19) are used

$$\mathscr{V}(\overline{x}) = \frac{\alpha_n + r_n}{m} W_I^{\frac{m}{\alpha_n + r_n}}(\overline{x}) + \frac{r_{n+1}}{m} \gamma_I |\zeta|^{\frac{m}{r_{n+1}}} + \gamma_{n-1} V_{n-1}(\overline{x}_{n-1}), \tag{2.24}$$

where $\zeta = x_{n+1}/k_n$, γ_{n-1} is an arbitrary positive constant and γ_I is a positive constant to be selected sufficiently large. Moreover,

$$W_I(\overline{x}) = Z(x_n, -\varphi(\overline{x})) = \frac{r_n}{\alpha_n + r_n} |x_n|^{\frac{\alpha_n + r_n}{r_n}} - \varphi(\overline{x}) x_n + \frac{\alpha_n}{\alpha_n + r_n} |\varphi(\overline{x})|^{\frac{\alpha_n + r_n}{\alpha_n}},$$

where Z is given in (2.22) and the function $\varphi(\overline{x})$ is defined as

$$\varphi(\overline{x}) = -k_{n-1}^{\frac{\alpha_n}{r_n}} \lceil \sigma_{n-1}(\overline{x}_{n-1}) \rfloor^{\frac{\alpha_n}{\alpha_{n-1}}} + \lceil \zeta \rfloor^{\frac{\alpha_n}{r_{n+1}}},$$

with $\sigma_{n-1}(\overline{x}_{n-1})$ given by (2.9). Furthermore, $V_{n-1}(\overline{x}_{n-1})$ is obtained from (2.10). Functions $\mathscr{V}(\overline{x})$, $W_I(\overline{x})$, $\varphi(\overline{x})$ are $\mathbf{r}$-homogeneous of degrees $m, \alpha_n + r_n, \alpha_n$, respectively, while functions $V_i(\overline{x}_i)$, $\sigma_i(\overline{x}_i)$, $W_i(\overline{x}_i)$ are $\mathbf{r}_i$-homogeneous of degrees m, α_i, $\alpha_i + r_i$, respectively. We note that $W_i(\overline{x}_i)$, defined in (2.11), can be written as

$$W_i(\overline{x}_i) = Z\left(x_i, k_{i-1}^{\frac{\alpha_{i-1}}{r_i}} \sigma_{i-1}(\overline{x}_{i-1})\right),$$

with $\rho_1 = r_i$, $\rho_2 = \alpha_{i-1}$, $\delta = \alpha_i + r_i$, and it follows from Lemma 2.2 that functions W_i and W_I are positive semidefinite and $\mathscr{C}^1$. As a consequence, for every positive values of γ_i, γ_I, and k_i, the function $\mathscr{V}(\overline{x})$ is continuously differentiable and positive semidefinite. Furthermore, $\mathscr{V}(\overline{x})$ can only vanish at the point $\bar{x} = 0$, and therefore it is positive definite and, due to its homogeneity [8], it is decrescent and radially unbounded.

2.5.2 Auxiliary Functions and Some Important Relations

From Lemma 2.2, we can also obtain that

$$\begin{aligned} \partial_{x_n} W_I(\overline{x}) &= \sigma_I(\overline{x}) \triangleq \lceil x_n \rfloor^{\frac{\alpha_n}{r_n}} - \varphi(\overline{x}) \\ \partial_{\varphi} W_I(\overline{x}) &= -s_I(\overline{x}) \triangleq -\left(x_n - \lceil \varphi(\overline{x}) \rfloor^{\frac{r_n}{\alpha_n}}\right), \end{aligned} \tag{2.25}$$

and for $i = 2, \ldots, n-1$

$$\begin{aligned}
\partial_{x_i} W_i\left(\overline{x}_i\right) &= \sigma_i\left(\overline{x}_i\right) \\
\partial_{\sigma_{i-1}} W_i\left(\overline{x}_i\right) &= \frac{\alpha_i}{\alpha_{i-1}} k_{i-1}^{\frac{\alpha_i}{r_i}} \left|\sigma_{i-1}\left(\overline{x}_{i-1}\right)\right|^{\frac{\alpha_i-\alpha_{i-1}}{\alpha_{i-1}}} s_i\left(\overline{x}_i\right),
\end{aligned} \tag{2.26}$$

where $\sigma_i\left(\overline{x}_i\right)$ is given by (2.9) and

$$s_i\left(\overline{x}_i\right) \triangleq x_i + k_{i-1}\left\lceil \sigma_{i-1}\left(\overline{x}_{i-1}\right)\right\rfloor^{\frac{r_i}{\alpha_{i-1}}}. \tag{2.27}$$

Moreover,

$$\begin{gathered}
\sigma_I\left(\overline{x}\right) = 0 \Leftrightarrow x_n = \left\lceil \varphi\left(\overline{x}\right)\right\rfloor^{\frac{r_n}{\alpha_{n-1}}} \Leftrightarrow W_I\left(\overline{x}\right) = 0 \Leftrightarrow s_I\left(\overline{x}\right) = 0, \\
s_I\left(\overline{x}\right)\sigma_I\left(\overline{x}\right) > 0, \text{ if } \sigma_I\left(\overline{x}\right) \neq 0,
\end{gathered}$$

and for $i = 2, \ldots, n-1$

$$\begin{gathered}
\sigma_i\left(\overline{x}_i\right) = 0 \Leftrightarrow x_i = -k_{i-1}\left\lceil \sigma_{i-1}\left(\overline{x}_{i-1}\right)\right\rfloor^{\frac{r_i}{\alpha_{i-1}}} \Leftrightarrow W_i\left(\overline{x}_i\right) = 0 \Leftrightarrow s_i\left(\overline{x}_i\right) = 0, \\
s_i\left(\overline{x}_i\right)\sigma_i\left(\overline{x}_i\right) > 0, \text{ if } \sigma_i\left(\overline{x}_i\right) \neq 0.
\end{gathered}$$

Note that functions $\varphi\left(\overline{x}\right)$, $\sigma_I\left(\overline{x}\right)$, and $\sigma_i\left(\overline{x}_i\right)$ are $\mathscr{C}^1$, while functions $s_I\left(\overline{x}\right)$ and $s_i\left(\overline{x}_i\right)$ are continuous ($\mathscr{C}^0$) but, in general, not differentiable.

From (2.10) and (2.26), it follows that

$$\partial_{x_i} V_i\left(\bar{x}_i\right) = W_i^{\frac{m-\alpha_i-r_i}{\alpha_i+r_i}}\left(\overline{x}_i\right)\sigma_i\left(\overline{x}_i\right), \tag{2.28}$$

and, similarly for each $j < i$

$$\begin{aligned}
\partial_{x_j} V_i\left(\bar{x}_i\right) = & \frac{\alpha_i}{\alpha_{i-1}} k_{i-1}^{\frac{\alpha_i}{r_i}} W_i^{\frac{m-\alpha_i-r_i}{\alpha_i+r_i}}\left(\overline{x}_i\right)\left|\sigma_{i-1}\left(\overline{x}_{i-1}\right)\right|^{\frac{\alpha_i-\alpha_{i-1}}{\alpha_{i-1}}} s_i\left(\overline{x}_i\right)\partial_{x_j}\sigma_{i-1}\left(\bar{x}_{i-1}\right) + \\
& \gamma_{i-1}\partial_{x_j} V_{i-1}\left(\bar{x}_{i-1}\right).
\end{aligned} \tag{2.29}$$

Note also that

$$\partial_{x_i} V_i\left(\bar{x}_i\right) = 0 \Leftrightarrow \sigma_i\left(\overline{x}_i\right) = 0.$$

Define also the $(\mathbf{r}_i,\ m+l)$-homogeneous functions

$$\begin{aligned}
H_1\left(\bar{x}_1\right) &\triangleq -k_1\partial_{x_1} V_1\left(\bar{x}_1\right)\left\lceil \sigma_1\left(\overline{x}_1\right)\right\rfloor^{\frac{r_2}{\alpha_1}} \\
H_i\left(\bar{x}_i\right) &\triangleq \sum_{j=1}^{i-1} \partial_{x_j} V_i\left(\bar{x}_i\right)\cdot x_{j+1} - k_i\partial_{x_i} V_i\left(\bar{x}_i\right)\cdot\left\lceil \sigma_i\left(\overline{x}_i\right)\right\rfloor^{\frac{r_{i+1}}{\alpha_i}}, \ i = 2, \ldots, n.
\end{aligned} \tag{2.30}$$

Using (2.28) and (2.29) in (2.30), we obtain

$$H_i(\bar{x}_i) = \frac{\alpha_i}{\alpha_{i-1}} k_{i-1}^{\frac{\alpha_i}{r_i}} W_i^{\frac{m-(\alpha_i+r_i)}{\alpha_i+r_i}}(\overline{x}_i) \left|\sigma_{i-1}(\overline{x}_{i-1})\right|^{\frac{\alpha_i-\alpha_{i-1}}{\alpha_{i-1}}} s_i(\overline{x}_i) \times$$
$$\sum_{j=1}^{i-1} \partial_{x_j}\sigma_{i-1}(\overline{x}_{i-1}) \cdot x_{j+1} - k_i \partial_{x_i} V_i(\bar{x}_i) \cdot \lceil \sigma_i(\overline{x}_i) \rfloor^{\frac{r_{i+1}}{\alpha_i}} +$$
$$\gamma_{i-1}\left(\sum_{j=1}^{i-2} \partial_{x_j} V_{i-1}(\overline{x}_{i-1}) \cdot x_{j+1} + \partial_{x_{i-1}} V_{i-1}(\overline{x}_{i-1}) \cdot x_i\right).$$

From (2.27), we get $x_i = s_i(\overline{x}_i) - k_{i-1}\lceil \sigma_{i-1}(\overline{x}_{i-1}) \rfloor^{\frac{r_i}{\alpha_{i-1}}}$ and, replacing this and (2.28) in the last equality, we obtain the recursive expression

$$H_i(\bar{x}_i) = \gamma_{i-1} H_{i-1}(\bar{x}_{i-1}) + s_i(\overline{x}_i)\Psi_i(\bar{x}_i) - k_i W_i^{\frac{m-\alpha_i-r_i}{\alpha_i+r_i}}(\overline{x}_i) \left|\sigma_i(\overline{x}_i)\right|^{\frac{r_{i+1}+\alpha_i}{\alpha_i}} \quad (2.31)$$
$$\Psi_i(\bar{x}_i) \triangleq \gamma_{i-1} W_{i-1}^{\frac{m-\alpha_{i-1}-r_{i-1}}{\alpha_{i-1}+r_{i-1}}}(\overline{x}_{i-1})\, \sigma_{i-1}(\overline{x}_{i-1}) +$$
$$\frac{\alpha_i}{\alpha_{i-1}} k_{i-1}^{\frac{\alpha_i}{r_i}} W_i^{\frac{m-(\alpha_i+r_i)}{\alpha_i+r_i}}(\overline{x}_i) \left|\sigma_{i-1}(\overline{x}_{i-1})\right|^{\frac{\alpha_i-\alpha_{i-1}}{\alpha_{i-1}}} \sum_{j=1}^{i-1} \partial_{x_j}\sigma_{i-1}(\overline{x}_{i-1}) \cdot x_{j+1}.$$

From (2.31), we conclude that in the set where $\sigma_i(\overline{x}_i) = 0$ the value of $H_i(\bar{x}_i) = \gamma_{i-1} H_{i-1}(\bar{x}_{i-1})$. Lemma 2.6 implies that if $H_{i-1}(\bar{x}_{i-1}) < 0$ then $H_i(\bar{x}_i)$ can be rendered negative definite by selecting a gain k_i sufficiently large. Since $H_1(\bar{x}_1) = -k_1 \partial_{x_1} V_1(\bar{x}_1) \lceil \sigma_1(\overline{x}_1) \rfloor^{\frac{r_2}{\alpha_1}} = -k_1 |x_1|^{\frac{m+r_2-r_1}{r_1}} < 0$ for $x_1 \neq 0$, it follows by induction that for every $i = 2, \ldots, n-1$ we can render $H_i(\bar{x}_i) < 0$ by appropriate selection of $k_1, \ldots, k_i$.

Lemma 2.3 *There exist gains $k_i > 0$ such that for all $i = 1, \ldots, n-1$ the functions $H_i(\bar{x}_i) < 0$ are negative definite.*

2.5.3 Derivative of the Lyapunov Function Candidate

We will show that the derivative of $\mathscr{V}(\overline{x})$ (2.24) along the trajectories of system (2.8), with $\vartheta_1(x)$ and $\vartheta_2(x)$ as in (2.18) (which corresponds to the general case), is negative definite for all $n \geq 1$ (when $n = 1$ we set $\sigma_0 = V_0 = 0$), for all positive values of $\gamma_i > 0$, every positive value of k_I (or $k_I > C$ when $l = -1$), and sufficiently large values of k_i and γ_I. From (2.24), and using (2.25), the time derivative of (2.24) along the closed-loop system (2.8) is ($\tilde{\gamma}_I \triangleq \gamma_I / k_n$)

$$\dot{\mathscr{V}}(\overline{x}) = W_I^{\frac{m-(\alpha_n+r_n)}{\alpha_n+r_n}}(\overline{x})\left(\sigma_I(\overline{x})\dot{x}_n - s_I(\overline{x})\dot{\varphi}(\overline{x})\right) + \tilde{\gamma}_I\lceil\zeta\rfloor^{\frac{m-r_{n+1}}{r_{n+1}}}\dot{x}_{n+1} + \gamma_{n-1}\dot{V}_{n-1}(\overline{x}_{n-1}) \,. \tag{2.32}$$

We consider each one of the terms in (2.32).

The Term $\dot{V}_{n-1}(\overline{x}_{n-1})$

From (2.27), we see that $x_n = s_n(\overline{x}_n) - k_{n-1}\lceil\sigma_{n-1}(\overline{x}_{n-1})\rfloor^{\frac{r_n}{\alpha_{n-1}}}$, and then, by a straightforward calculation, we get

$$\dot{V}_{n-1} = \sum_{j=1}^{n-1}\partial_{x_j}V_{n-1}\cdot x_{j+1} = H_{n-1}(\bar{x}_{n-1}) + \partial_{x_{n-1}}V_{n-1}\cdot s_n(\bar{x}_n) \,. \tag{2.33}$$

The Term $\dot{\varphi}$ and the Controller

Consider the term $\dot{\varphi}(\overline{x})$ in (2.32), which can be written as

$$\dot{\varphi}(\overline{x}) = -\frac{\alpha_n}{\alpha_{n-1}}k_{n-1}^{\frac{\alpha_n}{r_n}}\left|\sigma_{n-1}(\overline{x}_{n-1})\right|^{\frac{\alpha_n-\alpha_{n-1}}{\alpha_{n-1}}}\dot{\sigma}_{n-1}(\overline{x}_{n-1}) + \frac{\alpha_n}{r_{n+1}}\frac{1}{k_n}|\zeta|^{\frac{\alpha_n-r_{n+1}}{r_{n+1}}}\dot{x}_{n+1} \,. \tag{2.34}$$

Since $\sigma_{n-1}(\overline{x}_{n-1})$ is a $\mathscr{C}^1$ function, we have that $\dot{\sigma}_{n-1}(\overline{x}_{n-1})$ is a continuous homogeneous function.

We consider first the controllers (2.18) (or discontinuous controllers (2.13) when $l=-1$), so that

$$\dot{x}_n \in -k_n[K_m, K_M]\underbrace{\left(\lceil\sigma_n(x)\rfloor^{\frac{r_{n+1}}{\alpha_n}} - \zeta\right)}_{\sigma_d(\overline{x})}, \quad \dot{x}_{n+1} \in -k_I\underbrace{\left(\lceil\sigma_n(x)\rfloor^{\frac{r_{n+2}}{\alpha_n}} - \frac{1}{k_I}[-C, C]\right)}_{\Sigma_I(x)} \,.$$

Notice that $\sigma_I(\overline{x}) = 0 \Leftrightarrow \sigma_d(\overline{x}) = 0$ since

$$\begin{aligned}\sigma_I(\overline{x}) &= \lceil x_n\rfloor^{\frac{\alpha_n}{r_n}} - \varphi(\overline{x}) = \lceil x_n\rfloor^{\frac{\alpha_n}{r_n}} + k_{n-1}^{\frac{\alpha_n}{r_n}}\lceil\sigma_{n-1}(\overline{x}_{n-1})\rfloor^{\frac{\alpha_n}{\alpha_{n-1}}} - \lceil\zeta\rfloor^{\frac{\alpha_n}{r_{n+1}}} \\ &= \sigma_n(\overline{x}_n) - \lceil\zeta\rfloor^{\frac{\alpha_n}{r_{n+1}}} = 0 \Leftrightarrow \sigma_d(\overline{x}_e) = \lceil\sigma_n(\overline{x}_n)\rfloor^{\frac{r_{n+1}}{\alpha_n}} - \zeta = 0 \,.\end{aligned}$$

Moreover, $\sigma_I(\overline{x})\sigma_d(\overline{x}) > 0$ for $\sigma_I(\overline{x}) \neq 0$.

Using the control laws (2.18), and the relations (2.25), (2.28), (2.33), and (2.34), we can write $\dot{\mathscr{V}}(\overline{x})$ in (2.32) as

$$\dot{\mathscr{V}}(\overline{x}) \in -k_n[K_m, K_M]W_I^{\frac{m-(\alpha_n+r_n)}{\alpha_n+r_n}}(\overline{x})\sigma_I(\overline{x})\sigma_d(\overline{x}) - k_I\tilde{\gamma}_I\lceil\zeta\rfloor^{\frac{m-r_{n+1}}{r_{n+1}}}\Sigma_I(x) - s_I(\overline{x})W_I^{\frac{m-(\alpha_n+r_n)}{\alpha_n+r_n}}(\overline{x})\dot{\varphi}(\overline{x}) + \gamma_{n-1}\dot{V}_{n-1}(\overline{x}_{n-1}) \,. \tag{2.35}$$

If we consider that $C = 0$ when $l \in (-1, 0]$, or that $C \geq 0$ for $l = -1$, the r.h.s. of (2.35) is a homogeneous real-valued upper semicontinuous (u.s.) multivalued mapping with compact convex values. Lemma 2.4 provides a representation of such a multivalued function.

Lemma 2.4 ([13, Example 1.3]) *Let $\Omega \neq \varnothing$ be a subset of $\mathbb{R}^n$. Then the general upper semicontinuous real-valued multivalued map $F : \Omega \to 2^{\mathbb{R}} \setminus \varnothing$ with compact convex values is given by $F(\omega) = [\varphi(\omega), \psi(\omega)]$, where $\varphi : \Omega \to \mathbb{R}$ is lower semi-continuous (l.s.), $\psi : \Omega \to \mathbb{R}$ is u.s. and $\varphi(\omega) \leq \psi(\omega)$ on Ω.*

According to Lemma 2.4, the right-hand side of (2.35) can be upper bounded by a single-valued upper semicontinuous function. Therefore, in both the continuous and the discontinuous cases, we can use Lemma 2.6 (in Appendix).

The first term in (2.35), $-k_n W_I^{\frac{m-(\alpha_n+r_n)}{\alpha_n+r_n}} \sigma_I(\overline{x}) \sigma_d(\overline{x})$, is negative except when $\sigma_I(\overline{x}) = 0 \Leftrightarrow \lceil \sigma_n(\overline{x}_n) \rfloor^{\frac{r_{n+1}}{\alpha_n}} - \zeta = 0$. For the discontinuous case ($l = -1$ and $r_{n+2} = 0$), we assume that $k_I > C$, so that evaluating the second term in (2.35) on the set where $\sigma_I(\overline{x}) = 0$ (and $s_I(\overline{x}) = 0$), we obtain

$$-\tilde{\gamma}_I k_I \lceil \sigma_n(\overline{x}_n) \rfloor^{\frac{m-r_{n+1}}{\alpha_n}} \Sigma_I(\overline{x}) \leq -\tilde{\gamma}_I k_I \left(1 - \frac{C}{k_I}\right) |\sigma_n(\overline{x}_n)|^{\frac{m-r_{n+1}}{\alpha_n}} .$$

For the continuous case ($l > -1$ and $r_{n+2} > 0$), we assume that $k_I > 0$ and $C = 0$, so that evaluating the second term in (2.35) on the set where $\sigma_I(\overline{x}) = 0$, we obtain

$$-k_I \tilde{\gamma}_I \lceil \zeta \rfloor^{\frac{m-r_{n+1}}{r_{n+1}}} \lceil \sigma_n(\overline{x}_n) \rfloor^{\frac{r_{n+2}}{r_1}} = -\tilde{\gamma}_I k_I |\sigma_n(\overline{x}_n)|^{\frac{m-r_{n+1}+r_{n+2}}{\alpha_n}} .$$

And therefore, in both cases, the values of $\dot{\mathcal{V}}(\overline{x})$ restricted to the set where $\sigma_I(\overline{x}) = 0$, for which $\zeta = \lceil \sigma_n(\overline{x}_n) \rfloor^{\frac{r_{n+1}}{\alpha_n}}$, are given by

$$\Upsilon_I(\overline{x}_n) \triangleq \dot{\mathcal{V}}(\overline{x})\Big|_{\sigma_I(\overline{x})=0} \leq -\tilde{\gamma}_I k_I \Delta |\sigma_n(\overline{x}_n)|^{\frac{m-r_{n+1}+r_{n+2}}{\alpha_n}} + \gamma_{n-1} \dot{V}_{n-1}(\overline{x}_{n-1}) , \quad (2.36)$$

where $\Delta = 1$ in the continuous case and $\Delta = \left(1 - \frac{C}{k_I}\right) > 0$ in the discontinuous one. The first term in $\Upsilon_I(x)$, given by $-k_I \tilde{\gamma}_I \Delta |\sigma_n(\overline{x}_n)|^{\frac{m+r_{n+2}-r_{n+1}}{\alpha_n}}$, is negative, except on the set where $\sigma_n(\overline{x}_n) = 0 \Leftrightarrow s_n(\bar{x}_n) = 0$. The values of $\Upsilon_I(x)$ restricted to the set where $\sigma_n(\overline{x}_n) = 0$ are denoted by $\Upsilon_{I,\sigma}(\overline{x}_{n-1})$ and are given by (we use (2.33))

$$\Upsilon_{I,\sigma}(\overline{x}_{n-1}) \leq \gamma_{n-1} H_{n-1}(\bar{x}_{n-1}) < 0 ,$$

which is negative (recall Lemma 2.3). From these results, and using Lemma 2.6, $\Upsilon_I(x)$ can be made negative by selecting a sufficiently large value of $\tilde{\gamma}_I$ (for any positive value of k_I). Furthermore, applying Lemma 2.6 once more we conclude that $\dot{\mathcal{V}}(\overline{x})$ can be made negative definite by selecting a sufficiently large value of k_n. For this argumentation, it is important to note that $\tilde{\gamma}_I$ and k_n can be selected independently. Once they are fixed, one obtains the value of γ_I as $\gamma_I = \tilde{\gamma}_I k_n$. This is further discussed below in Sect. 2.5.3.2.

Now consider the rational controllers (2.19) (or quasi-continuous controllers (2.14) when $l = -1$). If $C = 0$, the same arguments presented before imply that $\dot{\mathscr{V}}(\bar{x})$ can be rendered negative definite. For $l = -1$, we want to show that $\dot{\mathscr{V}}(\bar{x})$ is still negative definite, despite the perturbation of size C. For this, we obtain from (2.35) and (2.37)

$$\dot{\mathscr{V}}(\bar{x}) \leq -\kappa \mathscr{V}^{\frac{m-1}{m}}(\bar{x}) + C\mathscr{H}(\bar{x}) ,$$
$$\mathscr{H}(\bar{x}) = \frac{1}{k_n}\left|\gamma_I \lceil \zeta \rfloor^{\frac{m-r_{n+1}}{r_{n+1}}} - \frac{\alpha_n}{r_{n+1}} s_I(\bar{x})\, W_I^{\frac{m-(\alpha_n+r_n)}{\alpha_n+r_n}}(\bar{x})\, |\zeta|^{\frac{\alpha_n-r_{n+1}}{r_{n+1}}}\right| ,$$

where $\mathscr{V}^{\frac{m-1}{m}}(\bar{x})$ and $\mathscr{H}(\bar{x})$ are continuous and homogeneous of the same homogeneity degree. On the unit sphere $-\kappa\mathscr{V}^{\frac{m-1}{m}}(\bar{x})$ is negative everywhere. Continuity of the functions and compactness of the unit sphere imply that $-\kappa\mathscr{V}^{\frac{m-1}{m}}(\bar{x}) + C\mathscr{H}(\bar{x})$ is negative for a sufficiently small value of C. Homogeneity implies that for small C, $\dot{\mathscr{V}}(\bar{x})$ is negative definite.

2.5.3.1 Type of Convergence and ISS

Lemma 2.7 implies that there exists a constant κ, depending on the gains k_i, k_I and C (when $l = -1$), such that

$$\dot{\mathscr{V}}(\bar{x}) \leq -\kappa \mathscr{V}^{\frac{m+l}{m}}(\bar{x}) . \tag{2.37}$$

This implies finite-time stability for $l \in [-1, 0)$ and exponential stability for $l = 0$. Moreover, for $l \in [-1, 0)$, using the differential inequality (2.37) and standard Lyapunov arguments [3, 5], we obtain inequality (2.20) as an estimation of the convergence time. Furthermore, for $l \in (-1, 0]$ and when $C \neq 0$, we obtain from (2.35) and (2.37)

$$\dot{\mathscr{V}}(\bar{x}) \leq -\kappa \mathscr{V}^{\frac{m+l}{m}}(\bar{x}) +$$
$$\left(\gamma_I \lceil \zeta \rfloor^{\frac{m-r_{n+1}}{r_{n+1}}} - \frac{\alpha_n}{r_{n+1}} s_I(\bar{x})\, W_I^{\frac{m-(\alpha_n+r_n)}{\alpha_n+r_n}}(\bar{x})\, |\zeta|^{\frac{\alpha_n-r_{n+1}}{r_{n+1}}}\right) \frac{1}{k_n}[-C,\, C] .$$

Again, Lemma 2.7 implies that there exist a constant μ, depending on the gains $k_1, \ldots, k_{n-1}, k_I$, such that $\dot{\mathscr{V}}(\bar{x}) \leq -\kappa\mathscr{V}^{\frac{m+l}{m}}(\bar{x}) + \mu\frac{C}{k_n}\mathscr{V}^{\frac{m-1}{m}}(\bar{x})$. This implies that $\dot{\mathscr{V}}(\bar{x}) \leq -\left(\kappa\mathscr{V}^{\frac{1+l}{m}}(\bar{x}) - \mu\frac{C}{k_n}\right)\mathscr{V}^{\frac{m-1}{m}}(\bar{x})$, which is negative for $c_2\,\|\bar{x}\|^m_{\mathbf{r}_{n+1},\,p} \geq \mathscr{V}(\bar{x}) > \left(\frac{\mu}{\kappa}\frac{C}{k_n}\right)^{\frac{m}{1+l}}$.

Remark 2.1 From the previous inequality, we have that by making k_n large enough the effect of the perturbation term can be arbitrarily reduced as well as by making $\kappa k_n > \mu C$ and selecting l close to -1. In this last case, the dynamical behavior becomes similar to the controllers with discontinuous integral term.

2.5.3.2 Gain Calculation

Note that values of the parameters k_i, for $i = 1, \ldots, n-2$, have to be calculated recursively such that H_i in (2.30) is negative definite. A method for choosing the values of the parameters k_i consists in fixing the parameters γ_i, to select $k_1 > 0$ and to calculate k_i for $i = 2, \ldots, n-1$ recursively as

$$k_i > \max_{\bar{x}_i \in S_i} \left\{ \frac{\sum_{j=1}^{i-1} \partial_{x_j} V_i(\bar{x}_i) \cdot x_{j+1}}{W_i^{\frac{m-\alpha_i-r_i}{\alpha_i+r_i}}(\bar{x}_i) \left|\sigma_i(\bar{x}_i)\right|^{\frac{\alpha_i+r_{i+1}}{\alpha_i}}} \right\}, \tag{2.38}$$

where $S_i = \{\bar{x}_i \in \mathbb{R}^i : \|\bar{x}_i\|_{\mathbf{r},p} = 1\}$ is the unit homogeneous sphere, which is a compact set. Further, we select $k_I > C$ in the discontinuous case and $k_I > 0$ in the continuous one. We choose $\tilde{\gamma}_I = \gamma_I / k_n$ such that in (2.36) $\Upsilon_I(\overline{x}_n) < 0$, i.e.,

$$\tilde{\gamma}_I > \frac{\gamma_{n-1}}{k_I \Delta} \max_{\bar{x}_n \in S_n} \left\{ \frac{\dot{V}_{n-1}(\overline{x}_{n-1})}{\left|\sigma_n(\overline{x}_n)\right|^{\frac{m-r_{n+1}+r_{n+2}}{\alpha_n}}} \right\},$$

and finally we choose k_n large enough such that in (2.35) $\dot{\mathscr{V}}(\overline{x}) < 0$. Notice that k_n is a function of k_{i-1}, $i = 2, \ldots, n$ and k_I, so that we can parameterize k_n in terms of k_1 and k_I. The previous maximizations are feasible, since the functions to be maximized have the following properties: (i) they are $(\mathbf{r}, 0)$-homogeneous, so that they achieve all their values on the unit sphere S_i and (ii) they are u.s., since they are continuous at the points where the denominator does not vanish, and when the denominator vanishes the numerator is negative, as shown above in the proof. It is well known that an u.s. function has a maximum on a compact set (see Lemma 2.7).

2.6 Convergence Proofs for the Passivity-Based I-Controllers

The original idea of this proof is given in [48] for $n = 2$. It has been nicely extended to arbitrary $n \geq 2$ in [28]. Here, we use a more constructive approach than in [28] and provide a novel passivity interpretation for the construction.

2.6.1 The Lyapunov Function Candidate

Assume first that system (2.8) has no perturbation, i.e., with $\frac{d}{dt}\theta(t, z) = 0$ and that the control coefficient $b(t, z) = b$ is constant (but possibly uncertain). For some $m \geq r_1 + \alpha_1$, consider the homogeneous, positive definite Lyapunov function candidate

$$\mathscr{W}(\overline{x}_{n+1}) = \frac{m}{2} V_n^{\frac{2}{m}}(\overline{x}_n) + \frac{1}{2\gamma} x_{n+1}^2 , \tag{2.39}$$

where $\gamma > 0$ is a positive constant and $V_n(\overline{x}_n)$ is obtained from a recursion given in (2.10) for all $i = 2, \ldots, n$. Since $r_{n+1} = 1$, $\mathscr{W}(\overline{x})$ is $\mathbf{r}$-homogeneous of degree 2 while $V_n(\overline{x}_n)$ is $\mathbf{r}_n$-homogeneous of degree m and continuously differentiable. Note that since (for $l < 0$) $\alpha_1 \geq r_1 > 1$ then $m > 2$ and $V_n^{\frac{2}{m}}(\overline{x}_n)$ is continuous everywhere, and continuously differentiable everywhere except at the origin $\bar{x}_n = 0$. $V_n^{\frac{2}{m}}(\overline{x}_n)$ is not Lipschitz continuous at $\bar{x}_n = 0$. Therefore, $\mathscr{W}(\overline{x}_{n+1})$ is differentiable everywhere except when $\overline{x}_n = 0$, and thus $\mathscr{W}(\overline{x}_{n+1})$ is not smooth and not Lipschitz continuous.[3] Its derivative exists everywhere except on the set $\mathscr{E} = \left\{\bar{x}_{n+1} \in \mathbb{R}^{n+1} | \bar{x}_n = 0\right\}$. Although classical Lyapunov theorems require the Lyapunov function to be at least locally Lipschitz, we can overcome the problem here using a similar argument as in the proof for the super-twisting algorithm provided in [49]. The basic idea is to show that the function $\mathscr{W}(\cdot)$ evaluated along the trajectory of the system $\varphi(t, \bar{x}_0)$, i.e., the composite function $\mathscr{W}(\varphi(t, \bar{x}_0))$ is an absolutely continuous (AC) function of time, and thus it is decreasing in time if $\dot{\mathscr{W}}(\varphi(t, \bar{x}_0)) < 0$ *almost everywhere*. Since $\varphi(t, \bar{x}_0)$ is AC, this is true at all points where $\mathscr{W}(\cdot)$ is locally Lipschitz continuous (in $\bar{x}$), i.e., everywhere except possibly at the points of the set $\mathscr{E}$. To conclude that $\mathscr{W}(\varphi(t, \bar{x}_0))$ is AC also at the points on $\mathscr{E}$, we need to show that all components $\varphi_i(t, \bar{x}_0)$, for $i = 1, \ldots, n$, are monotone. But this follows easily from the dynamics (2.6) evaluated at any point $\bar{x}_{n+1} = (0, x_{n+1}) \in \mathscr{E} \setminus \{0\}$.

The derivative $\dot{\mathscr{W}}$ (where it exists) along trajectories of system (2.8) is given by

$$\dot{\mathscr{W}}(\overline{x}_{n+1}) = V_n^{\frac{2-m}{m}}(\overline{x}_n)\, \dot{V}_n(\overline{x}_n) + \frac{1}{\gamma} x_{n+1}\dot{x}_{n+1} ,$$

which is an $\mathbf{r}$-homogeneous of degree $r_{n+1} + r_{n+2} = 2 + l$. To calculate the term $\dot{V}_n(\overline{x}_n)$, we note that

$$V_n(\overline{x}_n) = \frac{\alpha_n + r_n}{m} W_n^{\frac{m}{\alpha_n + r_n}}(\overline{x}_n) + \gamma_{n-1} V_{n-1}(\overline{x}_{n-1})$$

[3]This has been overseen in [28], where it is stated that $\mathscr{W}(\overline{x}_{n+1})$ is differentiable except at the origin, what is incorrect.

with

$$W_n(\overline{x}_n) = \frac{r_n}{\alpha_n + r_n} |x_n|^{\frac{\alpha_n + r_n}{r_n}} + k_{n-1}^{\frac{\alpha_n}{r_n}} \lceil \sigma_{n-1}(\overline{x}_{n-1}) \rfloor^{\frac{\alpha_n}{\alpha_{n-1}}} x_n + \frac{\alpha_n}{\alpha_n + r_n} k_{n-1}^{\frac{\alpha_n + r_n}{r_n}} |\sigma_{n-1}(\overline{x}_{n-1})|^{\frac{\alpha_n + r_n}{\alpha_{n-1}}} .$$

Therefore,

$$\begin{aligned}\dot{V}_n(\overline{x}_n) &= W_n^{\frac{m}{\alpha_n + r_n} - 1}(\overline{x}_n) \dot{W}_n(\overline{x}_n) + \gamma_{n-1} \dot{V}_{n-1}(\overline{x}_{n-1}) \\ &= W_n^{\frac{m}{\alpha_n + r_n} - 1}(\overline{x}_n) \dot{W}_n(\overline{x}_n) + \gamma_{n-1} H_{n-1}(\bar{x}_{n-1}) + \gamma_{n-1} \partial_{x_{n-1}} V_{n-1} \cdot s_n(\bar{x}_n)\end{aligned}$$

(where we have used (2.33)) and

$$\dot{W}_n(\overline{x}_n) = \sigma_n(\overline{x}_n) \dot{x}_n + \frac{\alpha_n}{\alpha_{n-1}} k_{n-1}^{\frac{\alpha_n}{r_n}} |\sigma_{n-1}(\overline{x}_{n-1})|^{\frac{\alpha_n}{\alpha_{n-1}} - 1} s_n(\overline{x}_n) \dot{\sigma}_{n-1}(\overline{x}_{n-1}) .$$

Since $\sigma_{n-1}(\overline{x}_{n-1})$ is a $\mathscr{C}^1$ function we have that $\dot{\sigma}_{n-1}(\overline{x}_{n-1})$ is a continuous homogeneous function. And thus (in the absence of perturbations)

$$\begin{aligned}\dot{\mathscr{W}}(\overline{x}_{n+1}) = {} & V_n^{\frac{2-m}{m}}(\overline{x}_n) W_n^{\frac{m}{\alpha_n + r_n} - 1}(\overline{x}_n) \sigma_n(\overline{x}_n) \dot{x}_n + \\ & V_n^{\frac{2-m}{m}}(\overline{x}_n) W_n^{\frac{m}{\alpha_n + r_n} - 1}(\overline{x}_n) \frac{\alpha_n}{\alpha_{n-1}} k_{n-1}^{\frac{\alpha_n}{r_n}} |\sigma_{n-1}(\overline{x}_{n-1})|^{\frac{\alpha_n}{\alpha_{n-1}} - 1} s_n(\overline{x}_n) \dot{\sigma}_{n-1}(\overline{x}_{n-1}) + \\ & V_n^{\frac{2-m}{m}}(\overline{x}_n) \left(\gamma_{n-1} H_{n-1}(\bar{x}_{n-1}) + \gamma_{n-1} \partial_{x_{n-1}} V_{n-1} \cdot s_n(\bar{x}_n)\right) + \frac{1}{\gamma} x_{n+1} \dot{x}_{n+1} \\ = {} & V_n^{\frac{2-m}{m}}(\overline{x}_n) W_n^{\frac{m}{\alpha_n + r_n} - 1}(\overline{x}_n) \sigma_n(\overline{x}_n) b \vartheta_1(x) + \\ & V_n^{\frac{2-m}{m}}(\overline{x}_n) W_n^{\frac{m}{\alpha_n + r_n} - 1}(\overline{x}_n) \frac{\alpha_n}{\alpha_{n-1}} k_{n-1}^{\frac{\alpha_n}{r_n}} |\sigma_{n-1}(\overline{x}_{n-1})|^{\frac{\alpha_n}{\alpha_{n-1}} - 1} s_n(\overline{x}_n) \dot{\sigma}_{n-1}(\overline{x}_{n-1}) + \\ & V_n^{\frac{2-m}{m}}(\overline{x}_n) \left(\gamma_{n-1} H_{n-1}(\bar{x}_{n-1}) + \gamma_{n-1} \partial_{x_{n-1}} V_{n-1} \cdot s_n(\bar{x}_n)\right) + \\ & x_{n+1} \left(\frac{1}{\gamma} \vartheta_2(x) + V_n^{\frac{2-m}{m}}(\overline{x}_n) W_n^{\frac{m}{\alpha_n + r_n} - 1}(\overline{x}_n) \sigma_n(\overline{x}_n) b\right) .\end{aligned}$$

Selecting $\vartheta_1(x)$ such that $\sigma_n(\overline{x}_n) \vartheta_1(x) < 0$ and $\vartheta_2(x)$ so that the last term in $\dot{\mathscr{W}}$ vanishes, e.g., as in (2.17) (note that there is freedom in selecting $\vartheta_1(x)$ but not on the selection of $\vartheta_2(x)$), i.e.,

$$\begin{aligned}\vartheta_1(x) &= -k_n \lceil \sigma_n(x) \rfloor^{\frac{r_{n+1}}{\alpha_n}} \\ \vartheta_2(x) &= -\gamma b V_n^{\frac{2-m}{m}}(\overline{x}_n) W_n^{\frac{m}{\alpha_n + r_n} - 1}(\overline{x}_n) \sigma_n(\overline{x}_n) ,\end{aligned}$$

where we have set $k_I = \gamma b$, then

$$\dot{\mathscr{W}}\left(\overline{x}_{n+1}\right) = -k_n b V_n^{\frac{2-m}{m}}\left(\overline{x}_n\right) W_n^{\frac{m}{\alpha_n+r_n}-1}\left(\overline{x}_n\right)\left|\sigma_n\left(x\right)\right|^{\frac{r_{n+1}}{\alpha_n}+1} +$$
$$V_n^{\frac{2-m}{m}}\left(\overline{x}_n\right) W_n^{\frac{m}{\alpha_n+r_n}-1}\left(\overline{x}_n\right)\frac{\alpha_n}{\alpha_{n-1}} k_{n-1}^{\frac{\alpha_n}{r_n}}\left|\sigma_{n-1}\left(\overline{x}_{n-1}\right)\right|^{\frac{\alpha_n}{\alpha_{n-1}}-1} s_n\left(\overline{x}_n\right)\dot{\sigma}_{n-1}\left(\overline{x}_{n-1}\right) +$$
$$V_n^{\frac{2-m}{m}}\left(\overline{x}_n\right)\left(\gamma_{n-1} H_{n-1}\left(\bar{x}_{n-1}\right) + \gamma_{n-1}\partial_{x_{n-1}} V_{n-1}\cdot s_n\left(\bar{x}_n\right)\right).$$

Note that the r.h.s. of the previous equality for $\dot{\mathscr{W}}\left(\overline{x}_{n+1}\right)$ is a function of $\overline{x}_n$ only, since it does not depend on x_{n+1}. The first term is negative definite (as a function of x), and by using Lemma 2.1, we can select k_n large enough to render $\dot{\mathscr{W}}$ negative semidefinite, and we can conclude that

$$\dot{\mathscr{W}}\left(\overline{x}_{n+1}\right) \le -\varepsilon\left\|x\right\|_{\mathbf{r},p}^{2+l}. \tag{2.40}$$

2.6.2 A Passivity Interpretation of the I-Controller

Consider the system (without perturbation and with constant control coefficient) (see (2.6))

$$\begin{aligned} \dot{x}_i &= x_{i+1}, & i &= 1, \ldots, n-1, \\ \dot{x}_n &= b\left(\vartheta_1\left(x\right) + u_1\right), \end{aligned}$$

with $\vartheta_1\left(x\right)$ as in (2.17). Then, since

$$\begin{aligned} \frac{d}{dt}\left\{\frac{m}{2} V_n^{\frac{2}{m}}\left(\overline{x}_n\right)\right\} &= V_n^{\frac{2-m}{m}}\left(\overline{x}_n\right) W_n^{\frac{m}{\alpha_n+r_n}-1}\left(\overline{x}_n\right)\left[\sigma_n\left(\overline{x}_n\right) b\vartheta_1\left(x\right) + \right. \\ &\quad \left. \frac{\alpha_n}{\alpha_{n-1}} k_{n-1}^{\frac{\alpha_n}{r_n}}\left|\sigma_{n-1}\left(\overline{x}_{n-1}\right)\right|^{\frac{\alpha_n}{\alpha_{n-1}}-1} s_n\left(\overline{x}_n\right)\dot{\sigma}_{n-1}\left(\overline{x}_{n-1}\right)\right] + \\ &\quad V_n^{\frac{2-m}{m}}\left(\overline{x}_n\right)\left(\gamma_{n-1} H_{n-1}\left(\bar{x}_{n-1}\right) + \gamma_{n-1}\partial_{x_{n-1}} V_{n-1}\cdot s_n\left(\bar{x}_n\right)\right) + \\ &\quad V_n^{\frac{2-m}{m}}\left(\overline{x}_n\right) W_n^{\frac{m}{\alpha_n+r_n}-1}\left(\overline{x}_n\right)\sigma_n\left(\overline{x}_n\right) b u_1 \\ &\le V_n^{\frac{2-m}{m}}\left(\overline{x}_n\right) W_n^{\frac{m}{\alpha_n+r_n}-1}\left(\overline{x}_n\right)\sigma_n\left(\overline{x}_n\right) b u_1, \end{aligned}$$

this system is passive with respect to the output $y = V_n^{\frac{2-m}{m}}\left(\overline{x}_n\right) W_n^{\frac{m}{\alpha_n+r_n}-1}\left(\overline{x}_n\right)\sigma_n\left(\overline{x}_n\right) b$ and with (a non-smooth) storage function $\frac{m}{2} V_n^{\frac{2}{m}}\left(\overline{x}_n\right)$. Thus, controller (2.17) results from the passive feedback interconnection of the previous subsystem and the integrator, which is also passive with storage function $\frac{1}{2\gamma} x_{n+1}^2$.

2.6.3 A Strong Lyapunov Function and the Robustness Issue

Function $\mathscr{W}$ in the previous paragraph is a weak (non-smooth) homogeneous Lyapunov function for the closed-loop system. We consider now as strong Lyapunov function candidate the function

$$\mathscr{U}\left(\bar{x}_{n+1}\right)=\mu\frac{2}{2-l}\mathscr{W}^{\frac{2-l}{2}}\left(\bar{x}_{n+1}\right)-x_{n}x_{n+1}\,,$$

which is continuous, $\mathbf{r}$-homogeneous of degree $r_n + r_{n+1} = 2 - l$, and positive definite for $\mu > 0$ sufficiently large. $\mathscr{U}\left(\bar{x}_{n+1}\right)$ is differentiable at the same points where $\mathscr{W}\left(\bar{x}_{n+1}\right)$ is differentiable. Consider now the derivative of $\mathscr{U}\left(\bar{x}_{n+1}\right)$ along the trajectories of system (2.8) with perturbation, which is given (where it exists) by

$$\dot{\mathscr{U}}\left(\bar{x}_{n+1}\right)=\mu\mathscr{W}^{\frac{-l}{2}}\left(\bar{x}_{n+1}\right)\dot{\mathscr{W}}-\dot{x}_{n}x_{n+1}-x_{n}\dot{x}_{n+1}\,.$$

With controller (2.17), we obtain that

$$\dot{x}_{n}=b\left(-k_{n}\left\lceil\sigma_{n}\left(x\right)\right\rfloor^{\frac{r_{n+1}}{\alpha_{n}}}+x_{n+1}\right),\quad\dot{x}_{n+1}\in-k_{I}\Sigma_{I}\left(x\right)\,,$$

where $\Sigma_I\left(\overline{x}\right)=V_n^{\frac{2-m}{m}}\left(\overline{x}_n\right)W_n^{\frac{m}{\alpha_n+r_n}-1}\left(\overline{x}_n\right)\sigma_n\left(\overline{x}_n\right)-\frac{1}{k_I}\left[-C,\,C\right]$. And thus

$$\begin{aligned}
\dot{\mathscr{U}}\in&-k_{n}\mu b\mathscr{W}^{\frac{-l}{2}}\left(\bar{x}_{n+1}\right)V_{n}^{\frac{2-m}{m}}\left(\overline{x}_{n}\right)W_{n}^{\frac{m}{\alpha_{n}+r_{n}}-1}\left(\overline{x}_{n}\right)\left|\sigma_{n}\left(\overline{x}_{n}\right)\right|^{\frac{r_{n+1}+\alpha_{n}}{\alpha_{n}}}+\\
&\mu\mathscr{W}^{\frac{-l}{2}}V_{n}^{\frac{2-m}{m}}W_{n}^{\frac{m}{\alpha_{n}+r_{n}}-1}\left(\overline{x}_{n}\right)\frac{\alpha_{n}}{\alpha_{n-1}}k_{n-1}^{\frac{\alpha_{n}}{r_{n}}}\left|\sigma_{n-1}\left(\overline{x}_{n-1}\right)\right|^{\frac{\alpha_{n}}{\alpha_{n-1}}-1}s_{n}\left(\overline{x}_{n}\right)\dot{\sigma}_{n-1}\left(\overline{x}_{n-1}\right)+\\
&\mu\mathscr{W}^{\frac{-l}{2}}\left(\bar{x}_{n+1}\right)V_{n}^{\frac{2-m}{m}}\left(\overline{x}_{n}\right)\left(\gamma_{n-1}H_{n-1}\left(\bar{x}_{n-1}\right)+\gamma_{n-1}\partial_{x_{n-1}}V_{n-1}\cdot s_{n}\left(\bar{x}_{n}\right)\right)+\\
&-\left(\frac{k_{I}}{\gamma}-b\right)\mu\mathscr{W}^{\frac{-l}{2}}\left(\bar{x}_{n+1}\right)V_{n}^{\frac{2-m}{m}}\left(\overline{x}_{n}\right)W_{n}^{\frac{m}{\alpha_{n}+r_{n}}-1}\left(\overline{x}_{n}\right)\sigma_{n}\left(\overline{x}_{n}\right)x_{n+1}+\\
&bk_{n}\left\lceil\sigma_{n}\left(x\right)\right\rfloor^{\frac{r_{n+1}}{\alpha_{n}}}x_{n+1}-bx_{n+1}^{2}+k_{I}x_{n}\Sigma_{I}\left(x\right)+\frac{1}{\gamma}\left[-C,\,C\right]\mu\mathscr{W}^{\frac{-l}{2}}\left(\bar{x}_{n+1}\right)x_{n+1}\\
\leq&-\mu\varepsilon\left\|x\right\|_{\mathbf{r},p}^{2+l}+bk_{n}\left\lceil\sigma_{n}\left(x\right)\right\rfloor^{\frac{r_{n+1}}{\alpha_{n}}}x_{n+1}-bx_{n+1}^{2}+k_{I}x_{n}\Sigma_{I}\left(x\right)+\\
&\frac{1}{\gamma}\left[-C,\,C\right]\mu\mathscr{W}^{\frac{-l}{2}}\left(\bar{x}_{n+1}\right)x_{n+1}\,,
\end{aligned}$$

where for the last inequality we have used the fact that $\frac{k_I}{\gamma}=b$ and that we can select k_n such that (2.40) is satisfied (see above).

Consider first the case that $C = 0$. We obtain the inequality

$$\dot{\mathscr{U}}\leq-\mu\varepsilon\left\|\overline{x}_{n}\right\|_{\mathbf{r},p}^{2+l}+bk_{n}\left\lceil\sigma_{n}\left(x\right)\right\rfloor^{\frac{r_{n+1}}{\alpha_{n}}}x_{n+1}-bx_{n+1}^{2}+k_{I}x_{n}V_{n}^{\frac{2-m}{m}}W_{n}^{\frac{m}{\alpha_{n}+r_{n}}-1}\sigma_{n}\left(\overline{x}_{n}\right)$$

with the right-hand side a homogeneous function. Since the first term is negative and vanishes on the set $\mathscr{S} = \left\{\bar{x}_{n+1} \in \mathbb{R}^{n+1} | \bar{x}_n = 0\right\}$ and in this set the value of $\dot{\mathscr{U}}\big|_{\mathscr{S}}$ is given by

$$\dot{\mathscr{U}}\big|_{\mathscr{S}} \leq -bx_{n+1}^2 ,$$

which is negative, then by Lemma 2.6 we conclude that selecting $\mu > 0$ large we can render $\dot{\mathscr{U}} < 0$, so that $\mathscr{U}$ is a strong Lyapunov function and the origin is globally finite-time stable for $-1 \leq l < 0$ (exponentially stable for $l = 0$).

When $C \neq 0$ we can use the same type of reasoning as in Sect. 2.5 to show that: (i) When $-1 < l \leq 0$ the I-controller $\vartheta_2(x)$ is continuous and the closed-loop system is ISS with respect to the perturbation or (ii) when $l = -1$ the I-controller $\vartheta_2(x)$ is discontinuous at $x = 0$ and if $k_I > C$ the origin of the closed-loop system is finite-time stable despite of the perturbation.

2.7 A Lyapunov Function Approach for the Partial State Integral Controller

In this section, we give a unified proof of both Theorems 2.1 and 2.2 for the discontinuous and the continuous partial state I-controllers.

Introducing as new state variables $\xi_1 = x_1 - \kappa_1^{-1}\left\lceil k_n^{-1} x_{n+1}\right\rfloor^{\frac{r_1}{r_{n+1}}}$, $\xi_2 = x_2, \ldots, \xi_n = x_n$, $\xi_{n+1} = k_n^{-1} x_{n+1}$, the dynamics of the closed-loop DI (2.8) becomes

$$\Sigma : \begin{cases} \dot{\xi}_1 \in \xi_2 - \kappa_1^{-1} \frac{r_1}{r_{n+1}} \left|\xi_{n+1}\right|^{\frac{r_1 - r_{n+1}}{r_{n+1}}} \dot{\xi}_{n+1} \\ \dot{\xi}_i = \xi_{i+1}, \quad i = 2, \ldots, n-1, \\ \dot{\xi}_n \in -k_n \left[K_m, \ K_M\right] \phi\left(\bar{\xi}_{n+1}\right) \\ \dot{\xi}_{n+1} \in -\tilde{k}_I \left(\psi\left(\bar{x}_n\right) - k_I^{-1} C\left[-1, \ 1\right]\right) , \end{cases} \tag{2.41}$$

where $\tilde{k}_I = \frac{k_I}{k_n}$ and

$$\phi\left(\bar{\xi}_{n+1}\right) := \left\lceil \sigma_n\left(\bar{\xi}_n\right) + \left\lceil \xi_{n+1} \right\rfloor^{\frac{r_1}{r_{n+1}}} \right\rfloor^{\frac{r_{n+1}}{r_1}} - \xi_{n+1} \tag{2.42}$$

and the **r**-homogeneous functions $\sigma_i\left(\bar{\xi}_i\right)$ are defined recursively as $\sigma_1\left(\bar{\xi}_1\right) = \xi_1$,

$$\sigma_i\left(\bar{\xi}_i\right) = \left\lceil \xi_i \right\rfloor^{\frac{r_1}{r_i}} + k_{i-1}^{\frac{r_1}{r_i}} \sigma_{i-1}\left(\bar{\xi}_{i-1}\right) = \left\lceil \xi_i \right\rfloor^{\frac{r_1}{r_i}} + \sum_{j=1}^{i-1} \prod_{\iota=j}^{i-1} k_\iota^{\frac{r_1}{r_{\iota+1}}} \left\lceil \xi_j \right\rfloor^{\frac{r_1}{r_j}} , \tag{2.43}$$

for $i = 2, \ldots, n$, and the constants k_ι are such that $\kappa_j = \prod_{\iota=j}^{n-1} k_\iota^{\frac{r_1}{r_{\iota+1}}}$ for $i = 1, \ldots, n-1$. Note that these values of $\sigma_i\left(\bar{\xi}_i\right)$ correspond to the ones defined by

(2.9), when the parameters α_i are selected as $\alpha_n = \cdots = \alpha_1 = r_1 = 1 - nl$. An important observation here is that

$$\phi\left(\bar{\xi}_{n+1}\right) \sigma_n\left(\bar{\xi}_n\right) > 0\,, \text{ if } \sigma_n\left(\bar{\xi}_n\right) \neq 0\,. \tag{2.44}$$

This is a consequence of the strict monotonicity of the function $\lceil z \rfloor^{\beta}$ for any $\beta > 0$, i.e., $\left(\lceil z \rfloor^{\beta} - \lceil y \rfloor^{\beta}\right)(z - y) > 0$ for all $z \neq y$.

The change of variables introduced $\bar{\xi} = T(\bar{x})$ is a diffeomorphism, since $r_1 \geq r_{n+1} = 1$, so that according to Theorem 1 in [15, Chap. 2, Paragraph 9], every solution of (2.41) corresponds to a solution of (2.8).

2.7.1 The LF Candidate for the I-Controller

To show that the equilibrium point $\bar{\xi}_{n+1} = 0$ of system (2.41) is asymptotically stable (with $C = 0$ in the continuous case), when the gains are appropriately selected, we will show that the continuously differentiable ($\mathscr{C}^1$), $(\mathbf{r}, m)$-homogeneous function,

$$V\left(\bar{\xi}_{n+1}\right) = \gamma_n V_n\left(\bar{\xi}_n\right) + \frac{r_{n+1}}{m}\left|\xi_{n+1}\right|^{\frac{m}{r_{n+1}}}\,,\ \gamma_n > 0 \tag{2.45}$$

is a (strong) LF for system (2.41), for every $n \geq 1$, $l \in [-1\,, 0]$, $m \geq r_1 + r_2 = 2 - (2n - 1)\,l$, and $\gamma_j > 0$. $V_n(\xi)$ is obtained from (2.10), fixing the parameters $\alpha_n = \cdots = \alpha_1 = r_1 = 1 - nl$. A similar analysis as the one performed in Sect. 2.5 shows that $V\left(\bar{\xi}_{n+1}\right)$ is $\mathscr{C}^1$, positive definite, decrescent, and radially unbounded.

2.7.2 Derivative of the LF Candidate

The derivative of the LF candidate (2.45) along the trajectories of (2.41) is given by

$$\dot{V} = -k_n F_0\left(\bar{\xi}_{n+1}\right) + F_1\left(\bar{\xi}_n\right) - \tilde{k}_I F_2\left(\bar{\xi}_{n+1}\right) + F_U\left(\bar{\xi}_{n+1}\right)\,, \tag{2.46}$$

where

$$\begin{aligned}
F_0\left(\bar{\xi}_{n+1}\right) &:= \gamma_n\left[K_m,\ K_M\right]\phi\left(\bar{\xi}_{n+1}\right)\partial_{\xi_n} V_n\left(\bar{\xi}_n\right)\,,\\
F_1\left(\bar{\xi}_n\right) &:= \gamma_n \sum_{j=1}^{n-1} \xi_{j+1}\partial_{\xi_j} V_n\left(\bar{\xi}_n\right)\,,\\
F_2\left(\bar{\xi}_{n+1}\right) &:= \left(\psi\left(\bar{x}_n\right) - \Gamma\left[-1,\ 1\right]\right) H\left(\bar{\xi}_{n+1}\right)\,,\\
H\left(\bar{\xi}_{n+1}\right) &:= \left|\xi_{n+1}\right|^{\frac{r_1 - r_{n+1}}{r_{n+1}}}\left(\lceil \xi_{n+1} \rfloor^{\frac{m - r_1}{r_{n+1}}} - \gamma_n \kappa_1^{-1}\frac{r_1}{r_{n+1}}\partial_{\xi_1} V_n\left(\bar{\xi}_n\right)\right)\,,
\end{aligned}$$

and

$$F_U\left(\bar{\xi}_{n+1}\right) = \begin{cases} 0 & \text{if } l = -1 \\ \tilde{\Delta}\left[-1,\ 1\right] H\left(\bar{\xi}_{n+1}\right) & \text{if } -1 < l \le 0 \end{cases},$$

where $\tilde{\Delta} = \frac{C}{k_n}$ and $\Gamma = \frac{C}{k_I}$ if $l = -1$ or $\Gamma = 0$ if $l > -1$.

From (2.28), we have $\partial_{\xi_n} V_n\left(\bar{\xi}_n\right) = W_n^{\frac{m-r_1-r_n}{r_1+r_n}}\left(\bar{\xi}_n\right)\sigma_n\left(\bar{\xi}_n\right)$, and therefore

$$\phi\left(\bar{\xi}_{n+1}\right)\partial_{\xi_n} V_n\left(\bar{\xi}_n\right) = W_n^{\frac{m-r_1-r_n}{r_1+r_n}}\left(\bar{\xi}_n\right)\phi\left(\bar{\xi}_{n+1}\right)\sigma_n\left(\bar{\xi}_n\right) \ge 0\,,$$

where we have used (2.44) to obtain the last inequality. We conclude that $F_0\left(\bar{\xi}_{n+1}\right) \ge 0$ and, moreover,

$$F_0\left(\bar{\xi}_{n+1}\right) \ge F_0^*\left(\bar{\xi}_{n+1}\right) := \gamma_n K_m W_n^{\frac{m-r_1-r_n}{r_1+r_n}}\left(\bar{\xi}_n\right)\phi\left(\bar{\xi}_{n+1}\right)\sigma_n\left(\bar{\xi}_n\right) \ge 0\,,$$

where $F_0^*\left(\bar{\xi}_{n+1}\right)$ is homogeneous of degree $m + l$ and it vanishes only on the set $\mathscr{S}_n = \left\{\bar{\xi}_n \in \mathbb{R}^n | \sigma_n\left(\bar{\xi}_n\right) = 0\right\}$.

As a consequence,

$$\dot{V} \le -k_n F_0^*\left(\bar{\xi}_{n+1}\right) + Q\left(\bar{\xi}_{n+1}\right) + F_U\left(\bar{\xi}_{n+1}\right)\,,$$

where $Q\left(\bar{\xi}_{n+1}\right) := F_1\left(\bar{\xi}_n\right) - \tilde{k}_I F_2\left(\bar{\xi}_{n+1}\right)$ is a (multivalued) homogeneous function of degree $m + l$. Note that F_1 is independent of the gains k_n and $\tilde{k}_I$. In the discontinuous case ($l = -1$) $F_U\left(\bar{\xi}_{n+1}\right) = 0$, and so it suffices to show that $-k_n F_0^* + Q < 0$. In the continuous case, ($0 \ge l > -1$) $F_U\left(\bar{\xi}_{n+1}\right)$ is of homogeneous degree $m - 1 < m + l$, and therefore when $-k_n F_0^* + Q < 0$ the origin is not an equilibrium point and only ultimately boundedness of the trajectories can be achieved.

Note that $\psi\left(\bar{x}_n\right) = \psi\left(\xi_1 + \kappa_1^{-1}\left\lceil\xi_{n+1}\right\rfloor^{\frac{r_1}{r_{n+1}}},\ \xi_2, \ldots,\ \xi_n\right)$, and that (using equality (2.33))

$$F_1\left(\bar{\xi}_n\right) = \gamma_n \sum_{j=1}^{n-1} \partial_{\xi_j} V_{n-1} \cdot \xi_{j+1} = \gamma_n H_{n-1}\left(\bar{\xi}_{n-1}\right) + \gamma_n \partial_{\xi_{n-1}} V_{n-1} \cdot s_n\left(\bar{\xi}_n\right).$$

And therefore, evaluating $F_1\left(\bar{\xi}_n\right)$ on the set $\mathscr{S}_n$, we get

$$F_1|_{\mathscr{S}_n} = \gamma_n H_{n-1}\left(\bar{\xi}_{n-1}\right)\,,$$

which, by selecting appropriately the gains k_i, can be rendered negative definite, i.e., $H_{n-1}\left(\bar{\xi}_{n-1}\right) < 0$ (see Lemma 2.3).

According to Lemma 2.6, since $F_0^* \ge 0$ and it vanishes only on the set $\mathscr{S}_n$, it suffices to show that Q evaluated at the set $\mathscr{S}_n$ (except at the origin) is negative, i.e., $Q\left(\bar{\xi}_{n+1}\right)\big|_{\mathscr{S}_n} < 0$, to conclude that $-k_n F_0^* + Q$ can be rendered negative definite selecting k_n sufficiently large. Note that because $F_1|_{\mathscr{S}_n} = \gamma_n H_{n-1}\left(\bar{\xi}_{n-1}\right)$ it follows that $Q\left(\bar{\xi}_{n+1}\right)\big|_{\mathscr{S}_n} = \gamma_n H_{n-1}\left(\bar{\xi}_{n-1}\right) - \tilde{k}_I\ F_2\left(\bar{\xi}_{n+1}\right)\big|_{\mathscr{S}_n}$. Now we consider the two possible cases, when $\psi\left(\bar{x}_n\right)$ is continuous and when it is discontinuous (multivalued).

2.7.2.1 $\psi\left(\bar{x}_n\right)$ Discontinuous ($l = -1$)

In this case,

$$Q|_{\mathscr{S}_n} = \gamma_n H_{n-1}\left(\bar{\xi}_{n-1}\right) - \tilde{k}_I\left(\psi\left(\bar{x}_n\right) - \Gamma[-1,\ 1]\right) H\left(\bar{\xi}_{n+1}\right)\big|_{\mathscr{S}_n}\ .$$

The first term $\gamma_n H_{n-1}\left(\bar{\xi}_{n-1}\right)$ is nonpositive and vanishes only on the set $\mathscr{S}_{n+1} = \left\{\bar{\xi}_{n-1} = 0\right\}$. Evaluating $Q\left(\bar{\xi}_{n+1}\right)\big|_{\mathscr{S}_n}$ on the set $\mathscr{S}_{n+1}$, we obtain

$$Q|_{\mathscr{S}_n \cap \mathscr{S}_{n+1}} = -\tilde{k}_I \varpi\left(\xi_{n+1}\right) \lceil \xi_{n+1} \rfloor^{\frac{m - r_{n+1}}{r_{n+1}}}\ ,$$

where $\varpi\left(\xi_{n+1}\right) := \psi\left(\kappa_1^{-1}\lceil \xi_{n+1} \rfloor^{\frac{r_1}{r_{n+1}}},\ 0, \ldots,\ 0\right) - \Gamma[-1,\ 1]$. From (2.16), we obtain that $\psi\left(x_1,\ 0, \ldots,\ 0\right) = \frac{\lceil x_1 \rfloor^{\frac{a + r_{n+2}}{r_1}}}{|x_1|^{\frac{a}{r_1}}} = \lceil x_1 \rfloor^0$. If $\Gamma < 1$ then the multivalued function satisfies

$$\varpi\left(\xi_{n+1}\right) = \begin{cases} 1 - [-1,\ 1]\Gamma > 0 & \text{if } \xi_{n+1} > 0 \\ [-1,\ 1] - [-1,\ 1]\Gamma & \text{if } \xi_{n+1} = 0, \\ -1 - [-1,\ 1]\Gamma < 0 & \text{if } \xi_{n+1} < 0, \end{cases}$$

and therefore $Q\left(\bar{\xi}_{n+1}\right)\big|_{\mathscr{S}_n \cap \mathscr{S}_{n+1}} < 0$ if $\xi_{n+1} \neq 0$. And thus, Lemma 2.6 implies that we can render $Q\left(\bar{\xi}_{n+1}\right)\big|_{\mathscr{S}_n} < 0$ selecting $\tilde{k}_I > 0$ sufficiently small. Consequently, selecting $k_n > 0$ sufficiently large we render $\dot{V}\left(\bar{\xi}_{n+1}\right) < 0$, and the origin is globally finite-time stable using Lyapunov's theorem for differential inclusions [3].

Remark 2.2 We note that to render $\dot{V}\left(\bar{\xi}_{n+1}\right) < 0$ we have to design appropriately three constants $\left(\Gamma,\ \tilde{k}_I,\ k_n\right)$: we select first $0 < \Gamma < 1$, and then we find $\tilde{k}_I$ (sufficiently small), and finally we determine a large enough value of k_n. It is important to note that these three constants can be assigned arbitrarily, and so the presented procedure is valid. Since the values of $\left(\Gamma,\ \tilde{k}_I,\ k_n\right)$ are in one-to-one relation with the values of $(C,\ k_I,\ k_n)$ through the relations $\Gamma = \frac{C}{k_I}$ and $\tilde{k}_I = \frac{k_I}{k_n}$, it follows that once $\left(\Gamma,\ \tilde{k}_I,\ k_n\right)$ are selected, we obtain the values of $(C,\ k_I,\ k_n) = \left(k_I\Gamma,\ k_n\tilde{k}_I,\ k_n\right)$. This procedure does not allow to select the value C arbitrarily, but it provides one for which stability is assured. This issue is discussed in more detail in Sect. 2.7.3.

2.7.2.2 $\psi(\bar{x}_n)$ Continuous ($-1 < l \leq 0$)

In this case,

$$Q|_{\mathscr{S}_n} = \gamma_n H_{n-1}\left(\bar{\xi}_{n-1}\right) - \tilde{k}_I\ \psi\left(\bar{x}_n\right) H\left(\bar{\xi}_{n+1}\right)\Big|_{\mathscr{S}_n}.$$

The first term $\gamma_n H_{n-1}\left(\bar{\xi}_{n-1}\right)$ is nonpositive and vanishes only on the set $\mathscr{S}_{n+1} = \left\{\bar{\xi}_{n-1} = 0\right\}$. Evaluating $Q\left(\bar{\xi}_{n+1}\right)\Big|_{\mathscr{S}_n}$ on the set $\mathscr{S}_{n+1}$, we obtain

$$Q|_{\mathscr{S}_n \cap \mathscr{S}_{n+1}} = -\tilde{k}_I \psi\left(\kappa_1^{-1}\left\lceil \xi_{n+1}\right\rfloor^{\frac{r_1}{r_{n+1}}},\ 0, \ldots, 0\right)\left\lceil \xi_{n+1}\right\rfloor^{\frac{m-r_{n+1}}{r_{n+1}}},$$

which is negative due to the hypothesis that $x_1\psi\left(x_1, 0, \ldots, 0\right) > 0$ for $x_1 \neq 0$. And thus, Lemma 2.6 implies that we can render $Q\left(\bar{\xi}_{n+1}\right)\Big|_{\mathscr{S}_n} < 0$ selecting $\tilde{k}_I > 0$ sufficiently small. Consequently, selecting $k_n > 0$ sufficiently large, we render $\dot{V}\left(\bar{\xi}_{n+1}\right) < 0$, and the origin is globally finite-time stable using Lyapunov's theorem, when $F_U = 0$, i.e., if $\Delta = 0$.

In the perturbed case, when $F_U \neq 0$, since the homogeneity degree of F_U is smaller than that of Q it follows (using standard Lyapunov arguments as in, e.g., [27]) that $\dot{V} \leq -k_n F_0^* + Q + F_U$ for every $\left\|\bar{\xi}_{n+1}\right\|_{\mathbf{r},p} \geq \mu\left(\tilde{\Delta}\right)$ for some $\mu\left(\tilde{\Delta}\right)$. And therefore, the trajectories of system are globally ultimately bounded (see [27, Sect. 4.8]), with an ultimate bound $\omega\left(\tilde{\Delta}\right)$ which is a monotone growing function of $\tilde{\Delta}$.

The scaling given in (2.21) can be obtained by performing a linear state transformation $z = Lx$ on the system (2.8) and noting that the resulting system has the same form with the scaled parameters. This implies that the stability of both systems is equivalent.

2.7.3 *Gain Calculation*

After fixing the parameters m, l, γ_i, and $\alpha_n = \cdots = \alpha_1 = r_1 = 1 - nl$, the values of the required k_i can be obtained recursively using the same procedure as in Sect. 2.5.3.2, finding values that fulfill inequalities (2.38). The gains κ_j of χ (2.15) are obtained from $\kappa_j = \prod_{\iota=j}^{n-1} k_\iota^{\frac{r_1}{r_{\iota+1}}}$ for $j = 1, \ldots, n-1$.

In the discontinuous (continuous) case, one first selects $\Gamma := \frac{C}{k_I} < 1$ ($\Gamma = 0$, respectively), then calculates $\tilde{k}_I > 0$, such that $Q\left(\bar{\xi}_{n+1}\right)\Big|_{\mathscr{S}_n} < 0$, i.e.,

$$\frac{1}{\tilde{k}_I} > \max_{\mathscr{S}_n}\left\{\frac{\left(\psi\left(\bar{x}_n\right) - \Gamma\left[-1,\ 1\right]\right) H\left(\bar{\xi}_{n+1}\right)\Big|_{\mathscr{S}_n}}{\gamma_n H_{n-1}\left(\bar{\xi}_{n-1}\right)}\right\},$$

and finally selects k_n large enough so that $\dot{V} < 0$, i.e.,

$$k_n > \max_{\bar{\xi}_{n+1} \in S_{n+1}} \left\{ \frac{Q\left(\bar{\xi}_{n+1}\right)}{F_0^*\left(\bar{\xi}_{n+1}\right)} \right\} .$$

Finally, in the discontinuous case, we calculate the values of $k_I = k_n \tilde{k}_I$, and of the size of the admissible perturbation $C^* = k_I \Gamma = k_n \tilde{k}_I \Gamma$. Note that this implies that the value of C cannot be freely given, but it is obtained from the calculation process. However, performing a scaling as in (2.21) will allow to stabilize for an arbitrary value of C by selecting L such that $LC^* \geq C$, where C is the desired size of the perturbation.

All maximizations are feasible by the same reasons described in Sect. 2.5.3.2.

2.8 Conclusions

We present in this chapter a discontinuous integral controller, which shares the properties of the classical PID control and the HOSM controllers: Similar to HOSM, it is able to fully compensate a Lipschitz perturbation or to track an (unknown) time-varying reference with bounded nth derivative, it has high precision due to the homogeneity properties, and it stabilizes the origin globally and in finite time. Similar to the PID control, it provides a continuous control signal. We present different types of integral controllers, all having the same basic features, but that are dissimilar in (a) the dependence on the full state or a part of the state of the function to be integrated, (b) the Lyapunov function required to show the convergence, and (c) the stability of the closed loop by increasing the gain of the integral term.

Some distinguishing features of our work are (i) we provide a whole family of smooth and non-smooth homogeneous Lyapunov functions of arbitrarily high homogeneity degree. (ii) The obtained functions ϑ_1 and ϑ_2 can be of "polynomial" or "rational" type (QC for ϑ_2). (iii) We propose a large family of possible controllers ϑ_1 and ϑ_2, and a procedure to calculate the gains associated with them using the proposed Lyapunov functions. (iv) Smooth Lyapunov functions allow assuring robustness properties of the algorithms such as input-to-state stability with respect to some external perturbations. Moreover, estimation of the convergence time or the input/output gains. However, in this work, we have not used the LFs for this explicit purpose, and it remains an open problem to perform these calculations. (v) We propose to use continuous homogeneous I-controllers of different homogeneity degrees to approximate the discontinuous algorithms in order to further increase the smoothness of the control signal $u(t)$. This approximation can be arbitrarily close to the discontinuous algorithm and we use a unified Lyapunov approach to study both the discontinuous and the continuous versions of the algorithms.

In order to achieve an output feedback scheme, an exact and finite-time differentiator can be used, which is a discontinuous algorithm. In the paper [46], for relative degree $n = 2$, a smooth observer is used instead, featuring some further interesting properties, as not requiring the boundedness of the perturbation θ, for achieving closed-loop stability. This remains an open problem for arbitrary relative degree.

Acknowledgements The authors would like to thank the financial support from PAPIIT-UNAM (Programa de Apoyo a Proyectos de Investigación e Innovación Tecnológica), project IN110719; Fondo de Colaboración II-FI UNAM, Project IISGBAS-100-2015; CONACyT (Consejo Nacional de Ciencia y Tecnología), project 241171; and SEP-PRODEP Apoyo a la Incorporación de NPTC project 511-6/18-9169UDG-PTC-1400.

Appendix: Some Technical Lemmas on Homogeneous Functions

We recall some useful lemmas needed for the development of our main results. Lemmas 2.6 and 2.7 are extensions of classical results for homogeneous continuous functions to semicontinuous ones [21, Theorems 4.4 and 4.1] (for their proofs see [12, 46]).

Lemma 2.5 ([19] Young's inequality) *For any positive real numbers $a > 0$, $b > 0$, $c > 0$, $p > 1$, and $q > 1$, with $\frac{1}{p} + \frac{1}{q} = 1$, the following inequality is always satisfied:*

$$ab \leq \frac{c^p}{p} a^p + \frac{c^{-q}}{q} b^q,$$

and equality holds if and only if $a^p = b^q$.

And the lemma is given as follows.

Lemma 2.6 *Let $\eta : \mathbb{R}^n \to \mathbb{R}$ and $\gamma : \mathbb{R}^n \to \mathbb{R}_+$ (resp. $\gamma : \mathbb{R}^n \to \mathbb{R}_-$) be two lower (upper) semicontinuous single-valued $\mathbf{r}$-homogeneous functions of degree $m > 0$. Suppose that $\gamma(x) \geq 0$ (resp. $\gamma(x) \leq 0$) on $\mathbb{R}^n$. If $\eta(x) > 0$ (resp. $\eta(x) < 0$) for all $x \neq 0$ such that $\gamma(x) = 0$, then there is a constant $\lambda^* \in \mathbb{R}$ and a constant $c > 0$ such that for all $\lambda \geq \lambda^*$ and for all $x \in \mathbb{R}^n \setminus \{0\}$,*

$$\eta(x) + \lambda\gamma(x) \geq c \left\| x \right\|_{\mathbf{r},p}^m , \quad \left(\text{resp. } \eta(x) + \lambda\gamma(x) \leq -c \left\| x \right\|_{\mathbf{r},p}^m\right) .$$

Lemma 2.7 *Let $\eta : \mathbb{R}^n \to \mathbb{R}$ be an upper semicontinuous, single-valued $\mathbf{r}$-homogeneous function, with weights $\mathbf{r} = [r_1, \ldots, r_n]^\top$ and degree $m > 0$. Then there is a point x_2 in the unit homogeneous sphere $S = \{x \in \mathbb{R}^n : \|x\|_{\mathbf{r},p} = 1\}$ such that the following inequality holds for all $x \in \mathbb{R}^n$:*

$$\eta(x) \leq \eta(x_2) \left\| x \right\|_{\mathbf{r},p}^m . \tag{2.47}$$

Under the same conditions, if η is lower semicontinuous, there is a point x_1 in the unit homogeneous sphere S such that the following inequality holds for all $x \in \mathbb{R}^n$:

$$\eta(x_1)\,\|x\|_{\mathbf{r},p}^m \leq \eta(x)\,. \tag{2.48}$$

References

1. Andrieu, V., Praly, L., Astolfi, A.: Homogeneous approximation, recursive observer design and output feedback. SIAM J. Control Optim. **47**(4), 1814–1850 (2008)
2. Astolfi, D., Praly, L.: Integral action in output feedback for multi-input multi-output nonlinear systems. IEEE Trans. Autom. Control **62**(4), 1559–1574 (2017). April
3. Bacciotti, A., Rosier, L.: Liapunov Functions and Stability in Control Theory, 2nd edn. Springer, New York (2005)
4. Bernuau, E., Efimov, D., Perruquetti, W., Polyakov, A.: On an extension of homogeneity notion for differential inclusions. In: European Control Conference, Zurich, Switzerland (2013)
5. Bernuau, E., Efimov, D., Perruquetti, W., Polyakov, A.: On homogeneity and its application in sliding mode control. J. Frankl. Inst. **351**(4), 1816–1901 (2014)
6. Bernuau, E., Efimov, D., Perruquetti, W., Polyakov, A.: Homogeneity of differential inclusions. In: Control, Robotics & Sensors, pp. 103–118. Institution of Engineering and Technology (2016)
7. Bernuau, E., Polyakov, A., Efimov, D., Perruquetti, W.: Verification of ISS, iISS and IOSS properties applying weighted homogeneity. Syst. Control Lett. **62**(12), 1159–1167 (2013)
8. Bhat, S.P., Bernstein, D.S.: Geometric homogeneity with applications to finite-time stability. Math. Control, Signals, Syst. **17**(2), 101–127 (2005)
9. Cruz-Zavala, E., Moreno, J.A.: Higher-order sliding mode control using discontinuous integral action. IEEE Trans. Autom. Control. (2019). https://doi.org/10.1109/TAC.2019.2956127
10. Cruz-Zavala, E., Moreno, J.A.: Levant's arbitrary order exact differentiator: a Lyapunov approach. IEEE Trans. Autom. Control **64**(7), 3034–3039 (2019). July
11. Cruz-Zavala, E., Moreno, J.A.: Lyapunov functions for continuous and discontinuous differentiators. In: 10th IFAC Symposium on Nonlinear Control Systems NOLCOS 2016; IFAC-PapersOnLine **49**(18), 660–665 (2016)
12. Cruz-Zavala, E., Moreno, J.A.: Homogeneous high order sliding mode design: a Lyapunov approach. Automatica **80**, 232–238 (2017)
13. Deimling, K.: Multivalued Differential Equations. Walter de gruyter, Berlin (1992)
14. Dorel, L., Levant, A.: On chattering-free sliding-mode control. In: 2008 47th IEEE Conference on Decision and Control, pp. 2196–2201 (2008)
15. Filippov, A.F.: Differential Equations with Discontinuous Righthand Side. Kluwer, Dordrecht (1988)
16. Fridman, L., Levant, A.: Sliding Mode Control in Engineering, chapter 3. Marcel Dekker Ink, New York (2002)
17. Fridman, L., Moreno, J.A., Bandyopadhyay, B., Kamal, S., Chalanga, A.: Continuous Nested Algorithms: The Fifth Generation of Sliding Mode Controllers, vol. 24, pp. 5–35. Springer International Publishing, Cham (2015)
18. Hahn, W.: Stability of Motion. Springer, Berlin (1967)
19. Hardy, G.H., Littlewood, J.E., Polya, G.: Inequalities. Cambridge University Press, London (1951)
20. Hermes, H.: Homogeneous coordinates and continuous asymptotically stabilizing feedback controls. In: Elaydi, S. (ed.) Differential Equations, Stability and Control. Lecture Notes in Pure and Applied Mathematics, vol. 127, pp. 249–260. Marcel Dekker Inc., New York (1991)
21. Hestenes, M.R.: Calculus of Variations and Optimal Control Theory. Wiley, New York (1966)

22. Isidori, A.: Nonlinear Control Systems, 3rd edn. Springer, Berlin (1995)
23. Isidori, A.: Nonlinear Control Systems II. Springer, London (1999)
24. Kamal, S., Chalanga, A., Moreno, J.A., Fridman, L., Bandyopadhyay, B.: Higher order super-twisting algorithm. In: 2014 13th International Workshop on Variable Structure Systems (VSS), pp. 1–5 (2014)
25. Kamal, S., Moreno, J.A., Chalanga, A., Bandyopadhyay, B., Fridman, L.M.: Continuous terminal sliding-mode controller. Automatica **69**, 308–314 (2016)
26. Khalil, H.K.: Universal integral controllers for minimum-phase nonlinear systems. IEEE Trans. Autom. Control **45**(3), 490–494 (2000). March
27. Khalil, H.K.: Nonlinear Systems, 3rd edn. Prentice-Hall, New Jersey (2002)
28. Laghrouche, S., Harmouche, M., Chitour, Y.: Higher order super-twisting for perturbed chains of integrators. IEEE Trans. Autom. Control **62**(7), 3588–3593 (2017)
29. Levant, A.: Sliding order and sliding accuracy in sliding mode control. Int. J. Control **58**(6), 1247–1263 (1993)
30. Levant, A.: Universal single-input single-output (siso) sliding-mode controllers with finite-time. IEEE Trans. Autom. Control **46**(9), 1447–1451 (2001)
31. Levant, A.: High-order sliding modes: differentiation and output-feedback control. Int. J. Control **76**(9), 924–941 (2003)
32. Levant, A.: Homogeneity approach to high-order sliding mode design. Automatica **41**, 823–830 (2005)
33. Levant, A.: Quasi-continuous high-order sliding-mode controllers. IEEE Trans. Autom. Control **50**(11), 1812–1816 (2005)
34. Levant, A.: Exact differentiation of signals with unbounded higher derivatives. In: Proceedings of the 45th IEEE Conference on Decision and Control, pp. 5585–5590 (2006)
35. Levant, A.: Principles of 2-sliding mode design. Automatica **43**(4), 576–586 (2007). April
36. Levant, A.: Globally convergent fast exact differentiator with variable gains. In: 2014 European Control Conference (ECC), pp. 2925–2930 (2014)
37. Levant, A., Alelishvili, L.: Integral high-order sliding modes. IEEE Trans. Autom. Control **52**(7), 1278–1282 (2007)
38. Levant, A., Livne, M.: Exact differentiation of signals with unbounded higher derivatives. IEEE Trans. Autom. Control **57**(4), 1076–1080 (2012)
39. Levant, A., Livne, M.: Weighted homogeneity and robustness of sliding mode control. Automatica **72**, 186–193 (2016)
40. Levant, A., Livne, M.: Globally convergent differentiators with variable gains. Int. J. Control **91**(9), 1994–2008 (2018)
41. Mendoza-Avila, J., Moreno, J.A., Fridman, L.: An idea for Lyapunov function design for arbitrary order continuous twisting algorithms. In: 2017 IEEE 56th Annual Conference on Decision and Control (CDC), vol. 2018, pp. 5426–5431 (2017)
42. Mercado-Uribe, A., Moreno, J.A.: Discontinuous integral action for arbitrary relative degree in sliding-mode control. Automatica. Submitted (2019)
43. Mercado-Uribe, A., Moreno, J.A.: Full and partial state discontinuous integral control. In: 2nd IFAC Conference on Modelling, Identification and Control of Nonlinear Systems MICNON 2018; IFAC-PapersOnLine, **51**(13), 573–578 (2018)
44. Mercado-Uribe, A., Moreno, J.A.: Output feedback discontinuous integral controller for siso nonlinear systems. In: 2018 15th International Workshop on Variable Structure Systems (VSS), vol. 2018, pp. 114–119 (2018)
45. Moreno, J.A.: Discontinuous integral control for mechanical systems. In: 2016 14th International Workshop on Variable Structure Systems (VSS), pp. 142–147 (2016)
46. Moreno, J.A.: Chapter 8: Discontinuous integral control for systems with relative degree two. New Perspectives and Applications of Modern Control Theory; in Honor of Alexander S. Poznyak, pp. 187–218. Springer International Publishing (2018)
47. Moreno, J.A.: Exact differentiator with varying gains. Int. J. Control **91**(9), 1983–1993 (2018)
48. Moreno, J.A., Osorio, M.: A Lyapunov approach to second-order sliding mode controllers and observers. In: 47th IEEE Conference on Decision and Control, pp. 2856–2861. Cancún, Mexico (2008)

49. Moreno, J.A., Osorio, M.: Strict Lyapunov functions for the super-twisting algorithm. IEEE Trans. Autom. Control **57**(4), 1035–1040 (2012)
50. Nakamura, H., Yamashita, Y., Nishitani, H.: Smooth Lyapunov functions for homogeneous differential inclusions. In: Proceedings of the 41st SICE Annual Conference, vol. 3, pp. 1974–1979 (2002)
51. Nakamura, N., Nakamura, H., Yamashita, Y., Nishitani, H.: Homogeneous stabilization for input affine homogeneous systems. IEEE Trans. Autom. Control **54**(9), 2271–2275 (2009)
52. Orlov, Y.: Finite time stability of homogeneous switched systems. In: Proceedings of 42nd IEEE Conference on Decision and Control, vol. 4, pp. 4271–4276 (2003)
53. Orlov, Y.V.: Discontinuous Systems: Lyapunov Analysis and Robust Synthesis Under Uncertainty Conditions. Springer, Berlin (2009)
54. Rosier, L.: Homogeneous Lyapunov function for homogeneous continuous vector field. Syst. Control Lett. **19**, 467–473 (1992)
55. Sanchez, T., Cruz-Zavala, E., Moreno, J.A.: An SOS method for the design of continuous and discontinuous differentiators. Int. J. Control **91**(11), 2597–2614 (2018)
56. Sanchez, T., Moreno, J.A.: Design of Lyapunov functions for a class of homogeneous systems: generalized forms approach. Int. J. Robust Nonlinear Control **29**(3), 661–681 (2019)
57. Shtessel, Y., Edwards, C., Fridman, L., Levant, A.: Sliding Mode Control and Observation. Birkhauser, Springer, New York (2014)
58. Torres-González, V., Fridman, L.M., Moreno, J.A.: Continuous twisting algorithm. In: 2015 54th IEEE Conference on Decision and Control (CDC), pp. 5397–5401 (2015)
59. Torres-González, V., Sanchez, T., Fridman, L.M., Moreno, J.A.: Design of continuous twisting algorithm. Automatica **80**, 119–126 (2017)
60. Zamora, C., Moreno, J.A., Kamal, S.: Control integral discontinuo para sistemas mecánicos. In: 2013 Congreso Nacional de Control Automático (CNCA AMCA), pp. 11–16, Ensenada, Baja California, Mexico (2013). Asociación de México de Control Automático (AMCA)
61. Zubov, V.I.: Methods of A. M. Lyapunov and Their Applications. P. Noordho Limited, Groningen (1964)

Part II
Properties of Continuous Sliding-Mode Algorithms

Chapter 3
Computing and Estimating the Reaching Time of the Super-Twisting Algorithm

Richard Seeber

Abstract The super-twisting algorithm is a second-order sliding-mode algorithm that may be used either for control or for observation purposes. An important performance characteristic of this algorithm is the so-called reaching or convergence time, the time it takes for the controller to reach the sliding surface or for the estimates to converge. In this chapter, three techniques are discussed to estimate, i.e., upper bound, and in some cases even compute this reaching time in the presence of additive perturbations, which are Hölder continuous in the state or Lipschitz continuous in the time. The first is obtained from an analytic computation of the unperturbed reaching time; the second is based on a family of quadratic Lyapunov functions; and the third is derived from a necessary and sufficient stability criterion. For each approach, the range of permissible perturbations, its asymptotic properties with respect to parameters and perturbation bounds, and, when applicable, the selection of parameters are discussed. Numerical comparisons illustrate the results obtained with each approach.

3.1 Introduction

The super-twisting algorithm [7] is a well-known second-order sliding-mode algorithm. It is given by the nonlinear system

$$\frac{\mathrm{d}x_1}{\mathrm{d}t} = -k_1\sqrt{|x_1|}\,\mathrm{sign}(x_1) + x_2 + \delta_1\sqrt{|x_1|}, \tag{3.1a}$$

$$\frac{\mathrm{d}x_2}{\mathrm{d}t} = -k_2\,\mathrm{sign}(x_1) + \delta_2 \tag{3.1b}$$

with state variables x_1 and x_2, and Lebesgue-integrable perturbation signals $\delta_1(t)$ and $\delta_2(t)$, which satisfy the bounds

R. Seeber (✉)
Christian Doppler Laboratory for Model Based Control of Complex Test Bed Systems, Institute of Automation and Control, Graz University of Technology, 8010 Graz, Austria
e-mail: richard.seeber@tugraz.at

M. Steinberger et al. (eds.), *Variable-Structure Systems and Sliding-Mode Control*, Studies in Systems, Decision and Control 271,
https://doi.org/10.1007/978-3-030-36621-6_3

$$|\delta_1(t)| \le K \qquad\qquad |\delta_2(t)| \le L, \tag{3.2}$$

for all t. It arises in the context of both, sliding-mode control and observation, see [7, 8]; this is briefly illustrated in the following.

To see its use as a control law, consider a sliding variable σ governed by the dynamics

$$\frac{\mathrm{d}\sigma}{\mathrm{d}t} = u + w_1 + w_2, \tag{3.3}$$

with perturbations w_1 and w_2, and a control input u. The perturbations are assumed to satisfy the bounds $|w_1| \le K\sqrt{|\sigma|}$ and $\left|\frac{\mathrm{d}w_2}{\mathrm{d}t}\right| \le L$ with nonnegative constants K and L. In order to drive σ to zero in spite of the perturbations w_1 and w_2, the control input is prescribed by the control law

$$u = -k_1\sqrt{|\sigma|}\,\mathrm{sign}(\sigma) + v, \tag{3.4a}$$

$$\frac{\mathrm{d}v}{\mathrm{d}t} = -k_2\,\mathrm{sign}(\sigma). \tag{3.4b}$$

With perturbation signals $\delta_1 := w_1\sqrt{|\sigma|}^{-1}$, $\delta_2 := \frac{\mathrm{d}w_2}{\mathrm{d}t}$ and state variables $x_1 = \sigma$, $x_2 = v + w_2$, the closed-loop dynamics are then described by the nonlinear system (3.1) with perturbation bounds (3.2).

In an observer context, the super-twisting algorithm is obtained when considering the task of differentiating a function $f(t)$, whose second time derivative is bounded by $\left|\frac{\mathrm{d}^2 f}{\mathrm{d}t^2}\right| \le L$, by means of the first-order[1] robust exact differentiator, see [8],

$$\frac{\mathrm{d}y_1}{\mathrm{d}t} = y_2 + k_1\sqrt{|f - y_1|}\,\mathrm{sign}(f - y_1), \tag{3.5a}$$

$$\frac{\mathrm{d}y_1}{\mathrm{d}t} = k_2\,\mathrm{sign}(f - y_1). \tag{3.5b}$$

With perturbations $\delta_1 := 0$, $\delta_2 := \frac{\mathrm{d}^2 f}{\mathrm{d}t^2}$ and states $x_1 := f - y_1$, $x_2 := \frac{\mathrm{d}f}{\mathrm{d}t} - y_2$, the observer error dynamics are described again by the nonlinear system (3.1) with perturbation bounds (3.2); the bound K is zero in this case.

An important performance characteristic of system (3.1) is the time it takes for the states x_1, x_2 to reach (or converge to) zero. This time is called the system's *reaching time* or *convergence time*. It is finite, when the system's origin is asymptotically stable, or, equivalently, *finite-time stable*. Depending on whether the super-twisting algorithm is considered in a control or a differentiator context, it is the time it takes for the controller to *reach* the sliding surface or for the differentiator error to *converge* to zero, respectively. In both contexts, computing upper bounds for the reaching time is desirable for the purpose of performance analysis and parameter tuning. In this chapter, three techniques to compute upper bounds for the reaching time of system (3.1) are described, summarizing and extending recent results from literature.

---

[1] The differentiator order is the order of the highest derivative it yields.

In particular, in Sects. 3.3 and 3.5, results obtained in [13, 15] are discussed and extended, and in Sect. 3.4, results due to Moreno et al. in [2, 10, 11] are partly summarized and further analyzed.

The first technique, which is proposed in [15], is based on an analytic computation of the reaching time in the absence of perturbations. This computation itself is mainly of theoretical interest, because disturbances typically are present in practice. It may, however, also be used to estimate the reaching time for the perturbed case, provided that the perturbations are not too large. This yields a reaching time bound, which is tight when the perturbations vanish, and thus is especially well suited for small to moderately large perturbations.

The second technique is proposed in [2] and uses a family of quadratic Lyapunov functions, which is proposed in [10, 11]. The estimate obtained this way depends on the particular Lyapunov function that is chosen from this family. A method to select this Lyapunov function by solving a semidefinite program, which is similar to the approach in [10], is thus shown. This technique yields more conservative bounds than the first technique, but permits for larger perturbations. Furthermore, it can be used to obtain reaching time bounds for an entire region of the state space instead of just a single initial state, which is difficult to do with the other approaches.

The third technique is based on a geometric stability proof for the super-twisting algorithm, which was originally proposed along with the algorithm itself in [7, 8]. Using this proof technique, a necessary and sufficient stability condition for the algorithm in analytic form has recently been proposed in [13]. The analytic expressions that are part of this condition may be used to estimate the reaching time for initial states that lie on the x_2-axis, i.e., where the initial value for x_1 is zero. When considering the super-twisting algorithm in a differentiator context, one can always have such an initial state by proper initialization of the differentiator's state y_1. Similar to the first approach, the obtained bound is tight when the perturbations vanish. At the same time, the range of permitted perturbations is much larger (in fact, as large as possible) with this approach, due to its being based on a necessary and sufficient stability condition. The approach is thus particularly well suited for analysis and design of differentiators.

Several other techniques are proposed in literature, see, e.g., [12, 18]. They are not described in detail here; for a concise description of them and a comparison to some of the techniques discussed here, the reader is referred to [15].

The chapter is structured as follows. Section 3.2 contains some preliminaries, such as notational conventions, the definitions of finite-time stability and the reaching time function, and some useful properties of the considered system. The three techniques for reaching time estimations are then discussed in Sects. 3.3, 3.4, and 3.5. Each of them is divided into subsections that study the estimate itself, as well as its asymptotic properties and the permitted range of perturbations. For the Lyapunov function family based estimate in Sect. 3.4, the choice of the Lyapunov function is additionally discussed. Important results are summarized in the form of theorems at the end of each subsection, and in some cases numerical examples are given to illustrate their application. Section 3.6, finally, presents numerical comparisons of the range of perturbations permitted by each approach, and of the estimates themselves. Some of the derivations that are too involved and would impair the readability of the chapter's main material are given in an appendix.

3.2 Preliminaries

In this section, some notational conventions are first discussed that are used throughout this chapter. Then, the notion of finite-time stability and the super-twisting algorithm's reaching time function, which is the main focus of the chapter, are introduced. Useful properties of this function are then shown that stem from the fact that the super-twisting algorithm is a homogeneous system. Finally, a quasi-linear representation of the considered system is discussed, which forms the basis for many considerations throughout the chapter.

3.2.1 Notation

Throughout this chapter, the following abbreviations and notational conventions are used. The sign-preserving power function is defined as

$$\lceil y \rfloor^p := |y|^p \operatorname{sign}(y), \tag{3.6}$$

and, in particular, $\lceil y \rfloor^0 = \operatorname{sign}(y)$. Note that for $y \neq 0$ and any real number p the useful relations

$$\frac{\mathrm{d}\lceil y \rfloor^p}{\mathrm{d}y} = p\,|y|^{p-1}, \qquad \frac{\mathrm{d}|y|^p}{\mathrm{d}y} = p\,\lceil y \rfloor^{p-1} \tag{3.7}$$

hold. The logarithm of a complex value y is defined as

$$\log y := \ln|y| + \mathrm{j}\arg y \qquad \text{with} \; -\pi \le \arg y < \pi. \tag{3.8}$$

Matrices and vectors are written as upper- and lowercase letters, respectively. The identity matrix is denoted by I. For a square matrix M, the matrix exponential function is denoted by e^M, i.e.,

$$\mathrm{e}^M := I + M + \frac{M^2}{2!} + \cdots = \sum_{\nu=0}^{\infty} \frac{M^\nu}{\nu!}. \tag{3.9}$$

The spectral norm of a (not necessarily square) matrix N, i.e., its largest singular value, is written as $\|N\|_2$ and the $\mathscr{H}_\infty$ norm of a transfer function matrix $G(s)$ is denoted by

$$\|G\|_{\mathscr{H}_\infty} := \sup_{\omega} \|G(\mathrm{j}\omega)\|_2 . \tag{3.10}$$

The Moore–Penrose pseudo-inverse of a matrix N is denoted by N^+.

For a symmetric matrix $R = R^\mathrm{T}$, the smallest and largest eigenvalues are denoted by $\lambda_{\min}(R)$ and $\lambda_{\max}(R)$, respectively. Furthermore, $R > 0$ or $R \geq 0$ mean that the matrix R is positive definite or positive semidefinite, i.e., that its smallest eigenvalue is positive or nonnegative, respectively.

Finally, the abbreviations

$$e_1 := \begin{bmatrix} 1 & 0 \end{bmatrix}^\mathrm{T}, \qquad e_2 := \begin{bmatrix} 0 & 1 \end{bmatrix}^\mathrm{T} \tag{3.11}$$

are used for the standard basis vectors $e_1, e_2 \in \mathbb{R}^2$, and the time derivative of a function $y(t)$ is denoted by

$$\dot{y}(t) := \frac{\mathrm{d}y}{\mathrm{d}t}. \tag{3.12}$$

Note that in each of Sects. 3.3, 3.4, and 3.5, the *same* symbol $\overline{T}_{K,L}$ denotes the upper reaching time bound in order to simplify notation, although the three approaches of course yield *different* values of this bound. Therefore, different relations or theorems involving $\overline{T}_{K,L}$ may only be combined or compared within the same section.

3.2.2 Considered System and Stability Notions

With the introduced notational conventions, the considered system (3.1) is given by

$$\dot{x}_1 = -k_1 \lceil x_1 \rfloor^{\frac{1}{2}} + x_2 + \delta_1 |x_1|^{\frac{1}{2}}, \tag{3.13a}$$

$$\dot{x}_2 = -k_2 \lceil x_1 \rfloor^{0} + \delta_2 \tag{3.13b}$$

with state variables x_1, x_2, which are aggregated in the state vector $x := [x_1 \quad x_2]^\mathrm{T}$, constant positive parameters k_1, k_2, and perturbations δ_1, δ_2. The perturbations are arbitrary, Lebesgue-integrable functions of time that satisfy

$$|\delta_1| \leq K, \qquad\qquad |\delta_2| \leq L. \tag{3.13c}$$

The initial state is denoted by $x_0 := x(t = 0)$. Mathematically, solutions of this system are not well defined in a classical sense due to the occurrence of the sign function in (3.13b). They are, therefore, understood in the sense of Filippov, i.e., as absolutely continuous functions that satisfy a differential inclusion that is obtained by replacing the sign function $\lceil x_1 \rfloor^0$ as well as the perturbations δ_1 and δ_2 by sets,[2] see, e.g., [7]. For the theoretical background, the mathematically inclined reader is referred to [4].

For the perturbed system (3.13), the following stability notion is used.

Definition 3.1 (*Finite-Time Stability*) The origin of the perturbed system (3.13) is called finite-time stable, if all its solutions converge to the origin in finite time, i.e., if for all[3] solutions $x(t)$, which are obtained with any initial condition x_0 and any admissible perturbations $\delta_1(t)$, $\delta_2(t)$, there exists a time instant τ such that $x(t) = 0$ holds for all $t \geq \tau$.

3.2.3 Reaching Time Function

The *reaching time* of system (3.13) is defined as the time it takes for *all* solutions starting from a given initial state x_0, i.e., solutions with any permitted perturbation, to reach the origin and stay there. For a *single* trajectory $x(t)$, the time it takes for the state to vanish is denoted by[4] τ_x; it may be formally defined as

$$\tau_x := \inf\left(\{\tau \in \mathbb{R} \mid x(t) = 0 \, \forall t \geq \tau\} \cup \{\infty\}\right), \tag{3.15}$$

i.e., as the first time instant after which $x(t) = 0$ holds for all times, or infinity, if no such time instant exists. For a given initial state x_0 and given perturbation bounds K and L, the system's reaching time is then given by the maximum of all individual trajectories' reaching times, i.e.,

$$T_{K,L}(x_0) := \sup\{\tau_x \mid x(t) \text{ is a solution of (3.13) with } x(0) = x_0\}. \tag{3.16}$$

The function $T_{K,L}$ is called the system's *reaching time function*. In addition to the perturbation bounds K and L, the function $T_{K,L}$ also depends on the parameters k_1 and k_2, but this dependence is suppressed for notational convenience.

Figure 3.1 illustrates this concept: It shows the unperturbed as well as three perturbed trajectories of system (3.13) along with the corresponding values of the unperturbed and the perturbed reaching time function.

---

[2]To be more specific, system (3.13) is understood as the inclusion

$$\dot{x}_1 \in \left[-k_1 \lceil x_1 \rfloor^{\frac{1}{2}} + x_2 - K |x_1|^{\frac{1}{2}}, -k_1 \lceil x_1 \rfloor^{\frac{1}{2}} + x_2 + K |x_1|^{\frac{1}{2}}\right], \tag{3.14a}$$

$$\dot{x}_2 \in \begin{cases} [-k_2 \lceil x_1 \rfloor^0 - L, -k_2 \lceil x_1 \rfloor^0 + L] & x_1 \neq 0 \\ [-k_2 - L, k_2 + L] & x_1 = 0. \end{cases} \tag{3.14b}$$

[3]In terms of the differential inclusion, this means all absolutely continuous functions $x_1(t)$, $x_2(t)$ that satisfy (3.14).

[4]Note that τ_x maps *functions* $x(t)$ rather than *vectors* x to real-valued time instants.

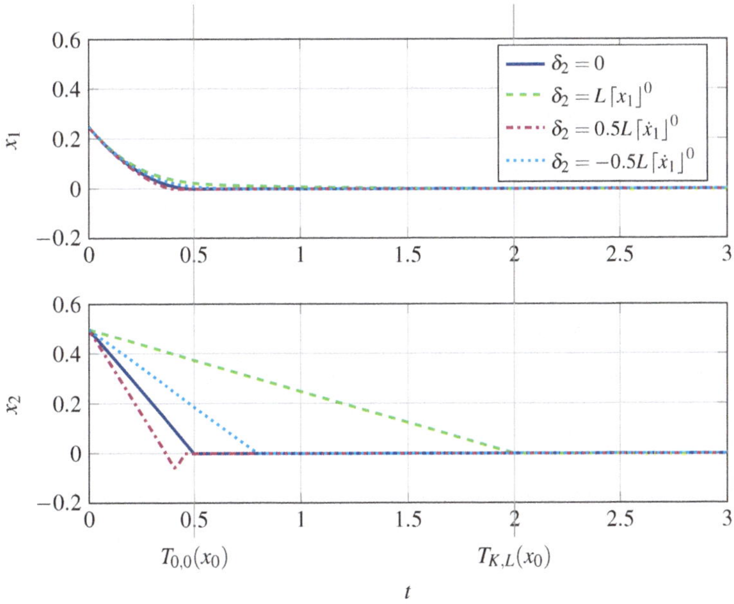

Fig. 3.1 One unperturbed and three perturbed trajectories of the super-twisting algorithm for parameter values $k_1 = 3$, $k_2 = 1$, perturbation bounds $K = 0$, $L = \frac{3}{4}$, and initial state $x_0 = \left[\frac{1}{4} \ \ \frac{1}{2}\right]$ with different perturbations, illustrating the unperturbed and perturbed reaching time functions $T_{0,0}$ and $T_{K,L}$, respectively

3.2.4 Homogeneity Properties

It is well known that (3.13) is a homogeneous system.[5] This means that any trajectory, after a suitable scaling of both the state and the time, is again a trajectory of the system. Due to this fact, the reaching time function also is a homogeneous function, i.e., suitably scaling the initial state yields a correspondingly scaled reaching time. Furthermore, as with linear systems, the system's behavior in the vicinity of the origin determines its behavior in the whole state space.

Due to homogeneity, the reaching time function $T_{K,L}$ has a number of useful scaling properties. To show them, consider the so-called dilation matrix

$$D_\varepsilon := \begin{bmatrix} \varepsilon^2 & 0 \\ 0 & \varepsilon \end{bmatrix} \tag{3.17}$$

with the positive parameter ε. New state variables $y := \left[\varepsilon^2 x_1 \ \ \varepsilon x_2\right]^{\mathrm{T}}$ and a new time coordinate τ are introduced by means of the combined state and time transformation

[5]For more information on homogeneity, the reader is referred to [1].

$$y = \begin{bmatrix} y_1 & y_2 \end{bmatrix}^{\mathrm{T}} = D_\varepsilon x, \qquad \tau = \varepsilon t. \tag{3.18}$$

Using (3.13), one finds that with respect to the time τ the transformed state variables y_1, y_2 are governed by the dynamics

$$\begin{aligned} \frac{\mathrm{d}y_1}{\mathrm{d}\tau} &= \frac{\mathrm{d}\varepsilon^2 y_1}{\mathrm{d}(\varepsilon t)} = \varepsilon \frac{\mathrm{d}x_1}{\mathrm{d}t} = -k_1 \left\lceil \varepsilon^2 x_1 \right\rfloor^{\frac{1}{2}} + \varepsilon x_2 + \delta_1 \left|\varepsilon^2 x_1\right|^{\frac{1}{2}} \\ &= -k_1 \lceil y_1 \rfloor^{\frac{1}{2}} + y_2 + \delta_1 |y_1|^{\frac{1}{2}}, \end{aligned} \tag{3.19a}$$

$$\begin{aligned} \frac{\mathrm{d}y_2}{\mathrm{d}\tau} &= \frac{\mathrm{d}\varepsilon x_2}{\mathrm{d}(\varepsilon t)} = \frac{\mathrm{d}x_2}{\mathrm{d}t} = -k_2 \left\lceil \varepsilon^2 x_1 \right\rfloor^{0} + \delta_2 \\ &= -k_2 \lceil y_1 \rfloor^{0} + \delta_2. \end{aligned} \tag{3.19b}$$

These are the same differential equations as those of the original system (3.13). Thus, if $x(t)$ is a solution of (3.13), then $y(\tau) = D_\varepsilon x(\varepsilon^{-1}\tau)$ is a solution of (3.19). Since these are the same equations as (3.13), one may replace τ again by t to see that $D_\varepsilon x(\varepsilon^{-1}t)$ is a solution of (3.13). Due to this time scaling, the reaching time function $T_{K,L}$ satisfies

$$T_{K,L}(D_\varepsilon x_0) = \varepsilon T_{K,L}(x_0), \tag{3.20}$$

i.e., the function $T_{K,L}$ is homogeneous of degree one with respect to the dilation D_ε.

One may show a similar scaling property with respect to the algorithm's parameters k_1, k_2 and perturbation bounds K, L. By considering instead of (3.18) the state transformation,

$$y = \varepsilon^2 x, \tag{3.21}$$

one obtains from (3.13) the differential equations

$$\begin{aligned} \frac{\mathrm{d}y_1}{\mathrm{d}t} &= \varepsilon^2 \frac{\mathrm{d}x_1}{\mathrm{d}t} = -\varepsilon k_1 \left\lceil \varepsilon^2 x_1 \right\rfloor^{\frac{1}{2}} + \varepsilon^2 x_2 + \varepsilon \delta_1 \left|\varepsilon^2 x_1\right|^{\frac{1}{2}}, \\ &= -\varepsilon^2 k_1 \lceil y_1 \rfloor^{\frac{1}{2}} + y_2 + \varepsilon \delta_1 |y_1|^{\frac{1}{2}}, \end{aligned} \tag{3.22a}$$

$$\begin{aligned} \frac{\mathrm{d}y_2}{\mathrm{d}t} &= \varepsilon^2 \frac{\mathrm{d}x_2}{\mathrm{d}t} = -\varepsilon^2 k_2 \left\lceil \varepsilon^2 x_1 \right\rfloor^{0} + \varepsilon^2 \delta_2 \\ &= -\varepsilon^2 k_2 \lceil y_1 \rfloor^{0} + \varepsilon^2 \delta_2. \end{aligned} \tag{3.22b}$$

By comparing with (3.13), one can see that, if $x(t)$ is a solution of the system with parameters k_1, k_2 and perturbation bounds K, L, then $\varepsilon^2 x(t)$ is a solution with parameters $\tilde{k}_1, \tilde{k}_2$ and perturbation bounds $\tilde{K}, \tilde{L}$, which are given by

$$\tilde{k}_1 = \varepsilon k_1, \qquad \tilde{k}_2 = \varepsilon^2 k_2, \tag{3.23a}$$

$$\tilde{K} = \varepsilon K, \qquad \tilde{L} = \varepsilon^2 L. \tag{3.23b}$$

The reaching time function $\tilde{T}_{\tilde{K},\tilde{L}}$ corresponding to the system with these scaled parameters thus satisfies

$$\tilde{T}_{\tilde{K},\tilde{L}}(\varepsilon^2 x_0) = T_{K,L}(x_0), \tag{3.24}$$

i.e., the reaching time is invariant with respect to a simultaneous scaling of initial state, parameters, and perturbation bounds.

Due to the latter fact, properties of the reaching time function are sometimes discussed or illustrated as a function of the parameter ratio $k_1 \sqrt{k_2}^{-1}$ and the normalized perturbation bounds Kk_1^{-1} and Lk_2^{-1}.

3.2.5 *Quasi-linear System Representation*

Several insights and derivations in this chapter are based on a quasi-linear representation of the system (3.13), which is presented in [10]. It is obtained by introducing new state variables $z := \begin{bmatrix} z_1 & z_2 \end{bmatrix}^{\mathrm{T}}$ by means of a the state transformation $z = g(x)$ with the function $g(x)$ given by

$$g(x) := \begin{bmatrix} \lceil x_1 \rfloor^{\frac{1}{2}} \\ x_2 \end{bmatrix}, \tag{3.25}$$

i.e., $z_1 = \lceil x_1 \rfloor^{\frac{1}{2}}$, $z_2 = x_2$. The transformed initial state is abbreviated as $z_0 := g(x_0)$. With respect to z, the system dynamics (3.13) can then formally be written in the form

$$\dot{z} = \frac{1}{|z_1|} A z + \frac{1}{2} e_1 \delta_1 + e_2 \delta_2, \tag{3.26}$$

where the matrix A is given by

$$A := \begin{bmatrix} -\frac{1}{2}k_1 & \frac{1}{2} \\ -k_2 & 0 \end{bmatrix}. \tag{3.27}$$

As one can see from its characteristic polynomial,

$$\det(sI - A) = s^2 + \frac{k_1}{2} s + \frac{k_2}{2}, \tag{3.28}$$

the matrix A is a Hurwitz matrix if and only if k_1 and k_2 are positive.

Note that $g(x)$ is not a diffeomorphism, because it is not differentiable for $x_1 = 0$. As a consequence, system (3.26) exhibits a singularity at $z_1 = 0$. Thus, existence of that system's solutions and their one-to-one correspondence to solutions of (3.13) are not immediately guaranteed. Despite this fact, all results obtained using (3.26) can also be proven formally, but some additional technical arguments are required.

These are only sketched briefly here; for details, the mathematically inclined reader is referred to [11, 15], as well as to [16] for an even more extensive version of these arguments applied to a different sliding-mode algorithm. Essentially, since $z(t)$ is the integral of the right-hand side in (3.26), zero crossings of $z_1(t)$ do not matter as long as they only occur at isolated time instants. It can be shown that such zero crossings only cluster at the system's reaching time $T_{K,L}$. This justifies the use of (3.26) instead of the original system (3.13) on any compact time interval of the form $[0, T]$ that does not include the reaching time, i.e., with $T < T_{K,L}$. The results are then obtained by taking the limit as T tends to $T_{K,L}$.

3.3 Estimation Based on the Reaching Time Function

In this section, the reaching time function of system (3.13) without perturbations δ_1, δ_2 is first computed. Based on this function, reaching time bounds for the *perturbed* system are derived by considering the *unperturbed* reaching time function as a Lyapunov function candidate. Contrary to other Lyapunov-based approaches proposed in literature and discussed later on, the resulting estimates tend to tight upper bounds for vanishing perturbations.

3.3.1 Unperturbed Reaching Time Function

The reaching time function $T_{0,0}(x_0)$ of system (3.13) without perturbations is now derived. For this purpose, the transformed system (3.26) is used. The reaching time function first is obtained in the form of an improper integral, and then an explicit analytic expression is computed for it.

3.3.1.1 Derivation in Integral Form

Consider the transformed system (3.26) without perturbations, i.e.,

$$\dot{z} = \frac{1}{|z_1|} A z. \tag{3.29}$$

One can see that it is linear except for the positive scaling factor $|z_1|^{-1}$, which amounts to a time scaling. To see this, a time transformation

$$\tau = \alpha(t) \tag{3.30}$$

with an absolutely continuous function α and a new time coordinate τ is introduced. In order for the time instant zero in original and in transformed time to coincide, the condition $\alpha(0) = 0$ is imposed. With respect to τ, the state z then satisfies

$$\frac{\mathrm{d}z}{\mathrm{d}\tau} = \dot{z}\frac{\mathrm{d}t}{\mathrm{d}\tau} = \frac{\dot{z}}{\dot{\alpha}} = \frac{1}{\dot{\alpha}\,|z_1|}Az. \tag{3.31}$$

If α satisfies the differential equation $\dot{\alpha}\,|z_1| = 1$, then z is governed by the linear dynamics

$$\frac{\mathrm{d}z}{\mathrm{d}\tau} = Az \tag{3.32}$$

with respect to τ. Note that $\alpha(t)$ depends on $z_1(t)$, and thus on the particular trajectory that is considered. It is thus impossible to obtain *one* function α such that the above considerations are valid for *all* initial states.

The function α is uniquely defined, however, for a given, fixed initial value z_0. In the following, a differential equation for α is derived. It is well known from linear systems theory that the solution of (3.32) with initial value $z(0) = z_0$ is given by

$$z(t) = \mathrm{e}^{A\tau}z_0 = \mathrm{e}^{A\alpha(t)}z_0. \tag{3.33}$$

In order to obtain a differential equation for α corresponding to the considered initial state z_0, this solution is substituted into $\dot{\alpha}\,|z_1(t)| = 1$; one thus obtains

$$\dot{\alpha}(t) = \frac{1}{|z_1(t)|} = \frac{1}{\left|e_1^{\mathrm{T}}z(t)\right|} = \frac{1}{\left|e_1^{\mathrm{T}}\mathrm{e}^{A\alpha(t)}z_0\right|}, \tag{3.34}$$

which is an autonomous differential equation for α with z_0 as a parameter. Solving this initial value problem by separation of variables, i.e., integrating both sides of the relation,

$$\left|e_1^{\mathrm{T}}\mathrm{e}^{A\alpha(t)}z_0\right|\mathrm{d}\alpha = \mathrm{d}t, \tag{3.35}$$

yields the implicit relation

$$\int_{\alpha(0)=0}^{\alpha(t)} \left|e_1^{\mathrm{T}}\mathrm{e}^{A\kappa}z_0\right|\mathrm{d}\kappa = t \tag{3.36}$$

for $\alpha(t)$. One can see that the left-hand side of this relation gives t as a function of $\alpha(t)$. It thus is the inverse function $\alpha^{-1}(\tau)$ of $\alpha(t)$, i.e.,

$$\alpha^{-1}(\tau) = \int_0^{\tau} \left|e_1^{\mathrm{T}}\mathrm{e}^{A\kappa}z_0\right|\mathrm{d}\kappa. \tag{3.37}$$

This expression yields the original time t as a function of the transformed time τ.

Inverting $\alpha^{-1}(\tau)$, i.e., computing the function $\alpha(t)$, can only be done numerically. Doing this inversion is not necessary, however, for the purpose of deriving the reaching time function. From (3.33), one can see that the state z of the linear system (3.32) tends to zero as τ tends to infinity. In original time, this corresponds to the time instant $\alpha^{-1}(\infty)$. By using (3.37) and recalling the fact that $z_0 = g(x_0)$, the reaching time function is thus obtained as

$$T_{0,0}(x_0) = \lim_{\tau\to\infty} \alpha^{-1}(\tau) = \int_0^\infty \left| e_1^{\mathrm{T}} \mathrm{e}^{A\alpha} g(x_0) \right| \mathrm{d}\alpha. \tag{3.38}$$

3.3.1.2 Explicit Analytic Expression

The value of the integral in (3.38) can be obtained analytically. For this purpose, the integration region is partitioned into intervals where the sign of the function $e_1^{\mathrm{T}} \mathrm{e}^{A\alpha} g(x)$ is constant. On each of these intervals, the integration may then be performed by using the indefinite integral

$$\int e_1^{\mathrm{T}} \mathrm{e}^{A\alpha} g(x)\, \mathrm{d}\alpha = e_1^{\mathrm{T}} A^{-1} \mathrm{e}^{A\alpha} g(x) = -\frac{1}{k_2} e_2^{\mathrm{T}} \mathrm{e}^{A\alpha} g(x). \tag{3.39}$$

This procedure leads to a rather lengthy computation, which is given in [15]. Its result is the reaching time function

$$T_{0,0}(x) = \frac{1}{k_2} \left| e_2^{\mathrm{T}} \left(I - \frac{2}{1-\lambda} \mathrm{e}^{A\alpha_1(x)} \right) g(x) \right| \tag{3.40}$$

with the abbreviations[6]

$$s_1 = -\frac{k_1}{4} + \frac{\sqrt{k_1^2 - 8k_2}}{4}, \qquad s_2 = -\frac{k_1}{4} - \frac{\sqrt{k_1^2 - 8k_2}}{4}, \tag{3.41a}$$

$$\alpha_1(x) := \begin{cases} 0 & x_1 = 0 \\ -\frac{2}{\lceil x_1 \rfloor^{-\frac{1}{2}} x_2 + 2s_1} & \lceil x_1 \rfloor^{-\frac{1}{2}} x_2 < -2\mathrm{Re} s_1,\ k_1^2 = 8k_2 \\ \frac{1}{s_1 - s_2} \log \frac{\lceil x_1 \rfloor^{-\frac{1}{2}} x_2 + 2s_2}{\lceil x_1 \rfloor^{-\frac{1}{2}} x_2 + 2s_1} & \lceil x_1 \rfloor^{-\frac{1}{2}} x_2 < -2\mathrm{Re} s_1,\ k_1^2 \neq 8k_2 \\ \frac{1}{s_1 - s_2} \left[2\pi \mathrm{j} + \log \frac{\lceil x_1 \rfloor^{-\frac{1}{2}} x_2 + 2s_2}{\lceil x_1 \rfloor^{-\frac{1}{2}} x_2 + 2s_1} \right] & \lceil x_1 \rfloor^{-\frac{1}{2}} x_2 \geq -2\mathrm{Re} s_1,\ k_1^2 < 8k_2 \\ \infty & \lceil x_1 \rfloor^{-\frac{1}{2}} x_2 \geq -2\mathrm{Re} s_1,\ k_1^2 \geq 8k_2, \end{cases} \tag{3.41b}$$

$$\lambda := \begin{cases} 0 & k_1^2 \geq 8k_2 \\ \exp\left(-\frac{k_1 \pi}{\sqrt{8k_2 - k_1^2}} \right) & k_1^2 < 8k_2. \end{cases} \tag{3.41c}$$

---

[6] Note that in the expression for $\alpha_1(x)$ the logarithm is used as defined in (3.8).

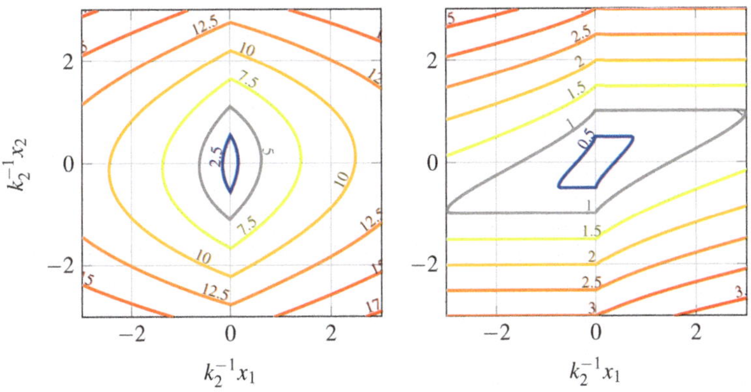

Fig. 3.2 Level lines of the unperturbed system's reaching time function $T_{0,0}(x)$ for $k_1 = 0.4\sqrt{k_2}$ (left) and $k_1 = 4\sqrt{k_2}$ (right)

The quantity α_1 is the first zero of the integrand, i.e, the smallest nonnegative value of α such that $e_1^\mathrm{T} \mathrm{e}^{A\alpha} g(x)$ vanishes. The values s_1 and s_2 are the eigenvalues of the matrix A, and the factor λ is nonzero if the integrand has multiple zeros, which is the case if the eigenvalues are complex-valued.

Figure 3.2 shows the level lines of the reaching time function $T_{0,0}(x)$ for the parameter settings $k_1 = 0.4\sqrt{k_2}$ and $k_1 = 4\sqrt{k_2}$. In the first case, the eigenvalues of A are complex-valued, while they are real-valued in the second case. One can see that this has a significant impact on the shape of the level lines: with real-valued eigenvalues, i.e., for $k_1^2 \geq 8k_2$, there are horizontal level-line segments, where the reaching time does not depend on x_1 locally. With complex-valued eigenvalues, this is not the case.

The following theorem summarizes the results of this section.

Theorem 3.1 (Unperturbed Reaching Time Function) *Let k_1, k_2 be positive and let the function $g(x)$ and the matrix $A \in \mathbb{R}^{2\times 2}$ be defined as in (3.25) and (3.27), respectively. Then, the reaching time function of system (3.13) without perturbation, i.e., with $K = L = 0$, is given by*

$$T_{0,0}(x) = \int_0^\infty \left| e_1^\mathrm{T} e^{A\alpha} g(x) \right| \mathrm{d}\alpha \tag{3.42}$$

or, equivalently, by

$$T_{0,0}(x) = \frac{1}{k_2} \left| e_2^\mathrm{T} \left(I - \frac{2}{1-\lambda} e^{A\alpha_1(x)} \right) g(x) \right| \tag{3.43}$$

with λ and $\alpha_1(x)$ as defined in (3.41).

Example 3.1 The use of Theorem 3.1 is shown in the following numerical example. Consider the unperturbed system (3.13) with parameter values $k_1 = 2$, $k_2 = 1$ and initial state $x_0 = e_2 = \begin{bmatrix}0 & 1\end{bmatrix}^{\mathrm{T}}$. In this case, one has according to (3.41)

$$s_1 = -\frac{1}{2}(1 - \mathrm{j}), \qquad \alpha(e_2) = 0, \tag{3.44a}$$

$$s_2 = -\frac{1}{2}(1 + \mathrm{j}), \qquad \lambda = \mathrm{e}^{-\pi}. \tag{3.44b}$$

Using (3.43) and the fact that $g(e_2) = e_2$, the unperturbed reaching time is obtained as

$$T_{0,0}(e_2) = \left| e_2^{\mathrm{T}} \left(I - \frac{2}{1 - \mathrm{e}^{-\pi}} I \right) e_2 \right| = \frac{1 + \mathrm{e}^{-\pi}}{1 - \mathrm{e}^{-\pi}} \approx 1.0903. \tag{3.45}$$

3.3.2 *Reaching Time Estimation*

To estimate the reaching time of the perturbed system, the unperturbed reaching time function $T_{0,0}(x)$ is considered as a Lyapunov function candidate. Along the trajectories of the *unperturbed* system, its time derivative is given by $\dot{T}_{0,0} = -1$. This may be verified by using its integral form (3.42) and (3.26) to obtain[7]

$$\begin{aligned}\dot{T}_{0,0} &= \int_0^\infty \left\lceil e_1^{\mathrm{T}} \mathrm{e}^{A\alpha} z \right\rfloor^0 e_1^{\mathrm{T}} \mathrm{e}^{A\alpha} \dot{z} \, \mathrm{d}\alpha = |z_1|^{-1} \int_0^\infty \left\lceil e_1^{\mathrm{T}} \mathrm{e}^{A\alpha} z \right\rfloor^0 e_1^{\mathrm{T}} \mathrm{e}^{A\alpha} A z \, \mathrm{d}\alpha \\ &= |z_1|^{-1} \int_0^\infty \frac{\mathrm{d}}{\mathrm{d}\alpha} \left(\left| e_1^{\mathrm{T}} \mathrm{e}^{A\alpha} z \right| \right) \mathrm{d}\alpha = - |z_1|^{-1} \left| e_1^{\mathrm{T}} z \right| = -1.\end{aligned} \tag{3.46}$$

For the *perturbed* system, one finds by using (3.13) that

$$\dot{T}_{0,0} = -1 + \frac{\partial T_{0,0}}{\partial x_1} |x_1|^{\frac{1}{2}} \delta_1 + \frac{\partial T_{0,0}}{\partial x_2} \delta_2 \tag{3.47}$$

holds. The partial derivatives in this expression are given by

$$\frac{\partial T_{0,0}}{\partial x_1} = \int_0^\infty \left\lceil e_1^{\mathrm{T}} \mathrm{e}^{A\alpha} g(x) \right\rfloor^0 e_1^{\mathrm{T}} \mathrm{e}^{A\alpha} \frac{\partial g}{\partial x_1} \mathrm{d}\alpha = \int_0^\infty \left\lceil e_1^{\mathrm{T}} \mathrm{e}^{A\alpha} g(x) \right\rfloor^0 \frac{e_1^{\mathrm{T}} \mathrm{e}^{A\alpha} e_1}{2 |x_1|^{\frac{1}{2}}} \mathrm{d}\alpha, \tag{3.48a}$$

$$\frac{\partial T_{0,0}}{\partial x_2} = \int_0^\infty \left\lceil e_1^{\mathrm{T}} \mathrm{e}^{A\alpha} g(x) \right\rfloor^0 e_1^{\mathrm{T}} \mathrm{e}^{A\alpha} \frac{\partial g}{\partial x_2} \mathrm{d}\alpha = \int_0^\infty \left\lceil e_1^{\mathrm{T}} \mathrm{e}^{A\alpha} g(x) \right\rfloor^0 e_1^{\mathrm{T}} \mathrm{e}^{A\alpha} e_2 \, \mathrm{d}\alpha. \tag{3.48b}$$

[7]As can be seen from Fig. 3.2, the unperturbed reaching time function $T_{0,0}(x)$ is not differentiable for $x_1 = 0$ or, equivalently, $z_1 = 0$. Thus, the time derivatives computed in the following are, strictly speaking, only defined for $z_1 \neq 0$. As pointed out in Sect. 3.2.5, however, this has no impact on the obtained results, and can thus essentially be ignored in the course of the derivations.

By taking the integrands' absolute values, these two expressions may be upper bounded by

$$2\,|x_1|^{\frac{1}{2}} \left|\frac{\partial T_{0,0}}{\partial x_1}\right| \le \int_0^\infty \left|e_1^{\mathrm{T}} \mathrm{e}^{A\alpha} e_1\right| \mathrm{d}\alpha = T_{0,0}(e_1), \tag{3.49a}$$

$$\left|\frac{\partial T_{0,0}}{\partial x_2}\right| \le \int_0^\infty \left|e_1^{\mathrm{T}} \mathrm{e}^{A\alpha} e_2\right| \mathrm{d}\alpha = T_{0,0}(e_2). \tag{3.49b}$$

It is noteworthy that both bounds can be expressed using the reaching time function $T_{0,0}(x)$ evaluated at $x = e_1$ and $x = e_2$, respectively. Taking into account the bounds (3.13c) on δ_1 and δ_2, one thus obtains

$$\begin{aligned}\dot{T}_{0,0} &\le -1 + \left|\frac{\partial T_{0,0}}{\partial x_1}\right| |x_1|^{\frac{1}{2}}\, |\delta_1| + \left|\frac{\partial T_{0,0}}{\partial x_2}\right| |\delta_2| \\ &\le -1 + K\frac{T_{0,0}(e_1)}{2} + L T_{0,0}(e_2).\end{aligned} \tag{3.50}$$

This expression's right-hand side is abbreviated as $-c$ in the following, i.e.,

$$c := 1 - K\frac{T_{0,0}(e_1)}{2} - L T_{0,0}(e_2). \tag{3.51}$$

If the time derivative's upper bound is negative, i.e., if $c > 0$ holds, then the unperturbed system's reaching time function $T_{0,0}$ is a Lyapunov function for the perturbed system. In this case, one may obtain a reaching time bound by integrating the differential inequality (3.50). For any time instant T, this yields

$$T_{0,0}(x(T)) = T_{0,0}(x_0) + \int_0^T \dot{T}_{0,0}\,\mathrm{d}t \le T_{0,0}(x_0) - cT. \tag{3.52}$$

When this expression's right-hand side is zero, the left-hand side and thus $x(T)$ need to be zero as well. The corresponding time instant T is thus upper bounded by

$$T \le \frac{T_{0,0}(x_0)}{c}, \tag{3.53}$$

which, therefore, corresponds to an upper bound $\overline{T}_{K,L}$ for the perturbed system's reaching time $T_{K,L}$, i.e.,

$$\overline{T}_{K,L}(x_0) := \frac{T_{0,0}(x_0)}{c} = \frac{T_{0,0}(x_0)}{1 - \frac{K}{2}T_{0,0}(e_1) - L T_{0,0}(e_2)} \ge T_{K,L}(x_0). \tag{3.54}$$

The following theorem summarizes the results of this section.

Theorem 3.2 (Reaching Time Function Based Estimate) *Let $T_{0,0}$ be the reaching time function of system (3.13) without perturbations as given in Theorem 3.1. If the nonnegative perturbation bounds K and L satisfy*

$$\frac{K T_{0,0}(e_1)}{2} + L T_{0,0}(e_2) < 1, \tag{3.55}$$

then the reaching time function $T_{K,L}(x)$ *of the perturbed system (3.13) is bounded by* $T_{K,L}(x) \le \overline{T}_{K,L}(x)$*, where* $\overline{T}_{K,L}$ *is given by*

$$\overline{T}_{K,L}(x) = \frac{T_{0,0}(x)}{1 - \frac{K T_{0,0}(e_1)}{2} - L T_{0,0}(e_2)}. \tag{3.56}$$

Example 3.2 As in Example 3.1, consider parameter values $k_1 = 2, k_2 = 1$ and initial state $x_0 = e_2$. A reaching time bound for the perturbed case with $L = 0.5$, $K = 0$ is computed using Theorem 3.2. Using the value of $T_{0,0}(e_2)$ computed in (3.45), the bound (3.56) is obtained as

$$\overline{T}_{0,0.5}(e_2) = \frac{T_{0,0}(e_2)}{1 - 0 T_{0,0}(e_1) - \frac{1}{2} T_{0,0}(e_2)} = \frac{\frac{1+\mathrm{e}^{-\pi}}{1-\mathrm{e}^{-\pi}}}{1 - \frac{1}{2}\frac{1+\mathrm{e}^{-\pi}}{1-\mathrm{e}^{-\pi}}} = \frac{2 + 2\mathrm{e}^{-\pi}}{1 - 3\mathrm{e}^{-\pi}} \approx 2.3972. \tag{3.57}$$

3.3.3 Range of Permitted Perturbations

One can see that Theorem 3.2 imposes upper bounds on K and L; a necessary condition for its applicability is that neither of the two terms on the left-hand side of (3.55) exceeds one. Introducing, thus, the abbreviations

$$\overline{K} := \frac{2}{T_{0,0}(e_1)}, \qquad \overline{L} := \frac{1}{T_{0,0}(e_2)}, \tag{3.58}$$

these necessary conditions are given by $K < \overline{K}$ and $L < \overline{L}$, and condition (3.55) may be written as

$$K\overline{K}^{-1} + L\overline{L}^{-1} < 1. \tag{3.59}$$

Using the reaching time function's integral form (3.42), an interesting theoretical insight may be gained into the nature of the two bounds $\overline{K}$ and $\overline{L}$. Introducing the transfer functions

$$\begin{aligned} G_1(s) &= \frac{1}{2} e_1^{\mathrm{T}} (sI - A)^{-1} e_1 & G_2(s) &= e_1^{\mathrm{T}} (sI - A)^{-1} e_2 \\ &= \frac{s}{2s^2 + k_1 s + k_2}, & &= \frac{1}{2s^2 + k_1 s + k_2}, \end{aligned} \tag{3.60}$$

and the associated impulse responses

$$g_1(t) = \frac{1}{2} e_1^{\mathrm{T}} \mathrm{e}^{At} e_1, \qquad\qquad g_2(t) = e_1^{\mathrm{T}} \mathrm{e}^{At} e_2, \tag{3.61}$$

one can write $\overline{K}$ and $\overline{L}$ as

$$\overline{K} = \frac{1}{\int_0^\infty |g_1(t)|\, \mathrm{d}t}, \qquad\qquad \overline{L} = \frac{1}{\int_0^\infty |g_2(t)|\, \mathrm{d}t}. \tag{3.62}$$

Note that the absolute integral of an impulse response corresponds to the $\mathscr{L}_\infty$-gain of the associated system. The bounds $\overline{K}$ and $\overline{L}$ can thus be seen to be reciprocals of the $\mathscr{L}_\infty$ gains of systems with transfer functions $G_1(s)$ and $G_2(s)$.

Using the definition (3.58) and the analytic form of the reaching time function (3.40), the values of $\overline{K}$ and $\overline{L}$ may also be computed analytically. In the course of a somewhat involved derivation, which is given in the appendix, one obtains the following theorem, which summarizes the obtained insights.

Theorem 3.3 (Reaching Time Function Based Perturbation Bounds) *Let $g_1(t)$ and $g_2(t)$ be the impulse responses corresponding to the transfer functions $G_1(s)$ and $G_2(s)$ defined in* (3.60)*, and let $T_{0,0}$ be the reaching time function of system (3.13) without perturbations as given in Theorem 3.1. Define the constants*

$$\overline{K} = \frac{1}{\int_0^\infty |g_1(t)|\, \mathrm{d}t} = \begin{cases} \sqrt{\frac{k_2}{2}} \left(\dfrac{\frac{k_1}{\sqrt{k_1^2 - 8k_2}} + 1}{\frac{k_1}{\sqrt{k_1^2 - 8k_2}} - 1} \right)^{\frac{k_1}{2\sqrt{k_1^2 - 8k_2}}} & k_1^2 > 8k_2 \\ \sqrt{\frac{k_2}{2}}\, e & k_1^2 = 8k_2 \\ \sqrt{\frac{k_2}{2}} \dfrac{2 \sinh \frac{\pi}{2} \frac{k_1}{\sqrt{8k_2 - k_1^2}}}{\exp\left(\frac{k_1}{\sqrt{8k_2 - k_1^2}} \arctan \frac{k_1}{\sqrt{8k_2 - k_1^2}} \right)} & k_1^2 < 8k_2, \end{cases} \tag{3.63a}$$

$$\overline{L} = \frac{1}{\int_0^\infty |g_2(t)|\, \mathrm{d}t} = \begin{cases} k_2 & k_1^2 \geq 8k_2 \\ k_2 \tanh \frac{\pi}{2} \frac{k_1}{\sqrt{8k_2 - k_1^2}} & k_1^2 < 8k_2. \end{cases} \tag{3.63b}$$

The condition of Theorem 3.2 is fulfilled if and only if *the perturbation bounds K, L satisfy the inequality*

$$K\overline{K}^{-1} + L\overline{L}^{-1} < 1. \tag{3.64}$$

If *this condition is fulfilled, then the reaching time function $T_{K,L}(x)$ of the perturbed system (3.13) is bounded by $T_{K,L}(x) \leq \overline{T}_{K,L}(x)$, where $\overline{T}_{K,L}(x)$ is given by*

$$\overline{T}_{K,L}(x) = \frac{T_{0,0}(x)}{1 - K\overline{K}^{-1} - L\overline{L}^{-1}}. \tag{3.65}$$

Example 3.3 The numerical computation of the maximum perturbation bounds $\overline{K}$ and $\overline{L}$ is illustrated for the parameter values $k_1 = 2$, $k_2 = 1$, which were already considered in previous examples. Since $k_1^2 < 8k_2$ holds in this case, one obtains

$$\overline{K} = \sqrt{\frac{1}{2}\frac{2\sinh\frac{\pi}{2}}{\exp(\arctan 1)}} = \frac{\sqrt{2}\sinh\frac{\pi}{2}}{e^{\frac{\pi}{4}}} \approx 1.4839, \tag{3.66a}$$

$$\overline{L} = \tanh\frac{\pi}{2} \approx 0.9172. \tag{3.66b}$$

3.3.4 Asymptotic Properties

The asymptotic behavior of the reaching time bound $\overline{T}_{K,L}$ is now considered. By construction, the bound is exact for vanishing perturbations, i.e.,

$$\lim_{K,L\to 0} \overline{T}_{K,L}(x) = \overline{T}_{0,0}(x) = T_{0,0}(x). \tag{3.67}$$

Furthermore, it will be shown that the bound is also asymptotically exact when the parameter k_1 tends to infinity. To see this, note that for fixed $\alpha > 0$, the matrix exponential $e^{A\alpha}$ satisfies

$$\lim_{k_1\to\infty} e^{A\alpha} = \lim_{k_1\to\infty} \exp\left(\begin{bmatrix} -\frac{k_1}{2} & \frac{1}{2} \\ -k_2 & 0 \end{bmatrix}\alpha\right) = \begin{bmatrix} 0 & 0 \\ 0 & 1 \end{bmatrix}. \tag{3.68}$$

While this is shown formally in the proof of [15, Theorem 5], it may also be seen intuitively using the corresponding linear system

$$\frac{dz_1}{d\alpha} = -\frac{k_1}{2}z_1 + \frac{1}{2}z_2 \tag{3.69a}$$

$$\frac{dz_2}{d\alpha} = -k_2 z_1. \tag{3.69b}$$

As k_1 grows, z_1 converges to an increasingly smaller neighborhood of zero with increasingly larger speed; thus, in the limit, z_1 tends to zero (pointwise) for any nonzero α, and z_2, therefore, stays constant and equal to its initial value.

In the course of evaluating the analytic expression (3.40), one finds that $\lambda = 0$ holds for sufficiently large k_1. Recalling the definition of $g(x)$ from (3.25), one thus finds

$$\lim_{k_1\to\infty} T_{0,0}(x) = \frac{1}{k_2}\left|e_2^{\mathrm{T}}\left(I - 2\begin{bmatrix} 0 & 0 \\ 0 & 1 \end{bmatrix}\right)g(x)\right| = \frac{1}{k_2}\left|-e_2^{\mathrm{T}}g(x)\right| = \frac{|x_2|}{k_2}. \tag{3.70}$$

Using (3.56), the asymptotic value of the perturbed reaching time bound is thus obtained as

$$\lim_{k_1\to\infty} \overline{T}_{K,L}(x) = \frac{\frac{|x_2|}{k_2}}{1 - \frac{K}{2}\frac{0}{k_2} - L\frac{1}{k_2}} = \frac{|x_2|}{k_2 - L}. \tag{3.71}$$

This last expression $(k_2 - L)^{-1} |x_2|$ is not only an upper but also a lower bound for the reaching time function of system (3.13). This can be seen by considering the time derivative of x_2 for the particular perturbation $\delta_2 = L \lceil x_1 \rfloor^0$, i.e.,

$$\dot{x}_2 = -(k_2 - L) \lceil x_1 \rfloor^0 . \tag{3.72}$$

One can see that the state x_2 changes with the rate $k_2 - L$ and thus cannot be reduced to zero in a time smaller than the one given in (3.71). Since the upper reaching time bound $\overline{T}_{K,L}(x)$ tends to a lower reaching time bound as k_1 tends to infinity, both bounds must coincide and be equal to the reaching time $T_{K,L}(x_0)$.

These asymptotic properties are summarized in the following theorem.

Theorem 3.4 (Asymptotics of Reaching Time Function Based Estimate) *The upper bound $\overline{T}_{K,L}(x)$ given in* (3.65) *of the reaching time function $T_{K,L}(x)$ of system (3.13) satisfies the asymptotic relations*

$$\lim_{K,L \to 0} \overline{T}_{K,L}(x) = \lim_{K,L \to 0} T_{K,L}(x), \tag{3.73a}$$

$$\lim_{k_1 \to \infty} \overline{T}_{K,L}(x) = \lim_{k_1 \to \infty} T_{K,L}(x) = \frac{|x_2|}{k_2 - L}, \tag{3.73b}$$

i.e., it is asymptotically exact when either the perturbation bounds K and L tend to zero or the parameter k_1 tends to infinity.

For certain values of the state x and vanishing perturbation δ_1, relation (3.73b) does not only hold asymptotically, but even for finite values of k_1. The following theorem states this fact; for the proof, the reader is referred to [15, Theorem 5].

Theorem 3.5 *Let the state $x = \begin{bmatrix} x_1 & x_2 \end{bmatrix}^{\mathrm{T}}$ be given and suppose that $x_2 \neq 0$ and $x_1 x_2 \geq 0$ hold, i.e., that either x_1 is zero or that x_1 and x_2 have the same sign. If the parameter k_1 satisfies $k_1 \geq \underline{k}_1(\lceil x_1 \rfloor^{-\frac{1}{2}} x_2)$ for $x_1 \neq 0$, or $k_1 \geq \underline{k}_1(\infty)$ for $x_1 = 0$ with the function $\underline{k}_1$ given by*

$$\underline{k}_1(c) = \begin{cases} \sqrt{8k_2} & c \geq \sqrt{2k_2} \\ \frac{2k_2}{c} + c & \sqrt{2k_2} > c > 0, \end{cases} \tag{3.74}$$

then, the reaching time function of system (3.13) for vanishing perturbation bound K is given by

$$T_{0,L}(x) = \frac{|x_2|}{k_2 - L}. \tag{3.75}$$

3.4 Estimation Based on a Lyapunov Function Family

In the previous section, the reaching time function was used as a Lyapunov function candidate to obtain reaching time bounds for the perturbed system. The same procedure may be used with different Lyapunov functions of system (3.13). One very popular family of Lyapunov functions that are based on the quasi-linear form (3.26) of the system is a family of quadratic functions, which are proposed in [10]. Reaching time estimates obtained with this Lyapunov function family are studied, e.g., in [2]. This section discusses that approach and studies the corresponding permitted perturbation bounds and asymptotic behavior. Furthermore, a method to choose the Lyapunov function from the considered family in an optimal way, such as to make the estimate as tight as possible, is proposed.

It is worth to point out that the approach discussed here is not the only one that is possible with the considered Lyapunov function family. The reader is referred to [14] for a more recent, alternative technique, which typically yields significantly tighter estimates, though at a higher computational cost of solving optimization problems constrained by linear matrix inequalities.

3.4.1 Quadratic Lyapunov Function Family

In [10], a family of quadratic Lyapunov functions of the form

$$V(x) = g(x)^{\mathrm{T}} P g(x) = z^{\mathrm{T}} P z \tag{3.76}$$

with a positive definite matrix $P \in \mathbb{R}^{2\times 2}$ is proposed for system (3.13). In the following, its time derivative along the trajectories of the unperturbed and the perturbed system is discussed.

3.4.1.1 Unperturbed System

Using the quasi-linear system representation (3.26), the time derivative $\dot{V}$ of V along the trajectories of the *unperturbed* system is found to be

$$\dot{V} = \dot{z}^{\mathrm{T}} P z + z^{\mathrm{T}} P \dot{z} = \frac{1}{|z_1|} z^{\mathrm{T}} (A^{\mathrm{T}} P + P A) z = -\frac{1}{|z_1|} z^{\mathrm{T}} Q z, \tag{3.77}$$

where the matrix $Q \in \mathbb{R}^{2\times 2}$ is defined as $Q := -A^{\mathrm{T}}P - PA$. Thus, $\dot{V} < 0$ holds for all nonzero values of z, if Q is positive definite. One can see that the matrices P and Q satisfy the Lyapunov equation

$$A^{\mathrm{T}}P + PA + Q = 0. \tag{3.78}$$

Provided that A is Hurwitz, i.e., that the parameters k_1 and k_2 are positive, a positive definite matrix P may be found for any given positive definite matrix Q. This yields a family of Lyapunov functions for the unperturbed system.

3.4.1.2 Perturbed System

For the perturbed system, the construction of such a Lyapunov function is slightly more involved. Along the perturbed system's trajectories, the time derivative of V as defined in (3.76) is given by

$$\dot{V} = z^{\mathrm{T}}(A^{\mathrm{T}}P + PA)z + \frac{1}{2}\delta_1(e_1^{\mathrm{T}}Pz + z^{\mathrm{T}}Pe_1) + \delta_2(e_2^{\mathrm{T}}Pz + z^{\mathrm{T}}Pe_2). \tag{3.79}$$

One can show that this expression may be upper bounded by

$$\dot{V} \leq -\frac{1}{|z_1|}z^{\mathrm{T}}Qz, \tag{3.80}$$

if there exist positive constants Θ_1, Θ_2 such that the positive definite matrix Q and the positive definite matrix P satisfy the Riccati inequality

$$A^{\mathrm{T}}P + PA + P\left(\frac{K}{4\Theta_1}e_1e_1^{\mathrm{T}} + \frac{L}{\Theta_2}e_2e_2^{\mathrm{T}}\right)P + (\Theta_1 K + \Theta_2 L)\,e_1e_1^{\mathrm{T}} + Q \leq 0. \tag{3.81}$$

A proof of this fact is given in the appendix.

3.4.2 *Reaching Time Estimation*

Using the Lyapunov function V, the reaching time may be estimated by finding an upper bound of $\dot{V}$ in terms of V. To that end, one may use the inequalities

$$|z_1| \leq \|z\|_2, \qquad z^{\mathrm{T}}Qz \geq \lambda_{\min}(Q)\,\|z\|_2^2, \qquad \lambda_{\max}(P)\,\|z\|_2^2 \geq z^{\mathrm{T}}Pz, \tag{3.82}$$

where $\lambda_{\min}(Q)$ and $\lambda_{\max}(P)$ denote the smallest eigenvalue of Q and the largest eigenvalue of P, respectively. By successively using each of these relations, the inequality

$$\dot{V} \leq -\frac{1}{|z_1|} z^{\mathrm{T}} Q z \leq -\frac{1}{\|z\|_2} z^{\mathrm{T}} Q z \leq -\lambda_{\min}(Q) \|z\|_2 \leq -\frac{\lambda_{\min}(Q)}{\sqrt{\lambda_{\max}(P)}} \sqrt{z^{\mathrm{T}} P z} \tag{3.83}$$

is obtained from (3.80). Introducing the abbreviation

$$\eta := \frac{\lambda_{\min}(Q)}{\sqrt{\lambda_{\max}(P)}} \tag{3.84}$$

one can see that V satisfies the differential inequality[8] $\dot{V} \leq -\eta\sqrt{V}$. Separating variables and integrating this inequality yield

$$\begin{aligned} \int_{V(x_0)}^{V(x(T))} \frac{1}{\sqrt{V}} \, \mathrm{d}V &\leq -\int_0^T \eta \, \mathrm{d}t \\ 2\sqrt{V(x(T))} - 2\sqrt{V(x_0)} &\leq -\eta T \end{aligned} \tag{3.85}$$

for any time instant T. Since $x(T)$ and, thus, also $V(x(T))$ are zero for T equal to the reaching time $T_{K,L}(x_0)$, one obtains the bound

$$T_{K,L}(x_0) \leq \frac{2}{\eta} \sqrt{V(x_0)}. \tag{3.86}$$

Substitution of η from (3.84) thus yields

$$T_{K,L}(x_0) \leq \frac{2\sqrt{\lambda_{\max}(P)}}{\lambda_{\min}(Q)} \sqrt{V(x_0)}. \tag{3.87}$$

The following theorem summarizes this result.

Theorem 3.6 (Lyapunov Function Family Based Estimate) *Let the function $g(x)$ and the matrix $A \in \mathbb{R}^{2\times 2}$ be defined as in (3.25) and (3.27), respectively. Suppose that for given perturbation bounds K and L, there exist positive constants Θ_1, Θ_2 and positive definite matrices $P, Q \in \mathbb{R}^{2\times 2}$, such that the matrix Riccati inequality*

$$A^{\mathrm{T}} P + PA + P\left(\frac{K}{4\Theta_1} e_1 e_1^{\mathrm{T}} + \frac{L}{\Theta_2} e_2 e_2^{\mathrm{T}}\right) P + (\Theta_1 K + \Theta_2 L) e_1 e_1^{\mathrm{T}} + Q \leq 0 \tag{3.88}$$

holds. Then, the reaching time function $T_{K,L}(x)$ of the perturbed system (3.13) is bounded by $T_{K,L}(x) \leq \overline{T}_{K,L}(x)$, where $\overline{T}_{K,L}(x)$ is given by

$$\overline{T}_{K,L}(x) = \frac{2\sqrt{\lambda_{\max}(P)}}{\lambda_{\min}(Q)} \sqrt{g(x)^{\mathrm{T}} P g(x)}. \tag{3.89}$$

[8] In [2], the same inequality is used, but with $\eta = \lambda_{\min}(Q)\sqrt{\lambda_{\max}(P)}^{-1}\sqrt{\lambda_{\min}(P)\lambda_{\max}(P)^{-1}}$. The resulting reaching time bounds are more conservative, i.e., larger, than those obtained here.

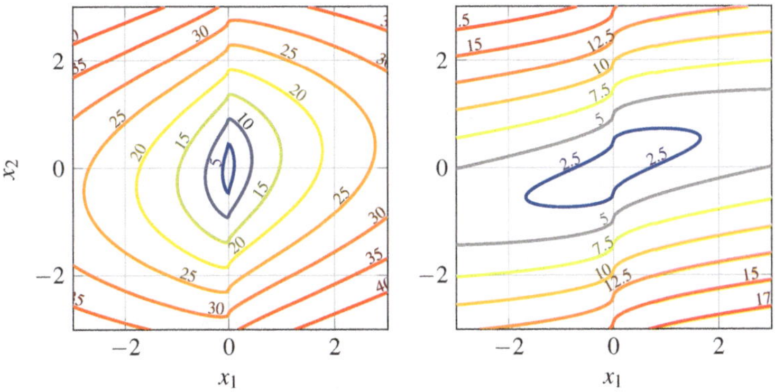

Fig. 3.3 Level lines of the unperturbed system's Lyapunov function family based reaching time estimate $\overline{T}_{0,0}(x)$ for $k_2 = 1, k_1 = 0.4\sqrt{k_2}$ (left) and $k_2 = 1, k_1 = 4\sqrt{k_2}$ (right)

Remark 3.1 Note that unlike the other two estimates considered in Sects. 3.3 and 3.5 of this chapter, this estimate does *not* exhibit the homogeneous scaling property (3.23), (3.24) described in Sect. 3.2.4 with respect to the parameters k_1, k_2, K, and L. Therefore, different reaching time bounds may be obtained by scaling the parameters according to (3.23) and computing the bound for those scaled parameters. For a more recent technique based on the family of quadratic Lyapunov functions, which does not exhibit this behavior and always yields the tightest possible bound, the reader is referred to [14].

Figure 3.3 shows level lines of the estimate $\overline{T}_{0,0}(x)$ for the unperturbed system[9] for $k_2 = 1$ with the two parameter settings $k_1 = 0.4\sqrt{k_2}$ and $k_1 = 4\sqrt{k_2}$, i.e., for real-valued and complex-valued eigenvalues of the matrix A, respectively. A numerical example to illustrate the computation of the reaching time bound given in this theorem, and to provide further insight on how to choose the Lyapunov function, i.e., the matrix P is given in Sect. 3.4.4.

3.4.3 Range of Permitted Perturbations

In the unperturbed case, a Lyapunov function exists for any values of the positive parameters k_1 and k_2. For the perturbed system, the existence of a Lyapunov function imposes additional conditions on the perturbation bounds K, L and the parameters k_1, k_2. As in Sect. 3.3.3, these conditions are connected to the transfer functions $G_1(s)$ and $G_2(s)$ introduced in (3.60), i.e.,

[9]The matrix P to compute the unperturbed estimate depicted in Fig. 3.3 was obtained by solving the Lyapunov equation (3.78) with $Q = I$.

$$G_1(s) = \frac{s}{2s^2 + k_1 s + k_2}, \qquad G_2(s) = \frac{1}{2s^2 + k_1 s + k_2}. \tag{3.90}$$

To see this connection, the Riccati inequality (3.88) may be rewritten as

$$A^{\mathrm{T}} P + PA + PBB^{\mathrm{T}} P + C^{\mathrm{T}} C + Q \leq 0 \tag{3.91}$$

with abbreviations $B \in \mathbb{R}^{2\times 2}$ and $C \in \mathbb{R}^{1\times 2}$ given by

$$B := \left[\sqrt{\tfrac{K}{\Theta_1}}\tfrac{1}{2}e_1 \; \sqrt{\tfrac{L}{\Theta_2}}e_2\right], \qquad C := e_1^{\mathrm{T}}\sqrt{\Theta_1 K + \Theta_2 L}. \tag{3.92}$$

It is well known from system theory, see [3, 19], that positive definite matrices P and Q satisfying this inequality exist if and only if the $\mathscr{H}_\infty$ norm of the transfer function matrix

$$\begin{aligned} G(s) &:= C(sI - A)^{-1}B \\ &= \sqrt{\Theta_1 K + \Theta_2 L}\left[\sqrt{\tfrac{K}{\Theta_1}}\tfrac{1}{2}e_1^{\mathrm{T}}(sI - A)^{-1}e_1 \; \sqrt{\tfrac{L}{\Theta_2}}e_1^{\mathrm{T}}(sI - A)^{-1}e_2\right] \\ &= \sqrt{\Theta_1 K + \Theta_2 L}\left[\sqrt{\tfrac{K}{\Theta_1}}G_1(s) \; \sqrt{\tfrac{L}{\Theta_2}}G_2(s)\right] \end{aligned} \tag{3.93}$$

is less than one, i.e., if and only if

$$\|G\|_{\mathscr{H}_\infty} < 1 \tag{3.94}$$

holds.

In the following, a sufficient condition as well as a necessary condition for the existence of positive scalars Θ_1, Θ_2 such that $G(s)$ satisfies this inequality is derived. Afterward, a necessary and sufficient condition for (3.94) to hold for given values of Θ_1 and Θ_2 is shown.

To obtain the sufficient condition, Θ_1 and Θ_2 are chosen as

$$\Theta_1 = \|G_1\|_{\mathscr{H}_\infty} = \sup_\omega |G_1(\mathrm{j}\omega)|, \qquad \Theta_2 = \|G_2\|_{\mathscr{H}_\infty} = \sup_\omega |G_2(\mathrm{j}\omega)|. \tag{3.95}$$

The $\mathscr{H}_\infty$ norm of $G(s)$ is then bounded by

$$\begin{aligned} \|G\|_{\mathscr{H}_\infty} = \sup_\omega \|G(\mathrm{j}\omega)\|_2 &= \sup_\omega \sqrt{\Theta_1 K + \Theta_2 L}\sqrt{\frac{K}{\Theta_1}|G_1(\mathrm{j}\omega)|^2 + \frac{L}{\Theta_2}|G_2(\mathrm{j}\omega)|^2} \\ &\leq \sqrt{\Theta_1 K + \Theta_2 L}\sqrt{\frac{K}{\Theta_1}\|G_1\|_{\mathscr{H}_\infty}^2 + \frac{L}{\Theta_2}\|G_2\|_{\mathscr{H}_\infty}^2} \\ &= \sqrt{\left(K\|G_1\|_{\mathscr{H}_\infty} + L\|G_2\|_{\mathscr{H}_\infty}\right)^2}. \end{aligned} \tag{3.96}$$

Thus, (3.94) is fulfilled if the perturbation bounds satisfy the inequality

$$K\overline{K}^{-1} + L\overline{L}^{-1} < 1 \tag{3.97}$$

with the abbreviations

$$\overline{K} := \frac{1}{\|G_1\|_{\mathscr{H}_\infty}} = \frac{1}{\sup_\omega |G_1(\mathrm{j}\omega)|}, \quad \overline{L} := \frac{1}{\|G_2\|_{\mathscr{H}_\infty}} = \frac{1}{\sup_\omega |G_2(\mathrm{j}\omega)|}. \tag{3.98}$$

Furthermore, similar to the conditions in Sect. 3.3.3, $K < \overline{K}$ and $L < \overline{L}$ are necessary conditions for (3.94) to be fulfilled for *any* choice of Θ_1, Θ_2. This can be seen from the fact that, if either of these two conditions is violated, then one of the relations

$$\|G\|_{\mathscr{H}_\infty} \geq \|G(\mathrm{j}\omega)\|_2 \geq \sqrt{\Theta_1 K}\sqrt{\frac{K}{\Theta_1}|G_1(\mathrm{j}\omega)|^2} = K\,|G_1(\mathrm{j}\omega)| = K\overline{K}^{-1} \geq 1, \tag{3.99}$$

$$\|G\|_{\mathscr{H}_\infty} \geq \|G(\mathrm{j}\omega)\|_2 \geq \sqrt{\Theta_2 L}\sqrt{\frac{L}{\Theta_2}|G_2(\mathrm{j}\omega)|^2} = L\,|G_2(\mathrm{j}\omega)| = L\overline{L}^{-1} \geq 1 \tag{3.100}$$

holds for some value of ω.

The values of $\overline{K}$ and $\overline{L}$ may be obtained analytically by means of a straightforward computation, which is found in the appendix. They are given, along with the obtained conditions, in the following theorem.

Theorem 3.7 (Lyapunov Function Family Based Perturbation Bounds) *Let the transfer functions $G_1(s)$ and $G_2(s)$ be defined as in* (3.60) *and define the constants*

$$\overline{K} = \frac{1}{\sup_\omega |G_1(j\omega)|} = k_1, \tag{3.101a}$$

$$\overline{L} = \frac{1}{\sup_\omega |G_2(j\omega)|} = \begin{cases} k_2 & k_1^2 \geq 4k_2 \\ \frac{k_1\sqrt{8k_2 - k_1^2}}{4} & k_1^2 < 4k_2. \end{cases} \tag{3.101b}$$

Positive constants Θ_1, Θ_2 and positive definite matrices $P, Q \in \mathbb{R}^{2\times 2}$ satisfying the conditions of Theorem 3.6 exist if *the perturbation bounds K and L fulfill*

$$K\overline{K}^{-1} + L\overline{L}^{-1} < 1 \tag{3.102}$$

and they exist only if *$K < \overline{K}$ and $L < \overline{L}$ hold.*

Note that this theorem's condition is necessary and sufficient when either K or L are zero. Furthermore, a necessary and sufficient condition may be derived for the existence of positive definite matrices P and Q that satisfy the Riccati inequality (3.88) for given constants Θ_1, Θ_2. This derivation is found in the appendix; it yields in the following theorem.

Theorem 3.8 *For given positive constants Θ_1 and Θ_2, there exist positive definite matrices $P, Q \in \mathbb{R}^{2\times 2}$ satisfying the conditions of Theorem 3.6* if and only if *the positive parameters k_1, k_2 and perturbation bounds K, L satisfy*

$$k_2^2 > L^2 + \frac{\Theta_1}{\Theta_2} K L, \tag{3.103a}$$

$$k_1^2 > K^2 + \frac{\Theta_2}{\Theta_1} K L + 4k_2 - 4\sqrt{k_2^2 - L^2 - \frac{\Theta_1}{\Theta_2} K L}. \tag{3.103b}$$

Remark 3.2 Note that the conditions of this theorem are independent of Θ_1 and Θ_2 if either K or L are zero; in this case, they are given by

$$k_2 > L, \qquad k_1 > 2\sqrt{k_2 - \sqrt{k_2^2 - L^2}}, \qquad \text{for } K = 0, \tag{3.104a}$$

$$k_2 > 0, \qquad k_1 > K, \qquad \text{for } L = 0. \tag{3.104b}$$

Example 3.4 To show the numerical computation of the bounds in Theorem 3.7 and to compare them with those previously computed in Example 3.3 for the reaching time function based approach, the parameter values $k_1 = 2$, $k_2 = 1$ are considered. Since $k_1^2 = 4k_2$ in this case, Theorem 3.7 yields the perturbation bounds

$$\overline{K} = k_1 = 2, \qquad \overline{L} = k_2 = 1. \tag{3.105}$$

For $K = 1$ and $L = 0.5$, for example, the existence of a Lyapunov function thus cannot be concluded from Theorem 3.7, because (3.102) is not satisfied. Nonetheless, a Lyapunov function exists in this case, because the conditions (3.103) of Theorem 3.8, i.e.,

$$1 = k_2^2 > 0.25 + \frac{\Theta_1}{\Theta_2} 0.5, \qquad 4 = k_1^2 > 1 + \frac{\Theta_2}{\Theta_1} 0.5 + 4 - 4\sqrt{1 - 0.25 - \frac{\Theta_1}{\Theta_2} 0.5}, \tag{3.106}$$

are fulfilled for $\Theta_2 = 2\Theta_1$, for example. Indeed, one finds that the Riccati inequality (3.88) is satisfied for, e.g.,

$$P = \begin{bmatrix} 44 & -10 \\ -10 & 24 \end{bmatrix}, \qquad Q = \begin{bmatrix} 1 & 0 \\ 0 & 1 \end{bmatrix}, \qquad \Theta_1 = 20, \qquad \Theta_2 = 40. \tag{3.107}$$

This shows that condition (3.102) of Theorem 3.7 is only sufficient, but not necessary.

3.4.4 Optimal Lyapunov Function Choice

Computing the reaching time bound (3.89) requires the choice of a Lyapunov function by selecting positive definite matrices P and Q as well as positive constants Θ_1 and Θ_2 that satisfy the Riccati inequality (3.88). The conservativeness of the estimate crucially depends on this choice, in particular, on P and Q, as the following example shows.

Example 3.5 Consider system (3.13) with parameter values $k_1 = 2$, $k_2 = 1$ and without perturbations, i.e., with $K = L = 0$. Two pairs of matrices $P^{(1)}$, $Q^{(1)}$ and $P^{(2)}$, $Q^{(2)}$ for this parameter setting that satisfy the Lyapunov equation (3.78), and thus also the Riccati inequality (3.88), are given by

$$P^{(1)} = \begin{bmatrix} 2 & -1 \\ -1 & 2 \end{bmatrix}, \qquad P^{(2)} = \begin{bmatrix} 11 & -10 \\ -10 & \frac{31}{2} \end{bmatrix}, \tag{3.108}$$

$$Q^{(1)} = \begin{bmatrix} 2 & 0 \\ 0 & 1 \end{bmatrix}, \qquad Q^{(2)} = \begin{bmatrix} 2 & 0 \\ 0 & 10 \end{bmatrix}. \tag{3.109}$$

Using (3.84), the corresponding values of $\eta = \lambda_{\min}(Q)\sqrt{\lambda_{\max}(P)}^{-1}$ are obtained as

$$\eta^{(1)} = \frac{1}{\sqrt{3}} \approx 0.5774, \qquad \eta^{(2)} = \frac{2\sqrt{2}}{\sqrt{47}} \approx 0.4126. \tag{3.110}$$

The estimate (3.89) for the initial state $x_0 = e_2$, i.e., $\overline{T}_{0,0}(e_2) = 2\eta^{-1}\sqrt{g(e_2)^{\mathrm{T}} P g(e_2)}$, is thus given by

$$\overline{T}^{(1)}_{0,0}(e_2) = \frac{2}{\eta^{(1)}}\sqrt{2} \approx 4.8990, \quad \overline{T}^{(2)}_{0,0}(e_2) = \frac{2}{\eta^{(2)}}\sqrt{\frac{31}{2}} \approx 19.0853. \tag{3.111}$$

The latter estimate is almost four times larger than the former one, despite the value of η being of similar magnitude. A suitable selection of the Lyapunov function is, therefore, crucial for obtaining good, i.e., small, upper bounds.

In the following, an optimization-based strategy for finding matrices P and Q is proposed. In particular, the problem of minimizing the upper reaching time bound for all initial states x_0 in a subset $\mathscr{R}(\varepsilon)$ of the state space of the form

$$\mathscr{R}(\varepsilon) := \{x_0 \in \mathbb{R}^2 : \|g(x_0)\|_2 \leq \varepsilon\} \tag{3.112}$$

with $g(x)$ as given in (3.25) is considered, where ε is a positive constant. The largest reaching time bound obtained for initial states from this set is denoted by $\hat{T}_{K,L}(\varepsilon)$ and is given by

$$\hat{T}_{K,L}(\varepsilon) = \max_{x_0 \in \mathscr{R}(\varepsilon)} \overline{T}_{K,L}(x_0) = \max_{\|z_0\|_2 \le \varepsilon} \frac{2\sqrt{\lambda_{\max}(P)}}{\lambda_{\min}(Q)} \sqrt{z_0^{\mathrm{T}} P z_0} = 2\varepsilon \frac{\lambda_{\max}(P)}{\lambda_{\min}(Q)}. \tag{3.113}$$

In the following, the matrices P and Q are to be selected such that this expression is minimized.[10]

In the unperturbed case, i.e., if P and Q fulfill the Lyapunov equation (3.78), it is well known that the minimum is obtained for $Q = I$, see [2, 6]. In the perturbed case, the minimum may be found by solving a semidefinite program. Recall that a matrix M given by

$$M = \begin{bmatrix} D & E^{\mathrm{T}} \\ E & F \end{bmatrix} \tag{3.114}$$

is positive semidefinite if and only if the matrix $(I - FF^{+})E$ is zero, and the matrices F and $D - E^{\mathrm{T}} F^{+} E$ are positive semidefinite,[11] see, e.g., [5]. The latter expression is the so-called generalized Schur complement of F in M. By applying this fact with matrices $D, E, F \in \mathbb{R}^{2\times 2}$ given by

$$D = -A^{\mathrm{T}} P - PA - (\Theta_1 K + \Theta_2 L) e_1 e_1^{\mathrm{T}} - Q, \tag{3.115a}$$

$$E = \begin{bmatrix} K e_1^{\mathrm{T}} P \\ L e_2^{\mathrm{T}} P \end{bmatrix}, \qquad F = \begin{bmatrix} 4\Theta_1 K & 0 \\ 0 & \Theta_2 L \end{bmatrix}, \tag{3.115b}$$

the nonlinear Riccati inequality (3.88) may be rewritten as the linear matrix inequality

$$M(P, Q, \Theta_1, \Theta_2) := \begin{bmatrix} -A^{\mathrm{T}} P - PA - (\Theta_1 K + \Theta_2 L) e_1 e_1^{\mathrm{T}} - Q & K P e_1 & L P e_2 \\ K e_1^{\mathrm{T}} P & 4\Theta_1 K & 0 \\ L e_2^{\mathrm{T}} P & 0 & \Theta_2 L \end{bmatrix} \ge 0 \tag{3.116}$$

with $M(P, Q, \Theta_1, \Theta_2) \in \mathbb{R}^{4\times 4}$. Note that both, this inequality and the objective function, are invariant to any scaling of the matrices P, Q and the scalars Θ_1, Θ_2 by a positive factor γ, because

$$M(\gamma P, \gamma Q, \gamma \Theta_1, \gamma \Theta_2) = \gamma M(P, Q, \Theta_1, \Theta_2), \tag{3.117a}$$

$$2\varepsilon \frac{\lambda_{\max}(\gamma P)}{\lambda_{\min}(\gamma Q)} = 2\varepsilon \frac{\gamma \lambda_{\max}(P)}{\gamma \lambda_{\min}(Q)} = 2\varepsilon \frac{\lambda_{\max}(P)}{\lambda_{\min}(Q)} \tag{3.117b}$$

holds. Thus, adding the constraint

$$\lambda_{\min}(Q) \ge 1 \tag{3.118}$$

does not change the attainable optimum, because one can always ensure that this inequality is satisfied by scaling Q along with all other variables by suitably choosing

---

[10] Note that a similar minimization problem is considered in [2], though with a value of η that is different from the one given in (3.84).

[11] The matrix F^{+} denotes the pseudo-inverse of the matrix F.

the value of γ. Introducing, additionally, an upper bound λ for the largest eigenvalue of P by adding

$$\lambda_{\max}(P) \leq \lambda \tag{3.119}$$

as a constraint, the expression to be minimized may be bounded from above as

$$2\varepsilon \frac{\lambda_{\max}(P)}{\lambda_{\min}(Q)} \leq 2\varepsilon \frac{\lambda}{\lambda_{\min}(Q)} \leq 2\varepsilon\lambda. \tag{3.120}$$

This bound is tight when equality holds in the constraints (3.118) and (3.119); therefore, minimizing λ subject to these two constraints in addition to (3.116) is equivalent to minimizing the original expression (3.113). Finding the optimal Lyapunov function is thus achieved by solving the semidefinite program given in the following theorem.

Theorem 3.9 *Let a scalar $\varepsilon > 0$ be given and let the vector-valued function g and the matrix-valued function M be defined as in* (3.25) *and* (3.116)*, respectively. Suppose that the positive definite matrices $P, Q \in \mathbb{R}^{2\times 2}$ and positive scalars Θ_1, Θ_2 are an optimal solution of the semidefinite program*

$$\hat{T}_{K,L}(\varepsilon) = \min_{P,Q,\Theta_1,\Theta_2,\lambda} 2\varepsilon\lambda$$
$$\text{subject to} \tag{3.121a}$$

$$P \geq 0, \qquad \lambda I \geq P, \qquad Q \geq I, \qquad M(P, Q, \Theta_1, \Theta_2) \geq 0. \tag{3.121b}$$

Then, P, Q, Θ_1, and Θ_2 fulfill the conditions of Theorem 3.6, the corresponding optimal value $\hat{T}_{K,L}(\varepsilon)$ is the largest value of the reaching time bound $\overline{T}_{K,L}(x)$ defined in (3.89) *in a region characterized by $\|g(x)\|_2 \leq \varepsilon$, i.e.,*

$$\hat{T}_{K,L}(\varepsilon) = \max_{\|g(x)\|_2 \leq \varepsilon} \overline{T}_{K,L}(x), \tag{3.122}$$

and this value $\hat{T}_{K,L}(\varepsilon)$ is minimal with respect to all matrices P and Q that fulfill the conditions of Theorem 3.6.

Example 3.6 The application of this theorem is illustrated by means of an example, using again the parameter values $k_1 = 2$, $k_2 = 1$ considered in Example 3.5. To that end, components of P and Q are denoted by $p_{11}, p_{12}, p_{22} \in \mathbb{R}$ and $q_{11}, q_{12}, q_{22} \in \mathbb{R}$, respectively, i.e.,

$$P = \begin{bmatrix} p_{11} & p_{12} \\ p_{12} & p_{22} \end{bmatrix}, \qquad Q = \begin{bmatrix} q_{11} & q_{12} \\ q_{12} & q_{22} \end{bmatrix}. \tag{3.123}$$

The matrix A defined in (3.27) is given by

$$A = \begin{bmatrix} -1 & \frac{1}{2} \\ -1 & 0 \end{bmatrix}, \tag{3.124}$$

and, thus, one obtains

$$A^{\mathrm{T}} P + P A = \begin{bmatrix} -2p_{12} - 2p_{11} & \frac{p_{11}}{2} - p_{22} - p_{12} \\ \frac{p_{11}}{2} - p_{22} - p_{12} & p_{12} \end{bmatrix}. \tag{3.125}$$

The matrix $M \in \mathbb{R}^{4\times 4}$ defined in (3.116) is then given by

$$M = \begin{bmatrix} 2p_{12} + 2p_{11} - \Theta_1 K - \Theta_2 L - q_{11} & p_{22} + p_{12} - \frac{p_{11}}{2} - q_{12} & K p_{11} & L p_{12} \\ p_{22} + p_{12} - \frac{p_{11}}{2} - q_{12} & -p_{12} - q_{22} & K p_{12} & L p_{22} \\ K p_{11} & K p_{12} & 4\Theta_1 K & 0 \\ L p_{12} & L p_{22} & 0 & \Theta_2 L \end{bmatrix}. \tag{3.126}$$

The optimization problem (3.121) was solved numerically for $\varepsilon = 1$ to obtain a bound for the unperturbed case $K = L = 0$ and one perturbed case with $L = 0.5$ and $K = 0$. To that end, the objective function 2λ was minimized subject to the constraints $\Theta_1, \Theta_2 \geq 0$ and the linear matrix inequalities $\lambda I \geq P \geq 0$, $Q \geq I$, and $M \geq 0$. For this purpose, YALMIP, see [9], with the solver SEDUMI [17] were used. For the unperturbed and the perturbed case, one obtains the numerical values

$$\hat{T}_{0,0}(1) \approx 5.2656 \quad \text{and} \quad \hat{T}_{0,0.5}(1) \approx 11.7277, \tag{3.127}$$

respectively, which in each case constitute an upper reaching time bound for all initial states x_0 satisfying $\|g(x_0)\|_2 \leq 1$. The corresponding optimal solutions of P, Q, and Θ_1, Θ_2 are

$$P \approx \begin{bmatrix} 1.5000 & -1.000 \\ -1.000 & 1.7500 \end{bmatrix}, \qquad Q \approx \begin{bmatrix} 1 & 0 \\ 0 & 1 \end{bmatrix}, \tag{3.128a}$$

with arbitrary $\Theta_1, \Theta_2 > 0$ for the unperturbed case $K = 0$, $L = 0$, and

$$P \approx \begin{bmatrix} 4.3302 & -1.8711 \\ -1.8711 & 3.5810 \end{bmatrix}, \quad Q \approx \begin{bmatrix} 1 & 0 \\ 0 & 1 \end{bmatrix}, \quad \Theta_2 \approx 7.3605 \tag{3.128b}$$

with any $\Theta_1 > 0$ for the perturbed case $K = 0$, $L = 0.5$. Using these two solutions to compute reaching time bounds for the initial state $x_0 = e_2$ by means of Theorem 3.6 yields

$$\overline{T}_{0,0}(e_2) \approx 4.2930 \quad \text{and} \quad \overline{T}_{0,0.5}(e_2) \approx 9.1649 \tag{3.129}$$

for the unperturbed and the perturbed case, respectively.

3.4.5 Asymptotic Properties

Similar to the considerations in Sect. 3.3.4, the Lyapunov-based estimate's asymptotic behavior for increasing k_1 is now studied. To that end, denote the components of P by p_{11}, p_{12}, p_{22} and introduce a vector v according to

$$P = \begin{bmatrix} p_{11} & p_{12} \\ p_{12} & p_{22} \end{bmatrix}, \qquad v := \begin{bmatrix} 1 \\ k_1 \end{bmatrix}. \tag{3.130}$$

From the Riccati inequality (3.88), one then obtains the inequalities

$$\begin{aligned} 0 &\geq e_2^{\mathrm{T}}(A^{\mathrm{T}}P + PA + PBB^{\mathrm{T}}P + C^{\mathrm{T}}C + Q)e_2 \\ &\geq e_2^{\mathrm{T}}(A^{\mathrm{T}}P + PA + Q)e_2 = p_{12}, \end{aligned} \tag{3.131}$$

$$\begin{aligned} 0 &\geq v^{\mathrm{T}}(A^{\mathrm{T}}P + PA + PBB^{\mathrm{T}}P + C^{\mathrm{T}}C + Q)v \\ &\geq v^{\mathrm{T}}(A^{\mathrm{T}}P + PA + \frac{L}{\Theta_2}Pe_2e_2^{\mathrm{T}}P + \Theta_2 Le_1e_1^{\mathrm{T}} + Q)v \\ &= -2k_2(k_1p_{22} + p_{12}) + L\left[\frac{(k_1p_{22} + p_{12})^2}{\Theta_2} + \Theta_2\right] + v^{\mathrm{T}}Qv \\ &\geq -2k_2(k_1p_{22} + p_{12}) + 2L(k_1p_{22} + p_{12}) + v^{\mathrm{T}}Qv. \end{aligned} \tag{3.132}$$

Combining them shows that

$$\begin{aligned} 0 &\geq -2(k_2 - L)(k_1p_{22} + p_{12}) + v^{\mathrm{T}}Qv \geq -2k_1(k_2 - L)p_{22} + \|v\|_2^2\lambda_{\min}(Q) \\ &= -2k_1(k_2 - L)e_2^{\mathrm{T}}Pe_2 + (1 + k_1^2)\lambda_{\min}(Q) \end{aligned} \tag{3.133}$$

holds, and one finds that the estimate $\overline{T}_{K,L}$ given in (3.89) is bounded from below by

$$\overline{T}_{K,L}(e_2) = \frac{2\sqrt{\lambda_{\max}(P)}}{\lambda_{\min}(Q)}\sqrt{e_2^{\mathrm{T}}Pe_2} \geq \frac{2e_2^{\mathrm{T}}Pe_2}{\lambda_{\min}(Q)} \geq \frac{2(1 + k_1^2)}{2k_1(k_2 - L)} = \frac{k_1 + \frac{1}{k_1}}{k_2 - L}. \tag{3.134}$$

Therefore, in the worst case, the estimate diverges with increasing k_1. The following theorem summarizes this property.

Theorem 3.10 (Asymptotics of Lyapunov Function Family Based Estimate) *For any positive definite matrices $P, Q \in \mathbb{R}^{2\times 2}$ and any positive scalars Θ_1, Θ_2 that fulfill the conditions of Theorem 3.6, the upper bound $\overline{T}_{K,L}(x)$ of the reaching time function of system (3.13) that is given in* (3.89) *satisfies*

$$\max_{\|x\|_2 \leq 1} \overline{T}_{K,L}(x) \geq \overline{T}_{K,L}(e_2) \geq \frac{k_1 + \frac{1}{k_1}}{k_2 - L}. \tag{3.135}$$

3.5 Estimation Based on a Majorant Curve

Another approach to estimate the reaching time is based on a geometric stability proof for the algorithm, which was originally proposed in [7, 8]. In [13], a necessary and sufficient stability condition in analytic form is derived from this proof. In this section, a reaching time estimate based on these results is proposed.

Compared to the previous approaches, however, an additional restriction is imposed: System (3.13) is considered only with initial states that lie on the x_2-axis, i.e., it is assumed that x_0 has the form

$$x_0 = \begin{bmatrix} 0 \; x_{0,2} \end{bmatrix}^{\mathrm{T}} \tag{3.136}$$

with $x_{0,2} \in \mathbb{R}$. Note that it is in principle possible to remove this restriction, but the required computations are quite tedious. Thus, they are thus not considered here. As discussed in Sect. 3.1, such an assumption is reasonable, however, when using the super-twisting algorithm in the form of the first-order robust exact differentiator, which is given in (3.5). To see this, recall that in such a case the super-twisting algorithm's state variables x_1, x_2 are given by the observer errors

$$x_1 = y_1 - f, \qquad\qquad x_2 = y_2 - \dot{f}, \tag{3.137}$$

where y_1, y_2 are the states of the differentiator (3.5) and f is the function to be differentiated. By choosing the initial state for y_1 as $y_1(0) = f(0)$, one can ensure that the initial state x_0 of system (3.13) has the required form (3.136).

3.5.1 *Geometric Stability Proof*

The geometric stability proof in [8, 13] is based on the computation of a so-called majorant curve of the system. This curve corresponds to a trajectory for a worst-case disturbance. In this section, it is first shown how to obtain this majorant curve. Then the resulting necessary and sufficient stability condition is given.

3.5.1.1 Majorant Curve

Loosely speaking, a majorant curve is found by selecting the perturbations δ_1, δ_2 such as to destabilize the system as much as possible. This yields a worst-case trajectory in the form of a curve that envelopes the trajectories obtained for all other perturbations. A curve with this property is called a majorant curve of the system. The system's origin is finite-time stable if and only if this worst-case trajectory converges to the origin in finite time.

To make this argument more formal, let $\overline{\delta}_1, \overline{\delta}_2$ denote the worst-case perturbation to be chosen. The normal vector $n \in \mathbb{R}^2$ of the corresponding trajectory, i.e., of the majorant curve, is obtained as

$$n := \begin{bmatrix} -\left.\dot{x}_2\right|_{\delta_2=\overline{\delta}_2} \\ \left.\dot{x}_1\right|_{\delta_1=\overline{\delta}_1} \end{bmatrix} = \begin{bmatrix} k_2 \lceil x_1 \rfloor^0 - \overline{\delta}_2 \\ -k_1 \lceil x_1 \rfloor^{\frac{1}{2}} + x_2 + \overline{\delta}_1 \end{bmatrix} = \begin{bmatrix} -\dot{x}_2 + (\delta_2 - \overline{\delta}_2) \\ \dot{x}_1 - (\delta_1 - \overline{\delta}_1) \left|x_1\right|^{\frac{1}{2}} \end{bmatrix}. \quad (3.138)$$

The scalar product of this normal vector n and the tangent vector of the trajectory $x(t)$, i.e., its time derivative $\dot{x}$, is given by

$$\begin{aligned} n^{\mathrm{T}}\dot{x} &= \begin{bmatrix} -\left.\dot{x}_2\right|_{\delta_2=\overline{\delta}_2} & \left.\dot{x}_1\right|_{\delta_1=\overline{\delta}_1} \end{bmatrix} \begin{bmatrix} \dot{x}_1 \\ \dot{x}_2 \end{bmatrix} \\ &= \begin{bmatrix} -\left.\dot{x}_2\right|_{\delta_2=\overline{\delta}_2} & \left.\dot{x}_1\right|_{\delta_1=\overline{\delta}_1} \end{bmatrix} \begin{bmatrix} \left.\dot{x}_1\right|_{\delta_1=\overline{\delta}_1} + (\delta_1 - \overline{\delta}_1) \left|x_1\right|^{\frac{1}{2}} \\ \left.\dot{x}_2\right|_{\delta_2=\overline{\delta}_2} + (\delta_2 - \overline{\delta}_2) \end{bmatrix} \\ &= -(\delta_1 - \overline{\delta}_1) \left|x_1\right|^{\frac{1}{2}} \left.\dot{x}_2\right|_{\delta_2=\overline{\delta}_2} + (\delta_2 - \overline{\delta}_2) \left.\dot{x}_1\right|_{\delta_1=\overline{\delta}_1} \\ &= (\delta_1 - \overline{\delta}_1) \lceil x_1 \rfloor^{\frac{1}{2}} (k_2 - \overline{\delta}_2 \lceil x_1 \rfloor^0) + (\delta_2 - \overline{\delta}_2) \left.\dot{x}_1\right|_{\delta_1=\overline{\delta}_1} . \end{aligned} \quad (3.139)$$

If $n^{\mathrm{T}}\dot{x} \leq 0$ holds, then the majorant curve envelopes all trajectories with the same initial value, because trajectories cannot cross the latter unless being—loosely speaking—directed "inwards". Since

$$k_2 - \overline{\delta}_2 \lceil x_1 \rfloor^0 \geq k_2 - L > 0 \quad (3.140)$$

holds, one can see from (3.139) that this condition is satisfied for all perturbations δ_1, δ_2 bounded by (3.13c) if and only if $\overline{\delta}_1, \overline{\delta}_2$ are selected as

$$\overline{\delta}_1 = K \lceil x_1 \rfloor^0, \quad (3.141a)$$

$$\overline{\delta}_2 = L \left\lceil \left.\dot{x}_1\right|_{\delta_1=\overline{\delta}_1} \right\rfloor^0 = L \left\lceil -(k_1 - K) \lceil x_1 \rfloor^{\frac{1}{2}} + x_2 \right\rfloor^0 . \quad (3.141b)$$

Note that $n^{\mathrm{T}}\dot{x}$ is maximal (and zero) if δ_1 and δ_2 are equal to these worst-case perturbations $\overline{\delta}_1$ and $\overline{\delta}_2$.

Due to the homogeneity properties of the system, it is sufficient to consider a single worst-case trajectory with initial value $x_0 = e_2$. This majorant curve is illustrated in Fig. 3.4 along with several perturbed trajectories in the x_1-x_2 plane. As shown in the figure, the magnitude of x_2 at the first intersection of the majorant curve and the x_2 axis is denoted by Φ. Due to the homogeneity properties, x_2 then has magnitudes $\Phi^2, \Phi^3, \Phi^4, \ldots$ at the following intersections. Since the time it takes to reach those intersections also is proportional to the magnitude of x_2, $\Phi < 1$ is necessary and sufficient for the convergence of this majorant curve to the origin in finite time. Therefore, the origin of system (3.13) is finite-time stable if and only if $\Phi < 1$ holds.

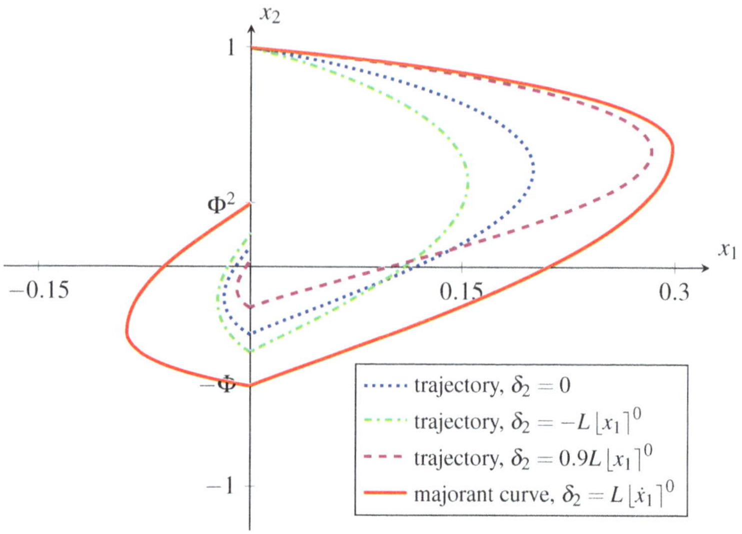

Fig. 3.4 Majorant curve and three trajectories up to their second intersection with the x_2-axis of system (3.13) for $k_1 = 1$, $k_2 = 1$, $K = 0$, $L = 0.5$ starting in $x_0 = e_2$

3.5.1.2 Necessary and Sufficient Stability Criterion

Computing the crucial quantity Φ requires a trajectory of system (3.13) with δ_1, δ_2 equal to the worst-case perturbations $\overline{\delta}_1, \overline{\delta}_2$ given in (3.141), i.e., of the system

$$\dot{x}_1 = -(k_1 - K)\lceil x_1 \rfloor^{\frac{1}{2}} + x_2, \tag{3.142a}$$

$$\dot{x}_2 = -k_2\lceil x_1 \rfloor^0 + L\lceil \dot{x}_1 \rfloor^0. \tag{3.142b}$$

For $K = 0$, this computation is done analytically[12] in [13]. The required computations are straightforward but quite lengthy; thus, they are not repeated here and the interested reader is referred to [13] instead. One can see from (3.142) that the same computations may also be used in the case $K \neq 0$ by replacing k_1 with $k_1 - K$. Thus, the following necessary and sufficient stability condition for system (3.13) is obtained from [13, Theorem 1].

Theorem 3.11 (Necessary and Sufficient Stability Criterion) *Let the logarithm of a complex number be defined as in* (3.8). *The origin of system (3.13) is finite-time stable if and only if its parameters k_1, k_2 and nonnegative perturbation bounds K, L satisfy the inequalities*

[12] Note that the system is considered in phase coordinates $w_1 = x_1$, $w_2 = \dot{x}_1$ in [13]. The w_2-axis considered there coincides with the x_2-axis considered here, however, because $x_2 = \dot{x}_1 = w_2$ if $w_1 = x_1 = 0$. Thus, the meaning of the quantity Φ, which is defined in [13] in terms of the intersection of trajectories with the w_2-axis, is the same.

$$k_1 > K, \qquad k_2 > L, \qquad 1 > \Phi(k_1 - K, k_2, L). \tag{3.143}$$

Therein, the function Φ is defined as

$$\Phi(\tilde{k}_1, k_2, L) = \sqrt{\frac{k_2 + L}{k_2 - L}} \exp\left(\frac{\tilde{k}_1(\alpha_1 - \alpha_2)}{4}\right) \tag{3.144}$$

for $\tilde{k}_1 \geq 0$, $k_2 > L$, with the real-valued quantities

$$\alpha_1 = \begin{cases} \frac{1}{s_{11}-s_{12}} \log \frac{s_{11}}{s_{12}} & s_{11} \neq s_{12} \\ \frac{1}{s_{11}} & s_{11} = s_{12}, \end{cases} \tag{3.145a}$$

$$\alpha_2 = \begin{cases} \frac{1}{s_{21}-s_{22}} \left[j2\pi + \log \frac{s_{21}}{s_{22}}\right] & s_{21}, s_{22} \notin \mathbb{R} \\ \infty & s_{21}, s_{22} \in \mathbb{R} \end{cases} \tag{3.145b}$$

and the possibly complex-valued quantities

$$s_{11} = \frac{1}{4}\left(-\tilde{k}_1 + \sqrt{\tilde{k}_1^2 - 8(k_2 - L)}\right), \quad s_{12} = \frac{1}{4}\left(-\tilde{k}_1 - \sqrt{\tilde{k}_1^2 - 8(k_2 - L)}\right), \tag{3.146a}$$

$$s_{21} = \frac{1}{4}\left(-\tilde{k}_1 + \sqrt{\tilde{k}_1^2 - 8(k_2 + L)}\right), \quad s_{22} = \frac{1}{4}\left(-\tilde{k}_1 - \sqrt{\tilde{k}_1^2 - 8(k_2 + L)}\right) \tag{3.146b}$$

being used as abbreviations.

Recalling the matrix $A \in \mathbb{R}^{2\times 2}$ defined in (3.27) it may be noted that the abbreviations introduced in this theorem are related to similarly defined matrices A_1, A_2 of the form

$$A_1 := \begin{bmatrix} -\frac{\tilde{k}_1}{2} & \frac{1}{2} \\ -k_2 + L & 0 \end{bmatrix}, \qquad A_2 := \begin{bmatrix} -\frac{\tilde{k}_1}{2} & \frac{1}{2} \\ -k_2 - L & 0 \end{bmatrix}. \tag{3.147}$$

In particular, s_{11}, s_{12} are the eigenvalues of A_1 and s_{21}, s_{22} are the eigenvalues of A_2, and α_1, α_2 are the zeros of certain matrix exponentials involving the two matrices. For details, the reader is referred to [13].

Example 3.7 The application of the criterion is briefly illustrated using the parameter values $k_1 = 2$, $k_1 = 1$. For the unperturbed case, i.e., for $K = L = 0$, one obtains

$$s_{11} = s_{21} = -\frac{1}{2} + \mathrm{j}\frac{1}{2}, \qquad s_{12} = s_{22} = -\frac{1}{2} - \mathrm{j}\frac{1}{2}, \tag{3.148}$$

and

$$\alpha_1 = \frac{1}{\mathrm{j}} \log \frac{1-\mathrm{j}}{1+\mathrm{j}} = \arg \frac{1-\mathrm{j}}{1+\mathrm{j}} = -\frac{\pi}{2}, \qquad \alpha_2 = \frac{1}{\mathrm{j}} \left(\mathrm{j}2\pi + \log \frac{1-\mathrm{j}}{1+\mathrm{j}} \right) = \frac{3\pi}{2}. \tag{3.149}$$

The value of Φ is thus obtained as

$$\Phi(2, 1, 0) = \mathrm{e}^{-\pi} \approx 0.0432. \tag{3.150}$$

For a perturbed case with $L = 0.5$ and $K = 0$, one finds

$$s_{21} = -\frac{1}{2} + \mathrm{j}\frac{1}{\sqrt{2}}, \qquad s_{11} = s_{12} = -\frac{1}{2}, \qquad s_{22} = -\frac{1}{2} - \mathrm{j}\frac{1}{\sqrt{2}} \tag{3.151}$$

and

$$\begin{aligned} \alpha_1 = -2, \quad \alpha_2 &= \frac{1}{\mathrm{j}\sqrt{2}} \left(\mathrm{j}2\pi + \log \frac{1-\mathrm{j}\sqrt{2}}{1+\mathrm{j}\sqrt{2}} \right) = \frac{2}{\sqrt{2}}\pi + \frac{1}{\sqrt{2}}(\arctan 2\sqrt{2} - \pi) \\ &= \frac{\pi + \arctan 2\sqrt{2}}{\sqrt{2}}, \end{aligned} \tag{3.152}$$

which yields

$$\Phi(2, 1, 0.5) = \sqrt{3} \exp\left(-1 - \frac{\pi + \arctan 2\sqrt{2}}{2\sqrt{2}} \right) \approx 0.1358. \tag{3.153}$$

One can see that in both cases α_1 is nonpositive, while α_2 is nonnegative; this is the case in general, i.e., for all values $\tilde{k}_1$, k_2, L, for which Φ is defined.

3.5.2 Reaching Time Estimation

The reaching time for the initial state $x_0 = x_{0,2} e_2$ given in (3.136) may be estimated by considering an upper bound on the total variation of x_2. This quantity is denoted by κ and is given by the integral

$$\kappa := \int_0^\infty |\dot{x}_2| \, \mathrm{d}t. \tag{3.154}$$

This integral may be split at the time instants t_1, t_2, t_3, …, at which $\dot{x}_2$ changes sign, i.e., at which the trajectory intersects the x_2-axis. Using the majorant curve to bound the values of x_2 at these time instants (see Fig. 3.4) yields the upper bound

$$\begin{aligned}\kappa &= \left|x_{0,2} - x_2(t_1)\right| + |x_2(t_1) - x_2(t_2)| + |x_2(t_2) - x_2(t_3)| + \cdots \\ &\le \left|x_{0,2}\right| \left[(1+\Phi) + (\Phi + \Phi^2) + (\Phi^2 + \Phi^3) + \cdots\right] \\ &= \left|x_{0,2}\right| \left(1 + 2\sum_{k=1}^{\infty} \Phi^k\right) = \left|x_{0,2}\right| \frac{1+\Phi}{1-\Phi}\end{aligned} \tag{3.155}$$

for the total variation κ.

Recall from Sect. 3.2.3 that $x_2(t)$ and thus also $\dot{x}_2(t)$ are zero for any $t \ge \tau_x$ and nonzero for $t < \tau_x$, where τ_x denotes the reaching time of the particular trajectory $x(t)$. Therefore, the inequality

$$\kappa = \int_0^{\infty} |\dot{x}_2| \,\mathrm{d}t = \int_0^{\tau_x} |\dot{x}_2| \,\mathrm{d}t \ge \int_0^{\tau_x} (k_2 - L) \,\mathrm{d}t = (k_2 - L)\tau_x \tag{3.156}$$

holds. Selecting the trajectory such that τ_x is equal to $T_{K,L}(x_{0,2}e_2)$ yields the reaching time estimate

$$T_{K,L}(x_{0,2}e_2) \le \frac{\kappa}{k_2 - L} = \frac{\left|x_{0,2}\right|}{k_2 - L} \frac{1+\Phi}{1-\Phi}. \tag{3.157}$$

The following theorem summarizes this result.

Theorem 3.12 (Majorant Curve Based Estimate) *Let the function Φ be defined as in Theorem 3.11 and suppose that the parameters k_1, k_2 and the nonnegative perturbation bounds K, L of system (3.13) satisfy*

$$k_1 > K, \qquad k_2 > L, \qquad 1 > \Phi(k_1 - K, k_2, L). \tag{3.158}$$

Then, for the special initial condition $x_0 = x_{0,2}e_2 \in \mathbb{R}^2$ with $x_{0,2} \in \mathbb{R}$, the system's reaching time $T_{K,L}(x_{0,2}e_2)$ is bounded by $T_{K,L}(x_{0,2}e_2) \le \overline{T}_{K,L}(x_{0,2}e_2)$, where $\overline{T}_{K,L}(x_{0,2}e_2)$ is given by

$$\overline{T}_{K,L}(x_{0,2}e_2) = \frac{\left|x_{0,2}\right|}{k_2 - L} \frac{1 + \Phi(k_1 - K, k_2, L)}{1 - \Phi(k_1 - K, k_2, L)}. \tag{3.159}$$

Example 3.8 To illustrate the theorem's application, the parameter values $k_1 = 2$, $k_2 = 1$ are considered. The corresponding values of Φ for the unperturbed and one perturbed case with $L = 0.5$, $K = 0$ have already been computed in (3.150) and (3.153) in Example 3.7. Using these results to evaluate (3.159), one obtains upper reaching time bounds for the initial condition $x_0 = e_2$ in these two cases as

$$\overline{T}_{0,0}(e_2) = \frac{1 + \Phi(2,1,0)}{1 - \Phi(2,1,0)} = \frac{1 + \mathrm{e}^{-\pi}}{1 - \mathrm{e}^{-\pi}} \approx 1.0903, \qquad \text{and} \tag{3.160a}$$

$$\overline{T}_{0,0.5}(e_2) = 2\frac{1+\Phi(2,1,0.5)}{1-\Phi(2,1,0.5)} = 2\frac{1+\sqrt{3}\exp\left(-1-\frac{\pi+\arctan 2\sqrt{2}}{2\sqrt{2}}\right)}{1-\sqrt{3}\exp\left(-1-\frac{\pi+\arctan 2\sqrt{2}}{2\sqrt{2}}\right)} \approx 2.6285, \tag{3.160b}$$

respectively.

3.5.3 Range of Permitted Perturbations

In this section, the permitted range of perturbation bounds K, L is studied. Due to the complicated dependence of the majorant curve's characteristic value Φ on these bounds, obtaining a condition like (3.64) is impossible. It is, however, possible to define upper bounds $\overline{K}$ and $\overline{L}$ such that the conditions $K < \overline{K}$ and $L < \overline{L}$ are necessary for using the approach, and also sufficient if either K or L is zero.

To obtain them, note first that the function $\Phi(k_1 - K, k_2, K)$ as defined in Theorem 3.11 is nondecreasing[13] with respect to K and L. While a formal proof of this fact based on the analytic expression is quite complicated, it is possible to see this intuitively from the way Φ is defined geometrically. To that end, consider system (3.142), whose trajectory is the majorant curve shown in Fig. 3.4. This curve envelopes trajectories obtained with any smaller perturbations, and thus it envelopes, in particular, any majorant curve obtained with smaller values of K and L. Thus, Φ cannot decrease when either K or L are decreased. Due to this fact, the range for either K or L is largest when L or K are zero, respectively.

For $L = 0$, one can see that the differential equations for the majorant curve (3.142) correspond to an unperturbed super-twisting algorithm with parameter values $\tilde{k}_1 = k_1 - K$ and $\tilde{k}_2 = k_2$. Thus, the curve converges, i.e.,$\Phi < 1$ holds, if and only if $k_2 > 0$ and

$$K < k_1 \tag{3.161}$$

holds. Thus, $\overline{K} := k_1$ is an upper bound for K.

For $K = 0$, on the other hand, one may define an upper bound $\overline{L}$ for L implicitly by means of the equation

$$\Phi(k_1, k_2, \overline{L}) = 1. \tag{3.162}$$

Since Φ is nondecreasing with respect to L, one has $\Phi(k_1, k_2, L) < 1$ if and only if $L < \overline{L}$ holds.

The following theorem summarizes these conditions.

Theorem 3.13 (Majorant Curve Based Perturbation Bounds) *Let positive parameters k_1, k_2 be given and let the function Φ be defined as in Theorem 3.11. Define furthermore the constant*

[13] This fact is also noted in [8].

$$\overline{K} = k_1, \tag{3.163}$$

and define the constant $\overline{L}$ *as the unique solution of*

$$\Phi(k_1, k_2, \overline{L}) = 1, \tag{3.164}$$

if it exists, or as $\overline{L} = k_2$ *otherwise. Then, the inequalities* $K < \overline{K}$ *and* $L < \overline{L}$ *are necessary for the conditions of Theorem 3.12 to be fulfilled. If either* K *or* L *are zero, they are also sufficient.*

Example 3.9 As an example, the upper perturbation bounds $\overline{K}$ and $\overline{L}$ as obtained from this theorem are given for the parameter values $k_1 = 2$, $k_2 = 1$. One can verify numerically that

$$\lim_{\overline{L} \to k_2} \Phi(k_1, k_2, \overline{L}) \approx 0.2984 < 1 \tag{3.165}$$

holds. Thus, (3.164) has no solution due the fact that Φ is nondecreasing with respect to k_2 and is defined only for $k_2 < L$. The constants $\overline{K}$ and $\overline{L}$ are, therefore, given by

$$\overline{K} = k_1 = 2, \qquad \overline{L} = k_2 = 1. \tag{3.166}$$

3.5.4 Asymptotic Properties

The bound (3.159) can be shown to possess even better asymptotic properties than the estimate based on the reaching time function discussed in Sect. 3.3. For any value $K < k_1$, it yields the actual reaching time rather than an upper bound for it provided either that the perturbation bound L is zero or that the parameter k_1 is sufficiently large.

To see that the bound is exact for $L = 0$, note that in this case equality holds in (3.156). Due to this fact, the derived upper bound (3.157) is also the reaching time of the majorant curve, and thus equal to the perturbed system's reaching time.

For the case that k_1 tends to infinity, this can be seen by looking at s_{21} and s_{22} defined in (3.146b). If $k_1 = \tilde{k}_1 + K$ satisfies the inequality

$$(k_1 - K)^2 \geq 8(k_2 + L), \tag{3.167}$$

then s_{21} and s_{22} are real-valued, and thus $\alpha_2 = \infty$ according to (3.145b). Therefore, (3.144) yields $\Phi = 0$, and according to (3.159) the estimate is given by

$$\overline{T}_{K,L}(x_2 e_2) = \frac{|x_2|}{k_2 - L} \tag{3.168}$$

for such parameter values, i.e., in particular for sufficiently large values of k_1. As argued in Sect. 3.3.4, this expression is both an upper and a lower bound for, and thus equal to, the system's reaching time.

The following theorems, which are analogous to those obtained in Sect. 3.3.4 for the reaching time function based estimate, summarize this result.

Theorem 3.14 (Asymptotics of Majorant Curve Based Estimate) *The upper bound* $\overline{T}_{K,L}(x_{0,2}e_2)$ *given in* (3.159) *of the reaching time* $T_{K,L}(x_{0,2}e_2)$ *of system (3.13) for the special initial condition* $x_0 = x_{0,2}e_2 \in \mathbb{R}^2$ *with* $x_{0,2} \in \mathbb{R}$ *satisfies the asymptotic relations*

$$\lim_{L\to 0} \overline{T}_{K,L}(x_{0,2}e_2) = \lim_{L\to 0} T_{K,L}(x_{0,2}e_2), \tag{3.169a}$$

$$\lim_{k_1\to \infty} \overline{T}_{K,L}(x_{0,2}e_2) = \lim_{k_1\to \infty} T_{K,L}(x_{0,2}e_2) = \frac{|x_{0,2}|}{k_2 - L}, \tag{3.169b}$$

i.e., it is asymptotically exact when either the perturbation bound L tends to zero or the parameter k_1 *tends to infinity.*

Remark 3.3 It should be highlighted that the result in this theorem for the majorant curve based estimate is *the same* as that obtained for the reaching time function based estimate in Theorem 3.4. Both yield the true reaching time for vanishing perturbation or large k_1.

Similar to Theorem 3.5, relation (3.169b) furthermore holds not only asymptotically, but already for finite, sufficiently large values of k_1.

Theorem 3.15 *Suppose that the parameters* k_1, k_2 *and the perturbation bounds* K, L *satisfy*

$$k_1 \geq K + \sqrt{8(k_2 + L)}. \tag{3.170}$$

Then, the reaching time $T_{K,L}(x_{0,2}e_2)$ *of system (3.13) for the special initial condition* $x_0 = x_{0,2}e_2 \in \mathbb{R}^2$ *with* $x_{0,2} \in \mathbb{R}$ *is given by*

$$T_{K,L}(x_{0,2}e_2) = \frac{|x_{0,2}|}{k_2 - L}. \tag{3.171}$$

3.6 Comparisons

In this section, numerical comparisons of the three discussed approaches are presented. First, the perturbation bounds permitted by each approach are illustrated. Then, numerical values of the obtained reaching time bounds are compared.

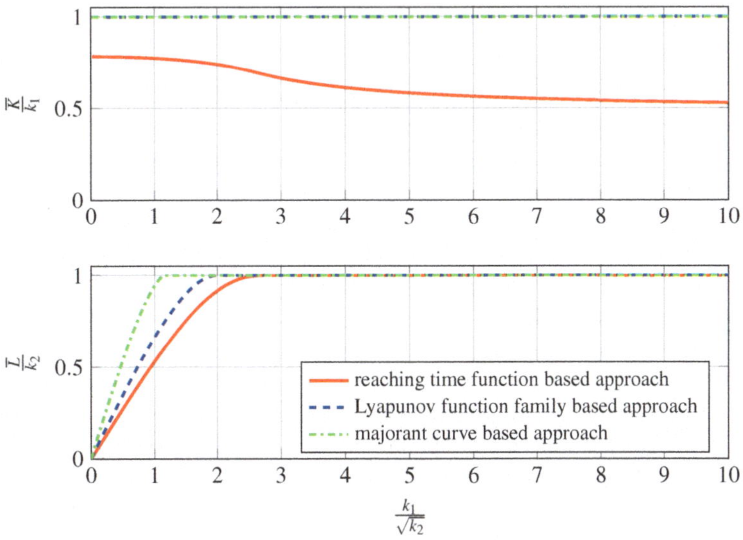

Fig. 3.5 Maximum perturbation bounds $\overline{K}$ and $\overline{L}$ of each presented approach

3.6.1 Perturbation Bounds

In Sects. 3.3.4, 3.4.3, and 3.5.3, upper bounds $\overline{K}$ and $\overline{L}$ for the perturbation bounds K and L are derived, such that $K < \overline{K}$ and $L < \overline{L}$ are necessary conditions for using each of the approaches. Furthermore, the condition

$$K\overline{K}^{-1} + L\overline{L}^{-1} < 1 \tag{3.172}$$

is necessary and sufficient for applying the reaching time function based approach described in Sect. 3.3 and sufficient for the Lyapunov function family based approach discussed in Sect. 3.4.

These upper perturbation bounds $\overline{K}, \overline{L}$ of each approach are compared in Fig. 3.5. Recall that the perturbation bounds for the reaching time function based approach in (3.63) are reciprocals of the $\mathscr{L}_\infty$ gain of systems with transfer functions $G_1(s)$ and $G_2(s)$ defined in (3.60). Similarly, the bounds for the Lyapunov function family based approach in (3.101) are reciprocals of the $\mathscr{L}_2$ gain of the same systems. Since the $\mathscr{L}_2$ gain of a system never exceeds the $\mathscr{L}_\infty$ gain, the latter approach permits for larger perturbations than the former. One can furthermore see that the majorant curve based approach permits the largest values of the bounds K and L, since it is based on a necessary and sufficient stability condition.

3.6.2 *Reaching Time Estimates*

Figures 3.6 and 3.7 compare the reaching time estimates obtained with each approach as functions of the perturbation bounds K and L for two values of k_1, and Fig. 3.8 illustrates the estimates' asymptotic behavior. In addition to the techniques discussed in this chapter, the plots also depict the alternative Lyapunov function family based approach [14], which is mentioned in Remark 3.1. To compute the estimates of the two Lyapunov function family based estimates, the semidefinite programs given in Theorem 3.9 and described in [14], respectively, were solved in MATLAB with YALMIP [9] using SEDUMI [17] as a solver.

The eigenvalues of the matrix A play an important role in each of the approaches. In order to illustrate their influence, the value of k_1 is selected to obtain real eigenvalues in Fig. 3.6 and complex eigenvalues in Fig. 3.7. In all plots, $k_2 = 1$ is chosen. Note, however, that all but the Lyapunov-based estimate discussed in Sect. 3.4 are independent of this parameter, provided that the initial state is scaled by k_2. This is due to the homogeneity properties of the reaching time function described in Sect. 3.2.4, which is shared by the estimates discussed in Sect. 3.3, Sect. 3.5, and [14].

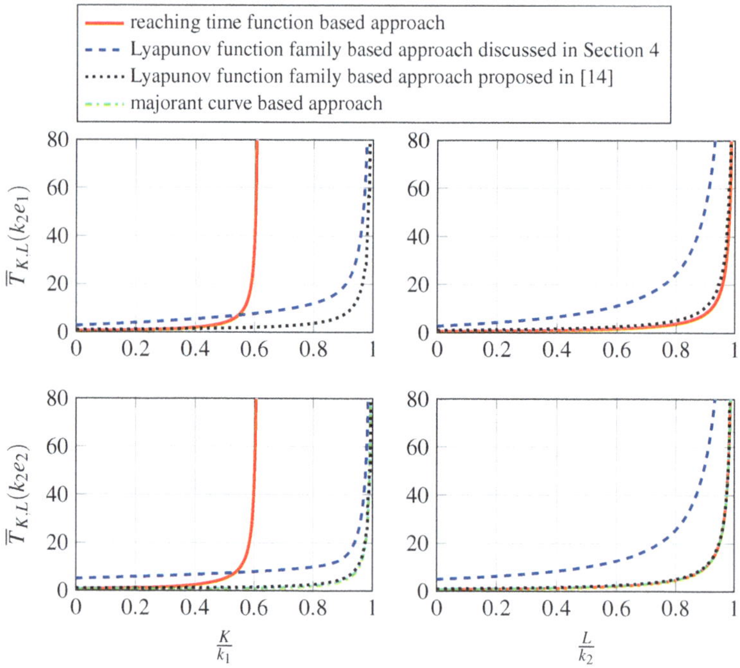

Fig. 3.6 Reaching time estimates for parameters $k_2 = 1$, $k_1 = 4\sqrt{k_2}$ and initial states $x_0 = k_2 e_1$ (top) and $x_0 = k_2 e_2$ (bottom) as a function of perturbation bounds K with $L = 0$ (left), and L with $K = 0$ (right)

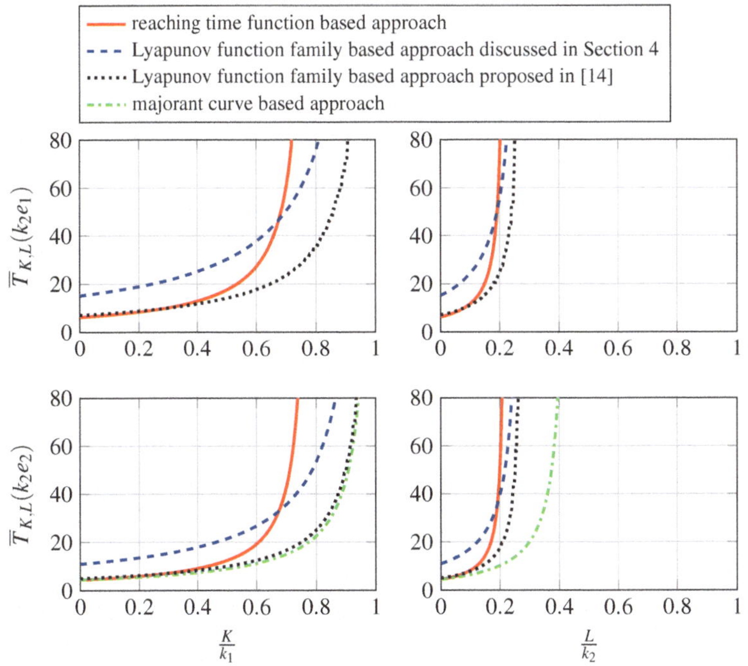

Fig. 3.7 Reaching time estimates for parameter values $k_2 = 1$, $k_1 = 0.4\sqrt{k_2}$ and initial states $x_0 = k_2e_1$ (top) and $x_0 = k_2e_2$ (bottom) as a function of perturbation bounds K with $L = 0$ (left), and L with $K = 0$ (right)

In Fig. 3.6, a value of $k_1 = 4\sqrt{k_2}$ is selected, which leads to real eigenvalues of the matrix A. One can see that the approaches based on the reaching time function and on the majorant curve yield the smallest (identical) bounds for any value of L, when K is zero. When the first component of the state is nonzero, on the other hand, the latter approach is not applicable. The former yields lower bounds for smaller values of K, but is overtaken by the Lyapunov function based approaches due to the larger value of $\overline{K}$.

A similar situation can be seen in Fig. 3.7, where k_1 is smaller, which leads to complex eigenvalues of the matrix A. Again, the lowest bounds are obtained with the majorant curve based technique when it is applicable. Bounds obtained with both the reaching time and the Lyapunov function family based approaches are larger, with the latter overtaking the former for large perturbations.

Figure 3.8, finally, illustrates the asymptotic behavior of the estimates with increasing value of the parameter k_1. One can see that the two approaches based on the reaching time function and on the majorant curve converge to the same constant value. This value has been shown to be not only a bound for, but, in fact, the true value of the perturbed system's reaching time. The Lyapunov function family based estimate discussed in Sect. 3.4, on the other hand, exhibits the diverging behavior shown in Sect. 3.4.5, whereas the approach from [14] appears to converge to a constant value for large values of k_1.

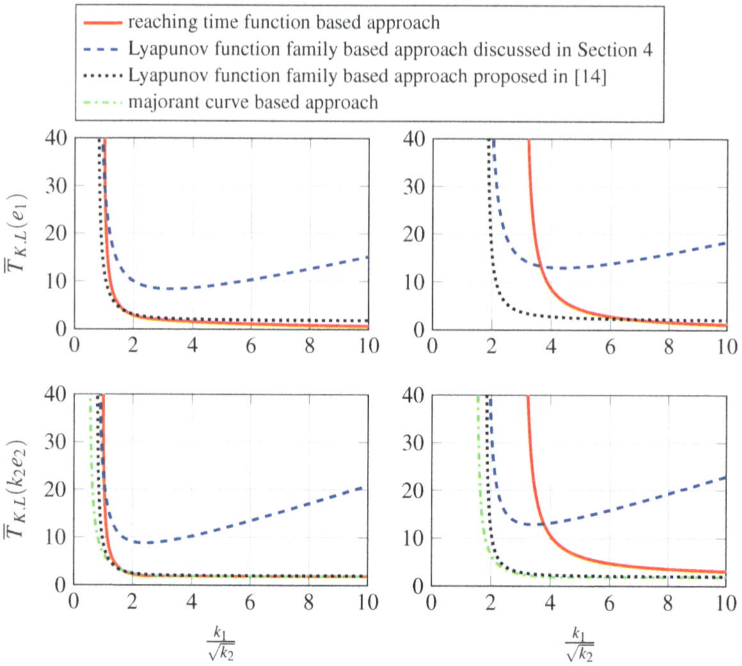

Fig. 3.8 Asymptotic behavior of the reaching time estimates for $k_2 = 1$ with the initial states $x_0 = k_2 e_1$ (top) and $x_0 = k_2 e_2$ (bottom) with $K = 0, L = 0.5k_2$ (left) and $K = 1, L = 0.5k_2$ (right) as a function of $k_1 \sqrt{k_2}^{-1}$

3.7 Conclusion

Three techniques for computing upper bounds for the super-twisting algorithm's reaching time have been discussed. Each of them has different advantages and shortcomings. The first technique, which is based on the unperturbed reaching time function, yields very good results when the perturbation bounds K and L are small or the parameter k_1 is large. The second one, which is obtained using a quadratic Lyapunov function family, diverges for large values of k_1, though this shortcoming is eliminated by a related, more recent approach in literature. Compared to the first technique, it is applicable to a larger class of perturbations, and performs well, in particular, for large values of the perturbation bound K. It is also the only technique, from which one may easily obtain a reaching time bound for a whole region in the state space. The third method, which is based on a majorant curve, yields the best results in the cases where it is applicable. Since it is based on a necessary and sufficient stability condition, it also permits for the largest class of perturbations. It is, however, restricted to initial states that lie on the x_2-axis, which restricts its use mainly to the analysis of the super-twisting algorithm in a differentiator context.

Acknowledgements The author is indebted to Prof. Nicolaos Dourdoumas, Graz University of Technology, for suggesting several improvements, and to Prof. Jaime Moreno, Universidad Nacional Autónoma de México, for several useful suggestions and for originally proposing the quasi-linear system representation, which forms the starting point for most considerations presented in this chapter. The financial support by the Christian Doppler Research Association, the Austrian Federal Ministry for Digital and Economic Affairs, and the National Foundation for Research, Technology and Development is gratefully acknowledged.

Appendix: Derivations

In the following, several derivations are presented that were omitted in the above discussions. First, the perturbation bounds for the reaching time function based approach are calculated. Then, the time derivative of the quadratic Lyapunov function is computed. Finally, the perturbation bounds for the Lyapunov function family based approach are derived.

Perturbation Bounds for Reaching Time Function Approach

In this section, the maximal perturbation bounds $\overline{K}$ and $\overline{L}$ given in Theorem 3.3 are derived using their definition (3.58).

Consider first $\overline{L}$, i.e., evaluate $T_{0,0}(x)$ for $x = e_2$, i.e., for $x_1 = 0$ and $x_2 = 1$. Using (3.41b) one can see that $\alpha_1(e_2) = 0$. Since $g(e_2) = e_2$, one thus obtains from (3.40)

$$T_{0,0}(e_2) = \frac{1}{k_2}\left|e_2^{\mathrm{T}}\left(I - \frac{2}{1-\lambda}I\right)e_2\right| = k_2^{-1}\frac{1+\lambda}{1-\lambda}. \tag{3.173}$$

According to (3.41c), $\lambda = 0$ for $k_1^2 \geq 8k_2$, yielding $T_{0,0}(e_2) = k_2^{-1}$. Otherwise,

$$\begin{aligned} T_{0,0}(e_2) &= k_2^{-1}\frac{1+\exp\left(-\frac{k_1\pi}{\sqrt{8k_2-k_1^2}}\right)}{1-\exp\left(-\frac{k_1\pi}{\sqrt{8k_2-k_1^2}}\right)} = k_2^{-1}\frac{\exp\left(\frac{k_1\pi}{2\sqrt{8k_2-k_1^2}}\right)+\exp\left(-\frac{k_1\pi}{2\sqrt{8k_2-k_1^2}}\right)}{\exp\left(\frac{k_1\pi}{2\sqrt{8k_2-k_1^2}}\right)-\exp\left(-\frac{k_1\pi}{2\sqrt{8k_2-k_1^2}}\right)} \\ &= \frac{1}{k_2\tanh\frac{\pi}{2}\frac{k_1}{\sqrt{8k_2-k_1^2}}}. \end{aligned} \tag{3.174}$$

To compute $\overline{K}$, one finds

$$\alpha_1(e_1) = \begin{cases} -\frac{1}{s_1} & k_1^2 = 8k_2 \\ \frac{1}{s_1-s_2}\log\frac{s_2}{s_1} & k_1^2 \neq 8k_2 \end{cases} \tag{3.175}$$

and

$$T_{0,0}(e_1) = \frac{1}{k_2}\left|e_2^{\mathrm{T}}\left(I - \frac{2}{1-\lambda}\mathrm{e}^{A\alpha_1(e_1)}\right)e_1\right| = \frac{2k_2^{-1}}{1-\lambda}\left|e_2^{\mathrm{T}}\mathrm{e}^{A\alpha_1(e_1)}e_1\right|. \tag{3.176}$$

Consider first the case $k_1^2 \neq 8k_2$. In this case, the matrix exponential function is given by

$$\mathrm{e}^{A\alpha} = \frac{1}{2(s_1 - s_2)}\begin{bmatrix} 2s_1\mathrm{e}^{\alpha s_1} - 2s_2\mathrm{e}^{\alpha s_2} & \mathrm{e}^{\alpha s_1} - \mathrm{e}^{\alpha s_2} \\ 4s_1s_2\mathrm{e}^{\alpha s_2} - 4s_1s_2\mathrm{e}^{\alpha s_1} & 2s_1\mathrm{e}^{\alpha s_2} - 2s_2\mathrm{e}^{\alpha s_1} \end{bmatrix}. \tag{3.177}$$

Then, one has

$$e_2^{\mathrm{T}}\mathrm{e}^{A\alpha_1(e_1)}e_1 = 2s_1s_2\frac{\mathrm{e}^{\alpha_1(e_1)s_2} - \mathrm{e}^{\alpha_1(e_1)s_1}}{s_1 - s_2}. \tag{3.178}$$

Substituting $\alpha_1(e_1)$ then yields

$$\begin{aligned} e_2^{\mathrm{T}}\mathrm{e}^{A\alpha_1(e_1)}e_1 &= 2s_1s_2\frac{\left(\frac{s_2}{s_1}\right)^{\frac{s_2}{s_1-s_2}} - \left(\frac{s_2}{s_1}\right)^{\frac{s_1}{s_1-s_2}}}{s_1 - s_2} \\ &= 2s_2\left(\frac{s_2}{s_1}\right)^{\frac{s_2}{s_1-s_2}}\frac{1 - \left(\frac{s_2}{s_1}\right)^{\frac{s_1-s_2}{s_1-s_2}}}{1 - \frac{s_2}{s_1}} = 2s_2\left(\frac{s_2}{s_1}\right)^{\frac{s_2}{s_1-s_2}} \\ &= 2s_2\sqrt{\frac{s_1}{s_2}}\left(\frac{s_2}{s_1}\right)^{\frac{s_2}{s_1-s_2}+\frac{1}{2}} = -2\sqrt{s_1s_2}\left(\frac{s_2}{s_1}\right)^{\frac{1}{2}\frac{s_1+s_2}{s_1-s_2}}, \end{aligned} \tag{3.179}$$

and thus, using the fact that $s_1s_2 = \det A = \frac{k_2}{2}$, one has

$$T_{0,0}(e_1) = \frac{4}{1-\lambda}k_2^{-1}\sqrt{s_1s_2}\left(\frac{s_2}{s_1}\right)^{\frac{1}{2}\frac{s_1+s_2}{s_1-s_2}} = \frac{2}{1-\lambda}\sqrt{\frac{2}{k_2}}\left(\frac{s_2}{s_1}\right)^{\frac{1}{2}\frac{s_1+s_2}{s_1-s_2}}. \tag{3.180}$$

If the eigenvalues are real-valued, i.e., if $k_1^2 > 8k_2$ holds, then one has $\lambda = 0$ according to (3.41c) and substituting s_1 and s_2 into this expression yields

$$T_{0,0}(e_1) = 2\sqrt{\frac{2}{k_2}}\left(\frac{k_1 + \sqrt{k_1^2 - 8k_2}}{k_1 - \sqrt{k_1^2 - 8k_2}}\right)^{-\frac{1}{2}\frac{k_1}{\sqrt{k_1^2-8k_2}}} \tag{3.181}$$

from which $\overline{K}$ is obtained by means of (3.58). If the eigenvalues are complex-valued, on the other hand, i.e., if $k_1^2 < 8k_2$, then one may further simplify (3.180). Recall that for $y, q \in \mathbb{C}$, the relation

$$y^q = \mathrm{e}^{q \log y} \tag{3.182}$$

holds with the logarithm as defined in (3.8) and introduce the abbreviation

$$c := \frac{k_1}{\sqrt{8k_2 - k_1^2}}. \tag{3.183}$$

Furthermore, one has[14]

$$\log \frac{s_2}{s_1} = \log \frac{k_1 + \mathrm{j}\sqrt{8k_2 - k_1^2}}{k_1 - \mathrm{j}\sqrt{8k_2 - k_1^2}} = 2\,\mathrm{j} \arctan \frac{\sqrt{8k_2 - k_1^2}}{k_1} = \mathrm{j}\pi - 2\,\mathrm{j}\arctan c \tag{3.184}$$

$$\frac{s_1 + s_2}{s_1 - s_2} = -\frac{k_1}{\mathrm{j}\sqrt{8k_2 - k_1^2}} = \mathrm{j}c \tag{3.185}$$

and

$$\lambda = \exp\left(-\frac{k_1 \pi}{\sqrt{8k_2 - k_1^2}}\right) = \mathrm{e}^{-c\pi} \tag{3.186}$$

holds according to (3.41c). Substituting into (3.180) thus yields

$$\begin{aligned} T_{0,0}(e_1) &= \frac{2}{1 - \mathrm{e}^{-c\pi}} \sqrt{\frac{2}{k_2}} \exp\left(\frac{\mathrm{j}c}{2}(\mathrm{j}\pi - 2\,\mathrm{j}\arctan c)\right) \\ &= \frac{2\mathrm{e}^{-\frac{c\pi}{2}}}{1 - \mathrm{e}^{-c\pi}} \sqrt{\frac{2}{k_2}} \mathrm{e}^{c \arctan c} = \frac{2}{\sinh c\pi} \sqrt{\frac{2}{k_2}} \mathrm{e}^{c \arctan c}. \end{aligned} \tag{3.187}$$

By substituting c from (3.183) and using (3.58), one obtains $\overline{K}$.

For $k_1^2 = 8k_2$, the matrix exponential is given by

$$\mathrm{e}^{A\alpha} = \begin{bmatrix} \alpha s_1 + 1 & \frac{\alpha}{2} \\ -2\alpha s_1^2 & 1 - \alpha s_1 \end{bmatrix} \mathrm{e}^{\alpha s_1}. \tag{3.188}$$

Thus, one has

$$e_2^{\mathrm{T}} \mathrm{e}^{A\alpha_1(e_1)} e_1 = -2\alpha_1(e_1) s_1^2 \mathrm{e}^{\alpha_1(e_1)s_1} = 2\sqrt{s_1^2}\mathrm{e}^{-1}. \tag{3.189}$$

Again, noting that in this case $s_2^2 = \frac{k_2}{2}$, one has

$$T_{0,0}(e_1) = 4k_2^{-1}\sqrt{s_1^2}\mathrm{e}^{-1} = 2\sqrt{\frac{2}{k_2}}\mathrm{e}^{-1}. \tag{3.190}$$

[14] Note that $\arctan y + \arctan \frac{1}{y} = \frac{\pi}{2}$ for $y > 0$.

Upper bound of Quadratic Lyapunov Function's Time Derivative

In this section, the time derivative $\dot{V}$ of V given in (3.76) along the trajectories of the perturbed system (3.26) is computed. By substituting the expression $A^{\mathrm{T}}P + PA$ from (3.88), one obtains for $z_1 \neq 0$

$$
\begin{aligned}
|z_1|\,\dot{V} &= \frac{|z_1|}{2}\delta_1(e_1^{\mathrm{T}}Pz + z^{\mathrm{T}}Pe_1) + \delta_2\,|z_1|\,(e_2^{\mathrm{T}}Pz + z^{\mathrm{T}}Pe_2) + z^{\mathrm{T}}(A^{\mathrm{T}}P + PA)z \\
&\le \delta_1\,|z_1|\,e_1^{\mathrm{T}}Pz + 2\delta_2\,|z_1|\,e_2^{\mathrm{T}}Pz - z^{\mathrm{T}}Qz \\
&\quad - \Theta_1 K z^{\mathrm{T}}e_1e_1^{\mathrm{T}}z - \frac{K}{4\Theta_1}z^{\mathrm{T}}Pe_1e_1^{\mathrm{T}}Pz - \Theta_2 L z^{\mathrm{T}}e_1e_1^{\mathrm{T}}z - \frac{L}{\Theta_2}zPe_2e_2^{\mathrm{T}}Pz \\
&\le \delta_1\,|z_1|\,e_1^{\mathrm{T}}Pz + 2\delta_2\,|z_1|\,e_2^{\mathrm{T}}Pz - z^{\mathrm{T}}Qz \\
&\quad - \Theta_1 K z_1^2 - \frac{K}{4\Theta_1}z^{\mathrm{T}}Pe_1e_1^{\mathrm{T}}Pz - \Theta_2 L z_1^2 - \frac{L}{\Theta_2}zPe_2e_2^{\mathrm{T}}Pz \\
&\quad + \left(\sqrt{\frac{K}{4\Theta_1}}\,z^{\mathrm{T}}Pe_1 - \sqrt{\frac{\Theta_1}{K}}\,|z_1|\,\delta_1\right)^2 + \left(\sqrt{\frac{L}{\Theta_2}}\,z^{\mathrm{T}}Pe_2 - \sqrt{\frac{\Theta_2}{L}}\,|z_1|\,\delta_2\right)^2 \\
&= \Theta_1\left(\frac{\delta_1^2}{K} - K\right)z_1^2 + \Theta_2\left(\frac{\delta_2^2}{L} - L\right)z_1^2 - z^{\mathrm{T}}Qz \\
&\le -z^{\mathrm{T}}Qz.
\end{aligned}
\tag{3.191}
$$

Perturbation Bounds for Lyapunov Function Family Approach

In this section, the bounds $\overline{K}$ and $\overline{L}$ that are part of the sufficient condition in Theorem 3.7 and the necessary and sufficient condition given in Theorem 3.8 are derived.

Conditions of Theorem 3.7

To obtain the bounds $\overline{K}$ and $\overline{L}$ in Theorem 3.7, the $\mathcal{H}_\infty$ norms of $G_1(s)$ and $G_2(s)$ given in (3.60) have to be computed. To get $\overline{K}$, one has to find

$$
\begin{aligned}
\overline{K}^{-2} = \sup_{\omega}|G_1(\mathrm{j}\omega)|^2 &= \sup_{\omega}\frac{\omega^2}{k_1^2\omega^2 + (k_2 - 2\omega^2)^2} \\
&= \sup_{y\ge 0}\frac{y}{k_1^2 y + (k_2 - 2y)^2},
\end{aligned}
\tag{3.192}
$$

where the abbreviation $y := \omega^2$ is used. The first-order necessary condition for an extremum of this function is given by

$$0 = \frac{\mathrm{d}}{\mathrm{d}y} \frac{y}{k_1^2 y + (k_2 - 2y)^2} = \frac{k_1^2 y + (k_2 - 2y)^2 - y k_1^2 + 4y(k_2 - 2y)}{(k_1^2 y + (k_2 - 2y)^2)^2}$$
$$= \frac{(k_2 + 2y)(k_2 - 2y)}{(k_1^2 y + (k_2 - 2y)^2)^2}. \tag{3.193}$$

One can see that the only nonnegative solution y of this equation is given by $2y = k_2$. Substituting this value for y into (3.192) yields

$$\overline{K}^{-2} = \frac{1}{k_1^2}. \tag{3.194}$$

To obtain $\overline{L}$, the same procedure is applied to $G_2(s)$, i.e.,

$$\overline{L}^{-2} = \sup_{\omega} |G_2(\mathrm{j}\omega)|^2 = \sup_{\omega} \frac{1}{k_1^2 \omega^2 + (k_2 - 2\omega^2)^2}$$
$$= \sup_{y \geq 0} \frac{1}{k_1^2 y + (k_2 - 2y)^2}. \tag{3.195}$$

Again, the first-order optimality condition is given by

$$0 = \frac{\mathrm{d}}{\mathrm{d}y} \frac{y}{k_1^2 y + (k_2 - 2y)^2} = \frac{-y k_1^2 + 4y(k_2 - 2y)}{(k_1^2 y + (k_2 - 2y)^2)^2} = \frac{y(4k_2 - k_1^2 - 8y)}{(k_1^2 y + (k_2 - 2y)^2)^2}. \tag{3.196}$$

If $k_1^2 \geq 4k_2$, then the only nonnegative solution is $y = 0$; substitution into (3.195) yields

$$\overline{L}^{-2} = \frac{1}{k_2^2}. \tag{3.197}$$

Otherwise, the supremum is obtained for $2y = k_2 - 4^{-1} k_1^2$ as

$$\overline{L}^{-2} = \frac{16}{k_1^2 (8k_2 - k_1^2)}. \tag{3.198}$$

Conditions of Theorem 3.8

To see the conditions given in Theorem 3.8, note that (3.94), i.e.,

$$\left\| \sqrt{\Theta_1 K + \Theta_2 L} \left[\sqrt{\tfrac{K}{\Theta_1}} G_1 \; \sqrt{\tfrac{L}{\Theta_2}} G_2 \right] \right\|_{\mathscr{H}_\infty} < 1 \tag{3.199}$$

is fulfilled if and only if

$$\frac{K}{\Theta_1}|G_1(\mathrm{j}\omega)|^2 + \frac{L}{\Theta_2}|G_2(\mathrm{j}\omega)|^2 = \frac{\frac{K}{\Theta_1}|\mathrm{j}\omega|^2 + \frac{L}{\Theta_2}}{\left|-2\omega^2 + \mathrm{j}k_1\omega + k_2\right|^2} = \frac{\frac{K}{\Theta_1}\omega^2 + \frac{L}{\Theta_2}}{k_1^2\omega^2 + (k_2 - 2\omega^2)^2}$$
$$= \frac{\frac{K}{\Theta_1}\omega^2 + \frac{L}{\Theta_2}}{k_1^2\omega^2 + k_2^2 - 4k_2\omega^2 + 4\omega^4} < \frac{1}{\Theta_1 K + \Theta_2 L} \tag{3.200}$$

holds for all ω. For this to be the case, there must not be any ω, for which equality holds in this relation. Substituting $\omega^2 = y$, one thus obtains the condition that the equation

$$\begin{aligned} 0 &= -(\Theta_1 K + \Theta_2 L)\left(\frac{K}{\Theta_1}y + \frac{L}{\Theta_2}\right) + k_1^2 y + k_2^2 - 4k_2 y + 4y^2 \\ &= K^2 y + \frac{\Theta_2}{\Theta_1}KLy + \frac{\Theta_1}{\Theta_2}KL + L^2 + k_1^2 y + k_2^2 - 4k_2 y + 4y^2 \end{aligned} \tag{3.201}$$

or equivalently

$$y^2 + \frac{k_1^2 - K^2 - \frac{\Theta_2}{\Theta_1}KL - 4k_2}{4}y + \frac{k_2^2 - L^2 - \frac{\Theta_1}{\Theta_2}KL}{4} =: y^2 + a_1 y + a_0 = 0 \tag{3.202}$$

must not have any nonnegative, real-valued solutions y. The coefficient a_0 is the product of the two (possibly complex-valued) solutions y_1, y_2 of this equation. Thus, $a_0 > 0$ has to hold; otherwise, there is always one positive and one negative solution. Additionally, solutions have to be either complex-valued or have negative real part. The former is the case, if the discriminant $a_1^2 - 4a_0$ is negative, which yields the condition

$$2\sqrt{a_0} > a_1 > -2\sqrt{a_0}. \tag{3.203}$$

The latter is guaranteed, if the polynomial is a Hurwitz polynomial, i.e., if $a_1 > 0$ holds in addition to $a_0 > 0$. Combining all conditions yields

$$\frac{k_2^2 - L^2 - \frac{\Theta_1}{\Theta_2}KL}{4} = a_0 > 0, \tag{3.204a}$$

$$\frac{k_1^2 - K^2 - \frac{\Theta_2}{\Theta_1}KL - 4k_2}{4} = a_1 > -2\sqrt{a_0} = -\sqrt{k_2^2 - L^2 - \frac{\Theta_1}{\Theta_2}KL}. \tag{3.204b}$$

References

1. Bhat, S.P., Bernstein, D.S.: Geometric homogeneity with applications to finite-time stability. Math. Control Signals Syst. **17**(2), 101–127 (2005)
2. Dávila, A., Moreno, J.A., Fridman, L.: Optimal Lyapunov function selection for reaching time estimation of super twisting algorithm. In: Proceedings of the 48th IEEE Conference on Decision and Control (CDC), pp. 8405–8410 (2009)
3. Doyle, J.C., Glover, K., Khargonekar, P.P., Francis, B.A.: State-space solutions to standard $\mathscr{H}_2$ and $\mathscr{H}_\infty$ control problems. IEEE Transa. Autom. Control **34**(8), 831–847 (1989)
4. Filippov, A.F.: Differential Equations with Discontinuous Right-Hand Side. Kluwer Academic Publishing, Dortrecht (1988)
5. Horn, R.A., Zhang, F.: Basic properties of the Schur complement. In: Zhang, F. (ed.) The Schur Complement and Its Applications. Springer, Berlin (2006)
6. Khalil, H.K.: Nonlinear Systems, 3rd edn. Prentice Hall, Upper Saddle River (2002)
7. Levant, A.: Sliding order and sliding accuracy in sliding mode control. Int. J. Control **58**(6), 1247–1263 (1993)
8. Levant, A.: Robust exact differentiation via sliding mode technique. Automatica **34**(3), 379–384 (1998)
9. Löfberg, J.: Yalmip: a toolbox for modeling and optimization in MATLAB. In: Proceedings of the CACSD Conference. Taipei, Taiwan (2004)
10. Moreno, J.A.: Lyapunov approach for analysis and design of second order sliding mode algorithms. In: Fridman, L., Moreno, J.A., Iriarte, R. (eds.) Sliding Modes After the First Decade of the 21st Century, pp. 113–149. Springer, Berlin (2011)
11. Moreno, J.A., Osorio, M.: Strict Lyapunov functions for the super-twisting algorithm. IEEE Trans. Autom. Control **57**(4), 1035–1040 (2012)
12. Polyakov, A., Poznyak, A.: Reaching time estimation for "super-twisting" second order sliding mode controller via Lyapunov function designing. IEEE Trans. Autom. Control **54**(8), 1951–1955 (2009)
13. Seeber, R., Horn, M.: Necessary and sufficient stability criterion for the super-twisting algorithm. In: 15th International Workshop on Variable Structure Systems (VSS), pp. 120–125. IEEE (2018)
14. Seeber, R., Horn, M.: Optimal Lyapunov-based reaching time bounds for the super-twisting algorithm. IEEE Control Syst. Lett. **3**(4), 924–929 (2019)
15. Seeber, R., Horn, M., Fridman, L.: A novel method to estimate the reaching time of the super-twisting algorithm. IEEE Trans. Autom. Control **63**(12), 4301–4308 (2018)
16. Seeber, R., Reichhartinger, M., Horn, M.: A Lyapunov function for an extended super-twisting algorithm. IEEE Trans. Autom. Control **63**(10), 3426–3433 (2018)
17. Sturm, J.F.: Using SeDuMi 1.02, A Matlab toolbox for optimization over symmetric cones. Optim. Methods Softw. **11**(1-4), 625–653 (1999)
18. Utkin, V.: On convergence time and disturbance rejection of super-twisting control. IEEE Trans. Autom. Control **58**(8), 2013–2017 (2013)
19. Zhou, K., Doyle, J.C., Glover, K.: Robust and Optimal Control. Prentice Hall, Upper Saddle River (1996)

Chapter 4
Saturated Feedback Control Using Different Higher-Order Sliding-Mode Algorithms

Mohammad Ali Golkani, Stefan Koch, Markus Reichhartinger, Martin Horn and Leonid Fridman

Abstract In this chapter, a second-order system, which is subject to different disturbances, is considered. Feedback control laws adopting the continuous twisting and the twisting algorithms are designed such that continuous and saturated control signals are introduced to the system. In the case that a state variable is not available for measurement, estimate information obtained through the robust exact differentiator is incorporated into the design. For the closed loops of the aforementioned techniques, global stability properties are established. Effectiveness of the saturated continuous twisting approach using a comparison with the standard continuous twisting one is illustrated in a real-world application.

4.1 Introduction

Sliding-mode control approaches have been successfully applied to systems affected by particular classes of plant uncertainties and external disturbances [1]. Conventional sliding-mode control, i.e., first-order sliding-mode approach, can be just

M. A. Golkani (✉) · M. Reichhartinger
Institute of Automation and Control, Graz University of Technology, Graz, Austria
e-mail: golkani@tugraz.at

M. Reichhartinger
e-mail: markus.reichhartinger@tugraz.at

S. Koch · M. Horn
Christian Doppler Laboratory for Model Based Control of Complex Test Bed Systems, Graz University of Technology, Graz, Austria
e-mail: stefan.koch@tugraz.at

M. Horn
e-mail: martin.horn@tugraz.at

L. Fridman
Department of Control Engineering and Robotics, Division of Electrical Engineering, Engineering Faculty, National Autonomous University of Mexico, México D.F., Mexico
e-mail: lfridman@unam.mx

M. Steinberger et al. (eds.), *Variable-Structure Systems and Sliding-Mode Control*, Studies in Systems, Decision and Control 271,
https://doi.org/10.1007/978-3-030-36621-6_4

employed in the case that the relative degree of the system with respect to a defined sliding variable is one. It guarantees a saturated and discontinuous control input. Having designed the aforementioned sliding function, a second-order sliding-mode technique such as the twisting as well as super-twisting algorithm provides a continuous control signal. In general, these high-order sliding-mode algorithms improve the sliding accuracy of the conventional sliding mode under discrete-time measurements. They are able to counteract perturbations, which are Lipschitz continuous, and recorded in the literature as the chattering reduction strategies in the case that the actuator dynamics are fast enough [2, 3].

For a perturbed double integrator system, the twisting algorithm contributes to a finite-time convergence of the state variables to the origin. There is no need to design a sliding function. A saturated and discontinuous control signal is ensured [4]. The continuous twisting controller introduced in [5] provides a continuous actuating signal for the aforementioned system. It also enjoys the advantage that both of the states converge to zero in a finite time. As a result of adopting this algorithm, perturbations, which are Lipschitz continuous, can be theoretically exactly compensated. For systems with fast actuator dynamics, the chattering effect can be reduced. Furthermore, under discrete-time measurements, higher precision is achieved compared to the standard twisting controller if the sampling interval is small enough.

For systems with saturating actuators, the control inputs provided by the standard super-twisting and continuous twisting controllers may exceed given saturation bounds. The continuous element of the super-twisting algorithm (the square root of the sliding variable absolute value) as well as the continuous elements of the continuous twisting algorithm (the cube and square roots of the states absolute values) will not be within the bounds for every initial condition. As a result of going beyond the limits, the windup effect is produced. This is due to the discontinuous integral actions of the controllers. In [6], a domain of attraction for such a system under the standard super-twisting control is computed such that the generated signal of the controller remains within the saturation limit of the actuator. The finite-time stability within this domain is proved and there is no windup effect if the initial condition of the closed-loop system belongs to this domain. The satisfactory closed-loop performance, however, may be degenerated when the initial values are outside this domain.

To attenuate the windup effect, a second-order sliding-mode control scheme is introduced in [4], which contributes to a continuous and bounded input. A suboptimal second-order sliding-mode controller is modified in [7] in order to ensure that the sliding variable converges to the origin in a finite time despite the fact that the actuator is saturated. In both of the control laws presented in [4, 7], high-frequency switching between two control strategies based on the saturation bounds may occur. Owing to the limitation on the switching frequency, some undesirable oscillations in the control signals as well as zigzag motions in the system trajectories appear.

Saturated super-twisting algorithms handling the problem of high-frequency switching are proposed in [8–10]. At most, one switch between two different sliding-mode algorithms based on a predefined neighborhood of the origin exists in [8, 9]. The fairly restrictive assumption, which is made on the bounds of perturbations in

[8], is relaxed in [9] applying a disturbance estimator. Furthermore, the convergence of the state to zero is speeded up removing the transient process of the super-twisting algorithm through the estimator. The control law designed in [10] is compact in the sense that neither switching from one algorithm to another one nor the estimator is required. Its implementation enjoys the advantage of a simple design and it contributes significantly to an improvement in the standard super-twisting performance in the case that the initial value is far away from the origin. However, these concepts need to be modified in order to be applicable to a system of order more than one.

This chapter deals with saturated feedback controls for perturbed second-order systems. The continuous twisting and the twisting algorithms are considered. A comprehensive saturated continuous twisting control is developed for a perturbed double integrator system. A nonredundant control law is presented in the sense that the disturbance estimator is not employed. In the case that a state variable is not measurable, a sliding-mode observer estimating the state is brought into play. A saturated and Lipschitz continuous actuating signal is introduced to the system through an observer-based control law applying the twisting algorithm. It is shown that a super-twisting controller based on a high-order sliding-mode observer cannot be implemented from the mathematical point of view in the case that the control input is multiplied by an unknown function. The proposed control laws are compact in the sense that switching between two different control algorithms is not incorporated into the design. The rest of this chapter is organized as follows: the problem and the objective are explained in Sect. 4.2. The saturated continuous twisting control law design is described in Sect. 4.3. The observer-based saturated feedback control using the twisting algorithm is discussed in Sect. 4.4. Experimental results comparing performances of the saturated continuous twisting and the standard continuous twisting algorithms are demonstrated in Sect. 4.5 followed by a conclusion given in Sect. 4.6.

4.2 Problem Statement

Consider a second-order system described by the model

$$\frac{dx_1}{dt} = x_2, \tag{4.1a}$$

$$\frac{dx_2}{dt} = b(t)u + a(t) , \tag{4.1b}$$

$$y = x_1 , \tag{4.1c}$$

where x_1 , $x_2 \in \mathbb{R}$ are the state variables, u is the scalar control input, and y is the output of the system. The time-varying function $a(t)$ denotes the effect of external disturbances. It is assumed to be globally bounded and Lipschitz continuous, i.e.,

$$|a(t)| \leq a_M \qquad \text{and} \qquad \left|\frac{da}{dt}\right| \leq L_a , \qquad \forall t > 0 , \tag{4.2}$$

where a_M and L_a are some known constants. In the first case addressed in this chapter, which is also known as a perturbed double integrator (PDI) system, the function $b(t)$ is without loss of generality[1] considered as $b = 1$. In the second case, in addition to perturbations $a(t)$, the input u is multiplied by unknown $b(t)$, which is a function of time t. In that case, it is assumed that

$$0 < b_m \leq b(t) \leq 1 \qquad \text{and} \qquad \left|\frac{db}{dt}\right| \leq L_b\,, \qquad \forall t > 0\,, \tag{4.3}$$

where the constant values b_m and L_b are also known and the upper bound of $b(t)$ without loss of generality[2] is one. For the aforementioned cases, investigations of two scenarios are conducted:

(i) Full state information: it is assumed that the system output y as well as the state x_2 is available for measurement.
(ii) Output feedback control: the output is just measurable and x_2 needs to be estimated.

The objective is to design feedback control laws for different cases and scenarios of system (4.1), such that

- both of the system states x_1 and x_2 tend to the origin despite the presence of perturbations and unknowns;
- the control signal is continuous and bounded for any initial condition $x_1(t=0) = x_{1,0} \in \mathbb{R}$ and $x_2(t=0) = x_{2,0} \in \mathbb{R}$, i.e.,

$$\sup |u(t)| \leq \rho\,, \qquad \forall t > 0\,, \tag{4.4}$$

where $\rho > \frac{a_M}{b_m}$ is a given constant value.

It is noted that ρ needs to be greater than $|\frac{a(t)}{b(t)}|$ in order to be able, in principle, to drive the system states to zero through the saturated control signal (otherwise $x_2 = 0$ cannot be an equilibrium state of system (4.1)).

The saturated continuous twisting algorithm presented in [11] can be employed in both of the scenarios of the PDI system. Saturated and continuous control inputs and finite-time convergence of the system states x_1 and x_2 are ensured. It is illustrated in Sect. 4.3 that no differentiator is required in scenario (i) and the first-order robust exact differentiator (RED) (see, e.g., [12]) just needs to be used in scenario (ii). Due to the fact that $a(t)$ is just reconstructed here through the controller, this control law becomes nonredundant. Furthermore, in this section, the sufficient condition for the control parameters imposed in [11] is relaxed by introducing a geometric proof for this algorithm.

In Sect. 4.4, having adopted the twisting algorithm as presented in [13], a Lipschitz continuous control signal with known maximum absolute value is introduced to the

[1] As long as the function $b(t) \neq 0$ is known, the discussed approach may be applied.

[2] For other known upper bounds, the proposed technique may still be applied.

system, where the relative degree of the system with respect to the sliding function is one. Therefore, the system states x_1 and x_2 drive to zero asymptotically. Since the time derivative of the sliding variable is incorporated into the control law design, estimate information of x_2 and its time derivative is required in scenario (ii). Whereas, in scenario (i), the time derivative of x_2 obtained through the first-order RED needs to be applied to the controller. It is noted that for the PDI system dealt with here, this control law is redundant since both the estimator and controller reconstruct disturbances $a(t)$. However, it is shown in this section that it absolutely makes sense to use this algorithm for the PDI system with multiplicative unknown $b(t)$.

4.3 Saturated Continuous Twisting Algorithm

In this section, the PDI system is taken into consideration. Assuming that the states x_1 and x_2 are available (full state information scenario), the proposed saturated and continuous control input is obtained through

$$u = -k_1 \mathrm{sat}_{\epsilon_1}\left(\lceil x_1 \rfloor^{\frac{1}{3}}\right) - k_2 \mathrm{sat}_{\epsilon_2}\left(\lceil x_2 \rfloor^{\frac{1}{2}}\right) + \nu \,, \tag{4.5a}$$

$$\frac{d\nu}{dt} = -k_3 \lceil x_1 \rfloor^0 - k_4 \nu \,, \qquad |\nu_0| \le \frac{k_3}{k_4} \,, \tag{4.5b}$$

where the initial value of ν denoted by ν_0 and positive constants k_1, ϵ_1, k_2, ϵ_2, k_3, and k_4 need to be selected appropriately. The sat_ϵ function is defined as

$$\mathrm{sat}_\epsilon\left(\lceil \cdot \rfloor^\gamma\right) = \begin{cases} \lceil \cdot \rfloor^\gamma & \text{for } |\cdot|^\gamma < \epsilon \,, \\ \epsilon \lceil \cdot \rfloor^0 & \text{for } |\cdot|^\gamma \ge \epsilon \,. \end{cases} \tag{4.6}$$

Therefore, it becomes evident that

$$\begin{aligned} |\mathrm{sat}_{\epsilon_1}\left(\lceil x_1 \rfloor^{\frac{1}{3}}\right)| &\le \epsilon_1 \,, \quad \forall x_1 \,, \\ |\mathrm{sat}_{\epsilon_2}\left(\lceil x_2 \rfloor^{\frac{1}{2}}\right)| &\le \epsilon_2 \,, \quad \forall x_2 \,. \end{aligned} \tag{4.7}$$

Sufficient conditions for choosing the control parameters are given later in the following subsection.

Remark 4.1 The term $\lceil x_2 \rfloor^0$ within the continuous twisting algorithm presented in [5] is omitted in (4.5). This is due to the fact that it does not contribute to the stability of the origin of the closed loop.

Lemma 4.1 *If the initial value ν_0 is chosen such that $|\nu_0| \le \nu_M$ is satisfied, where $\nu_M = \frac{k_3}{k_4}$, then the actuating signal u is bounded by*

$$|u(t)| \le k_1\epsilon_1 + k_2\epsilon_2 + \nu_M \,, \quad \forall t > 0 \,. \tag{4.8}$$

Proof Since (4.5b) is a linear differential equation with the state variable ν and the bounded input $\lceil x_1 \rfloor^0$, the supremum of $|\nu|$ is

$$\sup |\nu(t)| \leq \nu_M\,, \qquad \forall t > 0\,, \tag{4.9}$$

where the condition $|\nu_0| \leq \nu_M$ is fulfilled. The upper bound of $|u|$ in (4.8) is derived easily from inequalities (4.7) and (4.9).

If the state x_2 is not measurable (output feedback control scenario), it is replaced in (4.5a) with its estimate $\hat{x}_2$ as

$$u = -k_1 \mathrm{sat}_{\epsilon_1}\left(\lceil x_1 \rfloor^{\frac{1}{3}}\right) - k_2 \mathrm{sat}_{\epsilon_2}\left(\lceil \hat{x}_2 \rfloor^{\frac{1}{2}}\right) + \nu\,, \tag{4.10a}$$

$$\frac{d\nu}{dt} = -k_3 \lceil x_1 \rfloor^0 - k_4 \nu\,, \qquad |\nu_0| \leq \nu_M\,. \tag{4.10b}$$

In [14], the estimation obtained through the first-order as well as second-order RED is incorporated into the output feedback design of the continuous twisting algorithm. The closed-loop performances applying these two observers are compared therein and it is shown that under discrete-time measurements, the accuracy is improved in the case that the second-order RED is used. However, employing this order of the differentiator makes the continuous twisting control approach redundant since perturbations are reconstructed through the estimator as well as the controller. Therefore, for the saturated continuous twisting algorithm, the estimation is provided via a super-twisting observer designed as

$$e_1 = x_1 - \tilde{x}_1\,, \tag{4.11a}$$

$$\frac{d\tilde{x}_1}{dt} = \mu_1 \lceil e_1 \rfloor^{\frac{1}{2}} + \hat{x}_2\,, \tag{4.11b}$$

$$\frac{d\hat{x}_2}{dt} = u + \mu_2 \lceil e_1 \rfloor^0\,, \tag{4.11c}$$

where $\tilde{x}_1$ is an auxiliary variable and μ_1 and μ_2 are positive values to be chosen appropriately [12].

Remark 4.2 It is noted that the upper bound of the absolute value of the control input introduced in (4.10) remains the same as that one presented in Lemma 4.1.

4.3.1 Stability Analysis

For the closed loops, with and without observer, global finite-time stability properties are established in this subsection.

4.3.1.1 Full State Information

For system (4.1) with $b = 1$ under control law (4.5), the closed-loop dynamics reads as

$$\frac{dx_1}{dt} = x_2 \,, \tag{4.12a}$$

$$\frac{dx_2}{dt} = -k_1 \text{sat}_{\epsilon_1}\left(\lceil x_1 \rfloor^{\frac{1}{3}}\right) - k_2 \text{sat}_{\epsilon_2}\left(\lceil x_2 \rfloor^{\frac{1}{2}}\right) + x_3 \,, \tag{4.12b}$$

$$\frac{dx_3}{dt} = -k_3 \lceil x_1 \rfloor^0 - k_4 x_3 + \phi(t) \,, \tag{4.12c}$$

where the auxiliary variable x_3 within the state vector $x = \begin{bmatrix} x_1 & x_2 & x_3 \end{bmatrix}^{\mathrm{T}}$ is defined as $x_3 = \nu + a(t)$. Suppose that the inequalities in (4.2) hold, then $\phi(t) = k_4 a(t) + \frac{da}{dt}$ in the closed-loop system is bounded, i.e.,

$$|\phi(t)| \leq \phi_M = k_4 a_M + L_a \,, \quad \forall t > 0 \,. \tag{4.13}$$

Remark 4.3 It is worth mentioning that the sufficient condition set in [11] is relaxed in the following. This becomes feasible through a geometric scheme, which is explained later in the proof.

Proposition 4.1 *For any positive real value ϕ_M and any real values $x_{1,0}$ and $x_{2,0}$, the state vector x tends to zero within a finite time and the absolute value of the continuous control signal is bounded by the given value ρ for all $t > 0$ if the control gains k_1, k_2, and k_3 are selected appropriately, and furthermore the parameters ϵ_1, ϵ_2, and k_4 are chosen such that*

$$\begin{aligned} \frac{1}{2} k_1 \epsilon_1 = k_2 \epsilon_2 > a_M + \frac{k_3}{k_4} \,, \\ \rho \geq k_1 \epsilon_1 + k_2 \epsilon_2 + \frac{k_3}{k_4} \end{aligned} \tag{4.14}$$

hold.

Proof Even though system (4.12) is not a weighted homogeneous system (see, e.g., [15]) in the case that $|x_1|^{\frac{1}{3}} < \epsilon_1$ and $|x_2|^{\frac{1}{2}} < \epsilon_2$ due to the presence of the linear term $-k_4 x_3$, it is possible to use the Lyapunov function candidate considered in [5] based on the homogeneous generalized form function as

$$V_1(x) = \alpha_1 |x_1|^{\frac{5}{3}} + \alpha_2 x_1 x_2 + \alpha_3 |x_2|^{\frac{5}{2}} + \alpha_4 x_1 \lceil x_3 \rfloor^2 - \alpha_5 x_2 x_3^3 + \alpha_6 |x_3|^5 \,, \tag{4.15}$$

where the coefficient vector is defined as $\alpha = \begin{bmatrix} \alpha_1 & \cdots & \alpha_6 \end{bmatrix}^{\mathrm{T}} \in \mathbb{R}^6$. It is differentiable and homogeneous of degree five with the weights $r = \begin{bmatrix} 3 & 2 & 1 \end{bmatrix}^{\mathrm{T}}$. Taking the time derivative of V_1 in (4.15) along the trajectories of system (4.12) in the aforementioned case yields

$$\frac{dV_1}{dt} = -(W_1 + W_2)\,. \tag{4.16}$$

The function W_1, which is similar to the derived function for the unperturbed case in [5], reads as

$$\begin{aligned} W_1 = {} & \beta_1 |x_1|^{\frac{4}{3}} + \beta_2 x_1 \lceil x_2 \rfloor^{\frac{1}{2}} - \beta_3 \lceil x_1 \rfloor^{\frac{2}{3}} x_2 + \beta_4 \lceil x_1 \rfloor^{\frac{1}{3}} \lceil x_2 \rfloor^{\frac{3}{2}} \\ & + \beta_5 |x_2|^2 - \beta_6 x_1 x_3 + \beta_7 |x_1| |x_3| - \beta_8 \lceil x_1 \rfloor^{\frac{1}{3}} x_3^3 \\ & + \beta_9 \lceil x_1 \rfloor^0 \lceil x_3 \rfloor^4 - \beta_{10} \lceil x_2 \rfloor^{\frac{3}{2}} x_3 - \beta_{11} x_2 \lceil x_3 \rfloor^2 - \beta_{12} \lceil x_2 \rfloor^{\frac{1}{2}} x_3^3 \\ & - \beta_{13} \lceil x_1 \rfloor^0 x_2 |x_3|^2 + \beta_{14} |x_3|^4\,, \end{aligned} \tag{4.17}$$

where the coefficients are

$$\begin{aligned} & \beta_1 = \alpha_2 k_1\,, \quad \beta_2 = \alpha_2 k_2\,, \quad \beta_3 = \frac{5}{3}\alpha_1\,, \quad \beta_4 = \frac{5}{2}\alpha_3 k_1\,, \\ & \beta_5 = \frac{5}{2}\alpha_3 k_2 - \alpha_2\,, \quad \beta_6 = \alpha_2\,, \quad \beta_7 = 2\alpha_4 k_3\,, \quad \beta_8 = \alpha_5 k_1\,, \\ & \beta_9 = 5\alpha_6 k_3\,, \quad \beta_{10} = \frac{5}{2}\alpha_3\,, \quad \beta_{11} = \alpha_4\,, \quad \beta_{12} = \alpha_5 k_2\,, \\ & \beta_{13} = 3\alpha_5 k_3\,, \quad \beta_{14} = \alpha_5\,. \end{aligned}$$

The function W_2 in (4.16) is written as

$$W_2 = -\beta_{15} x_1 |x_3| + \beta_{16} x_2 |x_3|^2 - \beta_{17} \lceil x_3 \rfloor^4\,, \tag{4.18}$$

where the coefficients denote

$$\begin{aligned} & \beta_{15} = 2\alpha_4 \left(-k_4 \nu + \frac{da}{dt}\right)\,, \quad \beta_{16} = 3\alpha_5 \left(-k_4 \nu + \frac{da}{dt}\right)\,, \\ & \beta_{17} = 5\alpha_6 \left(-k_4 \nu + \frac{da}{dt}\right)\,. \end{aligned}$$

It is worth mentioning that both W_1 and W_2 are discontinuous homogeneous functions of degree four. In order to determine conditions of the control parameters and the coefficients α such that V_1 as well as $W_1 + W_2$ is positive definite, the Pólya's theorem as proposed in [5, 16] is applied. The coordinates are changed to $z = \begin{bmatrix} z_1 & z_2 & z_3 \end{bmatrix}^{\mathrm{T}}$ as

$$|x_1| = z_1^3\,, \qquad |x_2| = z_2^2\,, \qquad |x_3| = z_3\,. \tag{4.19}$$

Since the functions V_1, W_1, and W_2 are symmetric with respect to the origin, the following four sets out of eight sets only need to be considered:

$$\{x_1, x_2, x_3 \geq 0\}\,, \qquad \{x_2, x_3 \geq 0, x_1 \leq 0\}\,,$$
$$\{x_1, x_3 \geq 0, x_2 \leq 0\}\,, \qquad \{x_1, x_2 \geq 0, x_3 \leq 0\}\,.$$

For instance, $V_1(z)$ and $W_1(z)$ for the octant $\{x_1, x_2, x_3 \geq 0\}$ are given, respectively, as

$$V_{11}(z) = \alpha_1 z_1^5 + \alpha_2 z_1^3 z_2^2 + \alpha_3 z_2^5 + \alpha_4 z_1^3 z_3^2 - \alpha_5 z_2^2 z_3^3 + \alpha_6 z_3^5\,, \tag{4.20a}$$

and

$$\begin{aligned} W_{11}(z) = {} & \beta_1 z_1^4 + \beta_2 z_1^3 z_2 - \beta_3 z_1^2 z_2^2 + \beta_4 z_1 z_2^3 + \beta_5 z_2^4 \\ & - (\beta_6 - \beta_7) z_1^3 z_3 - \beta_8 z_1 z_3^3 - \beta_{10} z_2^3 z_3 \\ & - (\beta_{11} + \beta_{13}) z_2^2 z_3^2 - \beta_{12} z_2 z_3^3 + (\beta_9 + \beta_{14}) z_3^4. \end{aligned} \tag{4.20b}$$

It is noted that the coefficients of W_2 are bounded (according to (4.9) and (4.2)). As it is explained within the proof of [5, Lemma 8], in each octant, the positive definiteness of W_1 can be also achieved in the presence of small enough variations of its coefficients. $W_2(z)$ in the aforementioned octant can be written as

$$W_{21}(z) = -\bar{\beta}_{15} z_1^3 z_3 + \bar{\beta}_{16} z_2^2 z_3^2 - \bar{\beta}_{17} z_3^4 \tag{4.21}$$

with

$$\left[\bar{\beta}_{15}\ \bar{\beta}_{16}\ \bar{\beta}_{17}\right]^{\mathrm{T}} = \left[2\alpha_4\ 3\alpha_5\ 5\alpha_6\right]^{\mathrm{T}} \phi_M \tag{4.22a}$$

or

$$\left[\bar{\beta}_{15}\ \bar{\beta}_{16}\ \bar{\beta}_{17}\right]^{\mathrm{T}} = \left[2\alpha_4\ 3\alpha_5\ 5\alpha_6\right]^{\mathrm{T}} (-k_3 - L_a)\,. \tag{4.22b}$$

In order to consider the maximum deviation from the coefficients of W_{11}, the maximum and minimum values of the coefficients of W_2 in (4.18) are computed for this octant as given in (4.22a) and (4.22b), respectively. Since $x_3 \geq 0$ holds here, ν can be just greater than or equal to $-a_M$. From (4.13), it is derived that $a_M = \frac{\phi_M - L_a}{k_4}$. Thus, $-k_4\nu + \frac{da}{dt}$ is upper bounded by ϕ_M. It becomes evident that its lower bound in this octant can be obtained by setting $\nu = \nu_M$. It can be concluded that four polynomials for the Lyapunov function and eight polynomials for its time derivative need to be sum of squares at the same time in order to ensure that V_1 and $W_1 + W_2$ are positive definite. If the control gains k_1, k_2, and k_3 for $\phi_M = 1$ (e.g., $L_a = 0.8$) are selected exemplarily as

$$k_1 = \bar{k}_1 = 8\,, \qquad k_2 = \bar{k}_2 = 5\,, \qquad k_3 = \bar{k}_3 = 1.1\,, \tag{4.23}$$

the coefficients α can be computed employing a MATLAB toolbox, e.g., SOSTOOLS [17], as

$$\begin{aligned} &\alpha_1 = 27877\,, &&\alpha_2 = 7313\,, &&\alpha_3 = 1673\,,\\ &\alpha_4 = -660\,, &&\alpha_5 = 13.8\,, &&\alpha_6 = 1.1\,. \end{aligned} \tag{4.24}$$

For other values of ϕ_M, the aforementioned control parameters are scaled as

$$k_1 = \phi_M^{\frac{2}{3}}\bar{k}_1\,, \qquad k_2 = \phi_M^{\frac{1}{2}}\bar{k}_2\,, \qquad k_3 = \phi_M\bar{k}_3\,. \tag{4.25}$$

Furthermore, as mentioned in [5], the scaled coefficients of the Lyapunov function based on (4.24) are, respectively, $\alpha_1/\phi_M^{\frac{5}{3}}$, α_2/ϕ_M^2, $\alpha_3/\phi_M^{\frac{5}{2}}$, α_4/ϕ_M^3, α_5/ϕ_M^4, and α_6/ϕ_M^5. Please note that the other constants ϵ_1, ϵ_2, and k_4 do not play a role in the formulation of this sum of squares program and they come into play in the following.

In the case that $|x_1|^{\frac{1}{3}} \geq \epsilon_1$ and $|x_2|^{\frac{1}{2}} \geq \epsilon_2$, (4.12a) and (4.12b) read as

$$\frac{dx_1}{dt} = x_2\,, \tag{4.26a}$$

$$\frac{dx_2}{dt} = -k_1\epsilon_1\lceil x_1\rfloor^0 - k_2\epsilon_2\lceil x_2\rfloor^0 + x_3\,. \tag{4.26b}$$

The twisting algorithm is recovered since the defined variable x_3 is bounded (according to (4.9) and (4.2)) by

$$|x_3(t)| \leq \eta_M = \nu_M + a_M\,, \quad \forall t > 0\,. \tag{4.27}$$

Having selected the control parameters ϵ_1, ϵ_2, and k_4 such that (4.14) is satisfied, the states x_1 and x_2 drive to the origin (see, e.g., [2]), and therefore the previous case or one of the following cases occurs. For $|x_1|^{\frac{1}{3}} \geq \epsilon_1$ and $|x_2|^{\frac{1}{2}} < \epsilon_2$, the aforementioned closed-loop system is rewritten as

$$\frac{dx_1}{dt} = x_2\,, \tag{4.28a}$$

$$\frac{dx_2}{dt} = -k_1\epsilon_1\lceil x_1\rfloor^0 - k_2\lceil x_2\rfloor^{\frac{1}{2}} + x_3\,, \tag{4.28b}$$

where (4.27) holds. If $|x_1|^{\frac{1}{3}} < \epsilon_1$ and $|x_2|^{\frac{1}{2}} \geq \epsilon_2$, having considered the boundedness of x_3, the closed-loop system is represented as

$$\frac{dx_1}{dt} = x_2\,, \tag{4.29a}$$

$$\frac{dx_2}{dt} = -k_1\lceil x_1\rfloor^{\frac{1}{3}} - k_2\epsilon_2\lceil x_2\rfloor^0 + x_3\,. \tag{4.29b}$$

For these two cases, the following local Lyapunov functions are introduced in [11], which can guarantee the boundedness of the states. Therein, the proposed Lyapunov function candidate for (4.28) is

$$V_2(x_1, x_2) = k_1\epsilon_1 |x_1| + \frac{1}{2}x_2^2 + \frac{k_2}{\epsilon_2}x_1x_2 - x_1x_3 . \tag{4.30}$$

Since x_2 as well as x_3 is bounded in this case, this function is positive definite if (4.14) is satisfied. Differentiating V_2 along the trajectories of (4.28) gives

$$\frac{dV_2}{dt} \leq -\frac{k_2}{\epsilon_2}|x_1| \left(k_1\epsilon_1 + k_2\epsilon_2 \lceil x_1 \rfloor^0 \lceil x_2 \rfloor^0 - \left(\eta_M + \frac{(2k_3 + L_a)\,\epsilon_2}{k_2} \right) \lceil x_1 \rfloor^0 \right), \tag{4.31}$$

where $2k_3 + L_a$ is the upper bound of $|\frac{dx_3}{dt}|$ in (4.12c). It can be seen that $\frac{dV_2}{dt}$ is globally negative semidefinite if

$$k_1\epsilon_1 > k_2\epsilon_2 + \eta_M + \frac{(2k_3 + L_a)\,\epsilon_2}{k_2} \tag{4.32}$$

is fulfilled. Having compared the aforementioned condition with that one imposed in (4.14), it is revealed that satisfying the relaxed sufficient condition set here in Proposition 4.1 does not result in the global negative semidefiniteness of $\frac{dV_2}{dt}$. A Lyapunov function candidate for (4.29) is presented in [11] as

$$V_3(x_1, x_2) = |x_1|^{\frac{4}{3}} + \frac{2}{3k_1}x_2^2 . \tag{4.33}$$

It becomes evident that the radially unbounded function V_3 is globally positive definite. Its time derivative along the trajectories of (4.29) yields

$$\frac{dV_3}{dt} \leq -\frac{4}{3k_1}|x_2| \left(k_2\epsilon_2 - \eta_M \lceil x_2 \rfloor^0 \right) . \tag{4.34}$$

If (4.14) is met, the globally negative semidefiniteness of $\frac{dV_3}{dt}$ is ensured, which implies that x_2 is bounded.

The convergence of the states cannot be guaranteed through the Lyapunov functions V_2 and V_3. However, it is derived intuitively from the aforementioned arguments that the system trajectory in the corresponding cases cannot escape far away from the standard trajectory of the twisting algorithm. A mathematical explanation contributes here to the global stability proof of the closed-loop origin. Intersections of a majorant curve with a straight line parallel to the axis $x_1 = 0$ drawn at $x_1 = \epsilon_1^3$ as well as straight lines parallel to the axis $x_2 = 0$ passing through $x_2 = \pm\epsilon_2^2$ for $x_1 > 0$ (see Fig. 4.1) are computed. It is shown in the following that $|x_{2,M}| < |x_{2,0}|$ is always met if (4.14) holds, which implies the states convergence. The majorant

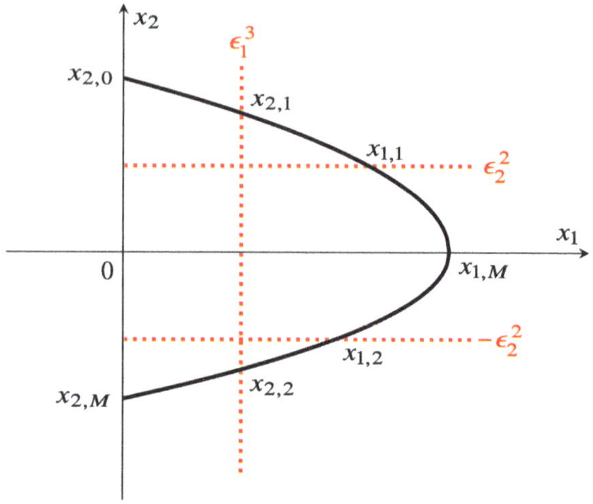

Fig. 4.1 A majorant trajectory for the saturated continuous twisting algorithm

trajectory of (4.28) for $x_1 > 0$ is pointed out in Fig. 4.1 by $x_{1,1}$, $x_{1,M}$, and $x_{1,2}$. It can be represented by

$$x_2 \frac{dx_2}{dx_1} = \begin{cases} -k_2 x_2^{\frac{1}{2}} - k_1 \epsilon_1 + \eta_M & \text{for} \quad x_2 > 0\,, \\ k_2 (-x_2)^{\frac{1}{2}} - k_1 \epsilon_1 - \eta_M & \text{for} \quad x_2 \leq 0\,, \end{cases} \tag{4.35}$$

where η_M is given in (4.27). The curve intersects the axis $x_2 = 0$ at the point $x_{1,M}$, which can be determined through the solution of (4.35) for $x_2 > 0$ based on $x_{1,1}$ (the intersection with $x_2 = \epsilon_2^2$) as

$$\begin{aligned} x_{1,M} = x_{1,1} &- \frac{2(k_1\epsilon_1 - \eta_M)^3 \ln(k_1\epsilon_1 - \eta_M + k_2\epsilon_2)}{k_2^4} + \frac{2(k_1\epsilon_1 - \eta_M)^2 \epsilon_2}{k_2^3} \\ &- \frac{(k_1\epsilon_1 - \eta_M)\epsilon_2^2}{k_2^2} + \frac{2\epsilon_2^3}{3k_2} + \frac{2(k_1\epsilon_1 - \eta_M)^3 \ln(k_1\epsilon_1 - \eta_M)}{k_2^4}\,. \end{aligned} \tag{4.36}$$

Having calculated the solution of (4.35) for $x_2 \leq 0$, $x_{1,2}$ (the intersection with $x_2 = -\epsilon_2^2$) based on $x_{1,M}$ is derived as

$$\begin{aligned} x_{1,2} = x_{1,M} &+ \frac{2(k_1\epsilon_1 + \eta_M)^3 \ln(k_1\epsilon_1 + \eta_M - k_2\epsilon_2)}{k_2^4} + \frac{2(k_1\epsilon_1 + \eta_M)^2 \epsilon_2}{k_2^3} \\ &+ \frac{(k_1\epsilon_1 + \eta_M)\epsilon_2^2}{k_2^2} + \frac{2\epsilon_2^3}{3k_2} - \frac{2(k_1\epsilon_1 + \eta_M)^3 \ln(k_1\epsilon_1 + \eta_M)}{k_2^4}\,. \end{aligned} \tag{4.37}$$

If the control parameters are selected as given in (4.14), substituting (4.36) into (4.37) gives

$$x_{1,2} = x_{1,1} - \frac{1}{k_2^4} q \tag{4.38}$$

with

$$q = 2\left(k_1^3\epsilon_1^3 + 3k_1\epsilon_1\eta_M^2\right)\ln(q_1) + 2\left(\eta_M^3 + 3k_1^2\epsilon_1^2\eta_M\right)\ln(q_2)$$
$$- \frac{28}{24}k_1^3\epsilon_1^3 - k_1\epsilon_1\eta_M^2 - \frac{1}{2}k_1^2\epsilon_1^2\eta_M\,.$$

The bounds of q_1 and q_2 are obtained as

$$2.77 < q_1 = \frac{\left(\frac{3}{2}k_1\epsilon_1 - \eta_M\right)\Big(k_1\epsilon_1 + \eta_M\Big)}{\Big(k_1\epsilon_1 - \eta_M\Big)\left(\frac{1}{2}k_1\epsilon_1 + \eta_M\right)} < 3\,,$$

$$0.75 < q_2 = \frac{\Big(k_1\epsilon_1 + \eta_M\Big)\Big(k_1\epsilon_1 - \eta_M\Big)}{\left(\frac{1}{2}k_1\epsilon_1 + \eta_M\right)\left(\frac{3}{2}k_1\epsilon_1 - \eta_M\right)} < 1.34\,.$$

Therefore, the lower bound of q is calculated as

$$q > 0.83k_1^3\epsilon_1^3 + 5k_1\epsilon_1\eta_M^2 - 0.58\eta_M^3 - 2.24k_1^2\epsilon_1^2\eta_M > 0\,.$$

This implies that $x_{1,2}$ is smaller than $x_{1,1}$. The points $x_{2,1}$ and $x_{2,2}$ of intersection with $x_1 = \epsilon_1^3$ can be obtained through the majorant trajectory of the twisting algorithm realized in (4.26). For $x_1 > 0$, this differential equation is rewritten as

$$\frac{d^2x_1}{dt^2} = \begin{cases} -k_1\epsilon_1 - k_2\epsilon_2 + \eta_M & \text{for} \quad x_2 > 0\,, \\ -k_1\epsilon_1 + k_2\epsilon_2 - \eta_M & \text{for} \quad x_2 \le 0\,, \end{cases} \tag{4.39}$$

where η_M, as mentioned above, is the upper bound of $|x_3|$. By solving (4.39), the aforementioned points are easily computed based on $x_{1,1}$ and $x_{1,2}$ as

$$x_{2,1}^2 = 2\left(x_{1,1} - \epsilon_1^3\right)\Big(k_1\epsilon_1 + k_2\epsilon_2 - \eta_M\Big) + \epsilon_2^4\,, \tag{4.40}$$

$$x_{2,2}^2 = 2\left(x_{1,2} - \epsilon_1^3\right)\Big(k_1\epsilon_1 - k_2\epsilon_2 + \eta_M\Big) + \epsilon_2^4\,. \tag{4.41}$$

Since $x_{1,2} < x_{1,1}$ holds, it can be concluded that $|x_{2,2}| < |x_{2,1}|$ is met if (4.14) is satisfied. Finally, the majorant trajectory of (4.29) for $x_1 > 0$ needs to be considered in order to determine the points $x_{2,0}$ and $x_{2,M}$. It can be denoted by

$$x_2 \frac{dx_2}{dx_1} = \begin{cases} -k_1 x_1^{\frac{1}{3}} - k_2 \epsilon_2 + \eta_M & \text{for} \quad x_2 > 0 \,, \\ -k_1 x_1^{\frac{1}{3}} + k_2 \epsilon_2 - \eta_M & \text{for} \quad x_2 \le 0 \,, \end{cases} \tag{4.42}$$

where x_3 is replaced with its bounds η_M and $-\eta_M$. The curve intersections with the axis $x_1 = 0$ are achieved based on $x_{2,1}$ and $x_{2,2}$ through the solution of (4.42) for $x_2 > 0$ and $x_2 \le 0$, respectively. They read as

$$x_{2,0}^2 = x_{2,1}^2 + 2\left(\frac{3}{4}k_1\epsilon_1 + k_2\epsilon_2 - \eta_M\right)\epsilon_1^{\frac{3}{2}} \,, \tag{4.43}$$

$$x_{2,M}^2 = x_{2,2}^2 + 2\left(\frac{3}{4}k_1\epsilon_1 - k_2\epsilon_2 + \eta_M\right)\epsilon_1^{\frac{3}{2}} \,. \tag{4.44}$$

Having fulfilled (4.14), $|x_{2,M}| < |x_{2,0}|$ is guaranteed owing to that $|x_{2,2}|$ is smaller than $|x_{2,1}|$. It can be carried out similarly for the half-plane $x_1 \le 0$.

This completes the global asymptotic stability proof of the origin $x = 0$ in the full state information scenario. Since the continuous function $-k_4 x_3 + \phi(t)$ within inhomogeneous system (4.12) is bounded (due to the boundedness of ν and $\frac{da}{dt}$), the conditions of the quasi-homogeneity principle [18, Theorem 4.2] are satisfied. Consequently, the finite-time convergence of the states is justified applying this principle. Furthermore, according to Lemma 4.1, it is ensured that the control input is bounded for any $x_{1,0},\ x_{2,0} \in \mathbb{R}$. In order to retain it within the given saturation bounds $-\rho$ and ρ, it is sufficient that the control constants are chosen as given in (4.14). □

4.3.1.2 Output Feedback Control

In the following, the stability analyses of the closed loop, in which the observer is incorporated into the design, are carried out.

Proposition 4.2 *Suppose that the assumptions in* (4.2) *are fulfilled. For system (4.1) with $b = 1$ under control law (4.10) employing observer (4.11), where $x_{1,0}$ and $x_{2,0}$ may be any real values, if the observer constant μ_2 is selected greater than a_M and μ_1 sufficiently large and the control parameters are chosen such that Proposition 4.1 is satisfied, then the origin $x = 0$ is globally finite-time stable and the continuous actuating signal remains bounded by the given saturation limit ρ.*

Proof Having considered observer dynamics (4.11) and defined the error variable $e_2 = x_2 - \hat{x}_2$, the overall closed-loop system is represented as

$$\frac{dx_1}{dt} = x_2 \,, \tag{4.45a}$$

$$\frac{dx_2}{dt} = -k_1 \text{sat}_{\epsilon_1}\left(\lceil x_1 \rfloor^{\frac{1}{3}}\right) - k_2 \text{sat}_{\epsilon_2}\left(\lceil x_2 - e_2 \rfloor^{\frac{1}{2}}\right) + x_3 \,, \tag{4.45b}$$

$$\frac{dx_3}{dt} = -k_3 \lceil x_1 \rfloor^0 - k_4 x_3 + \phi(t) \,, \tag{4.45c}$$

$$\frac{de_1}{dt} = -\mu_1 \lceil e_1 \rfloor^{\frac{1}{2}} + e_2 \,, \tag{4.45d}$$

$$\frac{de_2}{dt} = -\mu_2 \lceil e_1 \rfloor^{0} + a(t) \,. \tag{4.45e}$$

By taking into account the vector

$$\zeta = \left[\zeta_1 \; \zeta_2 \; \zeta_3\right]^{\mathrm{T}} = \left[x_1 \; x_2 - e_2 \; x_3\right]^{\mathrm{T}} , \tag{4.46}$$

the subsystem in the variables x_1, x_2, and x_3, i.e., (4.45a) to (4.45c), is rewritten as

$$\frac{d\zeta_1}{dt} = \zeta_2 + e_2 \,, \tag{4.47a}$$

$$\frac{d\zeta_2}{dt} = -k_1 \mathrm{sat}_{\epsilon_1}\left(\lceil \zeta_1 \rfloor^{\frac{1}{3}}\right) - k_2 \mathrm{sat}_{\epsilon_2}\left(\lceil \zeta_2 \rfloor^{\frac{1}{2}}\right) + \zeta_3 - \frac{de_2}{dt} \,, \tag{4.47b}$$

$$\frac{d\zeta_3}{dt} = -k_3 \lceil \zeta_1 \rfloor^{0} - k_4 \zeta_3 + \phi(t) \,. \tag{4.47c}$$

A possible choice of the observer constants as given in [12] is $\mu_2 = 1.1 a_M$ and $\mu_1 = 1.5\sqrt{a_M}$. As a result of choosing μ_1 and μ_2 properly, the observer errors e_1 and e_2 are bounded and converge to zero in the finite time T. It becomes evident that $\frac{de_2}{dt}$ is also bounded for all $t > 0$. A similar geometric approach as mentioned above can be used in order to show that within cascaded system (4.45), subsystem (4.47) is input-to-state stable with respect to the input $\left[e_2 \; -\frac{de_2}{dt} \; 0\right]^{\mathrm{T}}$. Therefore, it is ensured that the state vector ζ remains bounded for all $t > 0$. The finite-time convergence of the vector x, for $t \geq T$, is guaranteed as it is proved for the full state information scenario. □

4.4 Saturated Feedback Control Using the Twisting Algorithm

It becomes evident that in the case of the PDI system with an unknown $b(t)$, there is no chance to reconstruct either a or b, or both using an RED. Furthermore, for the scenario of output feedback control, the super-twisting algorithm cannot be applied from the mathematical point of view, which is explained in the following.

In [19], it is shown that the implementation of the standard super-twisting controller using the first-order RED as given in (4.11) to estimate x_2 does not have a mathematical justification. This is due to that, in the overall closed-loop system, the discontinuous element of this differentiator lies in the same channel as the continuous element of the controller, and therefore the second-order sliding mode does not exist. It is proposed therein that the estimation of x_2 should be obtained through a higher-order differentiator. Having employed the second-order RED, a state observer for system (4.1) is introduced as

$$e_1 = x_1 - \tilde{x}_1 \,, \tag{4.48a}$$
$$\frac{d\tilde{x}_1}{dt} = \mu_1 \lceil e_1 \rfloor^{\frac{2}{3}} + \hat{x}_2 \,, \tag{4.48b}$$
$$\frac{d\hat{x}_2}{dt} = u + \mu_2 \lceil e_1 \rfloor^{\frac{1}{3}} + \tilde{x}_2 \,, \tag{4.48c}$$
$$\frac{d\tilde{x}_2}{dt} = \mu_3 \lceil e_1 \rfloor^{0} \,, \tag{4.48d}$$

where $\hat{x}_2$ denotes the estimation of the state x_2, $\tilde{x}_1$ and $\tilde{x}_2$ are auxiliary variables, and the positive values μ_1, μ_2, and μ_3 are observer constants to be chosen appropriately. It may be noted that $u + \tilde{x}_2$ represents the estimate information of system's second channel (4.1b). By defining the error variables $e_2 = x_2 - \hat{x}_2$ and $\tilde{e}_2 = (b(t) - 1)\, u + a(t) - \tilde{x}_2$, the error dynamics is written as

$$\frac{de_1}{dt} = -\mu_1 \lceil e_1 \rfloor^{\frac{2}{3}} + e_2 \,, \tag{4.49a}$$
$$\frac{de_2}{dt} = -\mu_2 \lceil e_1 \rfloor^{\frac{1}{3}} + \tilde{e}_2 \,, \tag{4.49b}$$
$$\frac{d\tilde{e}_2}{dt} = -\mu_3 \lceil e_1 \rfloor^{0} + \psi \,, \tag{4.49c}$$

where ψ reads as

$$\psi = \frac{db}{dt} u + \Big(b(t) - 1\Big) \frac{du}{dt} + \frac{da}{dt} \,. \tag{4.50}$$

In order to achieve the convergence of e_1, e_2, and $\tilde{e}_2$ to zero in the finite time T, ψ needs to be globally bounded [12].

The estimation of the state x_2 is incorporated into the sliding function definition as

$$\sigma_1 = \hat{x}_2 + \lambda x_1 \,, \tag{4.51}$$

where λ is a positive constant value. A control law based on the super-twisting algorithm is designed as

$$u = -\lambda \hat{x}_2 - k_1 \lceil \sigma_1 \rfloor^{\frac{1}{2}} + \nu \,, \tag{4.52a}$$
$$\frac{d\nu}{dt} = -k_2 \lceil \sigma_1 \rfloor^{0} - \mu_3 \lceil e_1 \rfloor^{0} \,, \tag{4.52b}$$

where control parameters $k_1 > 0$ and $k_2 > 0$ need to be selected appropriately. It is noted that the control signal u is continuous, but it is not Lipschitz continuous. Consequently, ψ in (4.50) cannot be bounded for all $t > 0$ when $b(t)$ is unknown.

On this occasion, the twisting algorithm may be adopted to handle the problem. Since perturbations $a(t)$ cannot be estimated and compensated in a controller, this robust control technique becomes nonredundant. Moreover, providing a continuous control signal by using the twisting algorithm contributes significantly to avoidance of two dangerous chattering classes known as bounded and unbounded [20]. In order to introduce this continuous control input to the system, a sliding function needs to be defined such that the relative degree of the system with respect to this function is one. Due to the fact that the time derivative of the sliding variable is included in the control law design, in the scenario of full state information, the estimation of the time derivative of x_2 obtained through the first-order RED needs to be exploited to build up the controller. However, in the output feedback control scenario, estimates of x_2 and its time derivative obtained through the second-order RED mentioned in (4.48) need to be used. Since sliding functions and control laws will be similar in the scenarios of this case, the output feedback control scenario is considered here.

4.4.1 *Output Feedback Control*

Having considered sliding function (4.51), the proposed saturated and Lipschitz continuous control input is obtained through

$$\frac{du}{dt} = -k\left(\lceil\sigma_1\rfloor^0 + \frac{1}{2}\lceil\sigma_2\rfloor^0\right) - \mu_3\lceil e_1\rfloor^0 - \lambda u\,, \qquad u(t=0) = u_0\,, \tag{4.53}$$

where k is a positive constant to be selected and σ_2 is defined as

$$\sigma_2 = u + \tilde{x}_2 + \lambda\hat{x}_2\,. \tag{4.54}$$

In the following, it is outlined how the initial value u_0 of the control signal needs to be chosen.

Lemma 4.2 *The supremum of the control signal absolute value is*

$$\sup|u(t)| \le u_M = \frac{3k + 2\mu_3}{2\lambda}\,, \qquad \forall t > 0\,, \tag{4.55}$$

where the initial value u_0 is selected such that $|u_0| \le u_M$ holds.

Similar to the proof of Lemma 4.1, (4.55) can be easily derived since (4.53) is a linear differential equation with the state variable u and a bounded input. As a result of this lemma, the Lipschitz constant of the control input reads as

$$\left|\frac{du}{dt}\right| \le \frac{3k}{2} + \mu_3 + \lambda u_M = 2\lambda u_M\,. \tag{4.56}$$

Remark 4.4 It can be clearly seen that the proposed control law is compact in the sense that switching between two control strategies based on the saturation bound ρ is not required. This contributes considerably to alleviation of undesirable oscillations in the control signal as well as zigzag motions in the system trajectories, which may be caused by the structure of saturated control laws presented in [4, 7].

4.4.2 Stability Analysis

In the following, the stability properties of the closed loop, in which the observer error dynamics is considered, are investigated.

Proposition 4.3 *Suppose that the assumptions in* (4.2) *and* (4.3) *are satisfied. For system (4.1) with unknown* $b(t)$ *and any real values* $x_{1,0}$ *and* $x_{2,0}$ *under control law* (4.53)*, the system states* x_1 *and* x_2 *tend to zero asymptotically and the continuous control signal remains within the given saturation bounds, i.e.,* $u \in [-\rho, \rho]$*, if the observer gains* μ_1 *and* μ_2 *are selected large enough, and the observer gain* μ_3 *and the control parameters* k *and* λ *are chosen such that*

$$\begin{aligned} \mu_3 > \psi_M &= \Big(L_b + 2\lambda\,(1-b_m)\Big)u_M + L_a\,, \\ \rho &\geq \frac{3k+2\mu_3}{2\lambda}\,, \\ k &> 2\lambda\Big(u_M\,(1-b_m)+a_M\Big) \end{aligned} \tag{4.57}$$

hold.

Proof Substituting (4.53), (4.48c), (4.48d), and (4.49b) into the time derivatives of σ_1 and σ_2 gives

$$\frac{d\sigma_1}{dt} = \sigma_2 + \mu_2\,\lceil e_1 \rfloor^{\frac{1}{3}} + \lambda e_2\,, \tag{4.58a}$$

$$\frac{d\sigma_2}{dt} = -k\left(\lceil \sigma_1 \rfloor^{0} + \frac{1}{2}\lceil \sigma_2 \rfloor^{0}\right) + \lambda\left((b-1)\,u + a - \frac{de_2}{dt}\right). \tag{4.58b}$$

Having considered the boundedness of control input and its time derivative presented in (4.55) and (4.56), it can be concluded that the absolute value of ψ in (4.50) is globally bounded by the calculable constant value ψ_M. Hence, the observer errors e_1, e_2, and $\tilde{e}_2$ converge to zero in the finite time T if the positive gains μ_1, μ_2, and μ_3 are chosen properly. Exemplarily, as proposed in [21], they can be selected as $\mu_1 = 2\psi_M^{\frac{1}{3}}$, $\mu_2 = 1.5 \times \sqrt{2}\psi_M^{\frac{2}{3}}$, and $\mu_3 = 1.1\psi_M$. It becomes evident that e_1, e_2, as well as $\frac{de_2}{dt}$ is bounded for all $t > 0$. As proven in [22, Theorem 5.1], within the overall closed-loop system considering (4.58) and (4.49), the trajectories of driven subsystem (4.58) cannot scape to infinity in a finite time. Therefore, for $t < T$, the states σ_1 and σ_2 cannot become unbounded and afterward, for $t \geq T$, their finite-time convergence is guaranteed if the control constants k and λ are selected such that the inequalities in (4.57) are satisfied. This implies asymptotic stability of the system states x_1 and x_2. The boundedness of the actuating signal for any $x_{1,0}$, $x_{2,0} \in \mathbb{R}$ is justified applying Lemma 4.2. It becomes evident that its absolute value is bounded by the constant value ρ if $u_M \leq \rho$ is fulfilled. □

It is noted that a feasible region of the observer and controller gains μ_3 and k based on the inequalities given in (4.57) may be found numerically through a computer algebra software.

4.5 Experimental Implementation

In this section, a system, which represents an ideal platform to test the standard continuous twisting algorithm (Standard CTA) as well as saturated continuous twisting algorithm (Saturated CTA) in a real-world application, is considered. Since only its output is available for measurement, the output feedback control strategy is applied.

Hydraulic actuators offer a very high power-to-weight ratio, modular design, high precision, and durability. Typically they are used in industrial applications, which demand high forces or torques (e.g., in heavy equipment like earthmoving or forestry machines). Currently, there is a strong trend toward fully or at least partially automating such working machines. Automation requires advanced low-level control strategies allowing precise control of the hydraulic actuators. However, uncertainties like unknown load forces, external disturbances, and changing operating conditions render the control design a rather challenging task.

The synthesis of position control systems of hydraulic cylinders can be divided into two steps. The first step aims to design a controller for the nonlinear valve system, which, in most cases, relies on an exact linearization of the valve dynamics. Typically the valve dynamics are completely known and the parameters, which remain constant during operation, are mostly available in data-sheets. In the second step, an outer loop controller is designed for the moving piston. This mechanical subsystem is subject to external forces and possibly time-varying plant parameters (e.g., due to changing masses of the load). Therefore, usually a robust control strategy is essential.

The work presented in [23] forms the basis for the following study. Therein, a cascaded control structure was designed for the reference trajectory tracking of the piston rod of the hydraulic differential cylinder subject to an unknown load force. The proposed control law is implemented on the test bench which is equipped with industrial hydraulic components. It consists of two coupled hydraulic cylinders. One of these cylinders is regarded as the operating cylinder, while the other one is used to apply certain load profiles (disturbances). For evaluation purposes, a force sensor which provides real-time measurements of the external load force is installed at the test rig. The controller for the operating cylinder is composed of an inner loop, which aims to linearize the valve dynamics and an outer loop. Here, saturated continuous twisting controller for reference trajectory tracking is applied to the outer loop. The first-order RED is implemented to estimate the unmeasured state variable. The inner control loop also requires full state information. Hence, the information of the observer is provided to both loops. It is noted that the load force in the outer piston position control loop is not reconstructed through the estimator.

4.5.1 System Model

A schematic diagram of a hydraulic differential cylinder is depicted in Fig. 4.2. A typical feature of a differential cylinder is that the effective piston cross sections have different area. Due to this characteristic, depending on the direction, the cylinder moves at two different velocities at a constant flow rate Q_A and Q_B, respectively. The control goal is to make the position of the piston rod x track a certain reference profile despite the unknown load force F_{ext}. The flows Q_A and Q_B are regulated by a servo valve. A hydraulic pump supplies the valve with a substantially constant pressure, which is assumed to be independent on the external load force. The pressure and the volume in each chamber are denoted by p_A, p_B and V_A, V_B, respectively.

A mathematical model describing the dynamics of the piston movement is derived by applying Newton's second law of motion as

$$m\frac{d^2x}{dt^2} = F_h - F_r - F_{ext} \ , \tag{4.59}$$

where m is the total moving mass, i.e., the piston mass plus the mass of the hydraulic medium, and F_r represents the friction force. The hydraulic force reads as

$$F_h = (p_A - \alpha p_B)A_k \ , \tag{4.60}$$

where A_k is the so-called piston ring surface and α represents the ratio between the piston rod cross section and the piston ring surface. It is assumed that the valve is controlled and that the closed-loop dynamics is described by an integrator, i.e.,

$$\frac{dF_h}{dt} = u_h \ , \tag{4.61}$$

where u_h is considered as the system's input.

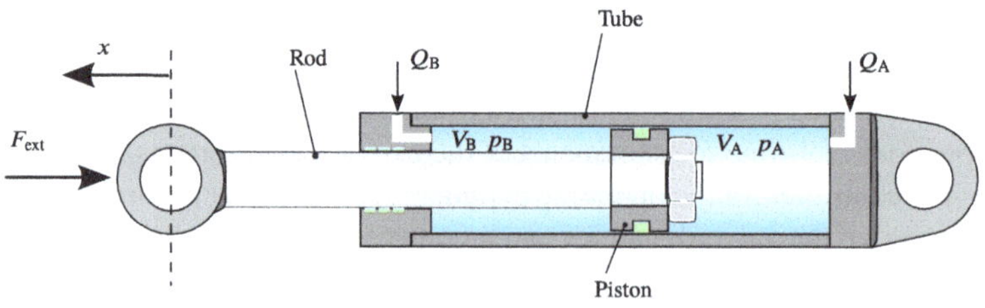

Fig. 4.2 A layout of the differential hydraulic cylinder

4.5.2 Control Design

The control input is obtained through

$$u_{\mathrm{h}} = k_0(F_{\mathrm{h,d}} - F_{\mathrm{h}}) \,, \tag{4.62}$$

where $F_{\mathrm{h,d}}$ denotes the desired piston force and k_0 is a positive constant. Hence, an inner force control loop with closed-loop dynamics

$$\frac{dF_{\mathrm{h}}}{k_0 dt} + F_{\mathrm{h}} = F_{\mathrm{h,d}} \tag{4.63}$$

is established. The choice $k_0 > 0$ ensures that the hydraulic force F_{h} asymptotically tracks a constant desired force $F_{\mathrm{h,d}}$. Having applied the saturated continuous twisting control law, a bounded desired piston force is introduced as

$$F_{\mathrm{h,d}} = u_{\mathrm{CTA}} + F_{\mathrm{r}} \,, \tag{4.64}$$

where u_{CTA} is identical to u given in (4.10) with $x_1 = x - x_{\mathrm{d}}$ and $\hat{x}_2 = \frac{dx}{dt} - \frac{dx_{\mathrm{d}}}{dt}$. It yields an outer feedback loop for reference trajectory tracking of the piston position and velocity. In the outer closed loop, x_3 reads as

$$x_3 = \nu - F_{\mathrm{ext}} - m\frac{d^2 x_{\mathrm{d}}}{dt^2} \,. \tag{4.65}$$

Please note that this control strategy does not require any information on the piston acceleration and external load. However, implementation of the outer control loop requires a model-based estimation of the friction force F_{r} as well as estimation of the piston velocity $\frac{dx}{dt}$. In order to achieve a perfect tracking in principle, a smooth function of time t to be twice differentiable and slow enough such that its second time derivative which is negligible is used as the reference position x_{d} (see Fig. 4.3). As a consequence, $-F_{\mathrm{ext}}$ is regarded as perturbations $a(t)$ taken into consideration in (4.1).

It is noted that, as it is explained in the previous section, it does not make sense to apply the twisting algorithm to this system since there is no multiplicative unknown $b(t)$. Thus, the closed-loop performance using the saturated CTA is compared here with the result achieved through the standard CTA. Having set $k_4 = 0$, $\epsilon_1 = \infty$, and $\epsilon_2 = \infty$ in (4.10), the standard CTA is recovered. The rest of the control parameters is well tuned as

$$k_1 = 1000 \,, \qquad k_2 = 800 \,, \qquad k_3 = 1500 \,.$$

In order to assess the effectiveness of the proposed saturated CTA, the aforementioned constants k_1, k_2, and k_3 are left the same in implementation of control law (4.10). The three other control gains are selected in this case such that $k_1\epsilon_1 + k_2\epsilon_2 + \frac{k_3}{k_4} \leq \rho$

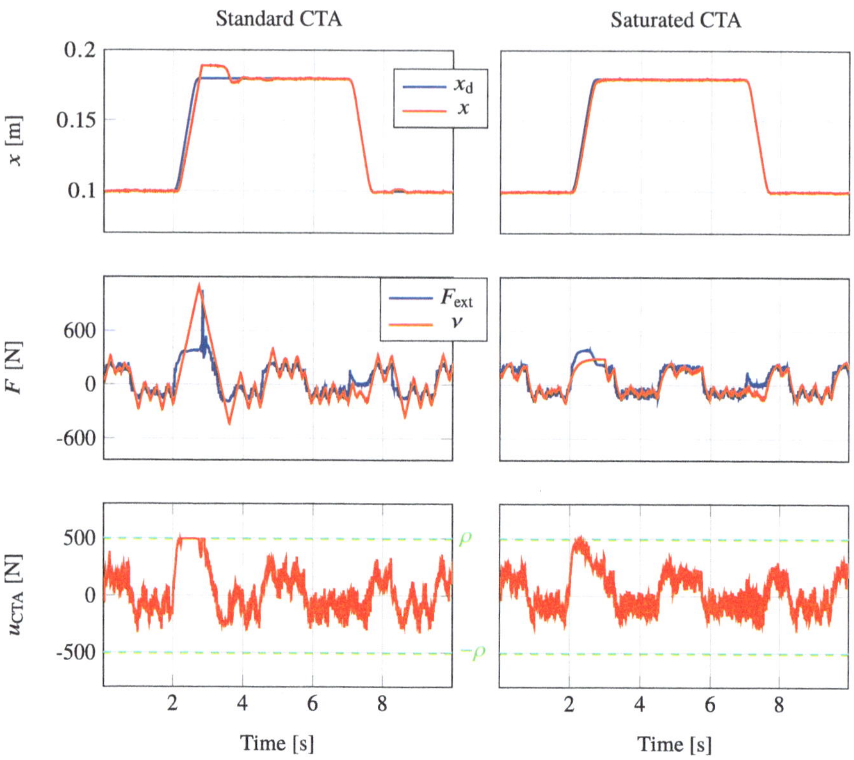

Fig. 4.3 Experimental response curves for output feedback control comparing the proposed saturated continuous twisting algorithm with the standard one. The position, external force, and control input are illustrated

holds, where ρ is assumed to be 500N. They can be

$$k_4 = 6\,, \qquad \epsilon_1 = 0.12\,, \qquad \epsilon_2 = 0.16\,.$$

Furthermore, the state observer is tuned such that the estimation error tends to zero in a finite time faster than the convergence of the system state.

The performance of both of the algorithms is depicted in Fig. 4.3. In the lower plot (right), it can be clearly seen that due to the structure of the saturated CTA, the produced actuating signal is bounded by the saturation limit specification. Whereas, the control input of standard CTA (left) introduced to the system is saturated through the actuator since the generated signal by the controller is got beyond the limit. As it is shown in the middle plots, both of the controllers can reconstruct the external force properly. A large overshoot is demonstrated in the response curve of piston position in the case of standard CTA. In contrast, the windup effect is significantly mitigated as well as a satisfactory performance is achieved in the case that saturated CTA is applied.

4.6 Conclusion

For both scenarios of the perturbed double integrator system, full state information and output feedback control, the saturated continuous twisting algorithm is applied. The Lyapunov function and the geometric scheme are incorporated into the global asymptotic stability proof of the closed-loop system origin. The finite-time convergence of the system states is realized based on the quasi-homogeneity principle. In the case that the control input is multiplied by a known function, the observer-based saturated feedback control adopting the twisting algorithm is employed. The finite-time convergence of the sliding function is guaranteed, which implies that the system states tend to the origin asymptotically. It is shown that the absolute values of the continuous actuating signals generated through the aforementioned strategies are bounded by the known constants. Therefore, the controllers can be tuned such that the signals remain within the saturation bounds. This contributes greatly to alleviation of the windup effect, which is confirmed through the experimental studies on a hydraulic differential cylinder.

Acknowledgements The authors would like to appreciate the financial support of the European Union's Horizon 2020 Research and Innovation Programme (H2020-MSCA-RISE-2016) under the Marie Sklodowska-Curie grant agreement No. 734832. The financial support of the Austrian Federal Ministry for Digital and Economic Affairs and the National Foundation for Research, Technology and Development is also gratefully acknowledged. Furthermore, L. Fridman would like to express his gratitude for the financial support of Consejo Nacional de Ciencia y Tecnología (CONACYT), grant No. 282013; Programa de Apoyo a Proyectos de Investigación e Innovación Tecnológica (PAPIIT-UNAM), grant No. IN 115419.

References

1. Li, S., Yu, X., Fridman, L., Man, Z., Wang, X.: Advances in Variable Structure Systems and Sliding Mode Control-Theory and Applications, vol. 115. Springer, Berlin (2017)
2. Shtessel, Y., Edwards, C., Fridman, L., Levant, A.: Sliding Mode Control and Observation. Springer, Berlin (2014)
3. Pérez-Ventura, U., Fridman, L.: When is it reasonable to implement the discontinuous sliding-mode controllers instead of the continuous ones? Frequency domain criteria. Int. J. Robust Nonlinear Control **29**(3), 810–828 (2019)
4. Levant, A.: Sliding order and sliding accuracy in sliding mode control. Int. J. Control **58**(6), 1247–1263 (1993)
5. Torres-González, V., Sanchez, T., Fridman, L.M., Moreno, J.A.: Design of continuous twisting algorithm. Automatica **80**, 119–126 (2017)
6. Behera, A.K., Chalanga, A., Bandyopadhyay, B.: A new geometric proof of super-twisting control with actuator saturation. Automatica **87**, 437–441 (2018)
7. Ferrara, A., Rubagotti, M.: A sub-optimal second order sliding mode controller for systems with saturating actuators. IEEE Trans. Autom. Control **54**(5), 1082–1087 (2009)
8. Castillo, I., Steinberger, M., Fridman, L., Moreno, J.A., Horn, M.: Saturated super-twisting algorithm: Lyapunov based approach. In: 14th International Workshop on Variable Structure Systems (VSS), pp. 269–273 (2016)

9. Castillo, I., Steinberger, M., Fridman, L., Moreno, J., Horn, M.: Saturated super-twisting algorithm based on perturbation estimator. In: IEEE 55th Conference on Decision and Control (CDC), pp. 7325–7328 (2016)
10. Golkani, M.A., Koch, S., Reichhartinger, M., Horn, M.: A novel saturated super-twisting algorithm. Syst. Control Lett. **119**, 52–56 (2018)
11. Golkani, M.A., Fridman, L.M., Koch, S., Reichhartinger, M., Horn, M.: Saturated continuous twisting algorithm. In: 15th International Workshop on Variable Structure Systems (VSS), pp. 138–143 (2018)
12. Levant, A.: Higher-order sliding modes, differentiation and output-feedback control. Int. J. Control **76**(9–10), 924–941 (2003)
13. Golkani, M.A., Fridman, L., Koch, S., Reichhartinger, M., Horn, M.: Observer-based saturated output feedback control using twisting algorithm. In: 14th International Workshop on Variable Structure Systems (VSS), pp. 246–250 (2016)
14. Sanchez, T., Moreno, J.A., Fridman, L.M.: Output feedback continuous twisting algorithm. Automatica **96**, 298–305 (2018)
15. Levant, A.: Homogeneity approach to high-order sliding mode design. Automatica **41**(5), 823–830 (2005)
16. Sanchez, T., Moreno, J.A.: A constructive Lyapunov function design method for a class of homogeneous systems. In: 53rd IEEE Conference on Decision and Control, pp. 5500–5505 (2014)
17. Papachristodoulou, A., Anderson, J., Valmorbida, G., Prajna, S., Seiler, P., Parrilo, P.A.: SOSTOOLS: Sum of squares optimization toolbox for MATLAB. Available from http://www.eng.ox.ac.uk/control/sostools and http://www.cds.caltech.edu/sostools (2013). arXiv:1310.4716
18. Orlov, Y.V.: Discontinuous Systems: Lyapunov Analysis and Robust Synthesis Under Uncertainty Conditions. Springer Science & Business Media, Berlin (2008)
19. Chalanga, A., Kamal, S., Fridman, L.M., Bandyopadhyay, B., Moreno, J.A.: Implementation of super-twisting control: super-twisting and higher order sliding-mode observer-based approaches. IEEE Trans. Ind. Electron. **63**(6), 3677–3685 (2016)
20. Levant, A.: Chattering analysis. IEEE Trans. Autom. Control **55**(6), 1380–1389 (2010)
21. Livne, M., Levant, A.: Proper discretization of homogeneous differentiators. Automatica **50**(8), 2007–2014 (2014)
22. Moreno, J.A.: A Lyapunov approach to output feedback control using second-order sliding modes. IMA J. Math. Control Inf. 291–308 (2012)
23. Koch, S., Reichhartinger, M.: Observer-based sliding mode control of hydraulic cylinders in the presence of unknown load forces. e & i Elektrotechnik und Informationstechnik **133**(6), 253–260 (2016)

Chapter 5
Constrained Sliding-Mode Control: A Survey

Massimo Zambelli and Antonella Ferrara

Abstract In this chapter, the robust control of input and state-constrained nonlinear systems is discussed from the perspective of sliding-mode control. The proposals currently available in the literature are summarized and some of them are briefly discussed, in particular cases providing also commented examples.

5.1 Introduction

In practical applications, the exertable control is always subject to constraints [1]. On the one hand, limitations exist due to physical reasons (e.g., on the torque produced by a motor, the force a car can apply on the ground through the tires, or the power a generator can provide to an electric circuit). The consequently appearing saturation effects can, in principle, disrupt performance and even stability. One example is the windup phenomenon that appears if an integral action is present in the controller (for instance, in PIDs [2, 3]). In that case, if the saturation zone is reached and held for a sufficiently long time, the actually applied control signal becomes appreciably different from that requested by the regulator, with obvious consequences in performance and effectiveness. On the other hand, it can sometimes also be useful to explicitly bound the control action in order to preserve the actuators themselves and/or the controlled plant (for instance, to reduce wear in mechanical systems, heat generation in electric circuits, and aging effects in lithium-ion cells).

Similarly, the states are often confined to a subset of the state space [4]. In some cases, this is imposed by structural reasons (i.e., they cannot physically exceed particular thresholds, as is the case, e.g., for a robot manipulator which is unable to reach points outside of its workspace), while in other situations it is desirable for safety or performance (cars in a platoon, for instance, must be constrained to keep a

M. Zambelli (✉) · A. Ferrara
University of Pavia, 27100 Pavia, Italy
e-mail: massimo.zambelli01@universitadipavia.it

A. Ferrara
e-mail: antonella.ferrara@unipv.it

M. Steinberger et al. (eds.), *Variable-Structure Systems and Sliding-Mode Control*, Studies in Systems, Decision and Control 271,
https://doi.org/10.1007/978-3-030-36621-6_5

satisfactorily large distance one from the successive in order to minimize the possibility of crashes). Actually, lots of other different constraints are typically present (among many others, input rate saturations, non-holonomicity, sensors reading rates) [5], but will not be explicitly accounted for in the present chapter.

Having a controller aware of the existing limitations, them being due to the system structure or imposed by design, is of crucial importance in order to enforce the desired behavior. Nevertheless, basic control techniques are not designed, in general, to effectively deal with limited actuators effort and state values. To this end, instead, one often resorts to model predictive control (MPC) [6–9], a well-known family of optimal control algorithms able to explicitly deal with state and input constraints. All of the flexibility offered by MPC comes, however, at the expense of computational complexity [10]. As a matter of fact, while specific formulations (e.g., those involving linear plant dynamics and constraints, and a quadratic cost function) can be efficiently run online due to the relatively low computational power requirements [11], when it comes to nonlinear systems (and/or constraints) the computational complexity grows very fast even in the presence of sufficiently tractable cost functions. The computational expense depends of course on the particular case, but most of the times it becomes quite hard to be able to implement such nonlinear MPC (NMPC) algorithms for online control, since the time required for the optimization program rapidly increases as the number of states and the length of the prediction horizon grow [12]. Furthermore, while for some simple cases a well-consolidated theory exists so that convergence and expected performance can be assessed a priori [13], for general nonlinear systems and cost functions, it becomes very hard to theoretically prove stability at least when constraints are active. Of course, linearization in the latter cases could be a viable option to reduce complexity and is often adopted in practice. However, it can be argued that such an approximation does not seriously harm performance and stability only in the presence of slow dynamics or quasi-linear system behaviors, which could be not the case most of the times. For the control of fast nonlinear dynamic systems, especially when inexpensive hardware implementations are required, one must try to resort to different, explicit control laws.

Additionally, it is important to underline the necessity for robustness, which is always a requirement in practical applications due to unavoidable modeling uncertainties and external disturbances [14–17]. In view of the receding horizon principle, most MPC formulations can be considered somehow robust by design in the presence of ideally precise state-feedback measurements (if in addition a sufficiently accurate model of the plant is provided). In case of disturbances on the input or on the measurements or large model uncertainty, however, robustness cannot be guaranteed with standard formulations, so that optimality has for sure to be dropped, while stability can also be at risk. Particular robust MPC algorithms exist in the literature (see, e.g., [18] and references therein), but come at the expense of an increased overall complexity.

Sliding-mode control (SMC) [19–22] is well suited for robust nonlinear control and allows for very lightweight implementations from a computational point of view even in the presence of highly complex systems. On the other hand, the classical theory does not enable for the consideration of input or state constraints. It appears then natural to try to preserve all the positive features of SMC (in particular, robustness and high performance) while explicitly taking into account saturations and state limitations.

To this end, some effective proposals couple SMC with MPC, to enforce robustness in the latter at very low additional expense. This allows to maintain optimality and constraints awareness, but unfortunately keeps also the computational issues previously discussed. In particular, works like [23–26] introduce an integral sliding-mode (ISM) correction term that aims at completely reject the matched uncertainty. The optimization over the prediction horizon can then be carried out relying on a nominal (completely known) description of the system. The concept has been later extended also to asynchronous networked systems [27] and the cases in which only output information is available [28].

In the following of the chapter, solutions inherently based on sliding-mode control will be presented and, for some of them, examples will be discussed. In particular, in Sect. 5.1.1 algorithms dealing with only input saturations will be considered. In Sect. 5.1.2, focus will be given to those including only state constraints and Sect. 5.1.3 will be devoted to describe some solutions which take into account both state and input limitations at the same time. To conclude, in Sect. 5.1.4 a brief recap with a summary of possible still open research questions in the field of constrained sliding-mode control is proposed.

5.1.1 SMC with Input Constraints

When dealing with input-constrained systems, plain first-order SMC laws of the kind

$$u(t) = -K sign(\sigma(t)) \tag{5.1}$$

with $\sigma(t)$ the sliding variable (defining the sliding manifold $\sigma(t) = 0$, to be reached in finite time and held afterward by means of (5.1)) and K a properly chosen gain [22], usually do not lead to issues. In fact, from (5.1) it is always verified that $|u(t)| \leq K$ and therefore it is sufficient that the choice of a K smaller than the input saturation value is suitable to guarantee finite-time convergence to the sliding manifold $\sigma(t) = 0$. In the opposite case, one can verify in advance the infeasibility of the solution without further computations or resort to particular formulations such as the ones in [29–36]. Solutions based on adaptive [37] or anti-windup SMC [38–40] (the latter being meaningful when an integral action is introduced in boundary layer methods [41] to enhance transient features and performance [42]) have also been presented for some specific applications. Similar considerations can be done when higher-order sliding mode (HOSM) control algorithms [21, 43–45] are employed to directly produce the system input.

When HOSM control is instead exploited for chattering alleviation purposes (see, e.g., [46, 47]), one has that the control signal actually fed to the actuators is continuous. It indeed corresponds to the pth-order integral of a bounded function, with

$p = k - r$, k being the order of the SMC law and r the relative degree of the pair $\{\sigma(t), u(t)\}$ [48]. In these cases, the magnitude of $u(t)$ cannot be bounded a priori within a given range and thus, in the presence of limitations on the exertable control, it can happen that the actual exerted action differs from the expected one. As a consequence, the convergence to the sliding set can either be slowed down or even never happen at all, with obvious consequences in the controlled system behavior [49]. In order to avoid such situations, one may want to resort to suitably designed HOSM algorithms able to guarantee convergence even in the presence of saturating actuators. To this end, works as [50–52] propose suitably modified versions of the super-twisting algorithm [53, 54]. In particular, the authors in [51] propose a Lyapunov-based desaturation scheme and enhance it in [50] with the addition of a perturbation estimator which enables to achieve better performance in terms of uncertainty rejection and convergence rate. A further extension is proposed in [55], where the assumptions on the disturbance amplitude are relaxed and a more easily implementable version of the control law is presented. In these proposals, first-order perturbed systems of the kind

$$\dot{x}(t) = u(t) + \phi(t) \tag{5.2}$$

with $\phi(t)$ a bounded unknown Lipschitz function, are considered. The task is to guarantee $|u(t)| < \rho$, $\rho > 0$ being a constant saturation threshold sufficiently high to allow for the enforcement of a sliding motion. The proposed law is based on a "behavior detector" which switches the structure of the controller between the conventional first-order (FOSM) form and the super-twisting strategy. One of the main features of the laws presented in the mentioned works is that of guaranteeing at most one switch throughout all the control time span, with the consequence of reducing high-frequency variations in the input value.

Another valuable result is that proposed in [49, 56], which constitutes in some sense an extension of the well-known suboptimal second-order sliding-mode (S-SOSM) control algorithm [47, 57] to input-saturated systems in the case of chattering alleviation. For a generic nonlinear system affine in the control

$$\dot{x} = a(x(t), t) + b(x(t), t)u(t), \tag{5.3}$$

where $x(t) \in \mathscr{X} \subsetneq \mathbb{R}^n$ is closed and bounded, $u(t) \in \mathbb{R}$ is a scalar input, while $a(x(t), t)$ and $b(x(t), t)$ are sufficiently smooth uncertain vector fields, let us define a differentiable sliding variable $\sigma(t)$. Provided that $r = 1$ holds for the pair $\{\sigma(t), u(t)\}$, by means of a proper diffeomorphism the following "auxiliary system" is obtained:

$$\begin{cases} \dot{z}_1(t) = f(z_1(t), z_2(t), t) + g(z_1(t), z_2(t), t)u(t) = z_2(t) \\ \dot{z}_2(t) = h(z_1(t), z_2(t), u(t), t) + g(z_1(t), z_2(t), t)w(t), \end{cases} \tag{5.4}$$

where $z_1(t) = \sigma(t)$ and $w(t) = \dot{u}(t)$. Assume that

$$|f(z_1(t), z_2(t), t)| \leq F, \quad |h(z_1(t), z_2(t), u(t), t)| \leq H, \\ 0 < G_1 \leq g(z_1(t), z_2(t), t) \leq G_2 \tag{5.5}$$

hold for known constants F, H, G_1, and G_2. According to the S-SOSM approach, the choice

$$w(t) = -\alpha(t)Wsign\left(\sigma(t) - \frac{\sigma_M}{2}\right), \tag{5.6}$$

where σ_M is the last value of $\sigma(t)$ for which $\dot{\sigma}(t) = 0$ (and is set equal to $\sigma(t_0)$ at the beginning), and

$$\alpha(t) = \begin{cases} \alpha^* & \text{if } \ (\sigma(t) - \frac{1}{2}\sigma_M)(\sigma_M - \sigma(t)) > 0 \\ 1 & \text{otherwise} \end{cases} \tag{5.7}$$

with

$$W > max\left(\frac{H}{\alpha^* G_1}, \frac{4H}{3G_1 - \alpha^* G_2}\right), \quad \alpha^* \in (0, 1] \cap \left(0, \frac{3G_1}{G_2}\right) \tag{5.8}$$

drives in a finite time the pair $(z_1(t), z_2(t)) = (\sigma(t), \dot{\sigma}(t))$ to the origin, enforcing a second-order sliding mode on the sliding set $\sigma(t) = \dot{\sigma}(t) = 0$.

Due to (5.6) and the fact that $w(t) = \dot{u}(t)$, the continuous input actually fed to the system can exceed the range $u(t) \in [-U, U]$ with

$$U > \frac{F}{G_1} \tag{5.9}$$

the saturation threshold. The strategy proposed in [49, 56] for desaturation, directly based on (5.6), is the following:

$$w(t) = \begin{cases} -\alpha(t)Wsign(\sigma(t) - \beta\sigma_M) & \text{if } \ |u(t)| < U \\ -Wsign(u(t)) & \text{if } \ |u(t)| \geq U. \end{cases} \tag{5.10}$$

The parameter β is taken at each time step in the interval $[0.5, \bar{\beta}]$, where $\bar{\beta} = 0.5$ at the beginning ($t = t_0$) and whenever a new peak is detected, and

$$\bar{\beta} = \frac{\sigma(t_{s_i})}{\sigma_M} \tag{5.11}$$

for each successive time instant t_{s_i} for which $|u(t_{s_i})| \geq U$.

Remark 5.1 Notice how, in the general case, the dynamics of $z_2(t)$ in (5.4) explicitly depend on $u(t)$. Therefore, boundedness of the control in the range $[-U, U]$ is useful to guarantee the respect of (5.5).

Remark 5.2 The value of β determines the overall behavior of the system during the reaching phase. In fact, if $\beta \equiv 0.5$, (5.10) corresponds to the usual suboptimal law (5.6) with an additional desaturation feature. With this choice, the system usually remains in a saturated condition for longer time, but the reaching phase lasts less than in the case in which $\beta \equiv \bar{\beta}$ that leads to a change of sign as soon as the saturation zone is hit. Intermediate values of β correspond to intermediate behaviors and can be chosen according to design specifications.

Example 5.1 In order to better clarify the effects that a saturation in the actuators can have in the finite-time convergence toward the sliding set, let us consider the system

$$\begin{cases} \dot{x}_1(t) = x_2(t) \\ \dot{x}_2(t) = \frac{1}{m}\left[u(t) + \phi(t) - k(1 + ax_1^2(t))x_1(t)\right], \end{cases} \tag{5.12}$$

which can be identified as a mass–spring–damper with a hardening spring and without friction [58], with values

$$m = 10, \quad k = 500, \quad a = 0.05. \tag{5.13}$$

The term $\phi(t)$ represents matched uncertainty, for instance, accountable to a malfunctioning of the force actuator. As for the sliding variable, a standard linear choice

$$\sigma(t) = x_1(t) + 0.5x_2(t) \tag{5.14}$$

is made for the sake of simplicity. The output (5.14) exhibits relative degree $r = 1$ with respect to the input $u(t)$, and therefore an S-SOSM control law can be used in order to alleviate the chattering effect. In particular, differentiating (5.14) in time and considering (5.13),

$$\begin{aligned} \dot{\sigma}(t) &= x_2(t) + \frac{1}{2m}\left[u(t) + \phi(t) - k(1 + ax_1^2)x_1\right] \\ &= \frac{1}{20}\left(20x_2(t) - 500x_1(t) - 25x_1^3(t) + \phi(t)\right) + \frac{1}{20}u(t) \end{aligned} \tag{5.15}$$

follows. Therefore, defining for convenience $z_1(t) := \sigma(t)$ and $z_2(t) := \dot{\sigma}(t)$, the auxiliary system

$$\begin{cases} \dot{z}_1(t) &= z_2(t) \\ \dot{z}_2(t) &= \frac{1}{20}\big(- 500x_2(t) - 75x_1^2(t)x_2(t) - 1000x_1(t) - 50x_1^3(t) \\ & \quad +2\phi(t) + \dot{\phi}(t) + 2u(t)\big) + \frac{1}{20}w(t) \\ & = h(t, u(t)) + gw(t) \end{cases} \tag{5.16}$$

is obtained differentiating again (5.15), with $w(t)$ an artificial input such that $\dot{u}(t) = w(t)$. Notice how the dependence of $h(t, u(t))$ upon the states $x_1(t)$ and $x_2(t)$ (here omitted for simplicity) is in fact also a dependence on $z_1(t)$, $z_2(t)$ in view of the diffeomorphism producing (5.16) from (5.12)–(5.14), while g is a constant in this particular case. In order to find a suitable control law, suppose that the state vector never exits the set

$$\mathscr{X} = \{(x_1, x_2) : |x_1| \le 2.5,\ |x_2| \le 10\} \tag{5.17}$$

and, for the uncertainty,

$$|\phi(t)| \le 10, \quad |\dot{\phi}(t)| \le 50 \tag{5.18}$$

holds. In order to highlight the effects of a saturated input, let us suppose to apply a classical suboptimal controller when the input force $u(t)$ is limited in the range $[-U, U] = [-50, 50]$ (which is actually less than what the sufficient condition (5.9) requires, but is chosen to give better evidence to the following discussion, it being nevertheless enough to enforce convergence in the considered case as will be evident in the following). In (5.16), one obtains

$$|h(t, u(t))| \le 660 \tag{5.19}$$

$$g = \frac{1}{20}, \tag{5.20}$$

and, then, picking $\alpha \equiv 1$, the gain W for the law (5.10) must be chosen such that $K > 26500$ in view of (5.8).

With the choice $W = 30000$, the required control amplitude as well as the saturated control $\underline{u}(t)$, which is actually applied to the plant, are reported in Fig. 5.1. The strongly limited exertable control, in this particular case, makes the convergence relatively slow as can be seen in Figs. 5.2 and 5.3, with the pair $(\sigma(t), \dot{\sigma}(t))$ cycling around the origin few times before convergence. As a matter of fact, the sliding set is not attained in the considered time interval, and no guarantee can be formally given that such a convergence will ever happen. If the desaturation strategy (5.10) is applied with $\beta = 0.5$, the obtained input lies in the allowed boundaries as can

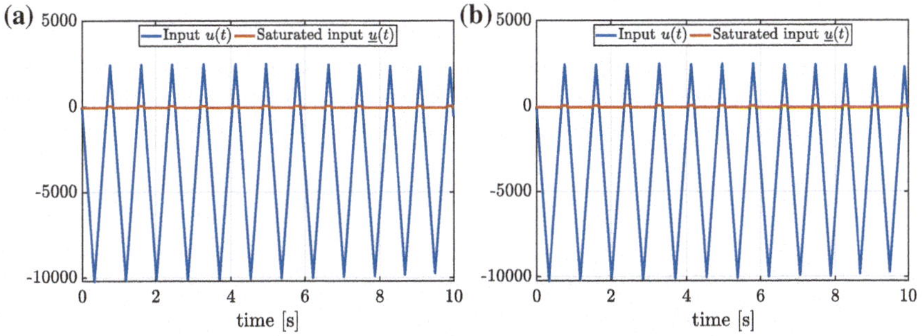

Fig. 5.1 *S-SOSM:* Computed and saturated input without matched uncertainty (i.e., $\phi(t) = 0$) (**a**) and in the presence of an actuator malfunctioning (**b**)

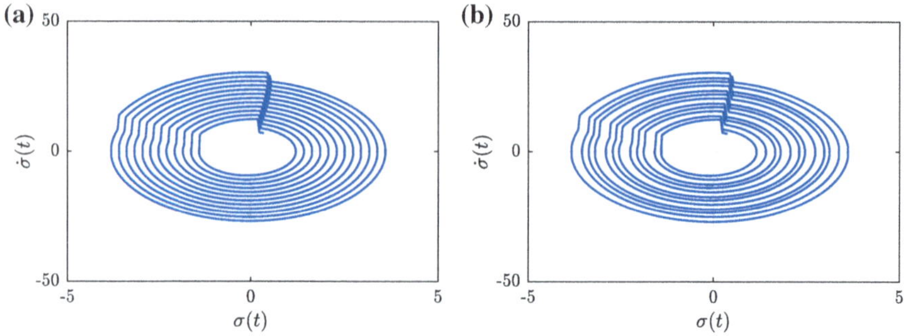

Fig. 5.2 *S-SOSM:* Sliding variable and its first time derivative without matched uncertainty (i.e., $\phi(t) = 0$) (**a**) and in the presence of an actuator malfunctioning (**b**)

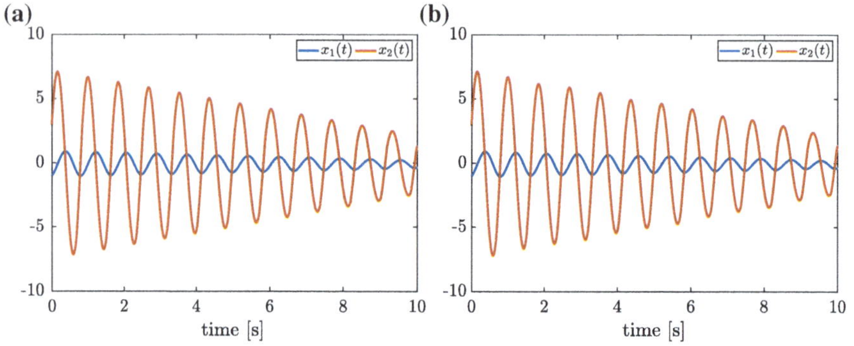

Fig. 5.3 *S-SOSM:* States evolution without matched uncertainty (i.e., $\phi(t) = 0$) (**a**) and in the presence of an actuator malfunctioning (**b**)

be seen in Fig. 5.4, and as an effect the speed of convergence to the sliding set is greatly enhanced (see Fig. 5.5). Notice, in particular, how the desaturation strategy is able to enforce fast convergence to the origin having at disposal a very low control power with respect to that required in a standard S-SOSM (see Fig. 5.1). This feature allows to design control systems with less powerful, and so cheaper, actuators without harming effectiveness. Once in sliding mode, the states converge asymptotically to the origin (Fig. 5.6) as expected, even in the presence of uncertainty (supposed here sinusoidal, as depicted in Fig. 5.7), highlighting the robustness of the proposed technique.

5.1.2 SMC with State Constraints

Even if one could, in principle, find a closed set in which the states lie once the sliding motion is established, at least during the reaching phase the states evolve according to

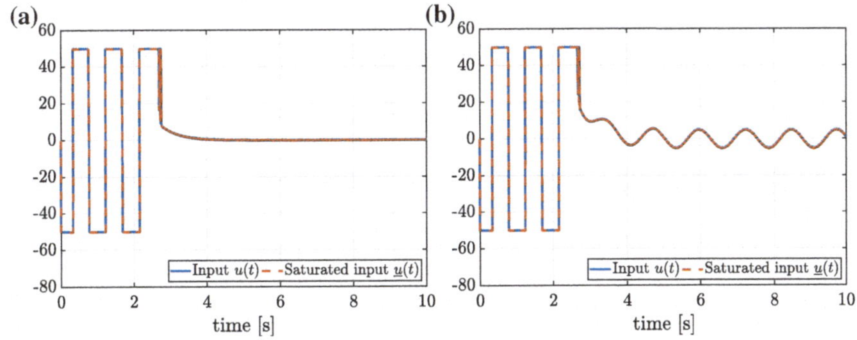

Fig. 5.4 *Desaturated S-SOSM:* Computed and saturated input without matched uncertainty (i.e., $\phi(t) = 0$) (**a**) and in the presence of an actuator malfunctioning (**b**)

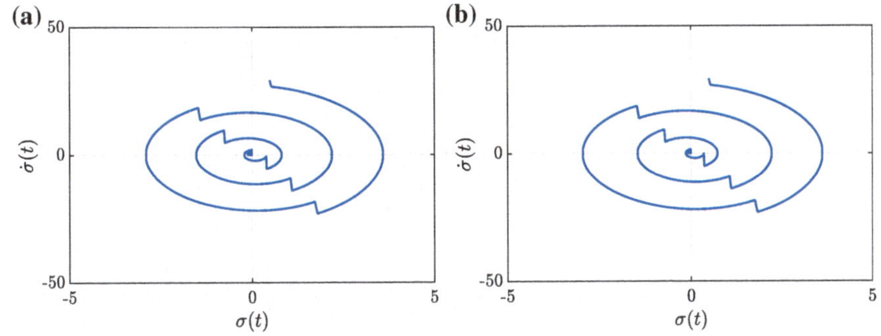

Fig. 5.5 *Desaturated S-SOSM:* Sliding variable and its first time derivative without matched uncertainty (i.e., $\phi(t) = 0$) (**a**) and in the presence of an actuator malfunctioning (**b**)

a complex combination of factors (such as the control magnitude, the plant dynamics, the disturbances, and uncertainties affecting the system), and therefore their trajectory cannot be ensured in a prescribed region. To do so, one must resort to particular versions of SMC laws, which are able to respect the given boundaries.

In [59], a strategy is proposed for the regulation of linear time-invariant systems subject to affine inequality constraints. State-constrained first-order SMC laws are proposed, for instance, in [60, 61], while a second-order algorithm is presented in [62]. In [63], the sub-optimal control law has been adapted in order to consider squared state constraints, while in [64] a third-order SMC is proposed with box state constraints, basing on the solution of the so-called Robust Fuller's problem [65] which gives rise to an optimal-time reaching law.

In particular, the work in [63] considers double integrators, i.e.,

$$\begin{cases} \dot{x}_1(t) = x_2(t) \\ \dot{x}_2(t) = f(x(t), t) + g(x(t), t)u(t), \end{cases} \tag{5.21}$$

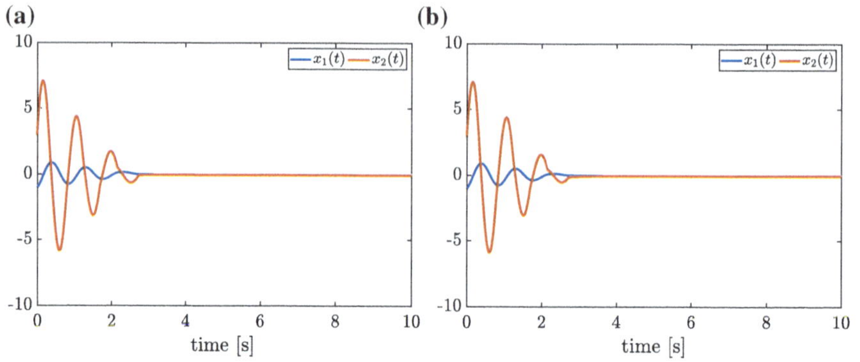

Fig. 5.6 *Desaturated S-SOSM:* States evolution without matched uncertainty (i.e., $\phi(t) = 0$) (**a**) and in the presence of an actuator malfunctioning (**b**)

Fig. 5.7 Matched uncertainty, introduced to emulate a faulty actuator in the proposed example

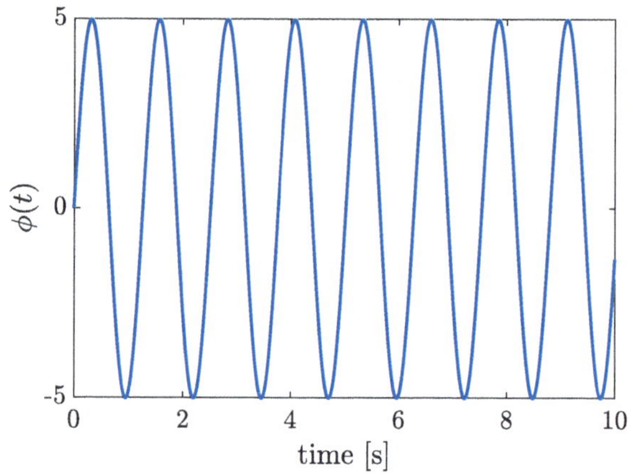

where $x(t) = [x_1(t)\, x_2(t)]^T \in \mathbb{R}^2$, and $f(x(t), t)$ and $g(x(t), t)$ are two sufficiently smooth unknown functions which can include uncertainty such that

$$\begin{aligned} |f(x(t), t)| &\leq F \\ 0 < G_1 \leq g(x(t), t) &\leq G_2 \end{aligned} \tag{5.22}$$

with F, G_1, and G_2 known constants. Let us assume that the states of (5.21) must be confined to a square region

$$R = \left\{(x_1(t), x_2(t)) : x_1^m \leq x_1(t) \leq x_1^M, x_2^m \leq x_2(t) \leq x_2^M\right\}. \tag{5.23}$$

Then, the proposed control law is

$$u(t) = \begin{cases} -K & \text{if} \quad x_1 \in S_1 \cup (S_3 \cap S_5) \cup S_7 \\ K & \text{if} \quad x_1 \in S_2 \cup (S_4 \cap S_6) \cup S_8, \end{cases} \tag{5.24}$$

where $K > \frac{F}{G_1}$ and

$$S_1 = \{(x_1(t), x_2(t)) : x_2(t) \geq x_2^M\} \tag{5.25}$$
$$S_2 = \{(x_1(t), x_2(t)) : x_2(t) \leq x_2^m\} \tag{5.26}$$
$$S_3 = \{(x_1(t), x_2(t)) : x_2(t) \leq x_2^M\} \tag{5.27}$$
$$S_4 = \{(x_1(t), x_2(t)) : x_2(t) \geq x_2^m\} \tag{5.28}$$
$$S_5 = \left\{(x_1(t), x_2(t)) : x_1(t) > -\frac{x_2(t)|x_2(t)|}{2U'}\right\} \tag{5.29}$$
$$S_6 = \left\{(x_1(t), x_2(t)) : x_1(t) < -\frac{x_2(t)|x_2(t)|}{2U'}\right\} \tag{5.30}$$
$$S_7 = \left\{(x_1(t), x_2(t)) : x_1(t) = -\frac{x_2(t)|x_2(t)|}{2U'} \text{ and } x_2(t) > 0\right\} \tag{5.31}$$
$$S_8 = \left\{(x_1(t), x_2(t)) : x_1(t) = -\frac{x_2(t)|x_2(t)|}{2U'} \text{ and } x_2(t) < 0\right\} \tag{5.32}$$

with $U' = G_1 K - F > 0$. The law (5.24) is guaranteed to establish finite-time convergence to the origin if the initial state lies in the invariant set

$$R' = R \setminus (S_9 \cup S_{10}) \tag{5.33}$$

with

$$S_9 = \left\{(x_1(t), x_2(t)) : x_1(t) > -\frac{x_2|x_2|}{2U'} + x_1^M \text{ and } x_2(t) > 0\right\} \tag{5.34}$$
$$S_{10} = \left\{(x_1(t), x_2(t)) : x_1(t) < -\frac{x_2|x_2|}{2U'} + x_1^m \text{ and } x_2(t) < 0\right\}. \tag{5.35}$$

The set (5.33) corresponds to the largest possible invariant region, which means that if the initial condition is outside of R', even if it lies into the constraints square R, no guarantee exists that the states trajectory will remain into it for all the subsequent time.

Example 5.2 Let us consider the need of regulating system (5.12), which is already in the form (5.21), to the origin in finite time. In this occasion, instead of the problem of input saturation, suppose it is required that the states do not exceed the limits

$$|x_1| \leq 2.5 \tag{5.36}$$
$$|x_2| \leq 5, \tag{5.37}$$

which has the corresponding physical meaning of a limited displacement of 2.5 m and a velocity which shall not exceed 5 m/s. The first can be necessary perhaps to preserve the spring, while the latter in order to maintain the system in a safe condition.

With reference to (5.23), we have

$$x_1^m = -2.5, \quad x_1^M = 2.5, \quad x_2^m = -5, \quad x_2^M = 5, \tag{5.38}$$

so that, looking at (5.22), we can safely assume, for instance,

$$F = 200 \tag{5.39}$$

$$G_1 = G_2 = \frac{1}{10}. \tag{5.40}$$

Therefore, we can pick $K = 2100$ (for the sake of conservativity) in view of (5.24) if we consider $|\phi(t)| \leq 15$, which is coherent with the observed values in Fig. 5.7. The obtained invariant region (5.33) is that depicted in black in Fig. 5.8. Supposing that the initial state is $x(0) = [-2\ 3]^T$ (which is inside of R'), the resulting motion is that reported in Fig. 5.9, which corresponds to the time evolution plotted in Fig. 5.10. From the results, it appears evident how the imposed constraints (5.38) are well

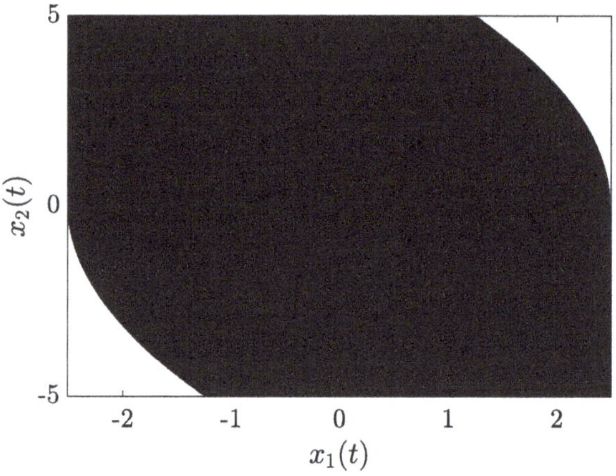

Fig. 5.8 Invariant region (5.33) for the system (5.12) and the control law design choices made

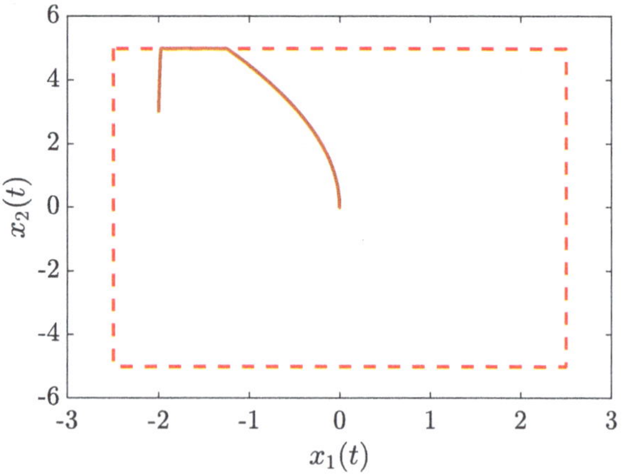

Fig. 5.9 Phase plane resulting from the application of (5.24) to the system under control, when $x(0) = [-2\ 3]^T$

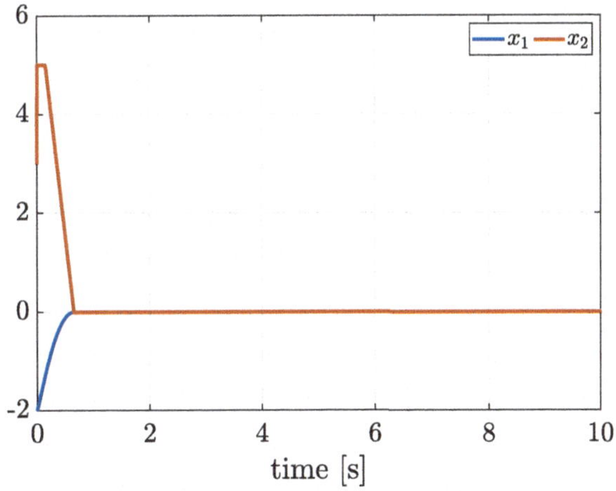

Fig. 5.10 Time evolution resulting from the application of (5.24) to the system under control, when $x(0) = [-2\ 3]^T$

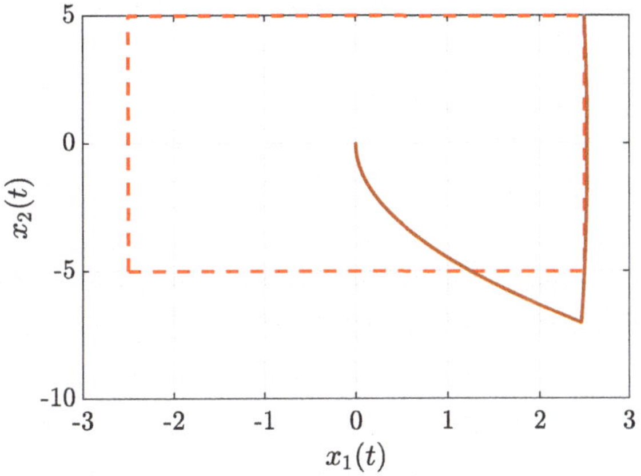

Fig. 5.11 Phase plane resulting from the application of (5.24) to the system under control, when $x(0) = [2.5\ 5]^T$

respected throughout all the reaching phase leading to the origin of the system under control, in spite of uncertainty. For the sake of completeness, let us try now to apply the same identical law when the initial condition lies in the constraints square region R, but not into R'. For the purpose, assume that $x(0) = [2.5\ 5]^T$. In Figs. 5.11 and 5.12 are reported the corresponding trajectories and time evolution, from which one can easily see that the constraints are no more respected.

Remark 5.3 A noticeable property of (5.24) (and, for instance, of the law proposed in [64], with some minor adjustments) is that also systems of higher order (i.e., with $n > 2$) can often be reduced to (5.21) by means of a proper diffeomorphism such that the states are partitioned into a second-order perturbed chain of integrators with states $x(t)$ as in (5.21) and an internal dynamics $\zeta(t) \in \mathbb{R}^{n-2}$ [58, 63]. In these cases, however, only $x(t)$ can be constrained by means of (5.24), while the zero dynamics evolves freely. This concept is at the basis of second-order SMC, where (5.21) (usually equipped with (5.22)) is referred to as the auxiliary system and is the object

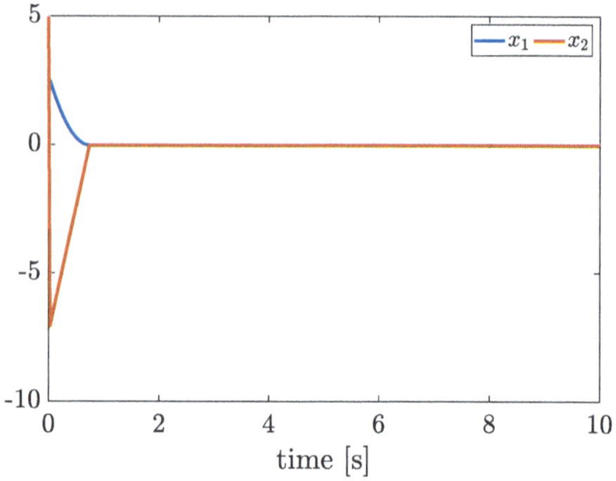

Fig. 5.12 Time evolution resulting from the application of (5.24) to the system under control, when $x(0) = [2.5\ 5]^T$

of the finite-time regulation to the origin (see, for instance, Eq. (5.16) in Example 5.1). The diffeomorphism generates the vector $x(t) = [\sigma(t)\,\dot{\sigma}(t)]^T$ correspondent to the sliding variable and its first time derivative, so that at the end the constraints are imposed on $\sigma(t)$ and $\dot{\sigma}(t)$.

Example 5.3 Given the generality of Remark 5.3, which can be extended to all the cases in which a constrained SMC law is able to deal with nth-order perturbed chains of integrators, an example is here reported to better clarify the concept. For the purpose, consider the model of a field-controlled DC motor with negligible shaft damping

$$\begin{cases} \dot{i}_f(t) = -ai_f(t) + u(t) \\ \dot{i}_a(t) = -bi_a(t) + k - ci_f(t)\omega(t) \\ \dot{\omega}(t) = \theta i_f(t)i_a(t), \end{cases} \tag{5.41}$$

where $i_f(t)$ and $i_a(t)$ are, respectively, the field and armature currents, and $\omega(t)$ is the angular velocity. The terms a, b, c, k, and θ are positive constants dependent on the particular considered motor, which in this example, for the sake of simplicity (the main purpose of this example being to illustrate the previously stated concept rather than propose an actual motor controller), are taken such that

$$a = b = c = k = \theta = 1. \tag{5.42}$$

The task is that of stopping the rotation of the motor in a finite time without violating prescribed state constraints. In particular, in view of Remark 5.3, let us consider the diffeomorphism

$$\Omega\left(\begin{bmatrix} i_f(t) \\ i_a(t) \\ \omega(t) \end{bmatrix}\right) = \begin{bmatrix} x_1(t) \\ x_2(t) \\ \zeta(t) \end{bmatrix} := \begin{bmatrix} \omega(t) \\ \theta i_f(t)i_a(t) \\ i_a(t) - \frac{k}{b} \end{bmatrix}, \tag{5.43}$$

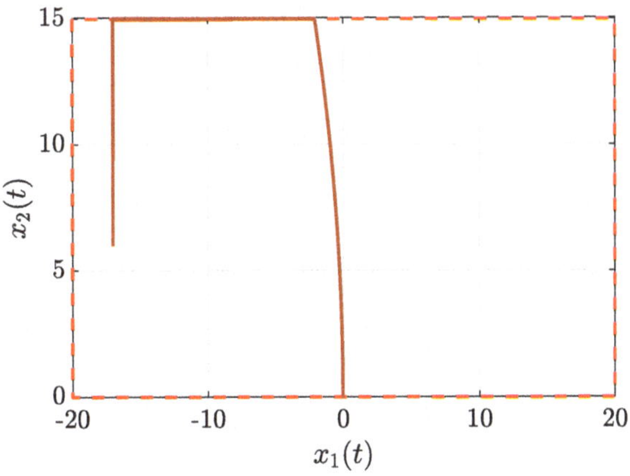

Fig. 5.13 Trajectory of the states $[x_1(t)\, x_2(t)]^T$ as in Eq. (5.44)

which is defined as long as $i_a(t) > 0$ (which we will suppose always true in this example). The obtained transformed system is then, taking into consideration (5.42),

$$\begin{cases} \dot{x}_1(t) &= x_2(t) \\ \dot{x}_2(t) &= -bx_2(t) + kb\frac{x_2(t)}{b\zeta(t)+k} - c\frac{x_2^2(t)}{\theta(\zeta(t)+k/b)^2}x_1(t) - ax_2(t) + \theta\left(\zeta(t) + \frac{k}{b}\right)u(t) \\ &= -2x_2(t) + \frac{x_2(t)}{\zeta(t)+1} - \frac{x_2^2(t)}{(\zeta(t)+1)^2}x_1(t) + (\zeta(t)+1)\,u(t) \\ \dot{\zeta}(t) &= -b\left(\zeta(t) + \frac{k}{b}\right) + k - cb\frac{x_1(t)x_2(t)}{\theta(b\zeta(t)+k)} \\ &= -\zeta(t) - \frac{x_1(t)x_2(t)}{\zeta(t)+1}. \end{cases} \tag{5.44}$$

As can be seen, the states $[x_1(t)\, x_2(t)]^T$ in (5.44) constitute a second-order chain of integrators (as a matter of fact, the choice $x_1(t) = \omega(t)$ corresponds to a relative degree 2 for (5.41)). The internal dynamics $\zeta(t) \to 0$ asymptotically, so that once $x_1(t)$ and $x_2(t)$ vanish, also $\zeta(t)$ approaches zero eventually. In fact,

$$\dot{\zeta}(t) = -b\zeta(t), \qquad x_1(t) = x_2(t) = 0. \tag{5.45}$$

Let us suppose that, according to design specifications, one would like the maximum velocity to be constrained as $|\omega(t)| = |x_1(t)| \leq 20$ [rad/s], while at the same time $|\theta i_f(t) i_a(t)| = |i_f(t) i_a(t)| = |x_2(t)| \leq 15$. With reference to (5.22) and supposing $0 < \zeta(t) < 25$,

$$F = 4545 \tag{5.46}$$

$$G_1 = 1, \quad G_2 = 26, \tag{5.47}$$

and therefore the gain $K = 4600$ is suitable for stabilizing (5.44) to the origin in finite time by means of (5.24).

The trajectories followed by the states $x_1(t)$ and $x_2(t)$ are reported in Fig. 5.13, where it is visible that the constraints are respected throughout the entire control interval. The respective time evolution is plotted in Fig. 5.14, along with that of $\zeta(t)$, that converges asymptotically to zero in consequence of the finite-time vanishing of $x_1(t)$ and $x_2(t)$. In the original coordinates, the time evolution of the states is that represented in Fig. 5.15, in which it is evident that $\omega(t)$ converges in finite time to zero, while $i_f(t) \rightarrow 0$ asymptotically. The state $i_a(t)$, instead, converges to $k/b = 1$. In fact, inverting the diffeomorphism (5.43) gives

$$\Omega^{-1}\left(\begin{bmatrix} x_1(t) \\ x_2(t) \\ \zeta(t) \end{bmatrix}\right) = \begin{bmatrix} i_f(t) \\ i_a(t) \\ \omega(t) \end{bmatrix} = \begin{bmatrix} \frac{x_2(t)}{(\zeta(t)+k/b)} \\ \zeta(t) + \frac{k}{b} \\ x_1(t) \end{bmatrix} = \begin{bmatrix} \frac{x_2(t)}{(\zeta(t)+1)} \\ \zeta(t) + 1 \\ x_1(t) \end{bmatrix}. \tag{5.48}$$

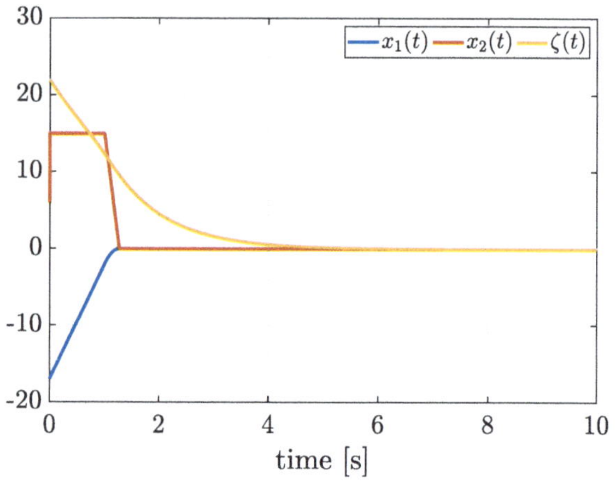

Fig. 5.14 Time evolution of the states of (5.44), transformed through (5.43)

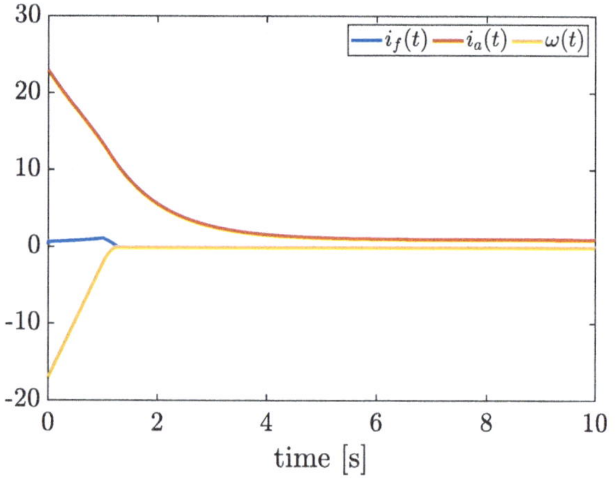

Fig. 5.15 Time evolution of the states of (5.41), in the original coordinates

5.1.3 SMC with Input and State Constraints

In real-world applications, in order to guarantee a proper functioning of the controlled systems, it would be valuable to be able to explicitly constrain both states and input. This need is often linked to safety and performance requirements, which must be fulfilled while taking into account an always limited available control effort.

Few proposals for the robust SMC of systems with state constraints and input saturations at the same time are present in the literature. In [66], an asymptotical sliding-mode approach is proposed for linear time-invariant SISO systems subject to linear constraints. The problem is then addressed in a rather general way in [67], where arbitrary-order HOSM laws are proposed for the fulfillment of both input saturation and state constraint requirements. Relying on the already mentioned results on minimum-time reaching phase, a suitable law is proposed for the stabilization of generic systems in the form (5.3). This work represents probably the most complete available result, which somehow encompasses many previously described algorithms while keeping at the same time a very low control law complexity.

The majority of the previously mentioned works, when dealing with state constraints, at the end address the problem from the point of view of the sliding variable, which is what is actually ensured in the prescribed region by means of the proposed control laws (more generally, the constraints are imposed on the system obtained after the chosen homeomorphism that transforms the original system in a chain of integrators and possibly a zero dynamics). To this end, a particular structure for $\sigma(t)$ must exist and be found in order to ensure the desired equivalent behavior and constraints respect at the same time (refer to Remark 5.3, for instance). Since it is not always the case, solutions which directly address the problem can be preferred sometimes. In [68], a first-order SMC law based on integral sliding-mode (ISM) control [69] is proposed for nonlinear systems in the form

$$\dot{x}(t) = f(x(t), t) + B(x)u(t) + \phi(t) \tag{5.49}$$

with an equal number n of states and inputs, $B(x)$ always invertible, and $|\phi(t)| \leq \Phi$ element-wise for a known vector Φ, with the aim of forcing the controlled system to follow a prescribed exponentially stable dynamics independently of possible matched uncertainty $\phi(t)$. The closed-loop behavior tracks that of a linear system, whose eigenvalues can be arbitrarily assigned, in such a way that the resulting control magnitude lies in the non-saturated region. Due to the properties of ISM, the reaching phase is completely avoided, thus ensuring the respect of constraints in the form

$$\begin{aligned} -x_m \leq x(t) \leq x_M, \quad x_m, x_M \in \mathbb{R}^n, \quad x_m \geq 0,\ x_M \geq 0 \\ -\infty \leq U_m(x) \leq u(t) \leq U_M(x) \leq \infty,\ U_m(x), U_M(x) \in \mathbb{R}^n,\ U_m(x) \leq 0 \leq U_M(x) \end{aligned} \tag{5.50}$$

from the beginning. As a matter of fact, the resulting exponential dynamics guarantees that if the initial state lies in the feasible region, no possibilities of escaping it exist during the sliding motion, which begins from the very first time instant. In particular, the proposed time-varying sliding variable is

$$\sigma(x,t) = b(x) - \int_0^t B^{-1}(x)Ax_f(t)\partial t, \tag{5.51}$$

where $b(x)$ is such that

$$B^{-1}(x) = \frac{\partial b(x)}{\partial x} \tag{5.52}$$

and A is a diagonal matrix

$$A = diag(a_1, \dots, a_n), \quad a_i = \lambda w_i \quad i = 1, \dots, n \tag{5.53}$$

establishing the dynamics of a fictitious system

$$\dot{x}_f(t) = Ax_f(t), \quad x_f(0) = x(0). \tag{5.54}$$

The diagonal terms in (5.53), which obviously correspond to the eigenvalues of the system (5.54), can be assigned, for instance, through the optimization of the parameter λ, such that the speed of convergence is maximum while the input constraints are satisfied considering the (possibly time-varying) weights w_i. In fact, the control law able to keep sliding mode on $\sigma(x,t) = 0$ is $u(x,t) = [u_1(x,t), \dots, u_n(x,t)]^T$, constructed such that

$$u_i(x,t) = -B_i^{-1}(x)(f(x,t) - Ax_f(t)) - K_i sign(\sigma_i(x,t)), \quad K_i > \Phi_{M_i} \tag{5.55}$$

with the index i referring to the ith row of the corresponding matrix and $\Phi_M(t)$ the matched uncertainty vector affecting the system (see Remark 5.4).

Remark 5.4 Due to the particular structure of the considered systems (5.49), it can be proved that all the uncertainty is always matched. In fact, the time-dependent component $\phi(t)$ can be decomposed as [70]

$$\phi(t) = B(x)B^+(x)\phi(t) + B'(x)B'^+(x)\phi(x), \tag{5.56}$$

where Γ' denotes the matrix which spans the null space of the generic matrix Γ, while Γ^+ its left pseudo-inverse. In the considered case, due to the invertibility of $B(x)$, $B'(x)$ does not exist and, in addition, $B^+(x) = B^{-1}(x)$. Therefore, defining $\phi_M(t) = B^{-1}(x)\phi(t)$, one obtains

$$\phi(t) = B(x)\phi_M(t) \tag{5.57}$$

and then

$$\dot{x} = f(x(t), t) + B(x)(u(t) + \phi_M(t)). \tag{5.58}$$

Additionally, in view of the assumption $|\phi(t)| \leq \Phi$, one has that also $\phi_M(t)$ is bounded element-wise. Hence, in all the cases in which the proposed procedure can be applied, the resulting motion is robust against any uncertainty affecting either the dynamics, the actuators, or both.

Two possible implementations of such an algorithm are suggested in [68], to which the reader is referred for the details. The first is an offline optimization switched policy, which consists in partitioning the states space into nonoverlapping regions, to each of which a particular set of eigenvalues, found through optimization, corresponds and is assigned during online control. The other requires online optimization in order to find the optimal eigenvalues set at a given time instant and throughout the subsequent control interval T, but allows for generally better performance.

Example 5.4 Among the huge amount of possible circumstances in which both states and inputs must be constrained, we will focus on a simple automotive example, namely, a stability control (yaw rate control) system (refer, e.g., to [71] for further details). The saturations on the inputs arise naturally due to the limited power (torque) availability and the friction coefficient guaranteed by the particular tire–road interaction conditions. The requirements on the states are imposed instead for safety and, secondarily, for driver and passengers' comfort. For example, let us start from a simplified planar description of the vehicle motion (see, e.g., [72]). For lateral stability, the hypothesis of pseudo-constant longitudinal velocity $v_x(t)$ is usually made (i.e., $\dot{v}_x(t) = 0$), so that the following two-state model is obtained:

$$\begin{cases} \dot{\beta} &= \frac{1}{mv_x}\left(F_{y,f} + F_{y,r} - mv_x\dot{\psi}\right) + \phi_1(t) \\ \ddot{\psi} &= \frac{1}{J_z}\left(l_f F_{y_f} + l_r F_{y,r} + M_\psi\right) + \phi_2(t), \end{cases} \tag{5.59}$$

where m is the vehicle mass, l_f and l_r are the distances between the center of mass and the front and rear axles, respectively, and J_z is the moment of inertia around the axis perpendicular to the considered plane. The two states β and $\dot{\psi}$ represent, respectively, the sideslip angle and the yaw rate. The former is defined as the angle present between the instantaneous velocity vector $v(t)$ and the longitudinal velocity component $v_x(t)$ (in the reference frame of the vehicle itself) and should then be kept ideally at zero during usual maneuvers. The latter, instead, must track the reference

$$\dot{\psi}_{ref}(t) = \frac{v(t)\delta_A(t)}{l_f + l_r}, \quad \delta_A(t) \sim \frac{1}{\rho(t)}, \tag{5.60}$$

where $\rho(t)$ is the (instantaneous) curvature radius of the turning, so that the resulting cornering is kinematic, and hence stability is ensured. The uncertainty functions $\phi_1(t)$ and $\phi_2(t)$ are introduced, for instance, to explicitly account for uncertainty in the description of the real vehicle behavior by means of the model (5.59) (here, such an uncertainty is supposed sinusoidal). For the purpose of this example, in (5.59), the lateral forces can be approximated as

$$F_{y,f} = C\left(\beta + \frac{l_f}{v}\dot{\psi} - \delta\right)$$
$$F_{y,r} = C\left(\beta - \frac{l_r}{v}\dot{\psi}\right) \tag{5.61}$$

with C the so-called cornering stiffness, which is associated with the tire–road characteristic and depends on the particular condition in which the turning is performed. Considering (5.59) and (5.61), one can act on the system through two inputs, namely, the steering angle $\delta(t) = u_1(t)$ and the moment $M_\psi(t) = u_2(t)$, having made the assumption that a sufficiently fast low-level controller can effectively apply a corresponding distribution of forces almost instantaneously.

The problem this time is that of regulating $\beta(t) \to 0$ and $\dot{\psi} \to \dot{\psi}_{ref}$ asymptotically. The error system must be then constructed, with $e_\beta(t) = \beta(t)$ and $e_{\dot{\psi}}(t) = \dot{\psi}(t) - \dot{\psi}_{ref}$, so that the overall problem becomes that of regulating the errors to zero, while keeping the constraints

$$|e_\beta(t)| \le 0.15 \quad \text{(rad)} \tag{5.62}$$
$$|e_{\dot{\psi}}(t)| \le 0.2 \quad \text{(rad/s)} \tag{5.63}$$
$$|\delta(t)| \le 0.45 \quad \text{(rad)} \tag{5.64}$$
$$|M_\psi(t)| \le 2 \cdot 10^4 \quad \text{(Nm)}. \tag{5.65}$$

The results obtained applying the described algorithm to a scenario with parameters as in Table 5.1 are reported here in Figs. 5.16 and 5.17 for the case in which the offline policy is adopted and Figs. 5.18 and 5.19 for the online one. In particular, the gains $K_1 = 0.0481$ and $K_2 = 350$ are used for conservativeness, since from Eq. (5.57) and the used parameters one can verify that $\Phi_{M_1} = -0.0476$ and $\Phi_{M_2} = 348$. Since the main task is that of making $\dot{\psi}(t) \to \dot{\psi}_{ref}$, the weight assigned to the convergence of $e_{\dot{\psi}}$ has been selected twice that of e_β (i.e., $w_2 = 2w_1$). For the offline procedure, the (error) state space has been partitioned in four nonoverlapping regions, depicted in Fig. 5.20, for each of which the optimal λ has been computed. In the same figure also the trajectory followed by the errors is presented, in order to evidence the resulting behavior. As a matter of fact, in the presented case, the

Table 5.1 Parameters of the scenario for Example 5.4

m	l_f	l_r	v_x	J_z
825 (Kg)	0.92 (m)	1.34 (m)	20 (m/s)	1750 (kg·m^2)
ρ	C	$\dot{\psi}_{ref}$	Φ_1	Φ_2
80 (m)	$5.2 \cdot 10^4$ (N/rad)	0.1106	0.2 ((kg · m)/s^2)	1.4 ((kg · m^2)/s^2)

associated eigenvalues grow in absolute value as the errors approach the origin and enter the innermost regions. In particular, with reference to Fig. 5.20, the eigenvalues λ_1 and λ_2 computed for each region X_i are, respectively, as follows:

$$X_1 : [-0.24, -0.48], X_2 : [-4.18, -8.36],$$
$$X_3 : [-16.18, -32.36], X_4 : [-39.62, -79.24]. \quad (5.66)$$

As a consequence, the convergence rate increases as time passes, resulting in the piecewise exponential shape in Fig. 5.16.

It is evident from the simulation that the online version corresponds to a faster convergence (compare Figs. 5.16 and 5.18), since the optimal λ are computed taking into account the actual followed trajectory. The resulting eigenvalue sequences, obtained with a selected sampling time of $T_s = 0.1$ s, are

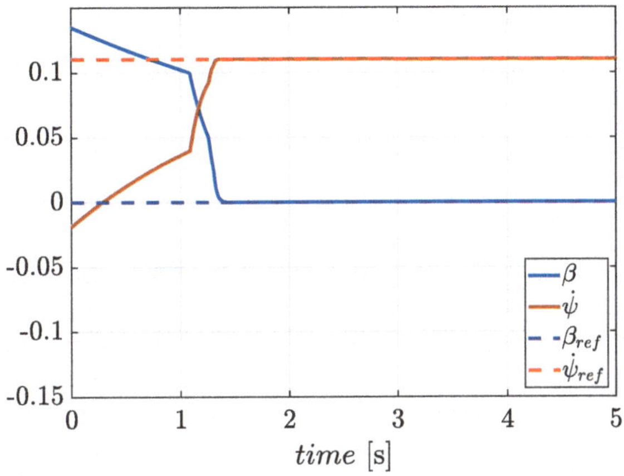

Fig. 5.16 *Offline switched policy:* states trajectory

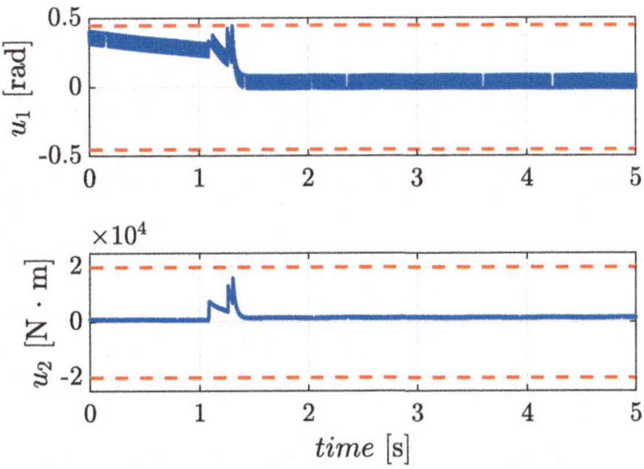

Fig. 5.17 *Offline switched policy:* control signals

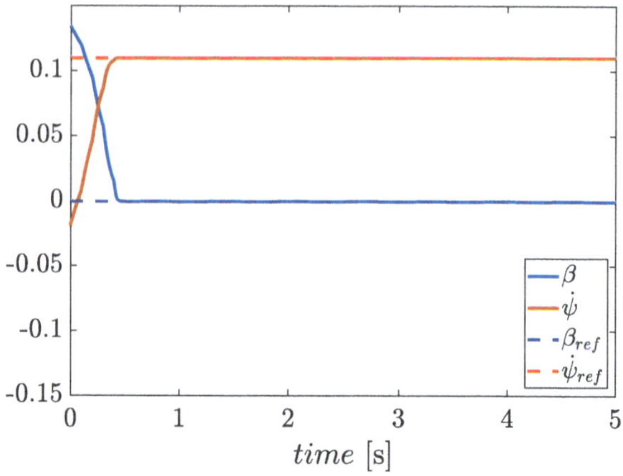

Fig. 5.18 *Online switched policy:* states trajectory

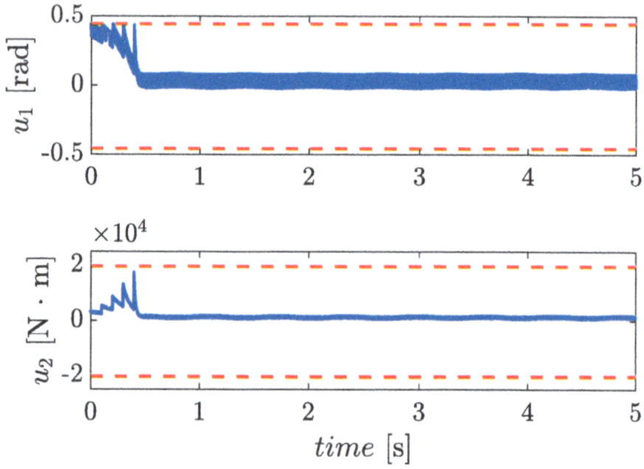

Fig. 5.19 *Online switched policy:* control signals

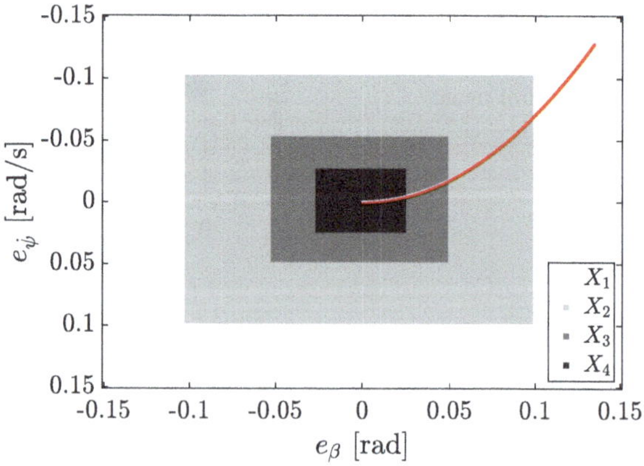

Fig. 5.20 *Offline switched policy:* error state-space partitioning and trajectory

$$
\begin{aligned}
\lambda_1 &: \{-1.18, -2.36, -5.08, -13.34, -69.58\} \\
\lambda_2 &: \{-2.36, -4.72, -10.16, -26.68, -139.16\}.
\end{aligned}
\tag{5.67}
$$

Notice that after the first 5 optimizations whose results appear in (5.67) (at $t = 0.5$), due to the convergence of the system to the origin which is obviously an invariant set for the linear dynamics, every value can be selected since such a choice is not affecting the dynamics anymore (in the specific case the last computed value was held). In the offline policy, such values are predetermined basing on a worst-case scenario for each of the considered partitions, which does not correspond necessarily to the real case. As expected, the piecewise exponential nature of the states trajectories translates in a natural fulfillment of the constraints while, at the same time, the control inputs never reach saturation (see Figs. 5.17 and 5.19). The "spikes" in the control action correspond to switches in the assigned eigenvalues (a new region entered in the case of offline optimization or a new optimization performed in the online case).

5.1.4 Conclusions

Due to all the reasons highlighted throughout the chapter and the effectiveness of the proposed techniques, sliding-mode control problems are gaining more and more interest from scientific research groups. Although the body of literature is not yet at an appreciably mature stage compared to other control fields, in the last years a great effort has been made in the direction of providing solid theoretical and practical results. Both in the case of input saturation and state constraints, available works prove to be effectively able to robustly deal with constrained control problems, as made also evident in the proposed case studies. Nonetheless, still open questions exist, which will likely constitute the main focus direction of future research in the field. Among them, one of the most important is the possibility of expressing the states constraints directly in the state space in the general case, instead of in the space of the sliding variable and its possible derivatives (refer, for instance, to [63, 64, 67] and Remark 5.3). On the one hand, this is expected to strongly broaden the applicability of state-constrained algorithms. On the other hand, making it easier and more intuitive to express the bounds according to the actual requirements, the gap in user-friendliness and so practical usability with respect to more widespread strategies as MPC would be filled. Additionally, until now only few works considered the multi-input case, which actually covers the great majority of the control problems. Extensions in this direction would probably constitute the major breakthrough imaginable nowadays for the topic, greatly widening the field of application and promoting thus the spread of such techniques as solid and effective control tools. Other possible research directions could also cover adaptations of input-saturated algorithms to the case of time-varying limits, and the same might be envisaged for the state constraints in the respective cases.

References

1. Hu, T., Lin, Z.: Control Systems with Actuator Saturation: Analysis and Design. Springer Science & Business Media, Berlin (2001)
2. Åström, K.J., Hägglund, T.: PID Controllers: Theory, Design, and Tuning, vol. 2. Instrument Society of America Research Triangle Park, NC (1995)
3. Åström, K.J., Hägglund, T., Astrom, K.J.: Advanced PID Control, vol. 461. ISA-The Instrumentation Systems, and Automation Society Research Triangle (2006)
4. Goodwin, G., Seron, M.M., De Doná, J.A.: Constrained control and estimation: an optimisation approach. Springer Science & Business Media, Berlin (2006)
5. Bak, M.: Control of systems with constraints. Ph.D. thesis, Technical University of Denmark, Department of Automation (2000)
6. Clarke, D., Scattolini, R.: Constrained receding-horizon predictive control. In: IEE Proceedings D (Control Theory and Applications), vol. 138, pp. 347–354. IET (1991)
7. Maciejowski, J.M.: Predictive Control with Constraints. Pearson Education, London (2002)
8. Mayne, D.Q., Rawlings, J.B., Rao, C.V., Scokaert, P.O.: Constrained model predictive control: stability and optimality. Automatica **36**(6), 789–814 (2000)
9. Qin, S.J., Badgwell, T.A.: A survey of industrial model predictive control technology. Control Eng. Pract. **11**(7), 733–764 (2003)
10. Garcia, C.E., Prett, D.M., Morari, M.: Model predictive control: theory and practice-a survey. Automatica **25**(3), 335–348 (1989)
11. Muske, K.R., Rawlings, J.B.: Model predictive control with linear models. AIChE J. **39**(2), 262–287 (1993)
12. Mayne, D.Q.: Model predictive control: recent developments and future promise. Automatica **50**(12), 2967–2986 (2014)
13. Rawlings, J.B., Muske, K.R.: The stability of constrained receding horizon control. IEEE Trans. Autom. Control **38**(10), 1512–1516 (1993)
14. Freeman, R., Kokotovic, P.V.: Robust Nonlinear Control Design: State-Space and Lyapunov Techniques. Springer Science & Business Media, Berlin (2008)
15. Hansen, L., Sargent, T.J.: Robust control and model uncertainty. Am. Econ. Rev. **91**(2), 60–66 (2001)
16. Wang, Y., Xie, L., De Souza, C.E.: Robust control of a class of uncertain nonlinear systems. Syst. Control Lett. **19**(2), 139–149 (1992)
17. Zhou, K., Doyle, J.C.: Essentials of Robust Control, vol. 104. Prentice Hall, Upper Saddle River (1998)
18. Bemporad, A., Morari, M.: Robust model predictive control: a survey. In: Robustness in Identification and Control, pp. 207–226. Springer, Berlin (1999)
19. Edwards, C., Fossas, E., Fridman, L.: Advances in Variable Structure and Sliding Mode Control. Birkh strä user, Basel (2008)
20. Edwards, C., Spurgeon, S.: Sliding Mode Control: Theory and Applications. CRC Press, Boca Raton (1998)
21. Shtessel, Y., Edwards, C., Fridman, L., Levant, A.: Sliding Mode Control and Observation. Birkhäuser, Basel (2014)
22. Utkin, V.I.: Sliding Modes in Control and Optimization. Springer, Berlin (1992)
23. Garcia-Gabin, W., Camacho, E.F.: Sliding mode model based predictive control for non minimum phase systems. In: 2003 European Control Conference (ECC), pp. 904–909 (2003)
24. Incremona, G.P., Ferrara, A., Magni, L.: Hierarchical model predictive/sliding mode control of nonlinear constrained uncertain systems. IFAC-PapersOnLine **48**(23), 102–109 (2015)
25. Raimondo, D., Rubagotti, M., Jones, C., Magni, L., Ferrara, A., Morari, M.: Multirate sliding mode disturbance compensation for model predictive control. Int. J, Robust Nonlinear Control (2014)

26. Rubagotti, M., Raimondo, D.M., Ferrara, A., Magni, L.: Robust model predictive control with integral sliding mode in continuous-time sampled-data nonlinear systems. IEEE Trans. Autom. Control **56**(3), 556–570 (2011)
27. Incremona, G.P., Ferrara, A., Magni, L.: Asynchronous networked MPC with ISM for uncertain nonlinear systems. IEEE Trans. Autom. Control **62**(9), 4305–4317 (2017)
28. Steinberger, M., Castillo, I., Horn, M., Fridman, L.: Model predictive output integral sliding mode control. In: 2016 14th International Workshop on Variable Structure Systems (VSS), pp. 228–233 (2016)
29. Corradini, M., Orlando, G.: Linear unstable plants with saturating actuators: Robust stabilization by a time varying sliding surface. Automatica **43**(1), 88–94 (2007)
30. Corradini, M.L., Cristofaro, A., Orlando, G.: On the robust stabilization of discrete-time SISO plants with saturating actuators. In: 2008 47th IEEE Conference on Decision and Control, pp. 1599–1604 (2008)
31. Corradini, M.L., Cristofaro, A., Orlando, G.: Robust stabilization of multi input plants with saturating actuators. IEEE Trans. Autom. Control **55**(2), 419–425 (2010)
32. Corradini, M.L., Orlando, G.: A result on the robust stabilization of MIMO plants with saturating actuators. In: 2007 46th IEEE Conference on Decision and Control, pp. 548–553 (2007)
33. Fulwani, D., Bandyopadhyay, B.: Design of Sliding Mode Controller with Actuator Saturation, pp. 207–219. Springer, Berlin (2013)
34. Ma, Z., Sun, G., Li, Z.: Dynamic adaptive saturated sliding mode control for deployment of tethered satellite system. Aerosp. Sci. Technol. **66**, 355–365 (2017)
35. Torchani, B., Sellami, A., M'hiri, R., Garcia, G.: Comparative analysis of the saturated sliding mode and lqr controllers applied to an inverted pendulum. In: 2011 International Conference on Communications, Computing and Control Applications (CCCA), pp. 1–6 (2011)
36. Wang, S., Gao, Y., Liu, J., Wu, L.: Saturated sliding mode control with limited magnitude and rate. IET Control Theory Appl. **12**(8), 1075–1085 (2018)
37. Xiao, B., Hu, Q., Zhang, Y.: Adaptive sliding mode fault tolerant attitude tracking control for flexible spacecraft under actuator saturation. IEEE Trans. Control Syst. Technol. **20**(6), 1605–1612 (2012)
38. Oliveira, C.M.R., Aguiar, M.L., Monteiro, J.R.B.A., Pereira, W.C.A., Paula, G.T., Almeida, T.E.P.: Vector control of induction motor using an integral sliding mode controller with anti-windup. J. Control Autom. Electr. Syst. **27**(2), 169–178 (2016)
39. Shi, X.J., Zhao, H.C., Xu, K.W.: Dynamic anti-windup design for missile overload control system. In: Advanced Technology for Manufacturing Systems and Industry. Applied Mechanics and Materials, vol. 236, pp. 273–277. Trans Tech Publications, Stafa (2012)
40. Yokoyama, M., Kim, G.N., Tsuchiya, M.: Integral sliding mode control with anti-windup compensation and its application to a power assist system. J. Vib. Control **16**(4), 503–512 (2010)
41. Slotine, J.J., Sastry, S.S.: Tracking control of non-linear systems using sliding surfaces with application to robot manipulators. In: 1983 American Control Conference, pp. 132–135 (1983). https://doi.org/10.23919/ACC.1983.4788090
42. Seshagiri, S., Khalil, H.K.: On introducing integral action in sliding mode control. In: Proceedings of the 41st IEEE Conference on Decision and Control, 2002, vol. 2, pp. 1473–1478 (2002). https://doi.org/10.1109/CDC.2002.1184727
43. Fridman, L., Levant, A., et al.: Higher order sliding modes. Sliding Mode Control Eng. **11**, 53–102 (2002)
44. Levant, A.: Universal single-input-single-output (SISO) sliding-mode controllers with finite-time convergence. IEEE Trans. Autom. Control **46**(9), 1447–1451 (2001)
45. Levant, A.: Quasi-continuous high-order sliding-mode controllers. In: 42nd IEEE International Conference on Decision and Control (IEEE Cat. No. 03CH37475), vol. 5, pp. 4605–4610. IEEE (2003)
46. Bartolini, G., Ferrara, A., Levant, A., Usai, E.: On Second Order Sliding Mode Controllers, pp. 329–350. Springer, London (1999)

47. Bartolini, G., Ferrara, A., Usai, E.: Chattering avoidance by second-order sliding mode control. IEEE Trans. Autom. Control **43**(2), 241–246 (1998)
48. Levant, A.: Sliding order and sliding accuracy in sliding mode control. Int. J. Control. **58**(6), 1247–1263 (1993)
49. Ferrara, A., Rubagotti, M.: A sub-optimal second order sliding mode controller for systems with saturating actuators. IEEE Trans. Autom. Control **54**(5), 1082–1087 (2009)
50. Castillo, I., Steinberger, M., Fridman, L., Moreno, J., Horn, M.: Saturated super-twisting algorithm based on perturbation estimator. In: 2016 IEEE 55th Conference on Decision and Control (CDC), pp. 7325–7328 (2016)
51. Castillo, I., Steinberger, M., Fridman, L., Moreno, J.A., Horn, M.: Saturated super-twisting algorithm: Lyapunov based approach. In: 2016 14th International Workshop on Variable Structure Systems (VSS), pp. 269–273 (2016)
52. Golkani, M.A., Koch, S., Reichhartinger, M., Horn, M.: A novel saturated super-twisting algorithm. Syst. Control Lett. **119**, 52–56 (2018)
53. Levant, A.: Robust exact differentiation via sliding mode technique. Automatica **34**(3), 379–384 (1998)
54. Levant, A.: Higher-order sliding modes, differentiation and output-feedback control. Int. J. Control **76**(9–10), 924–941 (2003)
55. Seeber, R., Horn, M.: Guaranteeing disturbance rejection and control signal continuity for the saturated super-twisting algorithm. IEEE Control Syst. Lett. (2019). https://doi.org/10.1109/LCSYS.2019.2917054
56. Ferrara, A., Rubagotti, M.: A sub-optimal second order sliding mode controller for systems with saturating actuators. In: 2008 American Control Conference, pp. 4715–4720 (2008)
57. Bartolini, G., Ferrara, A., Usai, E.: Output tracking control of uncertain nonlinear second-order systems. Automatica **33**(12), 2203–2212 (1997)
58. Khalil, H.K.: Noninear Systems. Prentice-Hall, New Jersey (1996)
59. Innocenti, M., Falorni, M.: State constrained sliding mode controllers. In: Proceedings of the 1998 American Control Conference, vol. 1, pp. 104–108 (1998)
60. Fan, Q.Y., Yang, G.H.: Nearly optimal sliding mode fault-tolerant control for affine nonlinear systems with state constraints. Neurocomputing **216**, 78–88 (2016). https://doi.org/10.1016/j.neucom.2016.06.063
61. Pietrala, M., Jaskuła, M., Leśniewski, P., Bartoszewicz, A.: Sliding mode control of discrete time dynamical systems with state constraints. In: Trends in Advanced Intelligent Control, Optimization and Automation, pp. 4–13 (2017)
62. Ding, S., Mei, K., Li, S.: A new second-order sliding mode and its application to nonlinear constrained systems. IEEE Trans. Autom. Control **64**, 2545–2552 (2018). https://doi.org/10.1109/TAC.2018.2867163
63. Rubagotti, M., Ferrara, A.: Second order sliding mode control of a perturbed double integrator with state constraints. In: Proceedings of the 2010 American Control Conference, pp. 985–990 (2010)
64. Ferrara, A., Incremona, G.P., Rubagotti, M.: Third order sliding mode control with box state constraints. In: 53rd IEEE Conference on Decision and Control, pp. 4727–4732 (2014)
65. Dinuzzo, F., Ferrara, A.: Higher order sliding mode controllers with optimal reaching. IEEE Trans. Autom. Control **54**(9), 2126–2136 (2009)
66. Tanizawa, H., Ohta, Y.: Sliding mode control under state and control constraints. In: 2007 IEEE International Conference on Control Applications, pp. 1173–1178 (2007)
67. Incremona, G.P., Rubagotti, M., Ferrara, A.: Sliding mode control of constrained nonlinear systems. IEEE Trans. Autom. Control **9286**(c), 1 (2016)
68. Zambelli, M., Ferrara, A.: Linearization-based integral sliding mode control for a class of constrained nonlinear systems. In: 2018 15th International Workshop on Variable Structure Systems (VSS), pp. 402–407 (2018)
69. Utkin, V., Shi, J.: Integral sliding mode in systems operating under uncertainty conditions. In: Proceedings of 35th IEEE Conference on Decision and Control, vol. 4, pp. 4591–4596 (1996)

70. Rubagotti, M., Estrada, A., Castanos, F., Ferrara, A., Fridman, L.: Integral sliding mode control for nonlinear systems with matched and unmatched perturbations. IEEE Trans. Autom. Control **56**(11), 2699–2704 (2011)
71. Kiencke, U., Nielsen, L.: Automotive Control Systems - For Engine, Driveline, and Vehicle. Springer, Berlin (2005)
72. Genta, G.: Motor Vehicle Dynamics - Modeling and Simulation. World Scientific, Singapore (1993)

Chapter 6
Analysis of Orbital Stability of Self-excited Periodic Motions in Lure System

Igor Boiko

Abstract Two criteria of orbital stability of self-excited periodic motions in a Lure system, the Loeb's criterion and the criterion based on the dynamic harmonic balance (the Boiko's criterion [17]), are reviewed in this book chapter. A relationship between these two criteria is established and investigated. An example of analysis is given.

6.1 Introduction

Analysis of orbital stability of self-excited periodic motions in dynamic systems is an important problem in science and engineering. Many practical problems require analysis of orbital stability of possible limit cycles to which the motion may converge. All phenomena and applications revealing oscillations require analysis of orbital stability of these oscillations, at least to some extent. Some notable applications of limit cycles in control engineering, which require analysis of orbital stability, are the relay feedback test [1] and various modifications of this test, limit-cycling servomechanisms, on–off, and phase-lock loop systems [2]. The problem of orbital stability requires its solution in the so-called hidden attractors and hidden oscillation analysis [3–5]. Exact periodic solution in dynamic systems and the subsequent analysis of orbital stability of this motion can be obtained through Poincare maps of the system [6]. However, Poincare maps can hardly be obtained for most systems that have limit cycles. An approximate approach to finding limit cycles is based on the describing function (DF) method [7, 8], and analysis of orbital stability, which involves DF, is proposed by Loeb [9]. The Loeb's criterion of orbital stability of periodic motions in nonlinear systems containing one nonlinearity is simple and convenient. Its extensions to the systems containing more than one nonlinearity, such as those that are found in second-order sliding-mode control algorithms, lead to relatively simple expressions too [10–12], which makes it a convenient tool for analysis of chattering in those systems, too.

I. Boiko (✉)
Khalifa University, Abu Dhabi, UAE
e-mail: igor.boiko@ku.ac.ae

M. Steinberger et al. (eds.), *Variable-Structure Systems and Sliding-Mode Control*, Studies in Systems, Decision and Control 271,
https://doi.org/10.1007/978-3-030-36621-6_6

However, the Loeb's criterion and other similar approaches [13, 14] are based on the assumption about the harmonic balance equation, obtained through the DF method, being valid not only at the point of the periodic motion but also for small perturbations in the vicinity of the periodic solution. Strictly speaking, this cannot be true, which is shown in the present book chapter. The formulation of the dynamic harmonic balance (DHB) principle [15, 16] allows for removing the above-noted assumption and producing a new criterion, which is based on the DHB. The present book chapter aims at finding conditions of orbital stability of periodic solutions in Lure systems, based on the DHB, which was first considered in [17], and comparison of these conditions with those of the Loeb's criterion. It is shown below that the proposed criterion, based on the DHB, eliminates the inherent contradiction of the Loeb's approach.

6.2 Limit Cycles in Lure System and Loeb's Criterion

Consider the following nonlinear system having a symmetric single-valued or hysteretic nonlinearity and linear dynamics (Lure system):

$$\dot{\mathbf{x}} = \mathbf{A}\mathbf{x} + \mathbf{B}u, \tag{6.1}$$

$$y = \mathbf{C}\mathbf{x}, \tag{6.2}$$

$$u = -f\left(y, sign\left(\dot{y}\right)\right), \tag{6.3}$$

where $\mathbf{x} \in R^n$ is the state vector, $y \in R^1$ is the system output, $u \in R^1$ is the control (which is the input to the linear part), $\mathbf{A} \in R^{n\times n}$, $\mathbf{B} \in R^{n\times 1}$, $\mathbf{C} \in R^{1\times n}$ are the state matrix, the input matrix, and the output matrix, respectively, the negative sign in Eq. (6.3) is attributed to the negative feedback in the closed-loop system, and the nonlinearity is a single-valued or hysteretic odd-symmetric nonlinearity that satisfies the condition $f\left(-y, sign\left(-\dot{y}\right)\right) = -f\left(y, sign\left(\dot{y}\right)\right)$. The linear part is given by (6.1) and (6.2) and can be described by the following transfer function $W(s) = \mathbf{C}(\mathbf{I}s - \mathbf{A})^{-1}\mathbf{B}$. We shall consider only the case of the transfer function $W(s)$ not having zeros:

$$W(s) = \frac{y(s)}{u(s)} = \frac{1}{Q(s)}, \tag{6.4}$$

where the denominator is the following polynomial of degree n: $Q(s) = a_0 + a_1 s + \cdots + a_n s^n$. This case can be extended to more complex situations of the transfer function with zeros and delay. But this is beyond the scope of the present book chapter.

A periodic solution in system (6.1)–(6.3), or self-excited periodic motions (subject to their stability), can be found from the solution of the harmonic balance (HB) equation:

$$N(A_0)W(j\Omega_0) = -1, \tag{6.5}$$

where A_0 and Ω_0 are the amplitude and the frequency of the self-excited periodic motions in system (6.1)–(6.3), function $N(a)$ is the describing function (DF) of the nonlinearity $f\,(y, sign\,(\dot{y}))$ [7, 8] given by:

$$N\,(a) = \frac{\omega}{\pi\ a}\int_0^{2\pi/\omega} u(t)\ \sin\omega t\ dt + j\,\frac{\omega}{\pi\ a}\int_0^{2\pi/\omega} u(t)\ \cos\omega t\ dt, \tag{6.6}$$

where a and ω are the amplitude and the frequency of the harmonic signal at the input to the nonlinearity.

According to the criterion proposed by Loeb [9], orbital stability of the periodic solution in (6.1)–(6.3) can be verified through the computation of the following quantity:

$$\Phi = \frac{\partial \mathrm{Re} W(j\omega)}{\partial\omega}\frac{\partial \mathrm{Im}\left(-N^{-1}(a)\right)}{\partial a} - \frac{\partial \mathrm{Im} W(j\omega)}{\partial\omega}\frac{\partial \mathrm{Re}\left(-N^{-1}(a)\right)}{\partial a}. \tag{6.7}$$

The necessary condition for the periodic solution to be orbitally stable is $\Phi > 0$. If $\Phi < 0$, the periodic solution is unstable. There is also another formulation of this criterion that involves magnitudes and phases of N and W, which is equivalent to (6.7).

Loeb's original proof proceeds from the assumption that the harmonic balance equation (6.7) remains valid also at some small deviations from the periodic solution (A_0, Ω_0), which is given by the following equation:

$$N(A_0 + \Delta a)W(j\Omega_0 + \sigma + j\Delta\omega) = -1, \tag{6.8}$$

where Δa is the deviation of the amplitude, $\Delta\omega$ is deviation of the frequency, and σ is the amplitude decay: $\sigma = \dot{a}/a$. The idea behind the proof is that if a positive deviation Δa results in a negative value of σ, and vice versa, then an increase of the amplitude above A_0 would result in its eventual decrease to A_0, and a decrease below A_0 would result in the amplitude increase up to A_0.

However, Eq. (6.8), which served in the Loeb's proof as a basis, cannot hold for any point other than the point of the periodic solution (A_0, Ω_0). Let us show that. At first let us consider the following theorem.

Theorem 6.1 *If the linear system given by (6.1), (6.2) is bounded-input bounded-output (BIBO) stable, then the zero-initial-state response to the input signal* $u(t) = e^{\sigma t}\sin\omega t$ *approaches the following function:*

$$y(t) = e^{\sigma t}\left|W\left(\sigma + j\omega\right)\right| \sin\left(\omega t + \arg W\left(\sigma + j\omega\right)\right), \tag{6.9}$$

as $t \to \infty$.

Proof A detailed proof of this theorem is given in [17]. A different and shorter proof is as follows. Zero-state response of a linear system is described by its transfer function. Because the Laplace transform of the sinusoidal signal of unity amplitude is $L\left[\sin\left(\omega t\right)\right]=\frac{\omega}{s^2+\omega^2}$, and given the property of the Laplace transform formulated as $L\left[e^{-at}f\left(t\right)\right]=F\left(s+a\right)$, where $F\left(s\right)=L\left[f\left(t\right)\right]$, the Laplace transform of the "decaying sinusoid" is $L\left[e^{\sigma t}\sin\left(\omega t\right)\right]=\frac{\omega}{(s-\sigma)^2+\omega^2}$. The output signal of the linear dynamics is given (in the Laplace domain) by $Y(s)=\frac{\omega}{(s-\sigma)^2+\omega^2}W(s)$. To analyze what this signal is, we introduce a "biased" Laplace variable $s'=s-\sigma$, which will yield the signal $y'(t)=L^{-1}\left[Y\left(s'\right)\right]$, and $Y(s'+\sigma)=\frac{\omega}{(s')^2+\omega^2}W(s'+\sigma)$. In accordance with the property of a linear system describing its reaction to a sinusoidal signal with $t\to\infty$, the reaction of a linear BIBO system approaches a sinusoid that has the frequency of ω, the amplitude of $|W(\sigma+j\omega)|$, and the phase of $\arg W(\sigma+j\omega)$. Accordingly, the output signal $y(t)=L^{-1}\left[Y\left(s\right)\right]$ can be produced from $y'(t)$ as $y(t)=e^{\sigma t}y'(t)$, which is a decaying sinusoid: $y(t)=e^{\sigma t}\left|W(\sigma+j\omega)\right|\sin\left(\omega t+\arg W(\sigma+j\omega)\right)$. ■

Lemma 6.2 was introduced and proved in [17].

Lemma 6.2 *If the input signal to the odd-symmetric nonlinearity (6.3) is given by* $y(t)=a_{0y}e^{\sigma t}\sin\omega t$ *then the output of the nonlinearity can be approximately (through the describing function method) found as* $u(t)=-a_{0u}e^{\sigma^* t}\sin\left(\omega t+\phi\right)$, *where* $\sigma^*\neq\sigma\neq 0$ *for all* $a_y=a_{0y}e^{\sigma t}$, *except possibly a few trivial values of the amplitude* a_y.

Proof Assuming, as in the Loeb's criterion proof, that $y(t)=a_{0y}e^{\sigma t}\sin\omega_0 t$, we find that the instantaneous amplitude of $y(t)$ is $a_y=a_{0y}e^{\sigma t}$. Therefore, the time derivative of the amplitude is $\dot{a}_y=a_{0y}\sigma e^{\sigma t}$, and the decay[1] of the amplitude is $\sigma=\dot{a}_y/a_y$. The output of the nonlinearity is, per the DF method, $u(t)=-a_{0y}e^{\sigma t}\sin\omega_0 t\ N(a_y)$, where $N(a_y)$ is the describing function of the nonlinearity. The amplitude of the signal $u(t)$ is $a_u=a_{0y}e^{\sigma t}\left|N(a_y)\right|=a_y\left|N(a_y)\right|$; the derivative of the amplitude is $\dot{a}_u=\dot{a}_y\left|N(a_y)\right|+a_y\frac{\partial\left|N(a_y)\right|}{\partial a_y}\dot{a}_y$. The decay of the amplitude of $u(t)$ can be found as

$$\sigma^*=\frac{\dot{a}_u}{a_u}=\frac{\dot{a}_y}{a_y}+\frac{1}{\left|N(a_y)\right|}\frac{\partial\left|N(a_y)\right|}{\partial a_y}\dot{a}_y=\sigma+\dot{a}_y\frac{\partial\ln\left|N(a_y)\right|}{\partial a_y}. \tag{6.10}$$

In accordance with (6.10), $\sigma^*=\sigma$ only if $\frac{\partial\left|N(a_y)\right|}{\partial a_y}=0$, which may hold only at certain points of minimum or maximum of function $N(a)$, or if the function $u=-f\left(y, sign\left(\dot{y}\right)\right)$ is linear. This completes the proof. ■

Theorem 6.1 and Lemma 6.2 enable us to show that the harmonic balance equation does not have a solution for $y(t)=a_{0y}e^{\sigma t}\sin\omega_0 t$ subject to $\sigma=\dot{a}_y/a_y\neq 0$. Indeed, write the harmonic balance equation for decaying oscillation, as assumed in the

[1] We shall refer to σ as decay, despite the fact that it corresponds to the decaying amplitude only if it is negative.

original Loeb's proof, that is, $W(\sigma + j\omega_0)N(a) = -1$. Write also the regeneration conditions $u(t) = -N(a_y)y(t)$ and $y(t) = W(\sigma + j\omega_0)u(t)$. They must be satisfied for $t \to \infty$ for the HB condition to hold. As per Theorem 6.1 and Lemma 6.2, these two equations may have a solution only if $\sigma^* = \sigma$. In this case, the exponents would be canceled out from both sides of both equations, and the equations would reduce to the equation of the harmonic balance given above. This, however, contradicts Lemma 6.2, which says that for $\sigma \neq 0$ the equality $\sigma^* = \sigma$ cannot hold. This implies that the assumption about the HB equation (6.8) that holds for the deviated amplitude and frequency is not correct. In fact, what is missing in the original proof is the component in the HB equation that would compensate for the balance of the magnitudes and phases due to the deviation from the point (A_0, Ω_0).

6.3 Dynamic Harmonic Balance

This missing part was covered with introduction of the dynamic harmonic balance (DHB) principle [15, 16]. The DHB allows one to find equations of a transient oscillatory motion in a vicinity of the periodic solution given by (6.5). Consequently, analysis of these transient oscillatory motions leads to the conditions of orbital stability. The DHB principle was formulated in [15]. Let us briefly review it.

We shall consider the Lure system given by (6.1)–(6.3), with the linear part satisfying the condition (6.4). Let us assume that the output signal can be described as a transient oscillation that has variable frequency and amplitude. The geometric representation of this motion is a *rotating phasor* (Fig. 6.1) which is a vector whose length corresponds to the amplitude and the angle to the phase of the output $y(t)$:

$$\bar{y}(t) = a(t)e^{j\Psi(t)}, \tag{6.11}$$

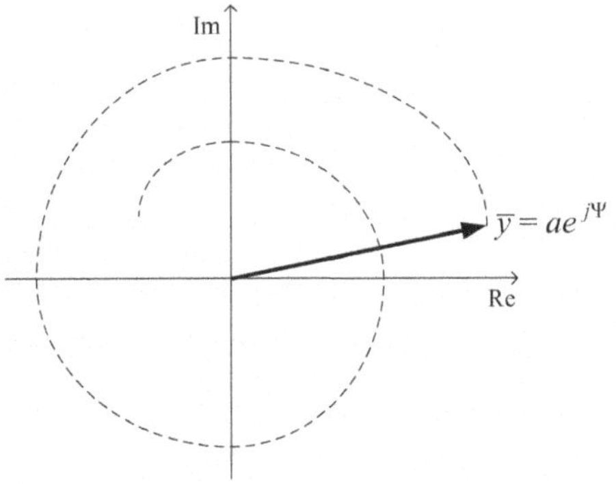

Fig. 6.1 Rotating phasor with variable length and speed of rotation

where a represents the length of the phasor and Ψ represents the angle between the phasor and the real axis. We can associate either the real or the imaginary part of the phasor $\bar{y}(t)$ with the real signal $y(t)$. Therefore, $a(t) = |\bar{y}(t)|$, $\Psi(t) = \arg \bar{y}(t)$.

Let us assume that the initial values of the instantaneous amplitude and the instantaneous phase angle are $a(0)$ and $\Psi(0)$, respectively, and introduce the variable $\Gamma(t)$ such that $a(t) = a(0)e^{\Gamma(t)}$. Prove the following lemma.

Lemma 6.3 *The variable $\Gamma(t)$ provides the change of the logarithm of the amplitude over the time interval [0, t], which can be found as the following integral:*

$$\Gamma(t) = \int_0^t \sigma(t)dt, \tag{6.12}$$

where $\sigma(t)$ is the decay of the amplitude.

Proof By taking the logarithms of the definition formula $a(t) = a(0)e^{\Gamma(t)}$, we find that

$$\Gamma(t) = \ln a(t) - \ln a(0), \tag{6.13}$$

which is the first part of the statement. Also, find the following integral by expressing the decay through the amplitude and its derivative as follows:

$$\int_0^t \sigma(t)dt = \int_0^t \frac{da/dt}{a}dt = \int_{a(0)}^{a(t)} \frac{da}{a} = \ln a|_{a(0)}^{a(t)} = \ln a(t) - \ln a(0). \tag{6.14}$$

Because the right-hand sides of (6.13) and (6.14) are the same, the variable $\Gamma(t)$ can, indeed, be found as per (6.12). ■

We can also write for the phase angle:

$$\Psi(t) = \Psi(0) + \int_0^t \omega(t)dt = \Psi(0) + \Delta\Psi(t), \tag{6.15}$$

where $\Delta\Psi(t)$ is the increment of the phase angle over the time interval [0, t]. In formula (6.15), we shall refer to the variable $\omega(t)$, as the instantaneous frequency, which reflects the rate of change of the phase angle: $\omega(t) = \dot{\Psi}(t)$.

Rewrite the formula (6.11) as follows:

$$\bar{y}(t) = a(0)e^{\Gamma(t)}e^{j\Psi(t)} = a(0)e^{\Gamma(t)+j\Psi(t)} = a(0)e^{j\Psi(0)}e^{\Gamma(t)+j\Delta\Psi(t)} = \bar{y}(0)e^{\Gamma(t)+j\Delta\Psi(t)}. \tag{6.16}$$

Therefore, the phasor vector at every time can be found as the product of the initial phasor by the exponent of the complex function of the amplitude and phase change over time t. It follows from (6.16) that the initial value of the phasor has no effect on the qualitative behavior of the system, and the unity vector aligned with the real axis can be used as the initial phasor. Denote

$$\Lambda(t) = e^{\Gamma(t)+j\Delta\Psi(t)} \tag{6.17}$$

and rewrite the formula (6.16) as

$$\bar{y}(t) = \bar{y}(0)\Lambda(t) \tag{6.18}$$

$\Lambda(t)$ represents a transformation of an initial phasor into the position of the phasor at time t. Let us also consider time derivatives of $\Lambda(t)$, which will be instrumental below:

$$\dot{\Lambda}(t) = \left(\dot{\Gamma}(t) + j\Delta\dot{\Psi}(t)\right)e^{\Gamma(t)+j\Delta\Psi(t)} = (\sigma(t) + j\omega(t))\,\Lambda(t) = \theta(t)\Lambda(t), \tag{6.19}$$

$$\ddot{\Lambda}(t) = \dot{\theta}(t)\Lambda(t) + \theta(t)\dot{\Lambda}(t) = \left(\dot{\theta}(t) + \theta^2(t)\right)\Lambda(t), \tag{6.20}$$

$$\dddot{\Lambda}(t) = \ddot{\theta}(t)\Lambda(t) + \dot{\theta}(t)\dot{\Lambda}(t) + 2\theta(t)\dot{\theta}(t)\Lambda(t) + \theta^2(t)\dot{\Lambda}(t) = \left(\ddot{\theta}(t) + 3\theta(t)\dot{\theta}(t) + \theta^3(t)\right)\Lambda(t), \tag{6.21}$$

where $\theta(t) = \sigma(t) + j\omega(t)$, and so on for higher derivatives. In a sense, the variable $\theta(t)$ is similar to the Laplace variable s. In fact, in a linear system it would be the same, because all transients would occur with constant decay and frequency. In this book chapter, we shall use a different notation to distinguish between these two variables.

Now, given the transfer function being considered, we can write for $y(t)$:

$$u = a_0 y + a_1\dot{y} + a_2\ddot{y} + \cdots + a_n y^{(n)}. \tag{6.22}$$

We can find the derivatives of $\bar{y}(t)$ through the multiplication of $\bar{y}(0)$ by the respective derivative of $\Lambda(t)$ (which follows from (6.18)):

$$\dot{\bar{y}}(t) = \bar{y}(0)\dot{\Lambda}(t) = \bar{y}(0)\theta\,\Lambda(t) = \theta\,\bar{y}, \tag{6.23}$$

$$\ddot{\bar{y}}(t) = \left[\dot{\theta} + \theta^2\right]\bar{y}.$$

Considering that at the point corresponding to a limit cycle the decay is zero, we disregard all derivatives of the decay and higher than first derivatives of the frequency in our model, which is used for analysis of the motions in the vicinity of the limit cycle. In fact, we are going to keep only four variables in the definition of the dynamic harmonic balance: the amplitude, the frequency, the derivative of the amplitude (or equivalently the decay), and the derivative of the frequency. These four would be enough for us to write and solve differential equations describing evolutions of the amplitude and the frequency. Therefore, $\dot{\theta} \approx j\dot{\omega}$ and we can rewrite the last formula as follows:

$$\ddot{\bar{y}}(t) \approx \left[(\sigma + j\omega)^2 + j\dot{\omega}\right]\bar{y} \tag{6.24}$$

and the formula for the third and fourth derivatives as follows:

$$\begin{aligned}\dddot{\bar{y}}(t) &\approx \left[2(\sigma + j\omega)(\dot{\sigma} + j\dot{\omega}) + j\ddot{\omega}\right]\bar{y} + \left[(\sigma + j\omega)^2 + j\dot{\omega}\right](\sigma + j\omega)\bar{y} \\ &\approx \left[j3\dot{\omega} + (\sigma + j\omega)^2\right](\sigma + j\omega)\bar{y} = \left[(\sigma + j\omega)^3 + j3\dot{\omega}(\sigma + j\omega)\right]\bar{y}\end{aligned} \tag{6.25}$$

$$\bar{y}^{(4)}(t) \approx \left[(\sigma + j\omega)^4 + j6\dot{\omega}(\sigma + j\omega) - 3\dot{\omega}^2\right]\bar{y}.$$

We can continue with taking further derivatives. It is worth noting that the formulas above are organized the way that they have a term $\theta = (\sigma + j\omega)$ to the respective power and the term which is the product of $\dot{\omega}$ and another multiplier. Therefore, we can write for $\bar{y}(t)$:

$$\bar{u} = \left[a_0 + a_1(\sigma + j\omega) + a_2(\sigma + j\omega)^2 + \cdots + a_n(\sigma + j\omega)^n\right]\bar{y} + S(\sigma, \omega, \dot{\omega})\,\bar{y}, \tag{6.26}$$

where $S(\sigma, \omega, \dot{\omega})$ includes all terms containing $\dot{\omega}$. This component can be accounted for as an additional feedback if represented by a system block diagram in the controllable canonical form.

If we introduce certain modified frequency response as

$$W_l^*(\sigma, \omega, \dot{\omega}) = \frac{\bar{y}}{\bar{u}} = \frac{1}{Q(\sigma + j\omega) + S(\sigma, \omega, \dot{\omega})} = \frac{W_l(\sigma + j\omega)}{1 + W_l(\sigma + j\omega)\,S(\sigma, \omega, \dot{\omega})}, \tag{6.27}$$

we can write the following complex equation that must hold during the transient oscillations:

$$N(a)W^*(\sigma, \omega, \dot{\omega}) = -1. \tag{6.28}$$

Obviously, Eq. (6.28) can be split into two equations: for real and imaginary parts, or for equations of the magnitude balance and phase balance.

Equation (6.28) must be complemented with an equation that relates the difference of the decays at the input and the output of the linear part and the frequency rate of change. We note that the amplitude of the control (first harmonic) is as follows: $a_u = a\,|Q(\sigma + j\omega) + S(\sigma, \omega, \dot{\omega})|$, and, therefore, its time derivative is $\dot{a}_u = \dot{a}\,|Q(\sigma + j\omega) + S(\sigma, \omega, \dot{\omega})| + a\frac{d|Q(\sigma + j\omega) + S(\sigma, \omega, \dot{\omega})|}{dt}$. The decay of signal $u(t)$ is computed as

$$\begin{aligned}\sigma_u = \frac{\dot{a}_u}{a_u} &= \frac{\dot{a}\,|Q(\sigma + j\omega) + S(\sigma, \omega, \dot{\omega})| + a\frac{d|Q(\sigma + j\omega) + S(\sigma, \omega, \dot{\omega})|}{dt}}{a\,|Q(\sigma + j\omega) + S(\sigma, \omega, \dot{\omega})|} \\ &= \sigma + \frac{d\ln|Q(\sigma + j\omega) + S(\sigma, \omega, \dot{\omega})|}{dt}.\end{aligned}$$

Because we disregard $\dot{\sigma}$ and $\ddot{\omega}$, we can rewrite the last formula as follows:

$$\sigma_u = \sigma + \frac{\partial \ln |Q(\sigma + j\omega) + S(\sigma, \omega, \dot{\omega})|}{\partial \omega}\dot{\omega} = \sigma - \frac{\partial \ln \left|W_l^*(\sigma, \omega, \dot{\omega})\right|}{\partial \omega}\dot{\omega}.$$

The derivative in the last formula defines the slope of the magnitude–frequency characteristic of $W_l^*(\sigma, \omega, \dot{\omega})$. Considering also the formula for the decay of $u(t)$ derived above through the DF (Eq. (6.10)), we can now write the condition of the balance of the decays in the closed-loop system as follows:

$$\frac{\partial \ln |W^*(\sigma, \omega, \dot{\omega})|}{\partial \ln \omega}\frac{\dot{\omega}}{\omega} = -\frac{\partial \ln |N(a)|}{\partial \ln a}\sigma. \tag{6.29}$$

Equation (6.28) comprises the conditions of *balance of the phases* and *balance of the magnitudes* [15]. And Eq. (6.29) is the equation of the *balance of the rates of change*. Therefore, if the transfer function is given as $W(s) = 1/Q(s) = 1/\left(a_0 + a_1 s + a_2 s^2 + \cdots + a_n s^n\right)$, then the modified frequency response can be found as

$$W^*(\omega, \dot{\omega}, \sigma) = \frac{1}{Q(\sigma + j\omega) + j\dot{\omega}R(\sigma, \omega, \dot{\omega})},$$

where R is a polynomial that can be computed through the rules that were established above for the polynomial S: for $n = 2$: $R = a_2$, for $n = 3$: $R = a_2 + 3a_3(\sigma + j\omega)$, for $n = 4$: $R = a_2 + 3a_3(\sigma + j\omega) + 6a_4(\sigma + j\omega) + 3ja_4\dot{\omega}$, etc.

The dynamic harmonic balance principle is formulated as follows [15]. *At every time during a transient oscillation being a single-frequency mode (not a complex multi-frequency mode), the oscillation can be described by the following variables: instantaneous frequency, instantaneous amplitude, instantaneous decay of the amplitude, and instantaneous rate of change (time derivative) of the frequency, which must satisfy Eqs. (6.28) and (6.29).*

6.4 Analysis of Orbital Stability Based on Dynamic Harmonic Balance

For deriving the stability conditions, let us assume that the system experiences a small deviation from the periodic motion (A_0, Ω_0). The initial instantaneous amplitude is $a = A_0 + \Delta a$, and the initial instantaneous frequency is $\omega = \Omega_0 + \Delta\omega$. We also assume that the motion would reveal certain decay of the amplitude $\sigma = \dot{a}/a \neq 0$ and a rate of change of the frequency $\dot{\omega} \neq 0$. The rates of change of the amplitude (decay) and frequency cannot simultaneously be zero because it would give a periodic motion, which contradicts the described scenario. Yet, if either the decay of the amplitude or the rate of change of the frequency is zero, the other rate should be zero too—according to Eqs. (6.28) and (6.29).

As in Loeb's approach, for the periodic motion to be orbitally stable, it is necessary that a positive value of Δa would lead to a negative value of σ and vice versa, or to the negative derivative $d\sigma/da$. In this case, $a = A_0$ would be a stable equilibrium point in the solution of the differential equation for the instantaneous amplitude $\dot{a}(t) = a(t)\sigma(t)$. Let us find the derivatives of both Eqs. (6.28) and (6.29) that define the DHB, and proceed with the derivation of the orbital stability conditions from here. We shall base the derivation on Theorem 2 from [17], with a more detailed proof provided in this chapter. We shall also use the advantage of logarithmic sensitivities over semi-logarithmic ones, which will allow us to further advance in the derivation. Let us formulate the orbital stability conditions as the following theorem.

Theorem 6.4 *For orbital stability of the periodic solution given by (6.5) in the system (6.1)–(6.3), it is necessary that the following inequality must hold:*

$$\frac{N_1}{D} < 0, \tag{6.30}$$

where $N_1 = c_{22}d_1 - c_{12}d_2$, $D = c_{11}c_{22} - c_{12}c_{21}$,

$$c_{11} = \mathrm{Re}\left[\frac{\partial \ln W}{\partial s}\right] - \Omega_0 \frac{S_a^{|N|}}{S_\omega^{|W|}} \frac{\mathrm{Re}Q \cdot \mathrm{Im}R - \mathrm{Im}Q \cdot \mathrm{Re}R}{|Q|^2},$$

$$c_{12} = -\mathrm{Im}\left[\frac{\partial \ln W}{\partial s}\right] = \frac{\partial \ln |W|}{\partial \omega},$$

$$c_{21} = \mathrm{Im}\left[\frac{\partial \ln W}{\partial s}\right] + \Omega_0 \frac{S_a^{|N|}}{S_\omega^{|W|}} \frac{\mathrm{Re}Q \cdot \mathrm{Re}R + \mathrm{Im}Q \cdot \mathrm{Im}R}{|Q|^2},$$

$$c_{22} = \mathrm{Re}\left[\frac{\partial \ln W}{\partial s}\right] = \frac{\partial \arg W}{\partial \omega}, \tag{6.31}$$

$$d_1 = -\frac{d \ln |N|}{da},$$

$$d_2 = -\frac{d \arg N}{da},$$

$S_a^{|N|} = \frac{\partial \ln |N|}{\partial \ln a}$, $S_\omega^{|W|} = \frac{\partial \ln |W|}{\partial \ln \omega}$, $s = j\omega$, *and all derivatives are taken at the point* (A_0, Ω_0).

Proof We start with finding a few derivatives that will be instrumental in the proof. Considering formula (6.30) of the modified frequency response, let us find the derivatives of W^* at the point $P := \{a = A_0, \omega = \Omega_0, \sigma = 0, \dot{\omega} = 0\}$ corresponding to the periodic solution:

$$\left.\frac{\partial \ln W^*}{\partial \ln \omega}\right|_P = \left.\frac{\partial \ln |W^*|}{\partial \ln \omega}\right|_P + j\left.\frac{\partial \arg W^*}{\partial \ln \omega}\right|_P = -\left.\frac{\partial \ln |Q + j\dot{\omega}R|}{\partial \ln \omega}\right|_P - j\left.\frac{\partial \arg(Q + j\dot{\omega}R)}{\partial \ln \omega}\right|_P$$

$$= -\frac{1}{|Q + j\dot{\omega}R|_P}\cdot\left.\frac{\partial |Q + j\dot{\omega}R|}{\partial \ln \omega}\right|_P - j\left.\frac{\partial \arg(Q + j\dot{\omega}R)}{\partial \ln \omega}\right|_P$$

$$= -\frac{1}{|Q|_P}\cdot\left.\frac{\partial |Q|}{\partial \ln \omega}\right|_P - j\left.\frac{\partial \arg Q}{\partial \ln \omega}\right|_P = -\left.\frac{\partial \ln |Q|}{\partial \ln \omega}\right|_P - j\left.\frac{\partial \arg Q}{\partial \ln \omega}\right|_P$$

$$= \left.\frac{\partial \ln |W|}{\partial \ln \omega}\right|_P + j\left.\frac{\partial \arg W}{\partial \ln \omega}\right|_P = \left.\frac{\partial \ln W}{\partial \ln \omega}\right|_P, \tag{6.32}$$

$$\left.\frac{\partial \ln W^*}{\partial \ln \sigma}\right|_P = \left.\frac{\partial \ln W}{\partial \ln \sigma}\right|_P = \left.\frac{\partial \ln |W|}{\partial \ln \sigma}\right|_P + j\left.\frac{\partial \arg W}{\partial \ln \sigma}\right|_P = -\left.\frac{\partial \ln |Q|}{\partial \ln \sigma}\right|_P - j\left.\frac{\partial \arg Q}{\partial \ln \sigma}\right|_P, \tag{6.33}$$

$$\left.\frac{\partial \ln |W^*|}{\partial \omega}\right|_P = \mathrm{Re}\left[\left.\frac{\partial \ln W}{\partial \omega}\right|_{\omega=\Omega_0}\right] = \mathrm{Re}\left[\left.\frac{\partial \ln W}{\partial s}\right|_{s=j\Omega_0}\cdot\frac{ds}{d\omega}\right]$$
$$= \mathrm{Re}\left[j\left.\frac{\partial \ln W}{\partial s}\right|_{s=j\Omega_0}\right] = -\mathrm{Im}\left[\left.\frac{\partial \ln W}{\partial s}\right|_{s=j\Omega_0}\right], \tag{6.34}$$

$$\left.\frac{\partial \ln |W^*|}{\partial \sigma}\right|_P = \mathrm{Re}\left[\left.\frac{\partial \ln W}{\partial \sigma}\right|_P\right] = \mathrm{Re}\left[\left.\frac{\partial \ln W}{\partial s}\right|_{s=j\Omega_0}\cdot\frac{ds}{d\sigma}\right] = \mathrm{Re}\left[\left.\frac{\partial \ln W}{\partial s}\right|_{s=j\Omega_0}\right], \tag{6.35}$$

$$\left.\frac{\partial \arg W^*}{\partial \omega}\right|_P = \mathrm{Im}\left[\left.\frac{\partial \ln W^*}{\partial \omega}\right|_P\right] = \mathrm{Im}\left[\left.\frac{\partial \ln W}{\partial \omega}\right|_{\omega=\Omega_0}\right] = \mathrm{Im}\left[\left.\frac{\partial \ln W}{\partial s}\right|_{s=j\Omega_0}\cdot\frac{ds}{d\omega}\right]$$
$$= \mathrm{Im}\left[j\left.\frac{\partial \ln W}{\partial s}\right|_{s=j\Omega_0}\right] = \mathrm{Re}\left[\left.\frac{\partial \ln W}{\partial s}\right|_{s=j\Omega_0}\right], \tag{6.36}$$

$$\left.\frac{\partial \arg W^*}{\partial \sigma}\right|_P = \mathrm{Im}\left[\left.\frac{\partial \ln W^*}{\partial \sigma}\right|_P\right] = \mathrm{Im}\left[\left.\frac{\partial \ln W}{\partial \sigma}\right|_P\right]$$
$$= \mathrm{Im}\left[\left.\frac{\partial \ln W}{\partial s}\right|_{s=j\Omega_0}\cdot\frac{ds}{d\sigma}\right] = \mathrm{Im}\left[\left.\frac{\partial \ln W}{\partial s}\right|_{s=j\Omega_0}\right], \tag{6.37}$$

$$\left.\frac{\partial \ln W^*}{\partial \dot{\omega}}\right|_P = -\left.\frac{\partial \ln(Q + j\dot{\omega}R)}{\partial \dot{\omega}}\right|_P = -\left.\frac{\partial \ln |Q + j\dot{\omega}R|}{\partial \dot{\omega}}\right|_P - j\left.\frac{\partial \arg(Q + j\dot{\omega}R)}{\partial \dot{\omega}}\right|_P$$

$$= -\frac{1}{|Q+j\dot{\omega}R|_P}\cdot\left.\frac{\partial\,|Q+j\dot{\omega}R|}{\partial\dot{\omega}}\right|_P - j\left.\frac{\partial \arctan\{\mathrm{Im}(Q+j\dot{\omega}R)/\mathrm{Re}(Q+j\dot{\omega}R)\}}{\partial\dot{\omega}}\right|_P$$

$$= -\frac{1}{|Q|}\cdot\left.\frac{\partial\sqrt{\mathrm{Re}^2(Q+j\dot{\omega}R)+\mathrm{Im}^2(Q+j\dot{\omega}R)}}{\partial\dot{\omega}}\right|_P - j\left.\frac{\partial \arctan\{\mathrm{Im}(Q+j\dot{\omega}R)/\mathrm{Re}(Q+j\dot{\omega}R)\}}{\partial\dot{\omega}}\right|_P$$

$$= -\frac{1}{|Q|}\cdot\frac{1}{2}\frac{1}{\sqrt{\mathrm{Re}^2Q+\mathrm{Im}^2Q}}\left(2\mathrm{Re}Q\cdot\mathrm{Re}(jR)+2\mathrm{Im}Q\cdot\mathrm{Im}(jR)\right)$$
$$- j\frac{\frac{\partial \mathrm{Im}(Q+j\dot{\omega}R)}{\partial\dot{\omega}}\mathrm{Re}Q - \frac{\partial \mathrm{Re}(Q+j\dot{\omega}R)}{\partial\dot{\omega}}\mathrm{Im}Q}{\left(1+[\mathrm{Im}Q/\mathrm{Re}Q]^2\right)\mathrm{Re}^2Q}$$

$$= -\frac{1}{|Q|^2}\left(-\mathrm{Re}Q\cdot\mathrm{Im}R+\mathrm{Im}Q\cdot\mathrm{Re}R\right) - j\frac{1}{|Q|^2}\cdot\left[\mathrm{Re}Q\cdot\mathrm{Im}(jR)-\mathrm{Im}Q\cdot\mathrm{Re}(jR)\right]$$
$$= \frac{1}{|Q|^2}\left\{(\mathrm{Re}Q\cdot\mathrm{Im}R-\mathrm{Im}Q\cdot\mathrm{Re}R) - j\left(\mathrm{Re}Q\cdot\mathrm{Re}R+\mathrm{Im}Q\cdot\mathrm{Im}R\right)\right\}. \quad (6.38)$$

Take a semi-logarithmic derivative at the point P of both sides of the DHB equation (6.10) with respect to a:

$$\frac{d\ln N}{da}+\frac{d\ln W^*(\omega,\dot{\omega},\sigma)}{da}=0. \quad (6.39)$$

Expand the derivatives in (6.39) considering them as composite functions:

$$\frac{d\ln|N|}{da}+j\frac{d\arg N}{da}+\frac{\partial\ln W^*}{\partial\sigma}\cdot\frac{d\sigma}{da}+\frac{\partial\ln W^*}{\partial\omega}\cdot\frac{d\omega}{da}+\frac{\partial\ln W^*}{\partial\dot{\omega}}\cdot\frac{d\dot{\omega}}{da}=0 \quad (6.40)$$

and take the real parts of both sides:

$$\frac{d\ln|N|}{da}+\frac{\partial\ln|W^*|}{\partial\sigma}\cdot\frac{d\sigma}{da}+\frac{\partial\ln|W^*|}{\partial\omega}\cdot\frac{d\omega}{da}+\frac{\partial\ln|W^*|}{\partial\dot{\omega}}\cdot\frac{d\dot{\omega}}{da}=0. \quad (6.41)$$

Substitution of (6.32)–(6.38) into (6.41) yields

$$\frac{d\ln|N|}{da}+\mathrm{Re}\left[\left.\frac{\partial\ln W}{\partial s}\right|_{s=j\Omega_0}\right]\frac{d\sigma}{da}-\mathrm{Im}\left[\left.\frac{\partial\ln W}{\partial s}\right|_{s=j\Omega_0}\right]\frac{d\omega}{da}+$$
$$+\frac{1}{|Q|^2}\left(\mathrm{Re}Q\cdot\mathrm{Im}R-\mathrm{Im}Q\cdot\mathrm{Re}R\right)\cdot\frac{d\dot{\omega}}{da}=0. \quad (6.42)$$

Taking the imaginary parts of (6.40) yields

$$\frac{d \arg N}{da} + \frac{\partial \arg W^*}{\partial \sigma} \cdot \frac{d\sigma}{da} + \frac{\partial \arg W^*}{\partial \omega} \cdot \frac{d\omega}{da} + \frac{\partial \arg W^*}{\partial \dot{\omega}} \cdot \frac{d\dot{\omega}}{da} = 0. \tag{6.43}$$

And substitution of (6.36)–(6.38) into (6.43) gives

$$\frac{d \arg N}{da} + \mathrm{Im}\left[\frac{\partial \ln W}{\partial s}\bigg|_{s=j\Omega_0}\right]\frac{d\sigma}{da} + \mathrm{Re}\left[\frac{\partial \ln W}{\partial s}\bigg|_{s=j\Omega_0}\right]\frac{d\omega}{da}$$

$$- \frac{1}{|Q|^2} \cdot (\mathrm{Re}Q \cdot \mathrm{Re}R + \mathrm{Im}Q \cdot \mathrm{Im}R)\,\frac{d\dot{\omega}}{da} = 0. \tag{6.44}$$

We now rewrite (6.42) and (6.44):

$$\mathrm{Re}\left[\frac{\partial \ln W}{\partial s}\right]\frac{d\sigma}{da} - \mathrm{Im}\left[\frac{\partial \ln W}{\partial s}\right]\frac{d\omega}{da} + \frac{\mathrm{Re}Q.\mathrm{Im}R - \mathrm{Im}Q.\mathrm{Re}R}{|Q|^2} \cdot \frac{d\dot{\omega}}{da} = -\frac{d \ln |N|}{da} \tag{6.45}$$

$$\mathrm{Im}\left[\frac{\partial \ln W}{\partial s}\right]\frac{d\sigma}{da} + \mathrm{Re}\left[\frac{\partial \ln W}{\partial s}\right]\frac{d\omega}{da} - \frac{\mathrm{Re}Q.\mathrm{Re}R + \mathrm{Im}Q.\mathrm{Im}R}{|Q|^2} \cdot \frac{d\dot{\omega}}{da} = -\frac{d \arg N}{da}. \tag{6.46}$$

Equations (6.45) and (6.46) are a system of three algebraic equations with three unknown variables: $\frac{d\sigma}{da}$, $\frac{d\omega}{da}$, and $\frac{d\dot{\omega}}{da}$. An additional equation is obtained through differentiation of the rates balance condition (6.29) with respect to a:

$$\left(\frac{\partial S_\omega^{|W^*|}}{\partial \omega}\frac{d\omega}{da} + \frac{\partial S_\omega^{|W^*|}}{\partial \sigma}\frac{d\sigma}{da} + \frac{\partial S_\omega^{|W^*|}}{\partial \dot{\omega}}\frac{d\dot{\omega}}{da}\right)\frac{\dot{\omega}}{\omega}$$

$$+ S_\omega^{|W^*|}\,\frac{\frac{d\dot{\omega}}{da}\omega - \frac{d\omega}{da}\dot{\omega}}{\omega^2} = -\left(\frac{\partial S_a^{|N|}}{\partial a} \cdot \sigma + S_a^{|N|} \cdot \frac{d\sigma}{da}\right), \tag{6.47}$$

where $S_\omega^{|W^*|} = \frac{\partial \ln |W^*|}{\partial \ln \omega}$, $S_\omega^{|W|} = \frac{\partial \ln |W|}{\partial \ln \omega}$, $S_a^{|N|} = \frac{\partial \ln |N|}{\partial \ln a}$.

Recall that all the derivatives are taken at the point $P := \{a = A_0, \omega = \Omega_0, \sigma = 0, \dot{\omega} = 0\}$. Therefore, formula (6.47) rewrites as

$$\frac{1}{\omega} S_\omega^{|W|} \cdot \frac{d\dot{\omega}}{da} = -S_a^{|N|} \cdot \frac{d\sigma}{da},$$

from which $\frac{d\dot{\omega}}{da}$ is expressed as follows:

$$\frac{d\dot{\omega}}{da} = -\Omega_0 \frac{S_a^{|N|}}{S_\omega^{|W|}}\frac{d\sigma}{da}. \tag{6.48}$$

Substitution of (6.48) into (6.45), (6.46) yields the set of two linear equations with two unknown variables:

$$\begin{cases} c_{11}\frac{d\sigma}{da} + c_{12}\frac{d\omega}{da} = d_1 \\ c_{21}\frac{d\sigma}{da} + c_{22}\frac{d\omega}{da} = d_2. \end{cases} \tag{6.49}$$

The coefficients for Eqs. (6.49) are given in the formulation of Theorem 6.4, in formulas (6.31). The derivative $\frac{d\sigma}{da}$ is, therefore, given by the left-hand side of (6.30). Because the decay is defined as $\sigma = \dot{a}/a$, the negative value of the derivative $\frac{d\sigma}{da}$ (at both positive and negative deviations of the amplitude from A_0) would imply convergence of the solution of $\dot{a}(t) = a(t)\sigma(t)$ to A_0. And the negative sign of this derivative is given by (6.30). Also, there are no specific requirements to the derivative of the frequency, except that it must be a finite value. However, it is always the case, because it may be infinite only if both $c_{12} = 0$ and $c_{22} = 0$, which would mean that at the point Ω_0 both the magnitude response and the phase response should have zero derivatives with respect to the frequency, which is impossible. ■

6.5 Comparison of the Loeb's Criterion and the Criterion Based on Dynamic Harmonic Balance

Analysis of the condition of orbital stability as per [17], which is given by (6.30), and its comparison with the Loeb's criterion [9] are as follows. One can notice that the numerator of (6.30) gives the Loeb's condition:

$$N_1 = c_{22}d_1 - c_{12}d_2 = -\frac{\partial \arg W}{\partial \omega}\frac{d\ln|N|}{da} + \frac{\partial \ln|W|}{\partial \omega}\frac{d\arg N}{da}. \tag{6.50}$$

Therefore, if the denominator of (6.30) is positive: $D = c_{11}c_{22} - c_{12}c_{21} > 0$, then the criterion (6.30) would be identical to Loeb's criterion. Let us find the conditions of $D > 0$. Find the formula for D:

$$\begin{aligned} D &= c_{11}c_{22} - c_{12}c_{21} \\ &= \mathrm{Re}^2\left[\frac{\partial \ln W}{\partial s}\right] - \frac{\partial \arg W}{\partial \omega}\Omega_0\frac{S_a^{|N|}}{S_\omega^{|W|}}\frac{\mathrm{Re}Q\cdot\mathrm{Im}R - \mathrm{Im}Q\cdot\mathrm{Re}R}{|Q|^2} \\ &+\mathrm{Im}^2\left[\frac{\partial \ln W}{\partial s}\right] + \frac{\partial \ln|W|}{\partial \omega}\Omega_0\frac{S_a^{|N|}}{S_\omega^{|W|}}\frac{\mathrm{Re}Q\cdot\mathrm{Re}R + \mathrm{Im}Q\cdot\mathrm{Im}R}{|Q|^2} \\ &= \left|\frac{\partial \ln W}{\partial s}\right|^2 + \frac{S_a^{|N|}}{S_\omega^{|W|}}\frac{\Omega_0}{|Q|^2}\left[\frac{\partial \ln|W|}{\partial \omega}(\mathrm{Re}Q\cdot\mathrm{Re}R + \mathrm{Im}Q\cdot\mathrm{Im}R)\right. \\ &\left. -\frac{\partial \arg W}{\partial \omega}(\mathrm{Re}Q\cdot\mathrm{Im}R - \mathrm{Im}Q\cdot\mathrm{Re}R)\right]. \end{aligned} \tag{6.51}$$

Considering that $S_{\omega}^{|W|} = \frac{\partial \ln |W|}{\partial \ln \omega} = \Omega_0 \frac{\partial \ln |W|}{\partial \omega}$ and $\frac{\partial \ln W}{\partial s} = j \frac{\partial \ln W}{\partial \omega}$, we rewrite (6.51) as

$$D = \left| \frac{\partial \ln W}{\partial s} \right|^2 + \frac{S_a^{|N|}}{S_{\omega}^{|W|} \, |Q|^2} \left[\frac{\partial \ln |W|}{\partial \ln \omega} (\mathrm{Re}Q \cdot \mathrm{Re}R + \mathrm{Im}Q \cdot \mathrm{Im}R) \right.$$

$$\left. - \frac{\partial \arg W}{\partial \ln \omega} (\mathrm{Re}Q \cdot \mathrm{Im}R - \mathrm{Im}Q \cdot \mathrm{Re}R) \right]$$

$$= \frac{1}{\Omega_0^2} \left| \frac{\partial \ln W}{\partial \ln s} \right|^2 + \frac{S_a^{|N|}}{S_{\omega}^{|W|} \, |Q|^2} \left[\frac{\partial \ln |W|}{\partial \ln \omega} (\mathrm{Re}Q \cdot \mathrm{Re}R + \mathrm{Im}Q \cdot \mathrm{Im}R) \right.$$

$$\left. - \frac{\partial \arg W}{\partial \ln \omega} (\mathrm{Re}Q \cdot \mathrm{Im}R - \mathrm{Im}Q \cdot \mathrm{Re}R) \right]$$

$$= \frac{1}{\Omega_0^2} \left| S_s^W \right|^2 + \frac{S_a^{|N|}}{\mathrm{Re}S_s^W \, |Q|^2} \left[\mathrm{Re}S_s^W (\mathrm{Re}Q \cdot \mathrm{Re}R + \mathrm{Im}Q \cdot \mathrm{Im}R) \right.$$

$$\left. - \mathrm{Im}S_s^W (\mathrm{Re}Q \cdot \mathrm{Im}R - \mathrm{Im}Q \cdot \mathrm{Re}R) \right]. \tag{6.52}$$

The first term in (6.52) is always positive and is expected to dominate in the whole sum, unless $\mathrm{Re}S_s^W$ has a small magnitude. The second and the third terms may be positive or negative—depending on the system. To analyze the possible situations, we need to note that at the point P, both $\mathrm{Re}R > 0$ and $\mathrm{Im}R > 0$. It follows from the way the polynomial R is designed (see above and [15]). Also, for the Lure systems, the real part at the point P is $\mathrm{Re}Q < 0$, and the imaginary part $\mathrm{Im}Q = 0$ for systems having single-valued nonlinearities (the periodic solution is on the real axis), $\mathrm{Im}Q > 0$ for system having hysteretic nonlinearities with positive hysteresis value (with the periodic solution in the third quadrant), and $\mathrm{Im}Q < 0$ for system having hysteretic nonlinearities with negative hysteresis value (with the periodic solution in the second quadrant). We further present the following expressions from (6.52) as real and imaginary parts of the dot product of two vectors:

$$\mathrm{Re}Q \cdot \mathrm{Re}R + \mathrm{Im}Q \cdot \mathrm{Im}R = \mathrm{Re}\left(\tilde{Q}R\right), \tag{6.53}$$

$$\mathrm{Re}Q \cdot \mathrm{Im}R - \mathrm{Im}Q \cdot \mathrm{Re}R = \mathrm{Im}\left(\tilde{Q}R\right), \tag{6.54}$$

where $\tilde{Q}$ is the complex conjugate of Q. To understand the location of vector $\tilde{Q}$, it is worth to note that it has the same phase angle as the frequency response W at the point of the periodic solution. Because vector R is located in the first quadrant, vector $\tilde{Q}R$ can be located in either third or fourth quadrants, with $\mathrm{Re}\left(\tilde{Q}R\right)$ that can be positive or negative, and $\mathrm{Im}\left(\tilde{Q}R\right) < 0$. Further, $\mathrm{Re}\left(\tilde{Q}R\right) / |Q|$ and $\mathrm{Im}\left(\tilde{Q}R\right) / |Q|$ are the real and the imaginary parts of the vector that is a result of rotation of vector R by the angle equal to the phase of $\tilde{Q}$ (phase of W at the point of the periodic solution). Denote

$$R^{+} = \mathrm{Re}\left(\tilde{Q}R\right)/|Q| + j\mathrm{Im}\left(\tilde{Q}R\right)/|Q|. \tag{6.55}$$

Therefore, R^{+} is a vector that is produced by the rotation of R by the angle equal to the phase of $W(j\Omega_0)$ at the point of the periodic solution. Rewrite (6.52) as follows:

$$D = \frac{1}{\Omega_0^2}\left|S_s^W\right|^2 + \frac{S_a^{|N|}}{\mathrm{Re}S_s^W\,|Q|}\left[\mathrm{Re}S_s^W\mathrm{Re}R^{+} - \mathrm{Im}S_s^W\mathrm{Im}R^{+}\right], \tag{6.56}$$

where $S_s^W = \frac{\partial \ln W}{\partial \ln s}$ is the sensitivity function, which is in fact a certain transfer function. For the considered $W(s)$, the sensitivity function is given by

$$S_s^W = -\frac{sQ'(s)}{Q(s)},$$

where $Q'(s)$ is a polynomial produced by differentiating the polynomial $Q(s)$ with respect to s:

$$Q'(s) = \sum_{i=1}^{n} a_i s^{i-1}.$$

Therefore, the sensitivity (transfer) function $\frac{sQ'(s)}{Q(s)}$ has relative degree zero. At low frequencies,

$$\lim_{s\to j0}\frac{sQ'(s)}{Q(s)} = \lim_{s\to j0}\frac{a_1 s}{a_0 + a_1 s} = 0,$$

and at high frequencies,

$$\lim_{s\to j\infty}\frac{sQ'(s)}{Q(s)} = \frac{s(n-1)a_n s^{n-1}}{a_n s^n} = n-1.$$

Under the assumptions of the monotonous phase characteristic, the phase of the sensitivity function $S_s^W(j\omega)$ changes from –90° to –180° when the frequency varies from 0 to ∞ (the phase of the function $\frac{sQ'(s)}{Q(s)} = -S_s^W$ changes from +90° to 0°). However, there may be other shapes of $S_s^W(j\omega)$ too. Rewrite now formula (6.46) as follows:

$$D = \frac{1}{\Omega_0^2}\left|S_s^W\right|^2 + \frac{S_a^{|N|}\mathrm{Re}\left(S_s^W R^{+}\right)}{|Q|\,\mathrm{Re}S_s^W}. \tag{6.57}$$

As per the above analysis, both vectors S_s^W and R^{+} are located in the third quadrant. Therefore, their product may be in the first or second quadrant. And $\mathrm{Re}\left(S_s^W R^{+}\right)$ may be positive or negative. This would depend on the magnitude and sign of $\mathrm{Re}S_s^W$. For example, at relatively low frequencies the phase of the sensitivity function $S_s^W(j\omega)$ is slightly below –90°, the magnitude $\left|S_s^W\right|$ is small, and $\mathrm{Re}\left(S_s^W R^+\right) < 0$ (third quadrant). Small magnitude of $\mathrm{Re}S_s^W$ can also occur in under-damped linear systems at frequencies close to the resonant frequency, where the slope of magnitude characteristic is low (Note that $\mathrm{Re}S_s^W = \frac{\partial \ln W(j\omega)}{\partial \ln \omega}$). However, the derivative must be taken at the frequency of the periodic solution in the nonlinear system, and therefore, the whole situation corresponds to the frequency of a periodic solution near the resonant frequency of the under-damped linear part. This is the only feasible scenario of the second term in (6.57) dominating, and the value of D becoming negative.

We now proceed with the type of Lure system that is found most frequently—the system with a single-valued nonlinearity. For a BIBO-stable linear part given by (6.1), (6.2) and a single-valued odd-symmetric nonlinearity given by

$$u = -f(y), \tag{6.58}$$

the derived conditions (6.50), (6.51) rewrite as (in this case $\arg N = 0$, $\mathrm{Re}Q = -Q$, $\mathrm{Im}Q = 0$, $\mathrm{Re}R^+ = -\mathrm{Re}R$)

$$D = \frac{1}{\Omega_0^2}\left|S_s^W\right|^2 - \frac{S_a^N \mathrm{Re}\left(S_s^W R\right)}{|Q|\,\mathrm{Re}S_s^W}, \tag{6.59}$$

$$N_1 = -\frac{\partial \arg W}{\partial \omega}\frac{d \ln N}{da}. \tag{6.60}$$

If D given by (6.59) is positive, then the criterion of orbital stability (6.30) is identical to Loeb's. Let us consider the situation of possible difference between the two criteria. If as stated above $S_a^N = A_0\frac{d\ln N}{da} < 0$, $\frac{\partial \arg W}{\partial \omega} < 0$, $\mathrm{Re}S_s^W < 0$ and $\mathrm{Re}\left(S_s^W R\right) > 0$ (the nontrivial case), then the second term in (6.59) is positive, and there is a possibility for $D < 0$.

However, all attempts to find examples of plants that would simultaneously satisfy the requirements of producing a periodic solution (6.5) and the denominator (6.59) $D < 0$ have failed. Plants that satisfy the condition of $D < 0$ have been found only among unstable dynamics, which do not provide solution to the harmonic balance equation (6.5).

Despite the noted deficiency in the proof of Loeb's criterion, there is no record of its falling short in establishing orbital stability. In that respect, the condition of $D >$ 0 can be treated as one more condition, which must be added to the Loeb's criterion.

6.6 Example of Analysis

Example. Consider an example of analysis of periodic solutions in the third-order dynamic system given by the linear part with the transfer function

$$W(s) = \frac{1}{a_0 + a_1 s + a_2 s^2 + a_3 s^3},$$

where the parameters of the denominator are $a_0 = 0.05$, $a_1 = 0.18$, $a_2 = 0.24$, $a_3 = 0.1$, and the nonlinearity shown in Fig. 6.2, with parameters $b = 1$, $c = 1$.

The DF of the nonlinearity can be derived per (6.6) as follows [8]:

$$N(a) = \frac{4c}{\pi a}\sqrt{1 - \frac{b^2}{a^2}}.$$

The plot of function $N(a)$ is presented in Fig. 6.3. One can see that $N(a)$ is not monotonous; it has a point of maximum at $a = \sqrt{2}b$ (this solution can be found by equating $dN(a)/da$ to zero). Therefore, the Nyquist plot of the transfer function $W(s)$ may have two intersections with negative reciprocal of the DF, because there may be two different points of intersection of the plot of $N(a)$ by a horizontal line. In the considered example, there are two periodic solutions found per (6.6): (A) $\Omega_0 = 1.3416$ and $A_{0A} = 1.05441$, and (B) $\Omega_0 = 1.3416$ and $A_{0B} = 3.1621$. It is worth noting that the frequency is the same in both periodic solutions.

Let us find the coefficients as per (6.25), (6.27). For the considered transfer function, polynomials Q and R at the point P are given by $Q = -ja_3\Omega_0^3 - a_2\Omega_0^2 + ja_1\Omega_0 + a_0$, $R = a_2 + j3a_3\Omega_0$, which results in $\mathrm{Re}\,Q = -0.3820$, $\mathrm{Im}\,Q =$

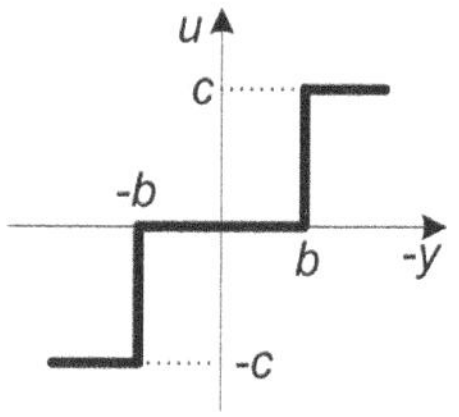

Fig. 6.2 Nonlinearity $u = -f(y)$ of the system in example

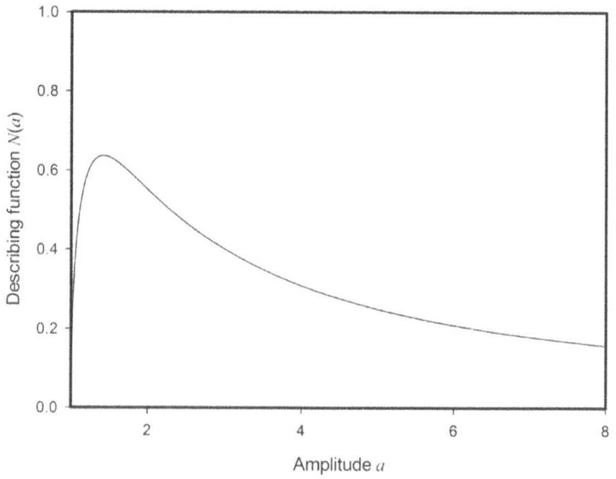

Fig. 6.3 Describing function $N(a)$ of nonlinearity in example

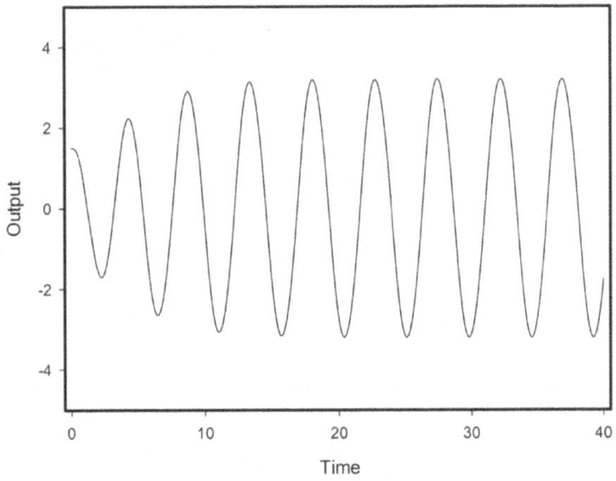

Fig. 6.4 Simulations of system in example; output $y(t)$

0, $|Q| = 0.3820$, $\mathrm{Re}R = 0.2400$, $\mathrm{Im}R = 0.4025$, $S_\omega^{|W|} = -2.2617$. For the considered nonlinearity, $\frac{d\ln|N|}{da} = \frac{1}{a}\frac{2b^2-a^2}{a^2-b^2}$, $S_a^{|N|} = \frac{\partial \ln|N|}{\partial \ln a}$, which gives for the periodic solution "A": $\frac{d\ln|N|}{da} = 7.5882$, $S_a^{|N|} = 7.9987$, and for the periodic solution "B", $\frac{d\ln|N|}{da} = -0.2811$, $S_a^{|N|} = -0.8889$. Computation of the value of the criterion given by (6.20) provides $\frac{c_{22}d_1 - c_{12}d_2}{c_{11}c_{22}-c_{12}c_{21}} = 0.531$ for the solution "A" and –0.100 for the solution "B". Therefore, the solution "B" is orbitally stable, and "A" is unstable. Simulations of the system totally support the results and conclusions: only oscillation of frequency $\Omega_0 = 1.3416$ and amplitude $A_{0B} = 3.1621$ can be produced (Fig. 6.4).

6.7 Conclusions

The chapter presents analysis of orbital stability through the concept of the dynamic harmonic balance and analysis of the relationship of this criterion with the Loeb's approach. It is shown that the DHB-based criterion requires an additional condition to be valid. However, the search for a system that would be stable according to one criterion and unstable according to the other was not successful. All produced examples of systems that give negative denominator of (6.30) also have unstable plants, with HB conditions (6.5) not satisfied.

Therefore, at this point, it would be reasonable to say that the DHB-based criterion provides the result matching to the Loeb's criterion, serves as a rigorous proof of the latter, and involves addition of the condition on the positiveness of the denominator of (6.30).

References

1. Astrom, K., Hagglund, T.: Automatic tuning of simple regulators. Automatica **20**(5), 645–651 (1984)
2. Gardner, F.: Phase-Lock Techniques. Wiley, New York (1966)
3. Leonov, G.A., Kuznetsov, N.V.: Hidden attractors in dynamical systems: from hidden oscillations in Hilbert–Kolmogorov, Aizerman, and Kalman problems to hidden chaotic attractor in Chua circuits. Int. J. Bifurc. Chaos **23**(1), 1–69 (2015)
4. Leonov, G.A., Kuznetsov, N.V., Solovyeva, E.P.: A simple dynamical model of hydropower plant: stability and oscillations. IFAC-PapersOnLine **48**(11), 656–661 (2015)
5. Bianchi, G., Kuznetsov, N.V., Leonov, G.A., Yuldashev, M.V., Yuldashev, R.V.: Limitations of PLL simulation: hidden oscillations in MatLab and SPICE. In: 2015 7th International Congress on Ultra Modern Telecommunications and Control Systems and Workshops (ICUMT), pp. 79–84 (2015)
6. Khalil, H.K.: Nonlinear Systems. Prentice Hall, Prentice (1996)
7. Gelb, A., Vander Velde, W.E.: Multiple-Input Describing Functions and Nonlinear System Design. McGraw-Hill, New York (1968)
8. Atherton, D.P.: Nonlinear Control Engineering -Describing Function Analysis and Design. Van Nostrand Company Limited, Workingham, Berks (1975)
9. Loeb, J.M.: Advances in nonlinear servo theory. In: Oldenburger, R. (ed.) Frequency Response, pp. 260–268. The Macmillan Company, New York (1956)
10. Boiko, I., Fridman, L.: Analysis of chattering in continuous sliding-mode controllers. IEEE Trans. Autom. Control **50**(9), 1442–1446 (2005)
11. Aguilar, L., Boiko, I., Fridman, L., Iriarte, R.: Generating self-excited oscillations via two-relay controller. IEEE Trans. Autom. Control **54**(2), 416–420 (2009)
12. Aguilar, L., Boiko, I., Fridman, L., Iriarte, R.: Self-oscillations in Dynamic Systems: A New Methodology via Two-relay Controllers. Birkhauser, Basel (2015)
13. Miller, R.K., Michel, A.N., Krenz, G.S.: On the stability of limit cycles in nonlinear feedback systems: analysis using describing functions. IEEE Trans. Circuits Syst. **CAS-30**(9), 684–696 (1983)
14. Miller, R.K., Michel, A.N., Krenz, G.S.: Stability analysis of limit cycles in nonlinear feedback systems using describing functions; improved results. IEEE Trans. Circuits Syst. **CAS-31**(6), 561–567 (1984)
15. Boiko, I.: Dynamic harmonic balance principle and analysis of rocking block motions. J. Frankl. Inst. **349**(3), 1198–1212 (2012)
16. Boiko, I.: Analysis of transient oscillations in Lure systems with delay, based on dynamic harmonic balance. Automatica **57**, 93–96 (2015)
17. Boiko, I.: Analysis of orbital stability in lure system, based on dynamic harmonic balance. J. Frankl. Inst. **354**, 4826–4837 (2017)

Chapter 7
Chattering Comparison Between Continuous and Discontinuous Sliding-Mode Controllers

Ulises Pérez-Ventura and Leonid Fridman

Abstract This chapter is devoted to comparison of chattering parameters in systems with stable actuators driven by discontinuous sliding-mode controllers (DSMC) and continuous sliding-mode controllers (CSMC), namely, by the discontinuous first-order sliding-mode controller (FOSMC) and the continuous super-twisting algorithm (STA). Taking into account the amplitude and frequency of fast oscillations (chattering), and the average power (AP) needed to keep the system into real sliding modes, proposed analysis allows to make the following conclusions: (a) for systems with slow actuators, the amplitude of oscillations and AP produced by the FOSMC be smaller than those one caused by the STA; (b) for bounded disturbances with fixed Lipschitz constant, there exist sufficiently fast actuators for which the amplitude of oscillations and AP produced by the FOSMC be greater than those one caused by the STA. On the other hand, a strategy to adjust the chattering in systems governed by STA is proposed, which consist of a proper selection of the controller gains. Simulations are presented to complement the results.

7.1 Introduction

The sliding-mode controllers (SMC) are efficiently used for exact compensation of matched disturbances in dynamical systems [36, 38]. Discontinuous sliding-mode controllers (DSMC), as the classical first-order sliding-mode controller (FOSMC) [38], were proposed to stabilize uncertain systems of relative degree one, in finite-time, and by means of bounded inputs with theoretically infinite switching frequency. However, only high-frequency commutation is feasible due to the presence

---

U. Pérez-Ventura (✉)
National Autonomous University of Mexico, C.P. 04510, Mexico City, Mexico
e-mail: ventury.sk8@gmail.com

L. Fridman
Engineering Faculty, National Autonomous University of Mexico, C.P. 04510, Mexico City, Mexico
e-mail: lfridman@unam.mx

M. Steinberger et al. (eds.), *Variable-Structure Systems and Sliding-Mode Control*, Studies in Systems, Decision and Control 271,
https://doi.org/10.1007/978-3-030-36621-6_7

of unmodeled (parasitic) dynamics which inevitably exists in *real* control loops, as actuators and sensors, delays and discretization, hysteresis, etc. Hence, the steady-state response describes high-frequency oscillations of finite amplitude and finite frequency; this phenomenon is well known as chattering [24]. The super-twisting algorithm (STA) [20], as well as continuous sliding-mode controllers (CSMC), was developed as an attempt to reduce the chattering in systems of relative degree one through the replacement of discontinuous inputs by continuous ones, and retaining some robustness properties and finite-time stabilization. For systems of relative degree two, the most popular DSMC are the twisting [10], the terminal [11], and the suboptimal [3] algorithms. Recently, CSMC for systems of relative degree two as the continuous terminal [14, 16], the continuous twisting [33], and the discontinuous integral controllers [28] were proposed. The quasi-continuous [23] and the nested [21] algorithms (see also [8, 9]) were designed to drive the output of the system and its $(r-1)$ successive time derivatives to zero in finite time by means of discontinuous control; these algorithms were also named as higher-order sliding-mode controllers (HOSMC). More recently, the higher-order super-twisting [16] and the high-order continuous twisting [26] algorithms were designed as extensions of CSMC for systems of arbitrary relative degree (see also [19]). CSMC ensure finite-time stabilization of the output and its r successive time derivatives by acting with a switching function on the $(r+1)$ time derivative of the sliding variable and using the same information as discontinuous HOSMC.

Professor Utkin proposed an example [37, 39] showing that the amplitude of chattering produced by the presence of parasitic dynamics (stable actuators) in some systems governed by the FOSMC is lower than one caused by the STA; this example was analyzed from the describing function (DF) approach [29, 37]. DF has been widely used to study the effects of the nonlinearities that compose DSMC/CSMC on the first-harmonic approximation of chattering [6, 13]. Harmonic balance (HB) methodology allows one to estimate (with high level of accuracy in *low-pass filter* systems) the parameters of the oscillations, amplitude, and frequency, caused by the presence of different kinds of parasitic dynamics, for example, actuators and sensors [4, 5, 13], time delays and hysteresis [15, 34], and discretization [18]. Stable actuators have been specially studied because their dynamics are usually neglected in the design step of the controller [6, 7]. The magnitude of chattering was estimated in terms of the actuator time constant (ATC) for DSMC of arbitrary order [24, 25, 32]. However, the problem of estimation of chattering caused by high-order CSMC remains open. Only the oscillations produced by the STA for the relative degree one systems in the presence of actuators were analyzed [5, 29, 30, 37].

The goal of this chapter is to analyze the parameters of the oscillations caused by the presence of fast-parasitic dynamics in systems of relative degree one, which are governed by the FOSMC or by the STA. By the comparison of the chattering parameters predicted by HB, it is possible to determinate when it is better to substitute the discontinuous input based on FOSMC by the continuous one based on STA. On the other hand, a strategy to adjust the chattering in systems with STA is proposed which consist of a proper selection of the controller gains. Simulation is presented to complement the results.

7.2 Preliminaries

7.2.1 *Ideal Sliding Modes*

Consider a dynamical SISO system such that a relative degree one output $x_1 \in \mathbb{R}$ can be chosen,

$$\dot{x}_1(t) = f(t) + \bar{u}(t)\,, \tag{7.1}$$

where $\bar{u} \in \mathbb{R}$ is the control input and $f(t)$ is a perturbation term. The input $\bar{u}(t)$ is based on a sliding-mode controller. For example, the discontinuous FOSMC,

$$\bar{u} = -c\text{sign}(x_1)\,, \tag{7.2}$$

or the continuous STA,

$$\begin{aligned} \bar{u} &= -k_1|x_1|^{1/2}\text{sign}(x_1) + \bar{v}\,, \\ \dot{\bar{v}} &= -k_2\text{sign}(x_1)\,. \end{aligned} \tag{7.3}$$

- System (7.1) in closed loop with the FOSMC (7.2) has the following properties:
  (a) The closed loop must be understood in the Filippov sense [12],

$$\dot{x}_1 = -c\text{sign}(x_1) + f\,, \tag{7.4}$$

  where the gain of the controller is chosen $c > |\delta|$ with the upper bound $|f| \le \delta$, in order to reject the perturbation.
  (b) The closed loop (7.4) is homogeneous of degree $q = -1$, with the dilation

$$d_\kappa : (t, x_1) \rightarrow (\kappa t, \kappa^{m_1} x_1)\,, \tag{7.5}$$

  where $m_1 = 1$ the homogeneity weight and $\kappa > 0$ (see more details about dilation and homogeneity in [2, 22]).
  (c) The FOSMC enforces the first-order sliding mode [38] in a finite time, i.e., there exist a time t_r such that

$$\forall t \ge t_r : x_1(t) = 0\,. \tag{7.6}$$

  Note that the vector field is directed toward the surface $x_1 = 0$.
- System (7.1) in closed loop with the STA (7.3) has the following properties:
  (a) Let the integral state $x_2 = f - \int_{t_0}^{t} k_2\text{sign}(x_1)\mathrm{d}\tau$, then the extended system can be written as

$$\begin{aligned} \dot{x}_1 &= -k_1|x_1|^{1/2}\text{sign}(x_1) + x_2\,, \\ \dot{x}_2 &= -k_2\text{sign}(x_1) + \dot{f}\,, \end{aligned} \tag{7.7}$$

where the gains of the controller are chosen $k_2 > \Delta$ with the upper bound $|\dot{f}| \leq \Delta$, while k_1 can be selected, for instance, to improve the setting time or to decrease the overshoot. Sufficient and necessary conditions for finite-time stabilization are provided in [27]. The most popular set of gain proposed by [20] was proved by a novel Lyapunov function in [35], i.e., $k_2 > \Delta$ then $k_1 > \sqrt{k_2 + \Delta}$.

(b) The closed loop (7.7) is homogeneous of degree $q = -1$ with the dilation

$$d_{\bar{\kappa}} : (t, x_1, x_2) \rightarrow (\bar{\kappa} t, \bar{\kappa}^{m_1} x_1, \bar{\kappa}^{m_2} x_2), \tag{7.8}$$

where $m_1 = 2, m_2 = 1$, are the homogeneity weights and $\bar{\kappa} > 0$.

(c) The STA enforces the second-order sliding mode [20] in a finite time, i.e., there exist a time t_r such that

$$\forall t \geq t_r : x_1(t) = x_2(t) = 0. \tag{7.9}$$

Note that the vector field is tangent to the surface defined by $x_1 = 0$, while it is directed toward the derivative $x_2 = 0$.

7.2.2 *Class of Disturbances*

DSMC are able to reject bounded disturbances (including discontinuous), while CSMC compensate Lipschitz disturbances (including not bounded). Then, a reasonable comparison between these algorithms should consider the class of disturbances: bounded and Lipschitz as shown in Fig. 7.1. Note that for the study of chattering only local effects of the disturbances must be considered for the computation of the upper bounds δ and Δ, because it is assumed that the sliding-mode controllers are able to keep the trajectories in a vicinity of (7.6) or (7.9), respectively.

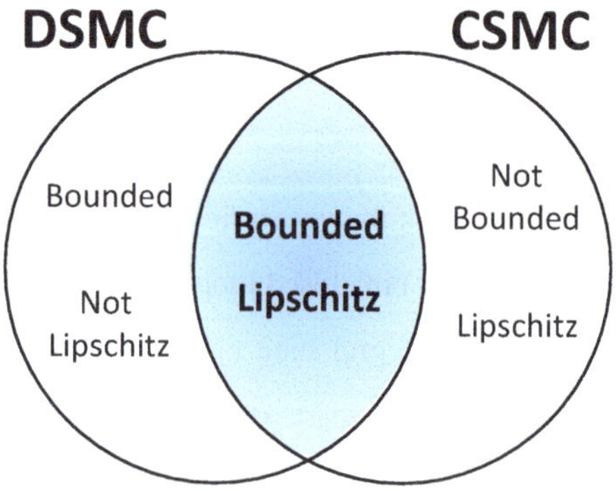

Fig. 7.1 Class of disturbances reasonable for comparison

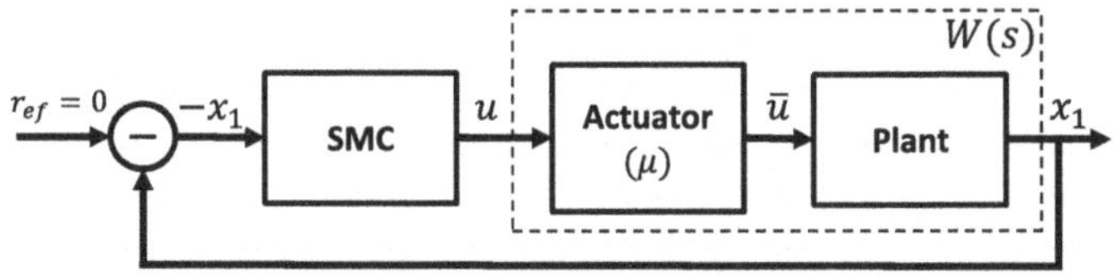

Fig. 7.2 Closed-loop system considered to study the chattering

7.2.3 *Real Sliding Modes*

The presence of unmodeled (parasitic) dynamics, as actuators and sensors, delays and discretization, hysteresis, and others, degrades the ideal properties of DSMC/CSMC so that high-frequency oscillations of so-called chattering appear. Any ideal sliding mode should be understood as a limit of the fast motions when switching imperfections vanish and the switching frequency tends to infinity (see [24, 31]). Let the parameter $\mu > 0$ that measures some imperfections, the sliding accuracy [21, 25] of any sliding-mode technique may be featured by the steady-state accuracy of the output as $\mu \to 0$. The real sliding modes in the presence of stable actuators were studied in detail because they are usually neglected for the design step of the controller [6]. As shown in Fig. 7.2, the input is dynamically affected by a stable actuator parametrized by the actuator time constant (ATC) $\mu \geq 0$ so that its effects can be assessed for different values,

$$\mu \dot{z} = h(z, u), \quad \bar{u} = \nu(z), \quad \dot{x}_1 = f(t) + \bar{u}(t), \tag{7.10}$$

where $z \in \mathbb{R}^m$ is the actuator state, $\bar{u} \in \mathbb{R}$ is the output of the actuator, and the input $u \in \mathbb{R}$ is now based on sliding-mode controllers. The functions $h(z, u)$ and $\nu(z)$ are assumed bounded and Lipschitz.

Remark 7.1 Any linearized system $G_a(s) = P(s)/Q(s)$ such that $G_a(0) = 1$ can be used as model of the actuator dynamics in (7.10). However, effects of delays, hysteresis, discretization, and other non-idealities can be studied through the methodology proposed in this chapter.

Dynamical systems driven by FOSMC and STA in the presence of fast actuators were analyzed in the frequency domain [5, 29, 37, 38]. The steady-state response of the system (7.10) is such that

$$|x_1| \leq \gamma_1 \mu, \tag{7.11}$$

for some $\gamma_1 > 0$, when the discontinuous FOSMC (7.2) is used. Moreover, the continuous STA (7.3) in the system (7.10) is finally bounded by

$$|x_1| \leq \bar{\gamma}_1 \mu^2, \quad |x_2| \leq \bar{\gamma}_2 \mu, \tag{7.12}$$

for some $\bar{\gamma}_1, \bar{\gamma}_2 > 0$. Note that the accuracy of the output and its derivative parametrized via ATC corresponds to their homogeneity weights into dilation (7.8).

7.2.3.1 Harmonic Balance Approach

The presence of the actuator dynamics in the closed loop (7.10) causes chattering. The total response is composed by slow motions (sliding-mode dynamics plus the effects of external perturbations) and fast motions caused by the actuator fast dynamics. Propagation of harmonic perturbations was analyzed through the *equivalent gain* in systems driven by FOSMC and STA [6]. In order to study only the fast motions caused by the actuator dynamics, the disturbance term is assumed $f(t) = 0$.

Describing function (DF) approach allows one to estimate the parameters of a possible periodic motion by considering the first harmonic of chattering,

$$x_1(t) \approx A \sin(\omega t)\,, \tag{7.13}$$

where A is the amplitude and ω is the frequency of fast oscillations. The expression (7.13) is a well approximation of the closed-loop behavior (after a transient process) if the linearized system $W(s)$ of (7.10) has *low-pass filter* characteristics [1, 15], i.e., $|W(j\omega)| >> |W(jn\omega)|$, for any $n = 2, 3, \ldots$. Parameters of a possible periodic motion may be found as an intersection point of the Nyquist plot of the frequency characteristic of $W(j\omega)$, and the negative reciprocal describing function $-N^{-1}(A, \omega)$ of the considered sliding-mode controller, which corresponds to the harmonic balance (HB) equation (see more details in [1, 15])

$$N(A, \omega)W(j\omega) = -1\,, \tag{7.14}$$

and whose solutions are predictions of the chattering parameters, amplitude A, and frequency ω of fast oscillations.

7.2.3.2 Estimation of the Amplitude and Frequency

The estimation of the chattering parameters, amplitude and frequency of the periodic motion (7.13), allows to compute the instantaneous power [17]

$$p(t) = \bar{u}(t)x_1(t) = \frac{A^2\omega}{2}\sin(2\omega t)\,. \tag{7.15}$$

Let us assume, for example, that system (7.1) models an electrical circuit where x_1 is the current and $\bar{u}$ is the voltage. Then, the average power needed to keep the trajectories into *real* sliding modes can be computed, for each period $T = \frac{2\pi}{\omega}$ predicted by HB, as

$$P = \frac{1}{T}\int_0^T |p(t)|dt = \frac{4A^2\omega}{\pi}\,. \tag{7.16}$$

It should be mentioned that the average power computation (7.16) can only be done taking into account the presence of parasitic dynamics, because there is no chattering in *ideal* sliding modes.

7.3 Comparison of Chattering Parameters for Systems of Relative Degree One with Fast-Parasitic Dynamics

7.3.1 Parameters of Chattering Caused by the First-Order Sliding-Mode Controller

Following [39], a critically damped model of actuator dynamics is considered. Thus, the linear system conformed by the cascade connection of the actuator and the plant (7.10) has the form

$$W(s) = \frac{1}{s(\mu s + 1)^2}, \tag{7.17}$$

whose dynamics are parametrized by the ATC, $\mu > 0$. Figure 7.3 shows the system (7.17) in closed loop with the nonlinearity (7.2). According to [38], the DF of (7.2) has the form

$$N(A) = \frac{4c}{\pi A}. \tag{7.18}$$

Then, the HB equation (7.14) for the system (7.17) and the DF (7.18) can be rewritten as

$$\frac{4c}{\pi A} = 2\,\mu\,\omega^2 + j\omega\,(\mu^2\,\omega^2 - 1), \tag{7.19}$$

whose solution are the parameters

$$A = \mu\left(\frac{2c}{\pi}\right), \tag{7.20}$$

$$\omega = \frac{1}{\mu}. \tag{7.21}$$

The AP (7.16) with the estimated parameters (7.20) and (7.21) has the form

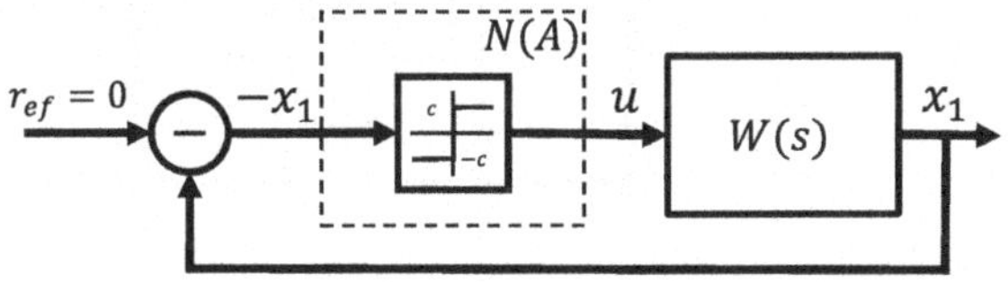

Fig. 7.3 System with first-order sliding-mode controller (7.2)

$$P = \mu \left(\frac{16c^2}{\pi^3} \right) . \tag{7.22}$$

Consider, for instance, $c = 1.1\delta$, then the chattering parameters predicted by HB, amplitude (7.20), frequency (7.21), and average power (7.22) are

$$A = 0.7003\mu\delta \,, \quad \omega = \tfrac{1}{\mu} \,, \quad P = 0.6244\mu\delta^2 \,. \tag{7.23}$$

The output of the linear system (7.17) in closed loop with FOSMC (7.2) has the following steady-state performance:

- The amplitude of the oscillations is proportional to μ and the upper bound δ.
- The frequency of the oscillations is inversely proportional to μ and is not affected by the upper bound δ.
- The average power is proportional to μ and the upper bound δ^2.

7.3.2 *Parameters of Chattering Caused by the Super-Twisting Algorithm*

Figure 7.4 shows the system (7.17) in closed loop with the nonlinearity (7.3). According to [5], the DF of (7.3) has the form

$$N(A, \omega) = \frac{3.496k_1}{\pi A^{1/2}} + \left(\frac{4k_2}{\pi A} \right) \left(\frac{1}{j\omega} \right) , \tag{7.24}$$

where an ideal integrator with $s = j\omega$ is considered as usually done in frequency-domain analysis. The HB equation (7.14) for the system (7.17) and the DF (7.24) can be rewritten as

$$\frac{3.496k_1}{\pi A^{1/2}} - j\frac{4k_2}{\pi A\omega} = 2\,\mu\,\omega^2 + j\omega\,(\mu^2\,\omega^2 - 1) \,, \tag{7.25}$$

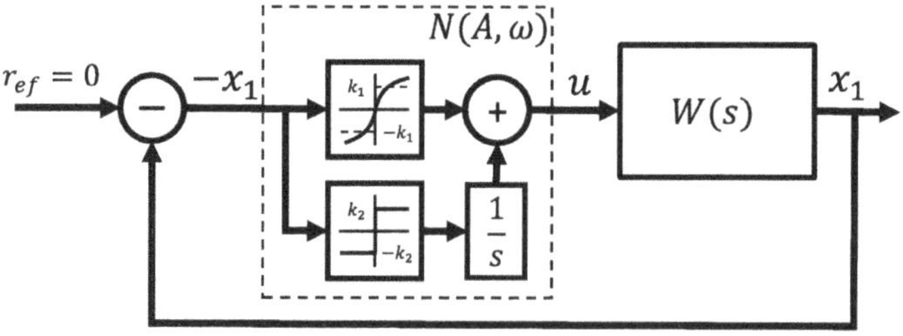

Fig. 7.4 System with super-twisting controller (7.3)

whose solution are the parameters

$$A = \mu^2 \left(\frac{(1.748k_1)^2 + 4\pi k_2}{1.748\pi k_1} \right)^2 , \tag{7.26}$$

$$\omega = \frac{1}{\mu} \left(\frac{(1.748k_1)^2}{(1.748k_1)^2 + 4\pi k_2} \right)^{1/2} . \tag{7.27}$$

The AP (7.16) with the estimated parameters (7.26) and (7.27) has the form

$$P = \mu^3 \left(\frac{4\left((1.748k_1)^2 + 4\pi k_2\right)^{7/2}}{\pi^5 (1.748k_1)^3} \right) . \tag{7.28}$$

Consider, for instance, $k_1 = 1.5\Delta^{1/2}$ and $k_2 = 1.1\Delta$, then the chattering parameters predicted by HB, amplitude (7.26), frequency (7.27), and average power (7.28) are

$$A = 6.3136\mu^2\Delta \, , \quad \omega = \tfrac{0.5763}{\mu} \, , \quad P = 29.251\mu^3\Delta^2 \, . \tag{7.29}$$

The output of the linear system (7.17) in closed loop with STA (7.3) has the following steady-state performance:

- The amplitude of the oscillations is proportional to μ^2 and the upper bound Δ.
- The frequency of the oscillations is inversely proportional to μ and is not affected by the upper bound Δ.
- The average power is proportional to μ^3 and the upper bound Δ^2.

Graphical solutions of the HB equations (7.19) and (7.25) are plotted in Fig. 7.5. Figure 7.6 shows the comparison of the chattering parameters predicted by HB as function of the ATC $\mu \to 0$, with the gains $c = 1.1\delta$, $k_1 = 1.5\Delta^{1/2}$, $k_2 = 1.1\Delta$, and the upper bounds $\delta = \Delta = 1$.

Remark 7.2 The frequency of oscillations produced by the FOSMC (7.2) is always greater than one caused by the STA (7.3), as can be seen in the HB graph (see Fig. 7.5).

Remark 7.3 There is a value of μ for which the amplitude of oscillations produced by the FOSMC (7.2) and by the STA (7.3) is the same,

$$\mu^* = \frac{2\pi c (1.748k_1)^2}{\left((1.748k_1)^2 + 4\pi k_2\right)^2} . \tag{7.30}$$

The expression (7.30) becomes

$$\mu^* = 0.1109 \frac{\delta}{\Delta} ;$$

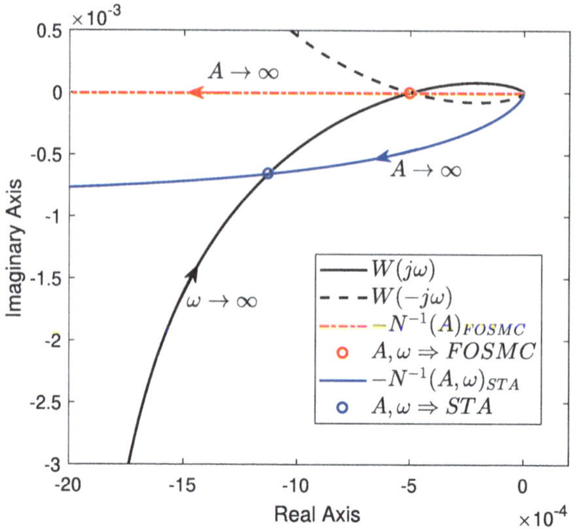

Fig. 7.5 Graphical solution of the HB equations (7.19) and (7.25) for the gains $c = 1.1\delta$, $k_1 = 1.5\Delta^{1/2}$, $k_2 = 1.1\Delta$, the ATC $\mu = 10^{-3}$, and the upper bounds $\delta = \Delta = 1$

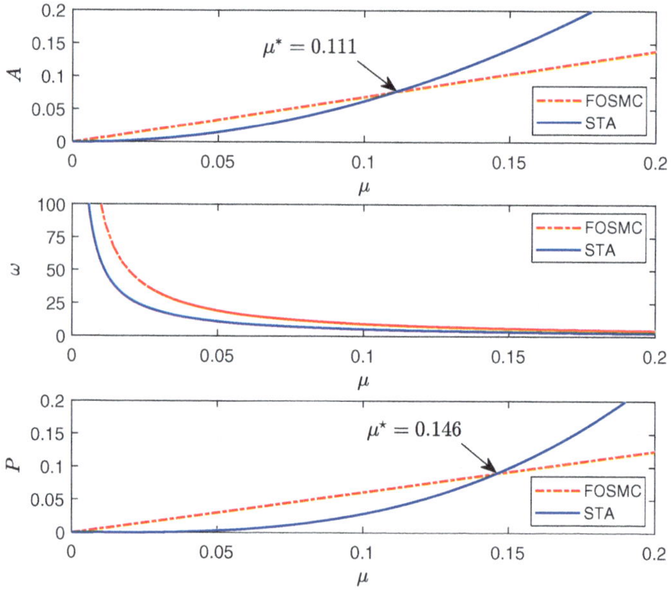

Fig. 7.6 Comparison of the chattering parameters as function of μ for the gains $c = 1.1\delta$, $k_1 = 1.5\Delta^{1/2}$, $k_2 = 1.1\Delta$, and the upper bounds $\delta = \Delta = 1$

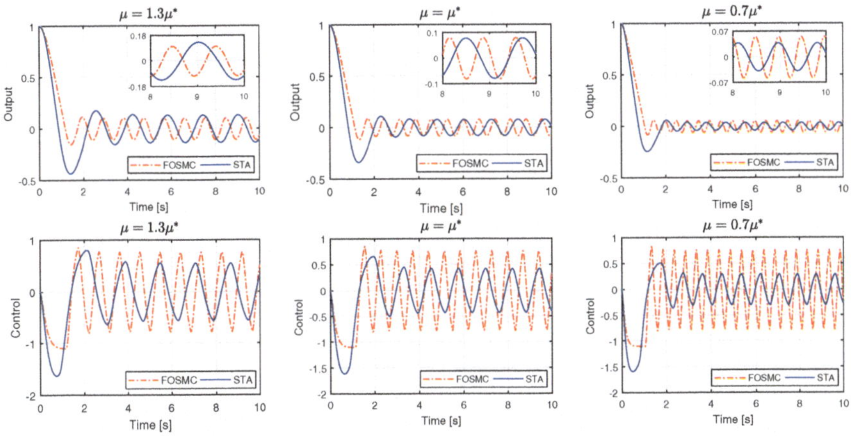

Fig. 7.7 Comparison of the magnitude of chattering regarding the critical value μ^* in accordance to (7.30) for the gains $c = 1.1\delta$, $k_1 = 1.5\Delta^{1/2}$, $k_2 = 1.1\Delta$, and the upper bounds $\delta = \Delta = 1$. The FOSMC is plotted in red-dashed line and the STA in blue-continuous line

for the considered gains, this value is shown in Fig. 7.6. We can also observe that

$$\mu > \mu^* \Rightarrow A_{\text{FOSMC}} < A_{\text{STA}}\,, \tag{7.31}$$

$$\mu < \mu^* \Rightarrow A_{\text{FOSMC}} > A_{\text{STA}}\,. \tag{7.32}$$

Remark 7.4 There is a value of μ for which the average power expended by the FOSMC (7.2) and by the STA (7.3) is the same,

$$\mu^\star = \frac{2\pi c(1.748k_1)^{3/2}}{\left((1.748k_1)^2 + 4\pi k_2\right)^{7/4}}\,. \tag{7.33}$$

The expression (7.33) becomes

$$\mu^\star = 0.1461\frac{\delta}{\Delta}\,;$$

for the considered gains, this value is shown in Fig. 7.6. We can also observe that

$$\mu > \mu^\star \Rightarrow P_{\text{FOSMC}} < P_{\text{STA}}\,, \tag{7.34}$$

$$\mu < \mu^\star \Rightarrow P_{\text{FOSMC}} > P_{\text{STA}}\,. \tag{7.35}$$

Note that STA produces oscillations of lower amplitude (average power) than the one caused by FOSMC when the actuator dynamics is *fast*, but also we must consider the ratio δ/Δ between the upper bounds [29].

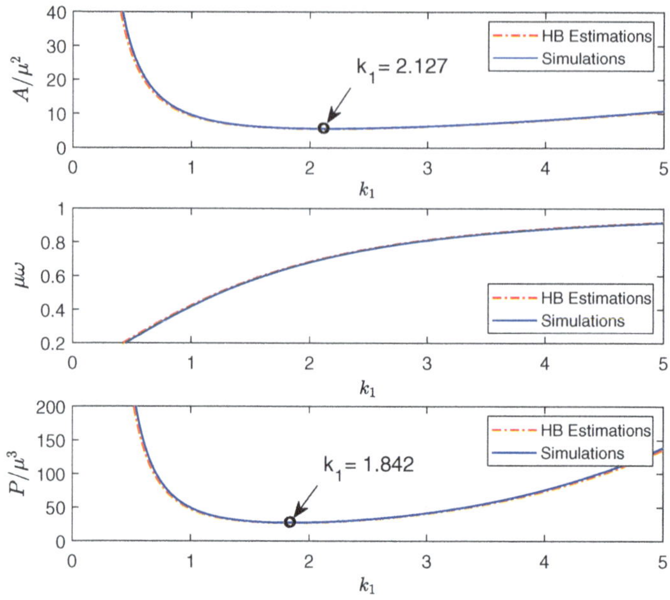

Fig. 7.8 Normalized parameters: amplitude (7.26), frequency (7.27), and average power (7.28), for several values of $k_1 \in (0 \quad 5]$ and fixing $k_2 = 1.1$

7.3.3 Simulations

Let the linear system (7.17) be in closed loop with the FOSMC (7.2) and the STA (7.3), respectively. Consider the gains $c = 1.1\delta$, $k_1 = 1.5\Delta^{1/2}$, $k_2 = 1.1\Delta$, and the upper bounds $\delta = \Delta = 1$. Also, the Euler's integration method with fixed step $\tau = 10^{-4}\,s$ is used. Figure 7.7 shows simulations for some values of ATC around of $\mu = \mu^*$, a larger value $\mu = 1.3\mu^*$ and a lower value $\mu = 0.7\mu^*$. It can be seen in the following:

$$\mu > \mu^* \Rightarrow A_{\text{FOSMC}} < A_{\text{STA}}\,, \tag{7.36}$$

$$\mu = \mu^* \Rightarrow A_{\text{FOSMC}} = A_{\text{STA}}\,, \tag{7.37}$$

$$\mu < \mu^* \Rightarrow A_{\text{FOSMC}} > A_{\text{STA}}\,. \tag{7.38}$$

7.4 Suboptimal Design of the Super-Twisting Gains for Systems of Relative Degree One with Fast-Parasitic Dynamics

Note that the chattering parameters (7.26) and (7.28) can be optimized from a suitable selection of the STA (7.3) gains. We can minimize the amplitude (or the average power) by fixing the gain $k_2 > \Delta$ and looking for the critical value of $k_1 > 0$.

Proposition 7.1 *Given the gain $k_2 > \Delta$ of the STA (7.3), the value*

$$k_1 = 2.028\sqrt{k_2} \tag{7.39}$$

minimizes the amplitude (7.26) of the output oscillations [30].

Proposition 7.2 *Given the gain $k_2 > \Delta$ of STA (7.3), the value*

$$k_1 = 1.756\sqrt{k_2} \tag{7.40}$$

minimizes the average power (7.28) [30].

Remark 7.5 The STA gains in Propositions 7.1 and 7.2 satisfy the sufficient conditions for finite-time stability proposed in [27].

7.4.1 Simulations

Consider the fixed value of time constant $\mu = 10^{-2}$, the upper bound $\Delta = 1$, and fixed gain $k_2 = 1.1$ of the STA (7.3). Then, the value of $k_1 > 0$ that minimizes the amplitude of chattering according to (7.39) is $k_1 = 2.127$. On the other hand, the value of $k_1 > 0$ that minimizes the average power from (7.40) is $k_1 = 1.842$. Figure 7.8 shows the normalized parameters (with respect to the time constant μ) estimated by HB in comparison with the measured from simulations. Several values of k_1 in the interval (0 5] are evaluated. It can be seen from Fig. 7.8 that the critical values of the gain $k_1 > 0$ predicted by HB to minimize the amplitude (at the top) and the average power (at the bottom) are well estimated in comparison with the parameters obtained by simulations.

7.5 Conclusions

The proposed methodology allows one to compare the chattering parameters in systems with stable actuators driven by the discontinuous first-order sliding-mode controller (FOSMC) and the continuous super-twisting algorithm (STA). Taking into account the amplitudes, frequencies of fast oscillations (chattering), and the average power (AP) needed to maintain the system into real sliding modes, the following conclusions are formulated from HB estimations and simulation results: (a) for systems with slow actuators, the amplitude of oscillations and AP produced by the FOSMC be smaller than those one caused by the STA; (b) for bounded disturbances with fixed Lipschitz constant, there exist sufficiently fast actuators for which the amplitude of oscillations and AP produced by the FOSMC be greater than those one caused by the STA. On the other hand, a strategy to adjust the chattering in systems with STA is presented which consist of a proper selection of the controller gains.

Acknowledgements The authors are grateful for the financial support of CONACyT (Consejo Nacional de Ciencia y Tecnología): CVU 631266; Project 282013; PAPIIT-UNAM (Programa de Apoyo a Proyectos de Investigación e Innovación Tecnológica): IN 115419; PASPA-UNAM (Programa de Apoyos para la Superación del Personal Académico de la UNAM).

References

1. Atherton Derek, P.: Nonlinear Control Engineering, p. 1975. Van Nostrand Reinhold, London (1975)
2. Bacciotti, A., Rosier, L.: Liapunov Functions and Stability in Control Theory. Springer Science & Business Media, Berlin (2006)
3. Bartolini, G., Ferrara, A., Usai, E.: Chattering avoidance by second-order sliding mode control. IEEE Trans. Autom. Control **43**(2), 241–246 (1998). IEEE
4. Boiko, I., Fridman, L., Castellanos, M.I.: Analysis of second-order sliding-mode algorithms in the frequency domain. IEEE Trans. Autom. Control **49**(6), 946–950 (2004). IEEE
5. Boiko, I., Fridman, L.: Analysis of chattering in continuous sliding mode controllers. IEEE Trans. Autom. Control **50**(9), 1442–1446 (2005). IEEE
6. Boiko, I.: Discontinuous Control Systems: Frequency-Domain Analysis and Design, p. 2009. Birkhäuser, Boston (2009)
7. Boiko, I.: On inherent gain margins of sliding-mode control systems. Advances in Variable Structure Systems and Sliding Mode Control Theory and Applications, pp. 133–147. Springer, Berlin (2018)
8. Cruz-Zavala, E., Moreno, J.A.: Homogeneous high order sliding mode design: a Lyapunov approach. Automatica **80**, 232–238 (2017). Elsevier
9. Ding, S., Levant, A., Li, S.: Simple homogeneous sliding-mode controller. Automatica **67**, 22–32 (2016). Elsevier
10. Emelianov, S.V., Korovin, S.K., Levantovskii, L.V.: Higher-order sliding modes in binary control systems. Sov. Phys. Dokl. **31**(4), 291–293 (1986)
11. Feng, Y., Yu, X., Man, Z.: Non-singular terminal sliding mode control of rigid manipulators. Automatica **38**(12), 2159–2167 (2002). Elsevier
12. Filippov, A.F.: Differential Equations with Discontinuous Right-Hand Side, p. 1988. Kluwer Academic, Dordrecht (1988)
13. Fridman, L.: An averaging approach to chattering. IEEE Trans. Autom. Control **46**(8), 1260–1265 (2001). IEEE
14. Fridman, L., Moreno, J.A., Bandyopadhyay, B., Kamal, S., Chalanga, A.: Continuous nested algorithms: the fifth generation of sliding mode controllers. Recent Advances in Sliding Modes: From Control to Intelligent Mechatronics, vol. 24, pp. 5–35. Springer International Publishing, Berlin (2015)
15. Gelb, A., Vander-Velde, W.E.: Multiple-Input Describing Functions and Nonlinear System Design. McGraw-Hill, New York (1968)
16. Kamal, S., Moreno, J.A., Chalanga, A., Bandyopadhyay, B., Fridman, L.: Continuous terminal sliding-mode controller. Automatica **69**, 308–314 (2016). Elsevier
17. Khalil, H.K.: Noninear Systems, p. 1992. MacMillan, New York (1992)
18. Koch, S., Reichhartinger, M., Horn, M., Fridman, L.: Sampled describing function analysis of second order sliding modes. In: 55th Conference on Decision and Control (CDC), pp. 7318–7324, Las Vegas, NV, USA. IEEE (2016). https://doi.org/10.1109/CDC.2016.7799399
19. Laghrouche, S., Harmouche, M., Chitour, Y.: Higher order super-twisting for perturbed chains of integrators. IEEE Trans. Autom. Control **62**(7), 3588–3593 (2017). IEEE
20. Levant, A.: Robust exact differentiation via sliding mode technique. Automatica **34**(3), 379–384 (1998). Elsevier

21. Levant, A.: Higher-order sliding modes, differentiation and output-feedback control. Int. J. Control **76**(9–10), 924–941 (2003). Taylor & Francis
22. Levant, A.: Homogeneity approach to high-order sliding mode design. Automatica **41**(5), 823–830 (2005). Elsevier
23. Levant, A.: Quasi-continuous high-order sliding-mode controllers. IEEE Trans. Autom. Control **50**(11), 1812–1816 (2005). IEEE
24. Levant, A.: Chattering analysis. IEEE Trans. Autom. Control **55**(6), 1380–1389 (2010). IEEE
25. Levant, A., Fridman, L.: Accuracy of homogeneous sliding modes in the presence of fast actuators. IEEE Trans. Autom. Control **55**(3), 810–814 (2010). IEEE
26. Mendoza-Avila, J., Moreno, J.A., Fridman, L.: An idea for Lyapunov function design for arbitrary order continuous twisting algorithms. In: 56th IEEE Conference on Decision and Control (CDC), pp. 5426–5431, Melbourne, Australia. IEEE (2017). https://doi.org/10.1109/CDC.2017.8264462
27. Moreno, J.A., Osorio, M.: Strict Lyapunov functions for the super-twisting algorithm. IEEE Trans. Autom. Control **57**(4), 1035–1040 (2012). IEEE
28. Moreno, J.A.: Discontinuous integral control for systems with relative degree two. New Perspectives and Applications of Modern Control Theory, pp. 187–218. Springer, Berlin (2018)
29. Pérez-Ventura, U., Fridman, L.: When is it reasonable to implement the discontinuous sliding-mode controllers instead of the continuous ones? Frequency domain criteria. Int. J. Robust Nonlinear Control **29**(3), 810–828 (2018). Wiley Online Library
30. Pérez-Ventura, U., Fridman, L.: Design of super-twisting control gains: a describing function based methodology. Automatica **99**, 175–180 (2019). Elsevier
31. Perruquetti, W., Barbot, J.P.: Sliding Mode Control in Engineering, pp. 53–101. CRC Press, New York (2002)
32. Rosales, A., Shtessel, Y., Fridman, L., Panathula, C.B.: Chattering analysis of HOSM controlled systems: frequency domain approach. IEEE Trans. Autom. Control **62**(8), 4109–4115 (2010). IEEE, 2017
33. Torres-González, V., Sanchez, T., Fridman, L., Moreno, J.A.: Design of continuous twisting algorithm. Automatica **80**, 119–126 (2017). Elsevier
34. Tsypkin, I.Z.: Relay Control Systems, p. 1984. Cambridge University Press, Cambridge (1984)
35. Seeber, R., Horn, M.: Necessary and sufficient stability criterion for the super-twisting algorithm. In: 15th International Workshop on Variable Structure Systems (VSS), pp. 120–125, Graz, Austria. IEEE (2017). https://doi.org/10.1109/VSS.2018.8460445
36. Shtessel, Y., Edwards, C., Fridman, L., Levant, A.: Sliding Mode Control and Observation. Birkhäuser, Boston (2014)
37. Swikir, A., Utkin, V.: Chattering analysis of conventional and super twisting sliding mode control algorithm. In: 14th International Workshop on Variable Structure Systems (VSS), pp. 98–102, Nanjing, China. IEEE (2016). https://doi.org/10.1109/VSS.2016.7506898
38. Utkin, V.: Sliding Modes in Optimization and Control Problems, p. 1992. Springer, New York (1992)
39. Utkin, V.: Discussion aspects of high-order sliding mode control. IEEE Trans. Autom. Control **61**(3), 829–833 (2016). IEEE

Part III
Usage of VSS Controllers for Solving Other Control Problems

Chapter 8
Sliding-Mode Stabilization of SISO Bilinear Systems with Delays

Tonametl Sanchez, Andrey Polyakov, Jean-Pierre Richard and Denis Efimov

Abstract In this chapter, we propose a sliding-mode controller for a class of scalar bilinear systems with delay in both the input and the state. Such systems have shown to be adequate for input–output modeling and control of a class of turbulent flow systems. Since the sliding dynamics is infinite dimensional and described by an integral equation, we show that the stability and robustness analysis is simplified by using Volterra operator theory.

8.1 Introduction

Reduction of fuel consumption is an important objective in transportation systems as aerial and ground vehicles [2, 10, 15]. *Aerodynamic drag* (also known as air resistance) is one of the main sources of energetic losses in such systems; moreover, the drag is significantly increased due to the flow separation phenomenon [2], see Fig. 8.1. Thus, active control of separated turbulent flows can produce very favorable effects in energy consumption and environmental impact [2, 3].

The performance of a model-based flow controller critically depends on its underlying model [16]. For the case of flow systems, the classical model is the set of Navier–Stokes equations; unfortunately, the complexity of such a model makes the control design process to be very complicated and the resultant controllers to be prac-

T. Sanchez (✉) · A. Polyakov · D. Efimov
Inria, Univ. Lille, CNRS, UMR 9189 - CRIStAL, 59000 Lille, France
e-mail: tonametl.sanchez_ramirez@inria.fr

A. Polyakov
e-mail: andrey.polyakov@inria.fr

D. Efimov
e-mail: denis.efimov@inria.fr

J.-P. Richard
Univ. Lille, École Centrale de Lille, Inria, CNRS, UMR 9189 - CRIStAL, 59000 Lille, France
e-mail: jean-pierre.richard@centralelille.fr

M. Steinberger et al. (eds.), *Variable-Structure Systems and Sliding-Mode Control*, Studies in Systems, Decision and Control 271,
https://doi.org/10.1007/978-3-030-36621-6_8

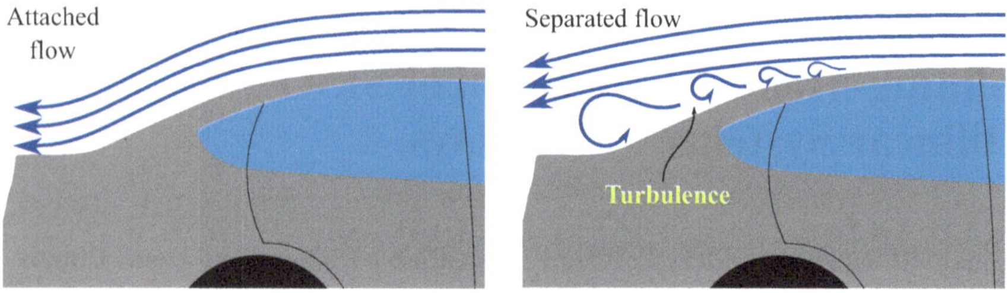

Fig. 8.1 Separation phenomenon in turbulent flows over a surface

tically impossible to implement [2]. Thus, models for practical flow control systems should exhibit a trade-off between accuracy and simplicity.

In [7], an input–output model for some separated flow systems was proposed; such a model consists in a bilinear differential equation with delays in the input and in the state. An attractive feature of the model is that, according to the experimental results, with few parameters the model reproduces the input–output behavior of the physical system with a *good precision*. The reasoning to adopt such a kind of equations as input–output models for flow systems was presented in [5]. We reproduce that reasoning as a motivational example in Sect. 8.2.

For a particular case of the model introduced in [7], a sliding-mode controller was proposed in [6]. That control technique was chosen due to the switching features of the actuators. A good experimental performance was obtained with such a controller. Initiated in the framework of a wing/flap experiment [5], the same model has been successfully used in another flow control situation (Ahmed body) in [3]. Hence, it is worth continuing with the study of the class of bilinear delayed systems and to develop general schemes for analysis and control design. It is important to mention that although there exist sliding-mode controllers for delayed systems (see, e.g., [17, 19], and the reference therein), and controllers for switched bilinear systems (see, e.g., [13]), to the best of our knowledge, there are no constructive procedures to design controllers that adjust well to the class of systems considered in this chapter.

In this chapter, we extend the description of the control design methodology presented in [23]. Such a methodology consists in the designing of a sliding-mode controller for scalar SISO bilinear systems with delays. The main features of the method are the following:

1. The sliding variable is proposed as an integral function of the system's states and the past values of the input. This is following the idea proposed in [6]. Nevertheless, the sliding variable presented in this chapter is slightly different from those in [6, 23]. This change permits a clearer reaching analysis with respect to that in [23].
2. The dynamics on the sliding surface is infinite dimensional and can be written as a Volterra integral equation. Thus, to avoid the analysis of an infinite-dimensional system in the frequency domain, we analyze the stability properties of such a dynamics by means of Volterra operator theory. This allows us to simplify the

analysis and to give simple conditions that guarantee asymptotic stability of the solutions.
3. Besides stability analysis, Volterra operator theory is also helpful to study robustness properties of the control scheme.

Chapter organization: In Sect. 8.3, a brief description of the control problem is given. The conditions that guarantee boundedness and nonnegativeness of the model's solutions are studied in Sect. 8.4. The design and analysis of the proposed controller are explained in Sect. 8.5. A robustness analysis is given in Sect. 8.6. A numerical example is shown in Sect. 8.7. Some final remarks are stated in Sect. 8.8.

Notation: For any real $p \geq 1$, $L^p(J)$ denotes the set of Lebesgue-measurable functions $x : J \subset \mathbb{R} \to \mathbb{R}$ with finite norm $\|x\|_{L^p(J)} = \left(\int_J |x(s)|^p \, ds\right)^{\frac{1}{p}}$, and $L^\infty(J)$ denotes the set of Lebesgue-measurable functions with finite norm $\|x\|_{L^\infty(J)} = \operatorname{ess\,sup}_{t \in J} |x(t)|$.

8.2 Motivational Example for the Bilinear Model

In this section, we reproduce the example given in [5] on how a bilinear delayed differential equation can be obtained as an input–output model for a controlled flow system.

Consider a flow over a surface as depicted in the scheme in Fig. 8.2. The actuator is an on–off air blower and the sensor is a hot film [5].

To describe the behavior of the flow, we consider the Burgers' equation

$$\frac{\partial \mathbf{v}(t,z)}{\partial t} + \mathbf{v}(t,z)\frac{\partial \mathbf{v}(t,z)}{\partial z} = \nu \frac{\partial^2 \mathbf{v}(t,z)}{\partial z^2}, \tag{8.1}$$

where $\mathbf{v} : \mathbb{R}^2 \to \mathbb{R}$ is the flow velocity field, $z \in \mathbb{R}$ is the spatial coordinate, and $\nu \in \mathbb{R}_+$ is the kinematic viscosity. Recall that (8.1) can be considered as a unidimensional approximation to the Navier–Stokes equations for an incompressible flow [4]. Assume that, for some $F \in \mathbb{R}_+$, $z \in [0, F]$ where $z = 0$ is the position of the actuator and $z = F$ is the position of the sensor. Suppose that $\mathbf{v}(t,z) = \bar{\mathbf{v}}(z - ct)$, i.e., the solution of (8.1) is a traveling wave with velocity $c \in \mathbb{R}_{\geq 0}$, it has been proven that (8.1) admits this kind of solutions [4].

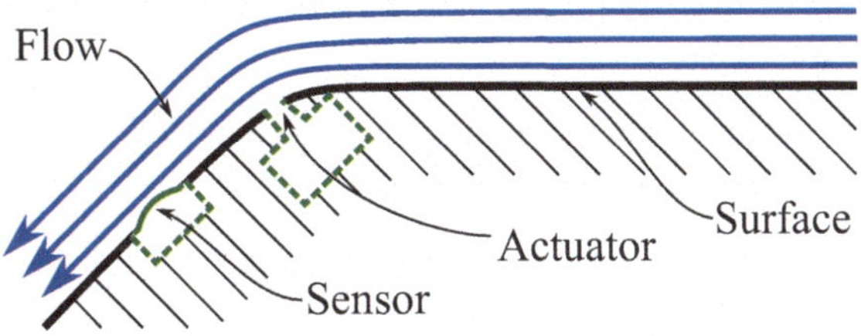

Fig. 8.2 Schematic diagram of a flow over an instrumented surface

An input–output model approximation of (8.1) can be obtained by discretizing (8.1) in the spatial coordinate. We use central finite differences to discretize the spatial derivatives with a mesh of three points and a step of $h = F/2$. Thus,

$$\frac{\partial \mathbf{v}(t, F/2)}{\partial t} + \frac{1}{F}\mathbf{v}(t, F/2)\left[\mathbf{v}(t, F) - \mathbf{v}(t, 0)\right] = \frac{4\nu}{F^2}\left[\mathbf{v}(t, F) - 2\mathbf{v}(t, F/2) + \mathbf{v}(t, 0)\right] . \tag{8.2}$$

Since $\mathbf{v}$ is assumed to be a traveling wave, it has a periodic pattern in space and time. In particular, note that $\mathbf{v}(t, F/2) = \bar{\mathbf{v}}(F/2 - ct) = \mathbf{v}(t + F/(2c), F) = \mathbf{v}(t - F/(2c), 0)$. With these relations and by defining $x(t) = \mathbf{v}(t, F)$, $u(t) = \mathbf{v}(t, 0)$, (8.2) can be rewritten as $\dot{x}(t + \varsigma) = -\frac{1}{F}u(t - \varsigma)x(t) + \frac{1}{F}x(t + \varsigma)u(t) + \frac{4\nu}{F^2}\left[x(t) - 2x(t + \varsigma) + u(t)\right]$, or equivalently,

$$\dot{x}(t) = -\frac{1}{F}x(t - \varsigma)u(t - 2\varsigma) + \frac{1}{F}x(t)u(t - \varsigma) + \frac{4\nu}{F^2}\left[x(t - \varsigma) - 2x(t) + u(t - \varsigma)\right] ,$$

where $\varsigma = F/(2c)$. Hence, in [5–7], the authors propose the more general equation

$$\dot{x}(t) = \sum_{i=1}^{N_1} a_i x(t - \tau_i) + \sum_{k=1}^{N_3}\left(\sum_{j=1}^{N_2} c_{j,k}x(t - \bar{\tau}_j) + b_k\right)u(t - \varsigma_k) , \tag{8.3}$$

as the input–output model for the separated flow control system shown in Fig. 8.2. The measured output is x and the control input is u. Observe that this approximating model still recovers two main features of the original flow model (8.1): first, it is nonlinear; and second, it is infinite dimensional. It is important to mention that for several experimental settings, this input–output model replicates with good accuracy the input–output behavior of the physical system, see [5] for more details.

8.3 Description of the Control Design Problem

We consider again the controlled system depicted in Fig. 8.2 and its input–output model given by (8.3). The two main concerns in the control design process of such a system are

1. the identification of the system's parameters and
2. the design of the control law.

These two tasks must be done taking into account the following physical restrictions of the system.

(a) The actuators are on–off air blowers; thus, the image of the input signal u is restricted to the set $\{0, 1\}$.

(b) A sensor measurement $x(t) = 0$ indicates that the flow is in a separated condition, and a measurement $x(t) > 0$ describes a reduction of the flow separation. Thus, the higher the sensor measurement, the less the flow separation. Hence,

   1. the solutions of the model must be nonnegative;
   2. without control action the solution must tend to zero as t increases;
   3. the solutions must be bounded for any actuation signal.

Remark 8.1 In this chapter, we focus on the control design; however, some conditions on the parameters of the system, which guarantee the properties 1–3 in (b), are given in Sect. 8.4. For suitable techniques of parameters identification, see, e.g., [7, 18, 21].

For control design purposes, we consider in this chapter the particular case of (8.3) with $N_1 = N_2 = 2$ and $N_3 = 1$, namely,

$$\dot{x}(t) = a_1 x(t - \tau_1) - a_2 x(t - \tau_2) + [c_1 x(t - \bar{\tau}_1) - c_2 x(t - \bar{\tau}_2) + b]\, u(t - \varsigma)\,, \tag{8.4}$$

where $x(t) \in \mathbb{R}$, and the parameters $a_1, a_2, c_1, c_2, b, \tau_1, \tau_2, \bar{\tau}_1, \bar{\tau}_2, \varsigma$ are nonnegative real constants. We assume that all the delays are bounded. We also assume that the initial conditions of (8.4) are

$$x(t) = 0\,, \quad u(t) = 0\,, \quad \forall t \leq 0\,. \tag{8.5}$$

The control objective: To drive the system's output $x(t)$ to a constant reference $x^* \in \mathbb{R}_+$.

8.4 Properties of the Model

As stated in Sect. 8.3, we require some features of the solutions of (8.4) to guarantee that it constitutes a suitable model for the physical system. In this section, we study the conditions on the parameters of (8.4) that guarantee nonnegativeness and boundedness of the solutions. Of course, existence and uniqueness of solutions must be guaranteed. To this aim we rewrite (8.4) as

$$\dot{x}(t) = a_1 x(t - \tau_1) + c_1 u(t - \varsigma) x(t - \bar{\tau}_1) - a_2 x(t - \tau_2) - c_2 u(t - \varsigma) x(t - \bar{\tau}_2) + b u(t - \varsigma)\,, \tag{8.6}$$

which can be seen as a linear delayed system with time-varying coefficients. The term $bu(t - \varsigma)$ is considered as the input. A locally absolutely continuous function that satisfies (8.6), for almost all $t \in [0, \infty)$, and its initial conditions for all $t \leq 0$ is called a solution of (8.6) [1]. Hence, if in addition to the assumptions in the previous section, we assume that $u : \mathbb{R} \to \{0, 1\}$ is a Lebesgue-measurable locally essentially bounded function, then the solution of (8.6) exists and it is unique, see Appendix 8.8.1. Such a definition of solution is adequate for the open-loop analysis made in this section. The framework for the closed-loop analysis is explained in Sect. 8.5.1, where we verify that the control signal u is indeed a Lebesgue-measurable bounded signal.

8.4.1 Nonnegative Solutions

We have said that the model has to be guaranteed to provide nonnegative solutions. Thus, we first search for some conditions that guarantee that the solutions of (8.6) are nonnegative. Below we recall a useful result to verify whether (8.6) has nonoscillatory solutions.[1] For the definition of the *fundamental function*, see Appendix 8.8.1.

Lemma 8.1 ([1], Corollary 3.13) *Consider* (8.6) *with* $b = 0$ *and initial conditions:* $x(t) = 0$, $u(t) = 0$ *for all* $t < 0$, *and* $x(0) = x_0$, *for some* $x_0 \in \mathbb{R}$. *Define*

$$P(t) = a_1 + c_1 u(t - \varsigma), \quad N(t) = a_2 + c_2 u(t - \varsigma). \tag{8.7}$$

If for any Lebesgue-measurable function $u : \mathbb{R} \to \{0, 1\}$ *the following holds: (1)* $\min(\tau_2, \bar{\tau}_2) \geq \max(\tau_1, \bar{\tau}_1)$*; (2)* $N(t) \geq P(t)$ *for all* $t \geq 0$ *and; (3) there exists* $\lambda \in (0, 1)$ *such that*

$$\limsup_{t\to\infty} \int_{t-\max(\tau_2,\bar{\tau}_2)}^{t-\min(\tau_1,\bar{\tau}_1)} (N(s) - \lambda P(s))\, \mathrm{d}s < \frac{\ln(1/\lambda)}{\mathrm{e}},$$

$$\limsup_{t\to\infty} \int_{t-\max(\tau_2,\bar{\tau}_2)}^{t} (N(s) - \lambda P(s))\, \mathrm{d}s < \frac{1}{\mathrm{e}}.$$

Then the fundamental function X *of* (8.6) *is such that* $X(t, s) > 0$, $t \geq s \geq 0$, *and* (8.6) *has an eventually positive solution with an eventually nonpositive derivative.*

Remark 8.2 Observe that a sufficient condition to satisfy (2) in Lemma 8.1 is

$$a_2 - a_1 + \min(0, c_2 - c_1) > 0. \tag{8.8}$$

Also note that (8.8) implies the inequalities $a_2 - a_1 > 0$ and $a_2 + c_2 - a_1 - c_1 > 0$. The integral conditions of the point 3) in Lemma 8.1 are satisfied with $\lambda = 1/\mathrm{e}$ if the following inequality holds:

---

[1] A continuous function $x : [t_0, \infty) \to \mathbb{R}$ is said to be nonoscillatory if there exists $t_1 > t_0$ such that $x(t) \neq 0$ for all $t > t_1$. Since x is continuous, if it is nonoscillatory it must be *eventually positive* or *eventually negative*. That is, there exists $T > t_0$ such that $x(t)$ is positive for all $t > T$ or it is negative for all $t > T$, see, e.g., [1, 12].

$$[a_2 - a_1/\mathrm{e} + \max(0,\ c_2 - c_1/\mathrm{e})]\max(\tau_2, \bar{\tau}_2) < 1/\mathrm{e}\,.$$

Now that we have nonoscillation conditions for (8.6); we can state the following corollary.

Corollary 8.1 *Consider* (8.6) *with* (8.5) *and* $b > 0$. *Suppose that the assumptions of Lemma 8.1 hold. If* $u(t) \geq 0$ *for all* $t \geq 0$, *then* $x(t) \geq 0$ *for all* $t \geq 0$.

The proof of Corollary 8.1 is straightforward by using the fundamental function and the solution representation of x given in (8.38), see Lemma 8.4 in Appendix 8.8.1.

8.4.2 Boundedness of Solutions

Observe that the nonoscillation conditions of Lemma 8.1 also guarantee the boundedness of the system's trajectories for $b = 0$. The case $b > 0$ is considered in the following lemma.

Lemma 8.2 *Consider* (8.6) *with initial conditions* (8.5). *If the hypotheses of Lemma 8.1 hold,* $b > 0$, *and condition* (8.8) *holds, then (1) for any Lebesgue-measurable function* $u : \mathbb{R} \to \{0, 1\}$ *there exists* $B \in \mathbb{R}_+$ *such that* $x(t) \leq B$ *for all* $t \geq 0$ *and; (2) for* $u(t) = 1\ \forall t > 0$,

$$\lim_{t\to\infty} x(t) = \bar{x}\,, \quad \bar{x} = \frac{b}{a_2 + c_2 - a_1 - c_1}\,. \tag{8.9}$$

Proof (1) First, let us consider (8.6) with $b = 0$, and $x(0) > 0$. According to Lemma 8.1, there exists $t_1 \geq 0$ such that $x(t) > 0$ and $\dot{x}(t) \leq 0$ for all $t \geq t_1$. This ensures that $x(t)$ is nonincreasing for all $t \geq t_1$. Hence, there exists $t_2 \geq t_1$ such that for all $t \geq t_2$ (denote $v = u(t - \varsigma)$),

$$\begin{aligned}
\dot{x}(t) &\leq a_1 x(t - \max(\tau_1, \bar{\tau}_1)) + v c_1 x(t - \max(\tau_1, \bar{\tau}_1)) - \\
&\qquad a_2 x(t - \min(\tau_2, \bar{\tau}_2)) - v c_2 x(t - \min(\tau_2, \bar{\tau}_2)) \\
&\leq (a_1 + v c_1) x(t - \max(\tau_1, \bar{\tau}_1)) - (a_2 + v c_2) x(t - \min(\tau_2, \bar{\tau}_2))\,, \\
&\leq -(a_2 - a_1 + v[c_2 - c_1]) x(t - \max(\tau_1, \bar{\tau}_1))\,.
\end{aligned} \tag{8.10}$$

Since $a_2 - a_1 + u(t - \varsigma)[c_2 - c_1] \geq a_2 - a_1 + \min(0, c_2 - c_1)$, conditions (8.8) and (8.10) imply the following:

(a) $\dot{x}(t) < 0$ for all $t \geq t_2$, consequently, $\lim_{t\to\infty} x(t) = 0$;
(b) the slowest decreasing rate of $x(t)$ (for all $t > t_1$) is achieved if u is constant, i.e., $u(t) = 1$ or $u(t) = 0$ (depending on the sign of $c_2 - c_1$).

When u is constant, (8.6) becomes a time-invariant system; thus, the asymptotic behavior of $x(t)$ guarantees that $x = 0$ is asymptotically stable; therefore, it is exponentially stable (see, e.g., [9, 12]). This guarantees that for $b = 0$ and any Lebesgue-measurable function $u : \mathbb{R} \to \{0, 1\}$ the solutions of (8.6) are upper bounded by a decreasing exponential. Therefore, the fundamental function $X(t, s)$ of (8.6) is exponentially bounded.

Now, let us consider the case $b > 0$. The solution of (8.6) can be written as follows (see Lemma 8.4 in Appendix 8.8.1):

$$x(t) = X(t, 0)x(0) + \int_0^t X(t, s)bu(s - \varsigma)\,\mathrm{d}s\,.$$

The result follows by noting that $bu(s - \varsigma) \leq b$ and $X(t, s)$ is upper bounded by a decreasing exponential.

(2) Lemma 8.1 ensures that the system's solution is nonoscillatory for $b = 0$. This and the analysis in 1) guarantee that for $b > 0$ and $u(t) \equiv 1$, $\lim_{t\to\infty} x(t)$ exists and it is some constant $\bar{x} \in \mathbb{R}_+$. Therefore,

$$\lim_{t\to\infty} \dot{x}(t) = 0 = -(a_2 + c_2 - a_1 - c_1)\bar{x} + b\,. \tag{8.11}$$

From this equality, we obtain the limit value in (8.9).

The following lemma gives a qualitative bound for the derivative of x. Such a bound will be used in the stability analysis of the controlled system.

Lemma 8.3 *If* (8.6), *with initial conditions* (8.5), *satisfies Lemma 8.2, and* $a_1\tau_1 + a_2\tau_2 + c_1\bar{\tau}_1 + c_2\bar{\tau}_2 < 1$, *then, for any* $T \in \mathbb{R}_+$, $\|\dot{x}(t)\|_{L^\infty([0,T])} \leq D(\|x(t)\|_{L^\infty([0,T])})$, *where*

$$D(s) = \frac{\max\big(b,\ (a_2 - a_1 + \max(0, c_2 - c_1))s\big)}{1 - (a_1\tau_1 + a_2\tau_2 + c_1\bar{\tau}_1 + c_2\bar{\tau}_2)}\,. \tag{8.12}$$

Proof The proof is divided into two cases. First, suppose that $\dot{x}(t) > 0$. From (8.6), we have that

$$\begin{aligned}
\dot{x}(t) &\leq a_1x(t - \tau_1) - a_2x(t - \tau_2) + c_1u(t - \varsigma)x(t - \bar{\tau}_1) - \\
&\quad c_2u(t - \varsigma)x(t - \bar{\tau}_2) + b\,, \\
&\leq \big(a_1 - a_2 + [c_1 - c_2]u(t - \varsigma)\big)x(t) + b - a_1\int_{t-\tau_1}^t \dot{x}(v)\,\mathrm{d}v + \\
&\quad a_2\int_{t-\tau_2}^t \dot{x}(v)\,\mathrm{d}v - u(t - \varsigma)\left(c_1\int_{t-\bar{\tau}_1}^t \dot{x}(v)\,\mathrm{d}v - c_2\int_{t-\bar{\tau}_2}^t \dot{x}(v)\,\mathrm{d}v\right).
\end{aligned}$$

Since $a_1 - a_2 + [c_1 - c_2]u(t-\varsigma) < 0$ and $x(t)$ is nonnegative, we have that

$$|\dot{x}(t)| \leq (a_1\tau_1 + a_2\tau_2 + c_1\bar{\tau}_1 + c_2\bar{\tau}_2)\,\|\dot{x}(t)\|_{L^\infty([0,T])} + b\,. \tag{8.13}$$

Now, suppose that $\dot{x}(t) < 0$. From (8.6), we have that

$$\begin{aligned}-\dot{x}(t) &\leq -a_1x(t-\tau_1) + a_2x(t-\tau_2) - c_1u(t-\varsigma)x(t-\bar{\tau}_1) + \\ &\quad c_2u(t-\varsigma)x(t-\bar{\tau}_2)\,,\\ &\leq \big(a_2 - a_1 + [c_2 - c_1]u(t-\varsigma)\big)x(t) + a_1\int_{t-\tau_1}^{t}\dot{x}(\nu)\,\mathrm{d}\nu - \\ &\quad a_2\int_{t-\tau_2}^{t}\dot{x}(\nu)\,\mathrm{d}\nu + u(t-\varsigma)\left(c_1\int_{t-\bar{\tau}_1}^{t}\dot{x}(\nu)\,\mathrm{d}\nu - c_2\int_{t-\bar{\tau}_2}^{t}\dot{x}(\nu)\,\mathrm{d}\nu\right);\end{aligned}$$

hence, we obtain the bound

$$\begin{aligned}|\dot{x}(t)| \leq (a_2 - a_1 + \max(0, c_2 - c_1))\|x(t)\|_{L^\infty([0,T])} + \\ (a_1\tau_1 + a_2\tau_2 + c_1\bar{\tau}_1 + c_2\bar{\tau}_2)\,\|\dot{x}(t)\|_{L^\infty([0,T])}\,.\end{aligned} \tag{8.14}$$

Since we are assuming that $a_1\tau_1 + a_2\tau_2 + c_1\bar{\tau}_1 + c_2\bar{\tau}_2 < 1$, the result is obtained from (8.13) and (8.14).

8.5 Sliding-Mode Controller

In this section, we present the sliding-mode control scheme for (8.4). The proposed control law is given by

$$u(t) = \begin{cases} 1, & \sigma(t) < 0, \\ 0, & \sigma(t) > 0, \end{cases} \tag{8.15}$$

where the sliding variable σ is given by

$$\sigma(t) = x(t) + T(t) - \sigma^*\,, \tag{8.16}$$

with $\sigma^* := (1 + \varsigma(a_2 + c_2 - a_1 - c_1))\,x^*$, and

$$T(t) := \int_{t-\varsigma}^{t}\Big[(u(\eta) - 1)\,[c_1x(\eta - \bar{\tau}_1 + \varsigma) - c_2x(\eta - \bar{\tau}_2 + \varsigma)] + bu(\eta)\Big]\,\mathrm{d}\eta\,.$$

Since the reference points x^* cannot be arbitrarily chosen, a suitable range for them must be specified. Observe that Lemma 8.2 gives an upper bound for the possible reference points, i.e., $x^* < \overline{x}$. However, the stability proof for the proposed controller imposes a lower bound for the reference points. Note that this is not problematic since, according to the physical problem, the flow separation is reduced by increasing the value of x^*.

Thus, for the controller (8.15), the set of reference points is the interval $(\underline{x}, \overline{x})$ where $\overline{x}$ is given by (8.9)

$$\underline{x} = \frac{\tilde{x} + \varsigma b}{1 + \varsigma \left(a_2 + c_2 - a_1 - c_1\right)}, \quad \tilde{x} = \frac{\left(a_1 \tau_1 + a_2 \tau_2 + c_1 \bar{\tau}_1 + c_2 \bar{\tau}_2\right) D(\overline{x})}{a_2 - a_1}, \tag{8.17}$$

with D given by (8.12). Note that it is necessary that the following inequality

$$\overline{x} > \underline{x} \tag{8.18}$$

holds to have a nonempty set of reference points.

To study the stability features in sliding regime, we have to define the function $K : \mathbb{R}_+ \cup \{0\} \to \mathbb{R}$ given by

$$K(r) = A_1(r) - A_2(r) + C_1(r) - C_2(r), \tag{8.19}$$

where

$$A_i(r) := \begin{cases} -a_i, & r \in [\tau_i, \tau_i + \varsigma] \\ 0, & r \notin [\tau_i, \tau_i + \varsigma] \end{cases}, \quad C_i(r) := \begin{cases} -c_i, & r \in [\bar{\tau}_i, \bar{\tau}_i + \varsigma] \\ 0, & r \notin [\bar{\tau}_i, \bar{\tau}_i + \varsigma] \end{cases}, \tag{8.20}$$

for $i = 1, 2$. Now we can give the sentence of the main result of this chapter.

Theorem 8.1 *If system* (8.4) *satisfies the conditions of Lemma 8.3, condition* (8.18), $a_2 > a_1 + c_1$, *and* $\varsigma \le \bar{\tau}_i$, *then*

(a) for any $x^* \in (\underline{x}, \overline{x})$, *the solution of* (8.4) *in closed loop with* (8.15) *establishes a sliding motion in finite time on the sliding surface described by* $\sigma(t) = 0$;
(b) if additionally, (8.19) *is such that*

$$\|K(r)\|_{L^1(\mathbb{R}_+ \cup \{0\})} < 1, \tag{8.21}$$

then in the sliding motion $x(t)$ *converges exponentially to* x^*.

The proof of this theorem is given in Sects. 8.5.2 and 8.5.3.

8.5.1 On the Closed-Loop Solutions

The time derivative of the sliding variable σ_i is given by

$$\begin{aligned} \dot{\sigma}(t) = {} & a_1 x(t - \tau_1) - a_2 x(t - \tau_2) + c_1 x(t - \bar{\tau}_1) - c_2 x(t - \bar{\tau}_2) - \\ & (1 - u(t))[c_1 x(t - \bar{\tau}_1 + \varsigma) - c_2 x(t - \bar{\tau}_2 + \varsigma)] + bu(t). \end{aligned} \tag{8.22}$$

Note that the right-hand side of (8.22) depends on x, and the right-hand side of (8.4) in closed-loop with (8.15) depends on σ. Moreover, the right-hand side of (8.22) is discontinuous in σ. Thus, (8.22), (8.4) constitute a system of discontinuous functional differential equations. Nevertheless, the existence and uniqueness of solutions of such a system can still be guaranteed by means of standard definitions as explained below.

First, let us rewrite (8.4) and (8.22) as Follows:

$$\dot{x}(t) = f\big(x(t), x_t, \sigma(t-\varsigma)\big), \tag{8.23}$$
$$\dot{\sigma}(t) = g\big(\sigma(t), x(t), x_t\big), \tag{8.24}$$

where f and g denote the right-hand side of (8.4) and (8.22), respectively. For a given t, x_t is the function given by $x_t(\theta) := x(t+\theta)$ with $\theta \in [-\tau_{\max}, 0)$, and $\tau_{\max}$ the maximum of all the delays in (8.4). Consider, for $t \in [t_0, t_0 + \tau_{\max}]$, an initial function ϕ_t given by $\phi_t := x(t+\theta)$ with $\theta \in [-\tau_{\max}, 0)$.

We can compute, with ϕ_t and (8.16), the initial function ψ given by $\psi := \sigma(\vartheta)$ with $\vartheta \in [t_0 - \varsigma, t_0)$.

To define the solution for the system (8.23)–(8.24), we use the *method of steps* (see, e.g., [14, p. 89]), i.e., for t in the intervals $[t_0, t_0 + \tau_{\min}]$, $[t_0 + \tau_{\min}, t_0 + 2\tau_{\min}]$, and so on, where $\tau_{\min}$ denotes the minimum of all nonzero delays in (8.4).

For t in the interval $[t_0, t_0 + \tau_{\min}]$, we rewrite (8.23)–(8.24) (by using the initial functions) as follows:

$$\dot{x}(t) = f\big(x(t), \phi_t, \psi(t-\varsigma)\big) := F(t, x), \tag{8.25}$$
$$\dot{\sigma}(t) = g\big(\sigma(t), x(t), \phi_t\big) := G(t, \sigma, x). \tag{8.26}$$

Observe that (8.25)–(8.26) is an ordinary differential equation. Moreover, F does not depend on σ, is linear in x, and is only discontinuous in t. Therefore, (8.25) can be solved as a Carathéodory differential equation and the solution is unique (see, e.g., [8, p. 3–6]). On the other hand, G is only discontinuous in σ and $x(t)$ is available from the solution of (8.25). Therefore, (8.26) can be solved by using the standard theory for differential equations with discontinuous right-hand side [8].[2]

The same reasoning is applied in the subsequent time intervals.

Note that, the solution x is an absolutely continuous function of t; hence, $\dot{x}$ exists almost everywhere and it is a Lebesgue-measurable function of t (see, e.g., [22, p. 109–110]). Thus, from (8.4) and by the properties of sum and product of Lebesgue-measurable functions (see, e.g., [22, p. 67]), we have that u (as a function of time t) is also a Lebesgue-measurable function.

[2]For (8.26), the three methods given in [8, p. 50–56] to construct the differential inclusion coincide, see also [20].

8.5.2 Reaching Phase Analysis

For the initial conditions (8.5), if $u(t) = 1$, and $t \geq t_0 + \varsigma$, we have that

$$\sigma(t) = x(t) + b\varsigma - \sigma^* .$$

Since $x(t) \to \overline{x}$ as $t \to \infty$ (see Lemma 8.2) and $x^* < \overline{x}$, there exists a finite $t_1 \in \mathbb{R}_+$ such that $x(t_1) = x^*$. Therefore,

$$\sigma(t_1) = x^* + \varsigma b - \sigma^* .$$

Now, from (8.9), $b = \overline{x}(a_2 + c_2 - a_1 - c_1)$, thus

$$\begin{aligned} \sigma(t_1) &= x^* + \varsigma\overline{x}(a_2 + c_2 - a_1 - c_1) - (1 + \varsigma(a_2 + c_2 - a_1 - c_1))\, x^* , \\ &= \varsigma(a_2 + c_2 - a_1 - c_1)(\overline{x} - x^*) . \end{aligned}$$

Hence, $\sigma(t_1) > 0$. This ensures the existence of a finite $t^* \in \mathbb{R}_+$ such that $\sigma(t^*) = 0$. Note that we still have to verify that $\sigma(t)$ grows as $x(t) \to x^*$. For $u(t) = 1$, the time derivative of σ is given by

$$\dot{\sigma}(t) = a_1 x(t - \tau_1) - a_2 x(t - \tau_2) + c_1 x(t - \bar{\tau}_1) - c_2 x(t - \bar{\tau}_2) + b ,$$

whose right-hand side corresponds to the dynamics of x, and thus σ remains growing as x is growing. This concludes the proof that the sliding surface is reached. Now, we have to verify that the system's trajectory remains on the sliding surface.

Once the sliding surface $\sigma = 0$ is reached, u switches from $u(t) = 1$ to $u(t) = 0$, and the dynamics of the sliding variable becomes

$$\begin{aligned} \dot{\sigma}(t) &= a_1 x(t - \tau_1) - a_2 x(t - \tau_2) + c_1 x(t - \bar{\tau}_1) - c_2 x(t - \bar{\tau}_2) - \\ &\quad c_1 x(t - \bar{\tau}_1 + \varsigma) + c_2 x(t - \bar{\tau}_2 + \varsigma) , \\ &= (a_1 - a_2)x(t) - a_1 \int_{t-\tau_1}^{t} \dot{x}(\nu)\,\mathrm{d}\nu + a_2 \int_{t-\tau_2}^{t} \dot{x}(\nu)\,\mathrm{d}\nu + \\ &\quad c_1 \int_{t-\bar{\tau}_1+\varsigma}^{t-\bar{\tau}_1} \dot{x}(\nu)\,\mathrm{d}\nu - c_2 \int_{t-\bar{\tau}_2+\varsigma}^{t-\bar{\tau}_2} \dot{x}(\nu)\,\mathrm{d}\nu . \end{aligned}$$

Hence, we can obtain the following bound:

$$\dot{\sigma}(t) \leq (a_1 - a_2)x(t) + (a_1\tau_1 + a_2\tau_2 + [c_1 + c_2]\varsigma)\|\dot{x}(t)\|_{L^\infty([0,t^*])} ,$$

thus, $\dot{\sigma}(t) < 0$ if $x(t) > (a_1\tau_1 + a_2\tau_2 + [c_1 + c_2]\varsigma)\|\dot{x}(t)\|_{L^\infty([0,t^*])}/(a_2 - a_1)$ (note that $a_2 > a_1$, see Remark 8.2). Hence, and by using (8.12), we obtain the value of $\tilde{x}$ given in (8.17). Now, we only have to verify that indeed $x(t) > \tilde{x}$ when the sliding surface is reached. Note that, when u switches from $u(t) = 1$ to $u(t) = 0$,

$u(t-\varsigma) = 1$. Thus, if $\sigma(t) = 0$, then

$$0 = x(t) + \varsigma b - \sigma^* .$$

From this equality, it is easy to verify that $x(t^*) > \tilde{x}$ if $x^* > \underline{x}$, where $\underline{x}$ is given in (8.17).

8.5.3 Sliding-Phase Analysis

To analyze the sliding dynamics, first observe that the sliding variable can be rewritten as follows:

$$\begin{aligned} \sigma(t) = x(t) - \int_{t-\varsigma}^{t} & \big(a_1 x(\eta-\tau_1) - a_2 x(\eta-\tau_2) + \\ & c_1 x(\eta-\bar{\tau}_1) - c_2 x(\eta-\bar{\tau}_2)\big)\, \mathrm{d}\eta - \sigma^* + \int_{t-\varsigma}^{t} \dot{\sigma}(\eta)\, \mathrm{d}\eta , \end{aligned} \tag{8.27}$$

or equivalently

$$\begin{aligned} \sigma(t-\varsigma) = x(t) - \int_{t-\varsigma}^{t} & \big(a_1 x(\eta-\tau_1) - a_2 x(\eta-\tau_2) + \\ & c_1 x(\eta-\bar{\tau}_1) - c_2 x(\eta-\bar{\tau}_2)\big)\, \mathrm{d}\eta - \sigma^* . \end{aligned} \tag{8.28}$$

Observe that, if the sliding surface is reached at $t = t^*$, the state variable $x(t)$ is not in sliding regime for $t \in [t^*, t^* + \varsigma)$; however, we know that it remains bounded. Thus, for all $t \geq t^* + \varsigma$, $\sigma(t-\varsigma) = 0$ and the dynamics of x is described by the integral equation

$$\begin{aligned} x(t) - \int_{t-\varsigma}^{t} & \big(a_1 x(\eta-\tau_1) - a_2 x(\eta-\tau_2) + \\ & c_1 x(\eta-\bar{\tau}_1) - c_2 x(\eta-\bar{\tau}_2)\big)\, \mathrm{d}\eta - \sigma^* = 0 . \end{aligned}$$

If we define the regulation error χ by means of the change of coordinates $\chi(t) = x(t) - x^*$, then the error dynamics is given by

$$\chi(t) - \int_{t-\varsigma}^{t} \big(a_1 \chi(\eta-\tau_1) - a_2 \chi(\eta-\tau_2) + c_1 \chi(\eta-\bar{\tau}_1) - c_2 \chi(\eta-\bar{\tau}_2)\big)\, \mathrm{d}\eta = 0 ,$$

or equivalently,

$$\chi(t) - \int_{t-\varsigma-\tau_1}^{t-\tau_1} a_1\chi(\eta)\,\mathrm{d}\eta + \int_{t-\varsigma-\tau_2}^{t-\tau_2} a_2\chi(\eta)\,\mathrm{d}\eta - \int_{t-\varsigma-\bar{\tau}_1}^{t-\bar{\tau}_1} c_1\chi(\eta)\,\mathrm{d}\eta + \int_{t-\varsigma-\bar{\tau}_2}^{t-\bar{\tau}_2} c_2\chi(\eta)\,\mathrm{d}\eta = 0\,.$$

Note that this equation can be rewritten as follows:

$$\chi(t) + \int_{t^*+\varsigma}^{t} \big(A_1(t-\eta) - A_2(t-\eta) + C_1(t-\eta) - C_2(t-\eta)\big)\chi(\eta)\,\mathrm{d}\eta = v(t;t^*+\varsigma)\,, \tag{8.29}$$

where the functions A_i, C_i are given by (8.20) (by replacing the parameter r with $t-\eta$). The function $v : [t^*+\varsigma, \infty) \to \mathbb{R}$ is given by

$$v(t;t^*+\varsigma) = -\int_{\rho_1}^{t^*+\varsigma} A_1(t-\eta)\chi(\eta)\,\mathrm{d}\eta - \int_{\rho_2}^{t^*+\varsigma} A_2(t-\eta)\chi(\eta)\,\mathrm{d}\eta - \int_{\rho_1}^{t^*+\varsigma} C_1(t-\eta)\chi(\eta)\,\mathrm{d}\eta - \int_{\rho_2}^{t^*+\varsigma} C_2(t-\eta)\chi(\eta)\,\mathrm{d}\eta\,, \tag{8.30}$$

with $\rho_i = \min(t^*+\varsigma,\ t-\varsigma-\tau_i)$, $i = 1, 2$, and $\rho_i = \min(t^*+\varsigma,\ t-\varsigma-\bar{\tau}_i)$. Note that $v(t;t^*+\varsigma) = 0$ for all $t \geq t^* + 2\varsigma + \max(\tau_1, \tau_2, \bar{\tau}_1, \bar{\tau}_2)$. Moreover, since $x(t)$ is bounded for all t, then $\chi(t)$ is also bounded. Hence, there exists $\bar{v} \in \mathbb{R}_+$ such that $\|v(t;t^*+\varsigma)\|_{L^1(\mathbb{R}_{\geq t^*+\varsigma})} \leq \bar{v}$.

Now, by considering (8.19), we can rewrite (8.29) as follows:

$$\chi(t) + \int_{t^*+\varsigma}^{t} K(t-\eta)\chi(\eta)\,\mathrm{d}\eta = v(t;t^*+\varsigma)\,, \tag{8.31}$$

which is a Volterra integral equation of the second kind with a kernel K of convolution type. Since K is measurable, according to Lemma 8.6 (see Appendix 8.8.2), condition (8.21) ensures that K is a kernel of type L^1; furthermore, it has a resolvent $\mathscr{R}$ of type L^1. Thus, since $v \in L^1$, Lemma 8.5 ensures the existence of a unique solution χ of (8.31) such that $\chi \in L^1$ and it is given by

$$\chi(t) = v(t;t^*+\varsigma) - \int_{t^*+\varsigma}^{t} \mathscr{R}(t-\eta)v(\eta;t^*+\varsigma)\,\mathrm{d}\eta\,.$$

Finally, since K is a convolution kernel, Lemma 8.7 guarantees that $\chi(t)$ tends to zero exponentially as t tends to infinity. Thus, on the sliding surface, $x(t) \to x^*$ exponentially.

8.6 Robustness

In this section, we analyze the effect of additive disturbances in (8.4). Consider the system

$$\begin{aligned}\dot{x}(t) &= a_1 x(t-\tau_1) - a_2 x(t-\tau_2) \\ &\quad + [c_1 x(t-\bar{\tau}_1) - c_2 x(t-\bar{\tau}_2) + b]\, u(t-\varsigma) + \delta(t), \end{aligned} \tag{8.32}$$

where $\delta : \mathbb{R}_+ \cup \{0\} \to \mathbb{R}$ is a Lebesgue-measurable essentially bounded function with $\|\delta(t)\|_{L^\infty(\mathbb{R}_+\cup\{0\})} = \Delta < b$, for some $\Delta \in \mathbb{R}_+$. We assume that the disturbances do not affect the nonnegativeness of the solutions of (8.32).

Theorem 8.2 *Consider* (8.32) *with* (8.5)*. Suppose that the assumptions of Theorem 8.1 hold. For any $\Delta \in \mathbb{R}_+$, the regulation error $x(t) - x^*$ is ultimately bounded, and its ultimate bound $\beta \in \mathbb{R}_+$ is such that $\beta \to 0$ as $\Delta \to 0$.*

Proof From the proof of Lemma 8.2, we can conclude that the solution of (8.32) is bounded for any essentially bounded input δ. This ensures ultimate boundedness of the regulation error $x(t) - x^*$ and the sliding variable σ.

The time derivative of the sliding variable is given by

$$\begin{aligned}\dot{\sigma}(t) &= a_1 x(t-\tau_1) - a_2 x(t-\tau_2) + c_1 x(t-\bar{\tau}_1) - c_2 x(t-\bar{\tau}_2) - \\ &\quad (1-u(t))[c_1 x(t-\bar{\tau}_1+\varsigma) - c_2 x(t-\bar{\tau}_2+\varsigma)] + bu(t) + \delta(t). \end{aligned} \tag{8.33}$$

Thus, from (8.28), the sliding variable can be rewritten as

$$\begin{aligned}\sigma(t-\varsigma) + \bar{\delta}(t) = x(t) - \int_{t-\varsigma}^{t} \big(& a_1 x(\eta-\tau_1) - a_2 x(\eta-\tau_2) + \\ & c_1 x(\eta-\bar{\tau}_1) - c_2 x(\eta-\bar{\tau}_2)\big)\, \mathrm{d}\eta - \sigma^*, \end{aligned} \tag{8.34}$$

where $\bar{\delta}(t) := \int_{t-\varsigma}^{t} \delta(\eta)\, \mathrm{d}\eta$. Note that $\|\bar{\delta}(t)\|_{L^\infty(\mathbb{R})} \le \Delta\varsigma$. Recall that the solution of (8.32) is given by

$$x(t) = X(t,0)x(0) + \int_0^t X(t,s)[bu(s-\varsigma) + \delta(s)]\, \mathrm{d}s.$$

Since $\overline{x} - x^* > 0$, there exists a sufficiently small Δ such that $x(t)$ increases and the sliding variable σ approximates from below the sliding surface $\sigma = 0$. Here, we have three cases: (1) the sliding surface is not reached; (2) the sliding surface is crossed; and (3) the sliding regime is established.

For the cases (1) and (2), we have already concluded ultimate boundedness. For the case (3), considering again the change of coordinates $\chi(t) = x(t) - x^*$, and by using the same procedure as in Sect. 8.5.3, we obtain from (8.34) the integral equation

$$\chi(t) + \int_{t_0}^{t} K(t-\eta)\chi(\eta)\,\mathrm{d}\eta = w(t)\,, \tag{8.35}$$

where $w(t) := \sigma(t-\varsigma) + \bar{\delta}(t) + v(t; t_0)$, and v is given by (8.30). Observe that, by using the same arguments as in Sect. 8.5.3, $v \in L^1 \cap L^\infty$, hence, $w \in L^\infty$. We have proven that the kernel K has a resolvent $\mathscr{R}$ of type L^1, and thus, according to Lemma 8.5, the solution χ of (8.35) is such that $\chi \in L^\infty$, and

$$\chi(t) = w(t) - \int_{t_0}^{t} \mathscr{R}(t-\eta)w(\eta)\,\mathrm{d}\eta\,.$$

If the sliding motion is established at $t = t^*$, then for $t \geq t^* + \varsigma$ we have that $w(t) = \bar{\delta}(t) + v(t; t^* + \varsigma)$, and

$$\chi(t) = \bar{\delta}(t) + v(t; t^* + \varsigma) - \int_{t^*+\varsigma}^{t} \mathscr{R}(t-\eta)(\bar{\delta}(\eta) + v(\eta; t^* + \varsigma))\,\mathrm{d}\eta\,.$$

Since $v(t; t^* + \varsigma) = 0$ for all $t \geq t^* + 2\varsigma + \max(\tau_1, \tau_2, \bar{\tau}_1, \bar{\tau}_2)$, and $\lim_{t\to\infty} \int_{t^*+\varsigma}^{t} \mathscr{R}(t-\eta)v(\eta; t^* + \varsigma)\,\mathrm{d}\eta = 0$, we have that

$$\chi(t) \to \bar{\delta}(t) - \int_{t^*+\varsigma}^{t} \mathscr{R}(t-\eta)\bar{\delta}(\eta)\,\mathrm{d}\eta\,,$$

as $t \to \infty$. But note that $\int_{t^*+\varsigma}^{t} \mathscr{R}(t-\eta)\bar{\delta}(\eta)\,\mathrm{d}\eta = \int_{t^*+\varsigma}^{t} \mathscr{R}(\eta)\bar{\delta}(t-\eta)\,\mathrm{d}\eta$, hence,

$$\bar{\delta}(t) - \int_{t^*+\varsigma}^{t} \mathscr{R}(t-\eta)\bar{\delta}(\eta)\,\mathrm{d}\eta \leq (1 + \|\mathscr{R}(t)\|_{L^1})\Delta\varsigma\,.$$

Therefore, the ultimate bound for χ is proportional to Δ.

8.7 Numerical Example

Consider (8.4) with the parameters $a_1 = 2$, $a_2 = 3$, $c_1 = 1$, $c_2 = 1$, $b = 2.5$, $\tau_1 = 0.05$, $\tau_2 = 0.06$, $\bar{\tau}_1 = 0.05$, $\bar{\tau}_2 = 0.07$, and $\varsigma = 0.03$. The values of these parameters were chosen of a similar order as those obtained in [6]. They satisfy all the conditions of Theorem 8.1 with $(\underline{x}, \overline{x}) = (1.86, 2.5)$.

The simulations were made with MATLAB® by using an explicit Euler integration method with a step of 0.1 ms, and by choosing the reference point $x^* = 2.2$.

In Fig. 8.3, we can observe the system's state in the nominal case. A more detailed behavior of the sliding phase can be seen in Fig. 8.4.

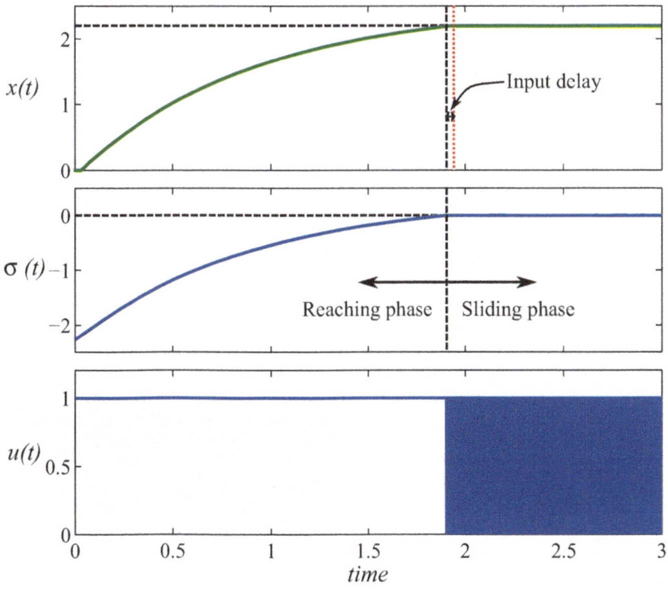

Fig. 8.3 System's signals in the nominal case

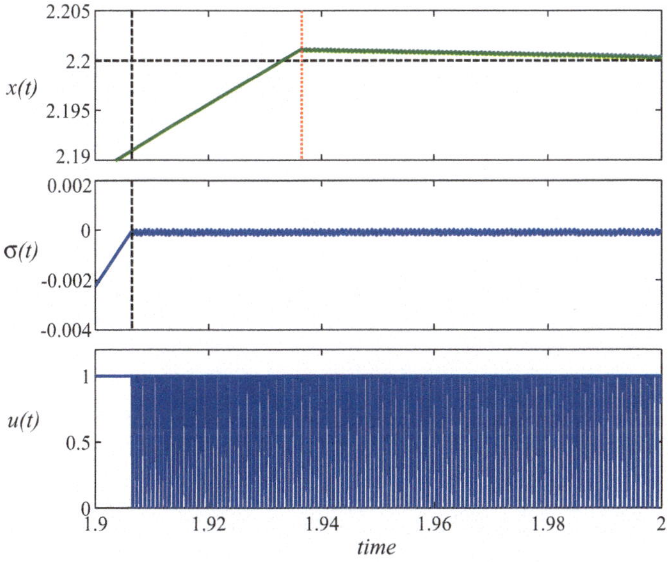

Fig. 8.4 Sliding phase in the nominal case

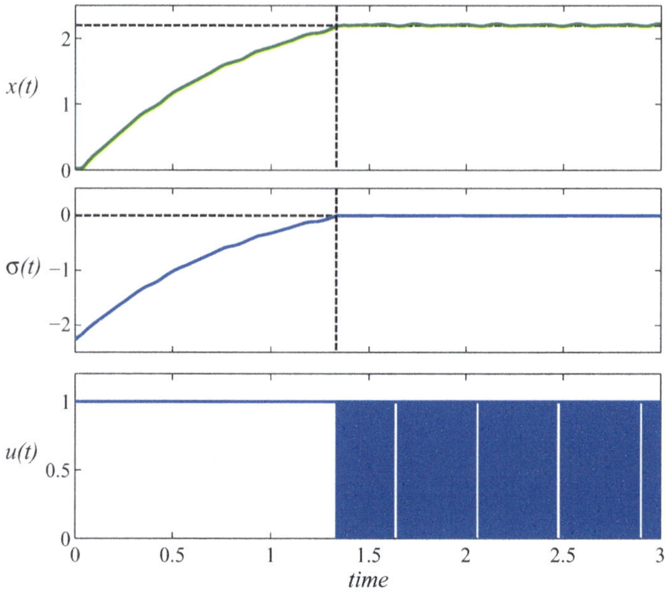

Fig. 8.5 System's signals in the disturbed case

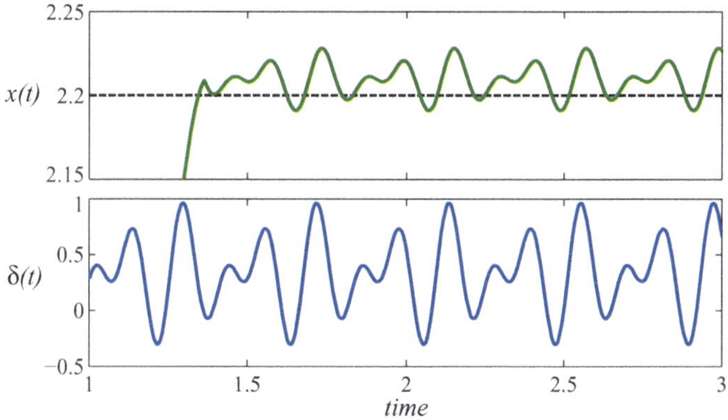

Fig. 8.6 Disturbance's effect on the state

In Fig. 8.5, we can see a simulation of the disturbed case, considering the disturbance $\delta(t) = (1 + \sin(30t) + \sin(45t))/3$. Figures 8.6 and 8.7 show in detail the effect of the disturbance on the state and on the sliding variable, confirming the robustness properties of the control scheme.

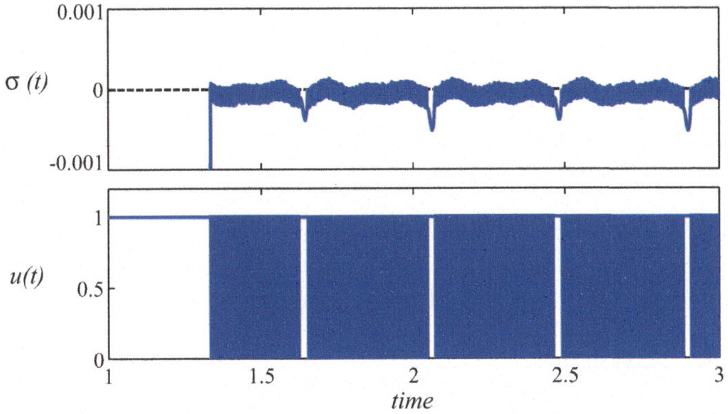

Fig. 8.7 Disturbance's effect on the sliding variable

8.8 Conclusions

We proposed a sliding-mode controller for a class of scalar bilinear systems with delays. We have shown that the Volterra operator theory is a useful tool to study the stability and robustness properties of the proposed control scheme. It is important to mention that, although the positivity conditions given in Sect. 8.4 are considered in the stability analysis of the closed loop, they are not related to the general methodology. This means that the same design methodology can be used by considering different positivity conditions.

Some ongoing and future directions in this research are (1) to extend the control scheme to the MIMO case; (2) to carry out experimental tests of the obtained controllers; (3) to analyze additional sources of uncertainty, e.g., uncertainty in the delays and; and (4) to study the case with multiple delays in the control input. These are challenging problems since the procedures available in the literature are not suitable to solve them, see, e.g., [24].

Acknowledgements This work was carried out within the framework of the CNRS Research Federation on Ground Transports and Mobility, in articulation with the ELSAT2020 project supported by the European Community, the French Ministry of Higher Education and Research, the Hauts de France Regional Council. The authors gratefully acknowledge the support of these institutions.

Appendix

8.8.1 Solutions of Delayed Differential Equations

The theory recalled in this section was taken from [1], see also [9, 12]. For some finite positive integer N, consider the system

$$\dot{x}(t) = \sum_{i=1}^{N} a_i(t)x(t - \tau_i), \quad x(t) \in \mathbb{R}. \tag{8.36}$$

Assumption 8.1 Each $\tau_i \in \mathbb{R}_+ \cup \{0\}$, and each a_i is a Lebesgue-measurable and locally essentially bounded function.

Definition 8.1 The function $X(t,s)$ that satisfies, for each $s \geq 0$, the problem $\frac{\mathrm{d}}{\mathrm{d}t}X(t,s) = \sum_{i=1}^{N} a_i(t)X(t - \tau_i, s)$, $\forall t \geq s$, $X(t,s) = 0$ for $t < s$, $X(s,s) = 1$, is called the fundamental function (or Cauchy function) of (8.36).

Now consider the system

$$\dot{x}(t) = \sum_{i=1}^{N} a_i(t)x(t - \tau_i) + f(t), \quad x(t) \in \mathbb{R}, \tag{8.37}$$

with initial conditions $x(t) = 0$ for all $t < 0$ and $x(0) = x_0$ for some $x_0 \in \mathbb{R}$.

Lemma 8.4 *Assume that a_i, τ_i satisfy Assumption 8.1, and f is a Lebesgue-measurable locally essentially bounded function, then there exists a unique solution of* (8.37) *and it can be written as*

$$x(t) = X(t,0)x_0 + \int_0^t X(t,s)f(s)\,\mathrm{d}s. \tag{8.38}$$

8.8.2 Volterra Equations

Most of the results recalled in this section can be found in [11]; some of them have been simplified for our particular needs.

Consider the integral equation $z(t) + \int_{t_0}^{t} k(t,s)z(s)\,\mathrm{d}s = f(t)$, where $t \in J = \{\tau \in \mathbb{R} : \tau \geq 0\}$, and $z, f : J \to \mathbb{R}^n$. The kernel k is assumed to be of convolution type, i.e., $k(t,s) = k(t - s)$; thus, k can be defined by means of the function $k : J \to \mathbb{R}$. This integral equation can be rewritten as follows

$$z(t) + (k * z)(t) = f(t), \tag{8.39}$$

where $k * z$ denotes the convolution map $t \mapsto \int_{t_0}^{t} k(t-s)z(s)\,\mathrm{d}s$. A function $r : J \to \mathbb{R}$ is called a resolvent of (8.39) if $z(t) = f(t) - (r * f)(t)$.

Lemma 8.5 ([11], Theorem 2–2.2, Theorem 9–3.6) *Suppose that $k \in L^1(J)$ is a convolution kernel that has a resolvent $r \in L^1(J)$. If $f \in L^1(J)$ ($f \in L^\infty(J)$, respectively), then* (8.39) *has a unique solution $z \in L^1(J)$ ($z \in L^\infty(J)$, respectively) given by $z(t) = f(t) - (r * f)(t)$. Moreover, $\|r * f\|_{L^1(J)} \leq \|r\|_{L^1(J)}\|f\|_{L^1(J)}$ ($\|r * f\|_{L^\infty(J)} \leq \|r\|_{L^1(J)}\|f\|_{L^\infty(J)}$, respectively).*

Lemma 8.6 (Corollary 9–3.10, [11]) *If* $k \in L^1(J)$ *is a convolution kernel such that* $\|k(t)\|_{L^1(J)} < 1$, *then it has a resolvent* $r \in L^1(J)$.

For the following lemma define $J = \mathbb{R}_+ \cup \{0\}$, and let us denote the Laplace transform of $k(t)$ as $\hat{k}(z)$, $z \in \mathbb{C}$.

Lemma 8.7 ([11], Theorem 2–4.1) *Let* $k \in L^1(J)$ *be a convolution kernel. The resolvent* r *is in* $L^1(J)$ *if and only if* $\det(I + \hat{k}(z)) \neq 0$ *for all* $z \in \mathbb{C}$ *such that* $\mathrm{Re}[z] \geq 0$.

References

1. Agarwal, R.P., Berezansky, L., Braverman, E., Domoshnitsky, A.: Nonoscillation Theory of Functional Differential Equations with Applications. Springer, New York (2012)
2. Brunton, S.L., Noack, B.R.: Closed-loop turbulence control: progress and challenges. ASME Appl. Mech. Rev. **67**(5), 1–48 (2015)
3. Chovet, C., Feingesicht, M., Plumejeau, B., Delprat, S., Lippert, M., Keirsbulck, L., Polyakov, A., Richard, J.-P., Kerhervé, F.: Consumption, reducing car, by means of a closed-loop drag control. In: VEHICULAR: The 7th IARIA International Conference on Advances in Vehicular Systems, Technologies and Applications, p. 2018 (2018)
4. Debnath, L.: Nonlinear Partial Differential Equations for Scientists and Engineers, 3rd edn. Birkhäuser, Basel (2012)
5. Feingesicht, M.: Nonlinear active control of turbulent separated flows: theory and experiments. Ph.D. thesis, Centrale Lille, France (2017)
6. Feingesicht, M., Polyakov, A., Kerhervé, F., Richard, J.-P.: SISO model-based control of separated flows: sliding mode and optimal control approaches. Int. J. Robust Nonlinear Control **27**(18), 5008–5027 (2017)
7. Feingesicht, M., Raibaudo, C., Polyakov, A., Kerhervé, F., Richard, J.-P.: A bilinear input-output model with state-dependent delay for separated flow control. In: 2016 European Control Conference (ECC), pp. 1679–1684 (2016)
8. Filippov, A.F.: Differential Equations with Discontinuous Righthand Sides. Kluwer, Dordrecht (1988)
9. Fridman, E.: Introduction to Time-Delay Systems. Birkhäuser, Basel (2014)
10. Gomez, O., Orlov, Y., Kolmanovsky, I.V.: On-line identification of SISO linear time-invariant delay systems from output measurements. Automatica **43**(12), 2060–2069 (2007)
11. Gripenberg, G., Londen, S.-O., Staffans, O.: Volterra Integral and Functional Equations. Cambridge University Press, Cambridge (1990)
12. Györi, I., Ladas, G.: Oscillation Theory of Delay Differential Equations with Applications. Oxford University Press, New York (1991)
13. Hetel, L., Defoort, M., Djemaï, M.: Binary control design for a class of bilinear systems: application to a multilevel power converter. IEEE Trans. Control Syst. Technol. **24**(2), 719–726 (2016)
14. Kolmanovskii, V., Myshkis, A.: Introduction to the Theory and Applications of Functional Differential Equations. Springer, Dordrecht (1999)
15. McCallen, R., Browand, F., Ross, J. (eds.): The Aerodynamics of Heavy Vehicles: Trucks, Buses, and Trains. Springer, Berlin (2004)
16. Noack, B.R., Morzyński, M., Tadmor, G. (eds.): Reduced-Order Modelling for Flow Control. Springer, Vienna (2011)
17. Oliveira, T.R., Cunha, J.P.V.S., Battistel, A.: Global stability and simultaneous compensation of state and output delays for nonlinear systems via output-feedback sliding mode control. J. Control, Autom. Electr. Syst. **27**(6), 608–620 (2016)

18. Orlov, Y., Belkoura, L., Richard, J.P., Dambrine, M.: Adaptive identification of linear time-delay systems. Int. J. Robust Nonlinear Control **13**(9), 857–872 (2003)
19. Polyakov, A.: Minimization of disturbances effects in time delay predictor-based sliding mode control systems. J. Frankl. Inst. **349**(4), 1380–1396 (2012). Special Issue on Optimal Sliding Mode Algorithms for Dynamic Systems
20. Polyakov, A., Fridman, L.: Stability notions and Lyapunov functions for sliding mode control systems. J. Frankl. Inst. **351**(4), 1831–1865 (2014)
21. Richard, J.-P.: Time-delay systems: an overview of some recent advances and open problems. Automatica **39**(10), 1667–1694 (2003)
22. Royden, H.L.: Real Analysis, 3rd edn. Macmillan Publishing Company, New York (1988)
23. Sanchez, T., Polyakov, A., Richard, J.-P., Efimov, D.: A robust Sliding Mode Controller for a class of SISO bilinear delayed systems. In: 15th International Workshop on Variable Structure Systems (VSS), pp. 126–131 (2018)
24. Tsubakino, D., Krstic, M., Oliveira, T.R.: Exact predictor feedbacks for multi-input LTI systems with distinct input delays. Automatica **71**, 143–150 (2016)

Chapter 9
Compensation of Unmatched Disturbances via Sliding-Mode Control

A Comparison of Classical Results and Recent Methods Using Integral and Higher-Order Sliding-Mode

Kai Wulff, Tobias Posielek and Johann Reger

Abstract We study the disturbance rejection properties of recent and classical sliding-mode approaches with respect to unmatched disturbances. We consider nonlinear systems with arbitrary relative degree subject to various configurations of unmatched disturbances. The disturbances are state-dependent, time-varying, or a combination of both. For our analysis, we choose to transform the system into Byrnes–Isidori form and discuss the impact of the disturbances on the relative degree and stability properties. We investigate the capability of the considered sliding-mode approaches to compensate the disturbances on a given output of interest. In a comprehensive case study, we illustrate the characteristics of each approach on various system configurations. Finally, we isolate several mechanisms that are responsible for the disturbance compensation in each case.

9.1 Introduction

Sliding-mode control techniques are well known for their robustness properties with regard to model uncertainties and external disturbances. In general, this feature applies to so-called matched disturbances, roughly speaking, disturbances (or uncertainties, or perturbations) that enter the system via the same input space as

K. Wulff (✉) · J. Reger
Control Engineering Group, Technische Universität Ilmenau, P.O. Box 10 05 65, 98684 Ilmenau, Germany
e-mail: kai.wulff@tu-ilmenau.de

J. Reger
e-mail: johann.reger@tu-ilmenau.de

T. Posielek
Institute of System Dynamics and Control, German Aerospace Center (DLR), Münchner Str. 20, 82234 Wessling, Germany
e-mail: tobias.posielek@dlr.de

M. Steinberger et al. (eds.), *Variable-Structure Systems and Sliding-Mode Control*,
Studies in Systems, Decision and Control 271,
https://doi.org/10.1007/978-3-030-36621-6_9

the control signal. This type of disturbance is completely rejected once the system is in sliding mode. There have been several propositions to extend this property to unmatched (or mismatched) disturbances, see, e.g., [1–4] and many others.

Conventional integral sliding-mode control (ISMC) designs choose some nominal control (usually continuous type) and an additional sliding-mode action to compensate matched disturbances. For the sliding-mode controller, a projection matrix can be chosen that minimizes the effect of the matched disturbance in the reduced dynamics, see [2] with extensions in [5]. This defines the dynamics of the integral state as well as the sliding surface which yields the desired nominal (reduced) dynamics. A different way to choose the integral state is proposed in [6]. Here, the integral state is chosen according to the design objective and uses the projection matrix to shape the reduced dynamics. This results in reduced dynamics of lower order than in conventional designs. Furthermore, the design is extended by conditions that allow to decouple the desired output from a certain class of unmatched disturbances. The decoupling is based on a state transformation of the system such that the disturbances are acting on the internal dynamics only. Similar decoupling approaches have been investigated in [7, 8].

In [3], an integral higher-order sliding-mode (HOSM) technique [9] is proposed that utilizes the so-called hierarchical quasi-continuous controller design [10], with further extensions given in [11] using a backstepping approach.

Another way to deal with unmatched disturbances in sliding-mode control is to find a transformation which results in an integrator chain system which does not contain unmatched disturbances. However, this requires the usage of an observer to estimate the disturbances [12, 13] or the knowledge of the transformed states [14]. Observer approaches can also be utilized to estimate the disturbance for subsequent compensation, e.g., [4, 15, 16].

The goal of this contribution is to analyze various sliding-mode control methods toward their capability to compensate unmatched disturbances. In particular, we consider classical first-order sliding-mode techniques, integral sliding mode as well as recent higher-order sliding-mode approaches. The chosen control methods are considered suitable to represent classical as well as modern techniques proposed to compensate unmatched disturbances. However, we are not providing a complete overview of sliding-mode control methods proposed for this problem.

We assume the control goal is specified via a given fixed output and shall investigate which design technique is capable of compensating different configurations of disturbances. In this spirit, we are considering various system classes with regard to their relative degree and stability of the internal dynamics. We provide a systematic analysis on the disturbance compensation capabilities of the control approaches for each system class and disturbance configuration.

This contribution is organized as follows. Section 9.2 gives a problem definition as well as the general system formulation that shall be considered. Note that the system class is refined in later sections to match the applicability of the various control methods considered in this contribution. Furthermore, we formulate conditions on the unmatched disturbance such that important system properties like relative degree or stability are unchanged. In Sects. 9.3–9.5, we recall the considered control methodologies in view of the previously defined system characteristics. In Sect. 9.6, we investigate and highlight the various cases via a simple simulation example. In Sect. 9.7, we identify several mechanisms that lead to the compensation of the disturbance. We discuss via an example which of these mechanisms compensates which type of disturbance in each configuration.

9.2 Problem Definition

In this contribution, we shall consider nonlinear systems of the form given below. For certain considerations and approaches, this system class will be refined in specific sections. Let

$$\dot{x} = f(x) + g(x)u + \phi(x,t) \tag{9.1a}$$
$$y = h(x)\,, \tag{9.1b}$$

where $x(t) \in \mathbb{R}^n$ denotes the state, $u(t) \in \mathbb{R}$ is the control input, and $y(t) \in \mathbb{R}$ is the output of interest. The vector fields f and g are sufficiently smooth and of matching dimensions, where $f(0) = 0$ and $g(x) \neq 0$ for all $\in \mathbb{R}^n$. The output function h is uniformly continuous, and the state vector x is assumed to be known. The function ϕ is an unknown bounded disturbance.[1] We shall distinguish cases where ϕ is state- or time-dependent or both. If ϕ is a time-invariant function of the states, $\phi(x)$, we shall call ϕ a (model) uncertainty.

The disturbance can be divided into a matched and an unmatched disturbance, ϕ_{m} and ϕ_{u}, respectively, with

$$\phi_{\mathrm{m}}(x,t) = g(x)g^{+}(x)\phi(x,t) \tag{9.2a}$$
$$\phi_{\mathrm{u}}(x,t) = g^{\perp}(x)g^{\perp+}(x)\phi(x,t)\,, \tag{9.2b}$$

where $g^{\perp}(x)$ is full-rank left annihilator of $g(x)$, i.e., a matrix with independent columns that spans the null space of $g(x)$, i.e., $g^{\perp}(x)g^{\top}(x) = 0$ and $\mathrm{rank}(g^{\perp}) = n-1$. Moreover, $g^{+}(x)$ is the left pseudo-inverse of $g(x)$, i.e., $g^{+}(x) = (g^{\top}(x)g(x))^{-1}g^{\top}(x)$.

Consider the nominal case where $\phi \equiv 0$. Then the relative degree r of the output $y = h(x)$ with respect to the input u at the point $x \in \mathbb{R}^n$ is r if

$$L_g L_f^k h(x) = 0\,, \quad \text{for} \quad k \in \{0, \dots r-2\} \tag{9.3a}$$
$$L_g L_f^{r-1} h(x) \neq 0\,, \tag{9.3b}$$

[1]Further requirements on the boundedness of various derivatives of ϕ may apply for some approaches considered and will be pointed out in the respective section.

where L denotes the Lie derivative. Note that the relative degree is a local property. If not stated otherwise, we consider the relative degree at the origin $x = 0$.

In case ϕ is state-dependent, the relative degree r of the nominal case may be altered (see Appendix 1 for an elementary example). To avoid this, we require that the relative degree r is uniform with respect to ϕ, i.e.,

$$L_g L^k_{f+\phi} h(x) = 0\,, \quad \text{for} \quad k \in \{0, \dots r-2\} \tag{9.4a}$$

$$L_g L^{r-1}_{f+\phi} h(x) \neq 0\,, \tag{9.4b}$$

where r denotes the relative degree of the nominal system with $\phi(x,t) = 0$. Note that $L^k_{f+\phi}$ is a sum of all possible compositions of L_f and L_ϕ of length k. Certainly, condition (9.4a) is fulfilled if all these compositions are within the kernel of L_g, while condition (9.4b) is fulfilled if all compositions but the nominal one are within the kernel of L_g.

Both of these conditions, (9.4a) and (9.4b), may be expressed as

$$L_g\big((L_f L_\phi)^{(\alpha,\beta)} h(x)\big) = 0, \quad \text{for} \quad k \in \{0, \dots r-1\}\,, \tag{9.5}$$

where $\alpha, \beta \in \{0, 1\}^k$, $|(\alpha, \beta)| = k$, and $\beta \neq 0$. The composition $(L_f L_\phi)^{(\alpha,\beta)}$ denotes permutations of Lie derivatives of length k; for the precise definition of this notation, see Appendix 1. This condition is satisfied by large system classes, e.g., if the system is in strict feedback form [17], c.f. Sect. 9.5.1.

We base our analysis on the Byrnes–Isidori form where the state space is decomposed into external and internal states obtained by a suitable state transformation, see [18]. Let the output y of the system (9.1) have relative degree r and consider the nominal case, i.e., $\phi \equiv 0$. Then, the first r elements of the state transformation $\tau(x)$ are given by

$$\xi_1 := \tau_1(x) = h(x)\,, \tag{9.6a}$$

$$\xi_i := \tau_i(x) = L_f^{i-1} h(x)\,, \quad \text{for} \quad i \in \{2, \dots, r-1\}\,, \tag{9.6b}$$

$$\xi_r := \tau_r(x) = L_f^{r-1} h(x)\,. \tag{9.6c}$$

The remaining components

$$\eta := \big(\tau_{r+1}(x), \dots, \tau_n(x)\big)^\top \tag{9.7}$$

are chosen such that τ is a diffeomorphism and $L_g \tau_j(x) = 0$ for $j \in \{r+1, \dots, n\}$. The states ξ are called external states, while η are called internal with respect to the output y and input u.

The resulting internal dynamics

$$\dot{\eta} = q(\xi, \eta) \tag{9.8}$$

with

$$q_i(\xi, \eta) := L_f \tau_i(x)|_{x=\tau^{-1}(\xi,\eta)} \tag{9.9}$$

for $i \in \{1, \ldots, n-r\}$ are called bounded-input bounded-state (BIBS) stable if η is bounded for every bounded input ξ.

We shall distinguish between systems with BIBS stable and unstable internal dynamics. In any case, we assume that the disturbance ϕ does not impair these stability properties. This is certainly given if the internal dynamics are not influenced at all by the disturbance, i.e.,

$$L_\phi \tau_i(x) = 0\,, \quad \text{for} \quad i \in \{r+1, \ldots, n\}\,.$$

A less strict assumption is to postulate the stability of the disturbed internal dynamics with components

$$\dot{\eta}_j = L_f \tau_{r+j}(x)|_{x=\tau^{-1}(0,\eta)} + L_\phi \tau_{r+j}(x)|_{x=\tau^{-1}(0,\eta)} \tag{9.10}$$

with $j \in \{1, \ldots, n-r\}$. Note if ϕ is time-varying, then $L_\phi \tau_i(x) = \frac{\partial \tau(x)}{\partial x}\phi(x,t)$ is also a time-varying expression.

The control objective throughout this contribution is to guarantee the existence of an asymptotically stable equilibrium and to compensate the various classes of disturbances ϕ at the output y, i.e.,

$$\lim_{t\to\infty} h(x(t)) = 0\,. \tag{9.11}$$

9.3 First-Order Sliding-Mode Control

This section revisits the classical first-order sliding-mode control in the context of unmatched disturbances and a system with internal dynamics. We shall distinguish between the conventional approach where only the system states are available and some extended approach that makes also use of (numerical) differentiation.

The idea for both approaches is to split the dynamics into three components. The dynamics of ξ_1 represent the external dynamics which shall be driven to zero by the sliding-mode control law. The sliding motion introduces new internal dynamics, represented by the states ζ, that are rendered stable by the choice of the sliding manifold σ. The order of these dynamics is $r-1$ where r denotes the relative degree of the original output y with respect to u. The remaining $n-r$ internal states η correspond to the original internal dynamics of the system as this is not shaped by design and thus is required to be stable.

9.3.1 Conventional First-Order Sliding Mode (FOSM)

System Class

Consider the system (9.1). Let the relative degree of the output y with respect to the input u be r and the internal dynamics be BIBS stable with respect to the output y.

Control Law

Conventional first-order sliding mode defines a switching surface σ

$$\sigma(x) = \sum_{i=0}^{r-1} a_i L_f^i h(x) \tag{9.12}$$

with the coefficients a_i of a Hurwitz polynomial with $a_{r-1} = 1$. This switching surface has relative degree 1 with respect to the input u. The transformed input w is defined as

$$w = -\alpha \operatorname{sign}(\sigma) \tag{9.13}$$

for $\alpha > |\sum_{i=0}^{r-1} a_i L_\phi L_f^i h(x)|$ in order to enforce the sliding mode. Finally, the input transformation

$$u = \frac{-\sum_{i=0}^{r-1} a_i L_f^{i+1} h(x) + w}{L_g L_f^{r-1} h(x)} \tag{9.14}$$

yields the control law. Note, that often the transformation (9.14) is omitted if $L_g L_f^{r-1} h(x) > 0$ for all x, as $\frac{-\sum_{i=0}^{r-1} a_i L_f^{i+1} h(x)}{L_g L_f^{r-1} h(x)}$ can be considered as a part of the matched disturbance and thus alters only the requirements on the gain α.

Transformation into Byrnes–Isidori Form

Consider the state transformation τ as in (9.6) and (9.7), but now with respect to the output σ such that

$$\tau(x) = \left(\tau_1(x), \ldots, \tau_n(x)\right)^\top = \left(\xi_1, \zeta_1, \ldots \zeta_{r-1}, \eta_1 \ldots \eta_{n-r}\right)^\top, \tag{9.15}$$

where

$$\xi_1 = \sum_{i=0}^{r-1} a_i L_f^i h(x), \tag{9.16a}$$

$$\zeta_1 = h(x), \tag{9.16b}$$

$$\zeta_i = L_f^{i-1} h(x)\,, \qquad i = 2, \ldots, r-2 \tag{9.16c}$$

$$\zeta_{r-1} = L_f^{r-2} h(x)\,, \tag{9.16d}$$

and the internal states $\eta = \left(\tau_{r+1}(x), \ldots, \tau_n(x)\right)^\top$ with respect to y chosen as in Eq. (9.7). Denote ξ_1 as the external and ζ and η as the internal states with respect to the fictitious output $\sigma = \xi_1$ and input u. For the conventional input–output transformation of the system (9.1) with respect to the output y, the internal states would only consist of η, while the states defined as ζ would be part of the external states as well. So the choice of this fictitious output σ allows seemingly less control authority due to the higher dimension of the internal dynamics.

Note that (9.16a) is equivalent to

$$\xi_1 - \sum_{i=0}^{r-2} a_i L_f^i h(x) = L_f^{r-1} h(x)\,. \tag{9.17}$$

In the remainder of this section, we write x although the expression should be read for $x = \tau^{-1}(\xi_1, \zeta, \eta)$. Substituting (9.17) into (9.16) yields

$$\dot{\xi}_1 = \sum_{i=0}^{r-2} a_i L_f^{i+1} h(x) + L_f^r h(x) + L_g L_f^{r-1} h(x) u + \sum_{i=0}^{r-1} a_i L_\phi L_f^i h(x) \tag{9.18a}$$

$$\dot{\zeta}_1 = \zeta_2 + L_\phi h(x) \tag{9.18b}$$

$$\dot{\zeta}_i = \zeta_{i+1} + L_\phi L_f^{i-1} h(x) \tag{9.18c}$$

$$\dot{\zeta}_{r-1} = \xi_1 - \sum_{i=0}^{r-2} a_i \zeta_{i+1} + L_\phi L_f^{r-2} h(x)\,. \tag{9.18d}$$

With (9.12)–(9.14), and the remaining internal dynamics defined by

$$\dot{\eta}_j = L_f \tau_{r+j}(x)|_{x=\tau^{-1}(\xi_1,\zeta,\eta)} + L_\phi \tau_{r+j}(x)|_{x=\tau^{-1}(\xi_1,\zeta,\eta)} =: q_j(\xi_1, \zeta, \eta)\,, \tag{9.19}$$

for $j = 1, \ldots, n-r$, we obtain

$$\dot{\xi}_1 = -\alpha \operatorname{sign}(\xi_1) + \sum_{i=0}^{r-1} a_i L_\phi L_f^i h(x) \tag{9.20a}$$

$$\dot{\zeta}_1 = \zeta_2 + L_\phi h(x) \tag{9.20b}$$

$$\dot{\zeta}_i = \zeta_{i+1} + L_\phi L_f^{i-1} h(x) \tag{9.20c}$$

$$\dot{\zeta}_{r-1} = \xi_1 - \sum_{i=0}^{r-2} a_i \zeta_{i+1} + L_\phi L_f^{r-2} h(x) \tag{9.20d}$$

$$\dot{\eta} = q(\xi_1, \zeta, \eta) . \tag{9.20e}$$

Note that the sliding-mode controller is able to compensate any disturbance entering into Eq. (9.20a), whereas the unmatched disturbance entering (9.20b)–(9.20d) will not be suppressed at the output $h(x) = \zeta_1$.

The decoupling of ξ_1 from η may be impaired by the disturbance. However, the external dynamics can indeed be decoupled form the internal dynamics if the following disturbance decoupling condition holds:

$$\frac{\partial}{\partial \eta} L_\phi L_f^{i-1} h(x)|_{x=\tau^{-1}(\xi_1,\zeta,\eta)} = 0 \quad \text{for} \quad i \in \{1, \dots, r\} . \tag{9.21}$$

Remark 9.1 If (9.21) holds, then only the disturbance acting on η is decoupled from the output. However, for the compensation of these disturbances using a first-order sliding-mode control in (9.20a), we only require the decoupling condition (9.21) to hold for $i \in \{1, \dots, r-1\}$.

9.3.2 FOSM with Numerical Derivatives (FOSM-D)

System Class

Consider the system (9.1) with BIBS stable internal dynamics and nontrivial unmatched disturbance ϕ_u. Let ϕ and its derivatives be sufficiently smooth and bounded. By defining new states that include the disturbance, it is possible to formulate a system where all disturbances are rendered a matched disturbance. Thus, first-order sliding mode is able to compensate this resulting matched disturbance. However, this procedure requires the use of output derivatives to be calculated numerically, e.g., by a sliding-mode differentiator as in Appendix 2.

Control Law

Similar to the conventional case (9.12), the switching function is defined by the output. However, instead of the nominal derivatives the numerical derivatives including the disturbance are used

$$\sigma(x) = \sum_{i=0}^{r-1} a_i L_{f+\phi}^i h(x) = \sum_{i=0}^{r-1} a_i h^{(i)}(x) \tag{9.22}$$

with the coefficients a_i of a Hurwitz polynomial with $a_{r-1} = 1$. Note that σ in (9.22) includes the disturbance ϕ since it involves the Lie derivatives with respect to the disturbance ϕ. Using numerical differentiation allows to incorporate this, by calculating the time derivatives of h.

In order to enforce the sliding mode, define the transformed input w as

$$w = -\alpha \operatorname{sign}(\sigma) \tag{9.23}$$

for $\alpha > |L^r_{f+\phi}h(x) - L^r_f h(x)|$. Finally, the input transformation

$$u = \frac{-\sum_{i=0}^{r-2} a_i h^{(i+1)}(x) - L^r_f h(x) + w}{L_g L_f^{r-1} h(x)} \tag{9.24}$$

yields the control law.

Transformation into Byrnes–Isidori Form

The state transformation (9.15) yields

$$\xi_1 = \sum_{i=0}^{r-1} a_i h^{(i)}(x) \tag{9.25a}$$

$$\zeta_1 = h(x) \tag{9.25b}$$

$$\zeta_i = h^{(i-1)}(x) \tag{9.25c}$$

$$\zeta_{r-1} = h^{(r-2)}(x)\,, \tag{9.25d}$$

where ξ_1 denotes the external state and the internal states $\eta = \left(\tau_{r+1}(x), \ldots, \tau_n(x)\right)^\top$ are chosen as in Eq. (9.7). Note that (9.25a) is equivalent to

$$\xi_1 - \sum_{i=0}^{r-2} a_i h^{(i)}(x) = h^{(r-1)}(x)\,. \tag{9.26}$$

Differentiation of (9.25) and substituting (9.26) yields

$$\dot{\xi}_1 = \sum_{i=0}^{r-2} a_i h^{(i+1)}(x) + h^{(r)}(x) \tag{9.27a}$$

$$= \sum_{i=0}^{r-2} a_i h^{(i+1)}(x) + L^r_f h(x) + L_g L_f^{r-1} h(x) u + (L^r_{f+\phi}h(x) - L^r_f h(x)) \tag{9.27b}$$

$$\dot{\zeta}_1 = \zeta_2 \tag{9.27c}$$

$$\dot{\zeta}_i = \zeta_{i+1} \tag{9.27d}$$

$$\dot{\zeta}_{r-1} = \xi_1 - \sum_{i=0}^{r-2} a_i \zeta_{i+1} \,. \tag{9.27e}$$

Note that we have stated the external dynamics of ξ_1 for a time-invariant uncertainty $\phi(x)$ to allow for a compact notation of the resulting matched uncertainty. For time-varying disturbances $\phi(x, t)$ we obtain very similar results; however, the choice of α must account for time derivatives of $\phi(x, t)$ included in the resulting matched disturbance.

With (9.22)–(9.24) and the internal dynamics

$$\dot{\eta}_j = L_f \tau_{j+r}(x)|_{x=\tau^{-1}(\xi_1,\zeta,\eta)} + L_\phi \tau_{j+r}(x)|_{x=\tau^{-1}(\xi_1,\zeta,\eta)} =: q_j(\xi_1, \zeta, \eta) \tag{9.28}$$

for $j = 1, \ldots, n-r$, we obtain

$$\dot{\xi}_1 = -\alpha \,\mathrm{sign}(\xi_1) + \left(L^r_{f+\phi} h(x) - L^r_f h(x)\right) \tag{9.29a}$$

$$\dot{\zeta}_1 = \zeta_2 \tag{9.29b}$$

$$\dot{\zeta}_i = \zeta_{i+1} \tag{9.29c}$$

$$\dot{\zeta}_{r-1} = \xi_1 - \sum_{i=0}^{r-2} a_i \zeta_{i+1} \tag{9.29d}$$

$$\dot{\eta} = q(\xi_1, \zeta, \eta) \,. \tag{9.29e}$$

Observe that all uncertainty ϕ is now accumulated in Eq. (9.29a) and thus can be compensated choosing α sufficiently large.

Remark 9.2 Note that any diffeomorphism that works for η from Sect. 9.3.1 can be also applied here to obtain the derived decomposition. However, the dynamics in (9.29e) differ from (9.20e) since ξ_1 and ζ are differently defined in both sections.

Remark 9.3 Different smoothness conditions for the disturbance arise in this section. By using the proposed transformation, derivatives up to the rth order of ϕ are necessary. In Sect. 9.3.1, differentiability of ϕ was not needed for the transformation.

9.3.3 Stability and Disturbance Compensation

We compare the approaches presented in this section and discuss the requirements for stability and compensation of the unmatched disturbances.

We consider the stability requirements for both approaches for different disturbances. Note that it is necessary that the internal dynamics of the system (9.1) are BIBS stable. First, suppose that ϕ consists only of a matched disturbance (9.2b). Then

$$L_\phi L_f^{i-1} h(x) = L_g L_f^{i-1} h(x) g^+(x) \phi(x,t) = 0, \quad i = 0, \ldots, r-2$$

due to the condition for the relative degree (9.3a) and the assumption that the uncertainty does not change the relative degree. Then, the closed-loop dynamics of both approaches (9.20) and (9.29) have the same form. Asymptotic stability of the origin can be ensured due to the convergence of ξ to zero in finite time, thanks to the choice of α [19], the asymptotic stability of ζ for $\xi = 0$, because of the choice of the coefficients a_i, and the stable zero dynamics of $\dot{\eta} = q(0, 0, \eta)$ due to the BIBS stability of the system (9.1).

Suppose ϕ is unmatched but fulfills the disturbance decoupling condition (9.21). In both approaches, the internal dynamics are stable due to the assumption of bounded-input bounded-state stability. Thus, for FOSM, it is necessary and sufficient to show stability of the system (9.20a)–(9.20d) to prove stability of (9.20). For FOSM-D, however, stability of (9.29) is guaranteed by design with the same reasoning as for matched disturbances.

Suppose ϕ is not matched and does not fulfill the decoupling condition (9.21). Then for FOSM the complete stability of the system (9.20) must be analyzed as the η-dynamics are not decoupled anymore. For FOSM-D, the stability is again guaranteed by design. The coupling from the η-dynamics into the ξ-dynamics (9.29a) occurs via a matched disturbance, which is compensated by the sliding-mode controller. In the time-invariant case, this matched uncertainty is given by $L_{f+\phi}^r h(x) - L_f^r h(x)$. The stability of the internal dynamics (9.29e) is ensured due to the assumption of BIBS stability of the internal dynamics (9.8).

9.4 Integral Sliding-Mode Control (ISMC)

Integral sliding-mode control has been introduced by Utkin in [20] as a general concept by introducing an additional integrator state. This allows to initialize the system exactly on the sliding manifold and therefore eliminate the reaching phase. Robustness regarding matched disturbances is guaranteed throughout the entire system response. This property makes an integral sliding-mode controller (ISMC) a powerful choice for systems with matched external disturbances or model uncertainties.

A classical design approach for ISMC is to augment a stable control loop with an ISMC in order to suppress matched disturbances, e.g., [2, 5, 21]. In [2], a method is presented that ensures minimal interference of the sliding-mode controller with the originally designed control law. In [22], an adaptive ISMC is proposed to control the longitudinal rotation of a tiltrotor aircraft. The ISMC guarantees robustness to bounded matched disturbances such as sensor noise. An ISMC for a two-wheeled mobile robot is given in [23] to completely annihilate the influence of the joint friction acting on the system.

As the sliding motion is completely insensitive to matched disturbances, but not to unmatched ones, minimizing the effects of the latter is part of designing an ISMC. In this section, we recall a recently presented approach to integral sliding-mode control [6] that focuses on the desired output and also allows for the compensation of certain unmatched uncertainties.

9.4.1 System Class

In this section, we shall limit the system class slightly. Let the system (9.1) be in regular form [24]:

$$\dot{x}_1 = f_1(x_1, x_2) + \phi_1(x_1, x_2) \tag{9.30a}$$

$$\dot{x}_2 = f_2(x_1, x_2) + g_2(x_1, x_2)u + \phi_2(t, x_1, x_2), \tag{9.30b}$$

where $x_1(t) \in \mathbb{R}^{n-1}$ and $x_2(t) \in \mathbb{R}$ denote the state, and the control input is $u(t) \in \mathbb{R}$. The system is subject to the uncertainty $\phi_1 : \mathbb{R}^{n-1} \times \mathbb{R} \to \mathbb{R}^{n-1}$ and the disturbance $\phi_2 : \mathbb{R} \times \mathbb{R}^{n-1} \times \mathbb{R} \to \mathbb{R}$ satisfying some boundary conditions given in the respective sections. The unmatched uncertainty ϕ_1 is time-invariant state-dependent, while the matched disturbance ϕ_2 may be time-varying.

9.4.2 Control Law

As proposed in [6] we define the integrator state v as per

$$\dot{v} = h(x_1, x_2) \tag{9.31}$$

in view of (9.11). We choose the switching function

$$\sigma(x_1, x_2, v) = s(x_1, x_2) + v, \tag{9.32}$$

where s is to be selected such that $\frac{\partial}{\partial x_2} s(x_1, x_2) \neq 0$ for all $(x_1, x_2) \in \mathbb{R}^n$. The implicit function theorem ensures that there is a function $l : \mathbb{R}^n \to \mathbb{R}$ such that

$$\sigma(x_1, x_2, v) = 0 \Leftrightarrow x_2 = l(x_1, v). \tag{9.33}$$

In order to obtain a more compact notation, we define

$$S(x) := \left[S_1(x)^\top \; S_2(x)\right] := \left[\frac{\partial s(x_1,x_2)}{\partial x_1} \; \frac{\partial s(x_1,x_2)}{\partial x_2}\right],$$

where $S_1(x) \in \mathbb{R}^{n-1}$ and $S_2(x) \in \mathbb{R}$. Then the control law that yields $\dot{\sigma} = -\rho \operatorname{sign}(\sigma)$ for $\phi \equiv 0$ and $\rho > 0$ is given by

$$u(x) = \frac{-1}{S_2(x)g_2(x)}\big(S(x)f(x) + h(x) + \rho\,\text{sign}(\sigma)\big). \tag{9.34}$$

Choosing the initial state $v(0)$ according to

$$v(0) = -s(x_1(0), x_2(0)) \tag{9.35}$$

guarantees sliding mode starting from $t = 0$. The reduced dynamics including the uncertainties take the shape

$$\dot{x}_1 = f_1(x_1, l(x_1, v)) + \phi_1(x_1, x_2) \tag{9.36a}$$
$$\dot{v} = h(x_1, l(x_1, v))\,. \tag{9.36b}$$

9.4.3 Stability and Comparison to Conventional ISMC

In conventional integral sliding-mode control as proposed in [20] with several design propositions, e.g., [2, 5, 21], a sliding manifold and integral state is designed based on a conventionally designed nominal control u_0 where

$$\dot{v} = -S(x)f(x) - S_2(x)g_2(x)u_0(x)\,, \tag{9.37}$$
$$\sigma(x_1, x_2, v) = s(x_1, x_2) + v\,. \tag{9.38}$$

Then, the reduced dynamics are the same as in the nominal case but without the influence of the matched disturbance. Stability and unmatched uncertainty compensation is completely determined by the choice of the nominal control u_0. For the proposed ISMC, sufficient stability conditions are given in [6]. These assume among others the existence of linear growth bounds on f, h, and ϕ.

We can cast our approach into the conventional scheme by identifying the continuous nominal control u_0 and the discontinuous control u_1 such that $u = u_0 + u_1$ in terms of

$$u_0(x) = \frac{-1}{S_2(x)g_2(x)}\big(S(x)f(x) + h(x)\big)\,,$$
$$u_1(x) = \frac{-1}{S_2(x)g_2(x)}\rho\,\text{sign}(\sigma)\,.$$

An overview of the two different design methods for an integral sliding-mode controller is given in Table 9.1.

In the conventional design, all effort is put into the choice of the nominal control u_0 to achieve the control objective (9.11). In particular, the integrator state is determined by the choice of u_0 and is not chosen directly to guarantee the control objective. Furthermore, the choice of s in the switching function does not influence the reduced dynamics in (x_1, x_2). In the proposed method, the integrator state is chosen directly

Table 9.1 Comparison of ISMC design methods for system (9.30). ©2018 IEEE. Reprinted, with permission, from T. Posielek, K. Wulff, and J. Reger. "Disturbance decoupling using a novel approach to integral sliding-mode". In: Proceedings of the IEEE International Workshop on Variable Structure Systems, pages 420–426, 2018

	Conventional ISMC design method	Proposed ISMC design method in [6]
Design parameter	Nominal control u_0 Switching function via s	Dynamics of the integrator state h Switching function via s
Switching function	$\sigma(x_1, x_2, v) = s(x_1, x_2) + v$ $\sigma(x_1, x_2, v) = 0 \Leftrightarrow x_2 = l(x_1, v)$	$\sigma(x_1, x_2, v) = s(x_1, x_2) + v$ $\sigma(x_1, x_2, v) = 0 \Leftrightarrow x_2 = l(x_1, v)$
Control law	$u = u_0 + u_1$ $u_1 = -[S_2(x)g_2(x)]^{-1}\rho\,\text{sign}(\sigma)$	$u = u_0 + u_1$ $u_0 = -[S_2(x)g_2(x)]^{-1}(S(x)f(x) + h(x))$ $u_1 = -[S_2(x)g_2(x)]^{-1}\rho\,\text{sign}(\sigma)$
Integrator state	$\dot{v} = -S(x)f(x) - S_2(x)g_2(x)u_0(x)$	$\dot{v} = h(x_1, x_2)$
Red. dyn. in (x_1, x_2)	$\dot{x}_1 = f_1(x) + \phi_1(x)$ $\dot{x}_2 = f_2(x) + g_2(x)u_0(x)$ $-[S_2(x)]^{-1}S_1(x)\phi_1(x)$	$\dot{x}_1 = f_1(x) + \phi_1(x)$ $\dot{x}_2 = -[S_2(x)]^{-1}\big(S_1(x)f_1(x) + h(x)\big)$ $-[S_2(x)]^{-1}S_1(x)\phi_1(x)$
Red. dyn. in (x_1, v), $(w = (x_1, l(x_1, v)))$	$\dot{x}_1 = f_1(x_1, l(x_1, v)) + \phi_1(x_1, l(x_1, v))$ $\dot{v} = -S(w)f(w) - S_2(w)g_2(w)u_0(w)$	$\dot{x}_1 = f_1(x_1, l(x_1, v)) + \phi_1(x_1, l(x_1, v))$ $\dot{v} = h(x_1, l(x_1, v))$
Condition on ρ	$\rho > \lVert S_1\phi_1 + S_2\phi_2 \rVert$	$\rho > \lVert S_1\phi_1 + S_2\phi_2 \rVert$

and the choice of s influences the reduced dynamics significantly and, thus, allows to shape the desired dynamics.

Furthermore, the order of the reduced dynamics in the conventional design is n plus the order needed for the nominal controller. Our approach results in reduced dynamics of order n. Thus, assuming a nominal control with integral action for the conventional design, our proposed approach will always yield reduced dynamics of lower order. This may lead to a simpler stability analysis and less restrictive requirements.

Note also, since our proposed method can be cast into the standard framework, results obtained for integral sliding-mode control also extend to our method. For example, the choice of $S(x)^\top = \begin{bmatrix} 0 & g_2(x) \end{bmatrix}$ minimizes the effect of an unmatched uncertainty in the sense of [5].

9.4.4 Decoupling of Unmatched Uncertainties

For unmatched uncertainties of a certain structure, the proposed ISMC in [6] allows the decoupling of the output from the uncertainty. Let (9.30) be structured as follows. Let $x_1(t)$ be split into $x_{11}(t) \in \mathbb{R}^{n-2}$ and $x_{12}(t) \in \mathbb{R}$ such that system (9.30) reads

$$\dot{x}_{11} = f_{11}(x_{11}, x_{12}, x_2) + \phi_{11}(x_{11}) \tag{9.39a}$$

$$\dot{x}_{12} = f_{12}(x_{11}, x_{12}, x_2) + \phi_{12}(x_{11}, x_{12}, x_2) \tag{9.39b}$$

$$\dot{x}_2 = f_2(x_1, x_2) + g_2(x_1, x_2)u + \phi_2(t, x_1, x_2) \tag{9.39c}$$

$$y = h(x_{11}). \tag{9.39d}$$

Theorem 9.1 ([6]) *Consider system (9.39) augmented by (9.31) with the integral sliding-mode control law (9.34) and initialization (9.35). Let the resulting system be locally asymptotically stable. Choose the integrator state (9.31) such that*

$$\dot{v} = h(x_{11}) \tag{9.40}$$

and the switching function (9.32) such that the first component of the reduced dynamics (9.36) is independent of x_{12}, i.e.,

$$f_{11}^{\mathrm{d}}(x_{11}, v) := f_{11}(x_{11}, x_{12}, l(x_{11}, x_{12}, v)) \,. \tag{9.41}$$

Then the effects of the matched uncertainty ϕ_2, and the unmatched uncertainty ϕ_{12} are compensated completely on h for all $t \geq 0$ in the closed-loop system.

Remark 9.4 If the reduced dynamics on the sliding manifold are BIBS stable for the input ϕ_{12} and state x_{12}, then ϕ_{12} may also be a time-varying disturbance without altering the achieved results.

Remark 9.5 The decoupling conditions (9.40), (9.41) may be achieved using other (nonintegral) sliding-mode techniques as shown in the numerical examples in [6].

Remark 9.6 Note that the approach taken in this section differs from the decomposition into external and internal dynamics as chosen in Sect. 9.3.1. Therefore, the relation of the decoupling conditions (9.41) and (9.21) is not drawn in a straightforward manner.

Remark 9.7 This ISMC approach may be augmented by the use of differentiators as in Sect. 9.3.2. Then the sliding manifold (9.32) may depend on derivatives of v up to the order r. This allows for the compensation of a larger class of disturbances as demonstrated and discussed in the Sects. 9.6 and 9.7.

9.5 Higher-Order Sliding-Mode Control (HOSM)

In this section, we discuss higher-order sliding-mode methods to compensate unmatched disturbances. We focus on the work presented in [3, 4, 25], where the main ideas for the compensation of unmatched disturbances are introduced. Both of these approaches use sliding-mode differentiators [26] and quasi-continuous higher order sliding mode (QC-HOSM) controllers [9], see Appendix 2 and 3 for a brief summary, respectively. The QC-HOSM controllers are chosen because, even when they are theoretically discontinuous, in the presence of switching delays, measurement noise, and singular perturbations, they produce a continuous control and produce less chattering than other HOSM controllers, c.f. [9].

9.5.1 Nested Quasi-continuous HOSM Control (nHOSM)

System Class

In [3], the system (9.1) is assumed to be in the following nonlinear block controllable form:

$$\dot{x}_1 = f_1(x_1) + g_1(x_1)x_2 + \phi_1(x_1, t) \tag{9.42a}$$
$$\dot{x}_i = f_i(\bar{x}_i) + g_i(\bar{x}_i)x_{i+1} + \phi_i(\bar{x}_i, t), \quad i = 2, \dots n-1 \tag{9.42b}$$
$$\dot{x}_n = f_n(x) + g_n(x)u + \phi_n(x, t) \tag{9.42c}$$
$$y = x_1 \tag{9.42d}$$

with $x(t) \in \mathbb{R}^n, x_i(t) \in \mathbb{R}$ and $\bar{x}_i = (x_1, \dots x_i)^\top, \bar{f}_i = (f_1, \dots f_i)^\top, \bar{g}_i = (g_1, \dots g_i)^\top$ and $u(t) \in \mathbb{R}$. The disturbances ϕ_i are bounded and unknown. Note that the original work [3] also allows for time-varying f_i, g_i. This is omitted here to be consistent with the considered system class (9.1).

The structure of this system class implies that the relative degree of y with respect to u is $r = n$. Thus, there are no internal dynamics. Note that the disturbances fulfill condition (9.5) and thus do not influence the relative degree.

Control Law

The design idea is to define virtual control laws w_i to control x_i to w_{i+1}. For $i = 1$ define

$$\sigma_1 = x_1 \tag{9.43a}$$
$$w_1(x_1, u_{1,1}) = g_1^{-1}(x_1)(-f_1(x_1) + u_{1,1}) \tag{9.43b}$$
$$\dot{u}_{1,1} = u_{1,2} \tag{9.43c}$$
$$\vdots$$
$$\dot{u}_{1,n-1} = -\alpha_1 \Psi_{n-1,n}(\sigma_1, \dot{\sigma}_1, \dots, \sigma_1^{(n-1)}), \tag{9.43d}$$

with Ψ defined as in Appendix 3. For $i = 2, \dots n-1$ define

$$\sigma_i = x_i - w_{i-1}(\bar{x}_i, u_{i,1}) \tag{9.44a}$$
$$w_i(\bar{x}_i, u_{i,1}) = g_i^{-1}(\bar{x}_i)(-f_i(\bar{x}_i) + u_{i,1}) \tag{9.44b}$$
$$\dot{u}_{i,1} = u_{i,2} \tag{9.44c}$$
$$\vdots$$
$$\dot{u}_{i,n-i} = -\alpha_i \Psi_{n-i,n-i+1}(\sigma_i, \dot{\sigma}_i, \dots, \sigma_i^{(n-i)}) \tag{9.44d}$$

and finally for $i = n$

$$\sigma_n = x_n - w_{n-1}(\bar{x}_{n-1}, u_{n-1,1}) \tag{9.45a}$$
$$u(x, u_{n,1}) = g_n^{-1}(x)(-f_n(x) + u_{n,1}) \tag{9.45b}$$
$$u_{n,1} = -\alpha_n \operatorname{sign}(\sigma_n) \tag{9.45c}$$

with $\alpha_n \geq |\phi_n^{\text{sup}}| + |\dot{w}_{n-1}(x)|$.

Remark 9.8 For the sake of compactness, we write $\dot{w}_{n-1}$ instead of

$$\dot{w}_{n-1}(x) = L_{\bar{f}_{n-1}} \bar{g}_{n-1}^{-1}(\bar{x}_{n-1})\big(-L_{\bar{f}_{n-1}} f_{n-1} - \alpha_{n-1}\Psi_{1,2}(\sigma_{n-1}, \dot{\sigma}_{n-1})\big).$$

Conditions for the remaining α_i are functions dependent on $w_{n-i}^{(i)}$ which will lead to even more complex expressions, see also Remark 9.12.

Remark 9.9 This method assumes that the exact derivatives of the sliding variable including the acting disturbances are available. If only the nominal derivatives are used, the method loses its ability to compensate unmatched disturbances.

Remark 9.10 Instead of using (9.45c), we may introduce $w = -\alpha_n \operatorname{sign}(\sigma_n)$ and use the first-order sliding-mode input transformation

$$u_{n,1} = L_{\bar{f}_{n-1}} \bar{g}_{n-1}^{-1}(\bar{x}_{n-1})\big(-L_{\bar{f}_{n-1}} f_{n-1} - \alpha_{n-1}\Psi_{1,2}(\sigma_{n-1}, \dot{\sigma}_{n-1})\big)$$

to compensate a part of the nominal dynamics. Otherwise, this part is treated as a matched disturbance.

9.5.2 *Nested Backward Compensation of Unmatched Disturbances via HOSM Observation (nFOSM)*

System Class

In [4, 15], the system (9.1) is considered to be linear and in block controllable form

$$\dot{x}_1 = A_1 x_1 + B_1(x_2 + \Gamma_1 \nu) \tag{9.46a}$$
$$\dot{x}_i = A_i \bar{x}_i + B_i(x_{i+1} + \Gamma_i \nu), \quad \text{for} \quad i = 2, \ldots, k-1 \tag{9.46b}$$
$$\dot{x}_k = A_k \bar{x}_k + B_r(u + \phi_n) \tag{9.46c}$$
$$y = x_1 \tag{9.46d}$$

with $x_i(t) \in \mathbb{R}^{n_i}$ and $\text{rk}(B_i) = \min\{n_i, n_{i+1}\}$ for $i = 1, \ldots k < n$ and $\sum_{i=0}^{k} n_i = n$. The unmatched disturbances $\phi_{\text{u},i} := B_i \Gamma_i \nu$ are parametrized by $\nu(t) \in \mathbb{R}^q$ where $q \leq n-1$. The last component ϕ_n is matched. Note that the relative degree may be less than n.

The control approach in [4, 15] uses an higher-order sliding-mode observer based on [27], in order to estimate the states as well as the disturbance ν. As a consequence, the states and the disturbances are assumed to be exactly known for the control.

Control Law

The control is designed in a nested fashion. For $i = 1$ define

$$\sigma_1 = x_1 \tag{9.47a}$$

$$w_1 = -\Gamma_1 \nu - B_1^{+}(A_1 x_1 - \hat{A}_1 x_1)\,, \tag{9.47b}$$

where the matrix $\hat{A}$ is Hurwitz. Further, for $i \in \{2, \dots, k-1\}$ define

$$\sigma_i = x_i - w_{i-1} \tag{9.48a}$$

$$w_i = -\Gamma_i \nu - B_i^{+}(A_i \bar{x}_i - \hat{A}_i(x_i - w_{i-1})) + X_{i-1}(x_{i-1} - w_{i-1}) - \dot{w}_{i-1} \tag{9.48b}$$

with $X_{i-1} = P_i^{-1} B_i^{\top} P_{i-1}$ and P_i such that $P_i \hat{A}_i + \hat{A}_i^{\top} P_i = -I$. Finally, for $i = k$ we use the control law

$$\sigma_k = x_k - w_{k-1} \tag{9.49a}$$

$$u = -B_k^{+}(A_k \bar{x}_k + \alpha \operatorname{sign} \sigma_k) + B_k^{+} \dot{w}_{k-1} - \phi_n\,. \tag{9.49b}$$

Remark 9.11 In [4], the control law is proposed without the final two terms in (9.49b). This is due to the fact that they can be considered as a matched disturbance and can thus be incorporated in the choice of a suitably large α. In [15], the first-order sliding-mode approach for the controller is replaced by a super-twisting type.

Remark 9.12 The expression of w_i in (9.48b) involves the derivative of the preceding w_{i-1} which can be obtained by differentiation of (9.48b) with shifted indices. Iteration yields w_2 which contains $\dot{w}_1$ given by the derivative of (9.47b). Thus w_{k-1} in (9.49a) needs $k-1$ iterations of this procedure and $\dot{w}_{k-1}$ in (9.49b) contains the derivative of order $k-1$ with respect to w_1. The resulting expression for w_{k-1} is rather involved and shall be omitted here.

Remark 9.13 Instead of using (9.49b) and similar to the nHOSM approach (c.f. Remark 9.10), we may introduce $w = -\alpha_n \operatorname{sign}(\sigma_n)$ and use the first-order sliding-mode input transformation

$$u_{n,1} = \dot{w}_{k-1}$$

to compensate a part of the nominal dynamics instead of compensating it as a matched disturbance.

9.5.3 Stability and Disturbance Compensation

For the nHOSM, stability is guaranteed by designing n sliding manifolds with virtual controls which are driven iteratively to zero. In each sliding manifold, the design of a higher-order sliding-mode controller is proposed by having $n - i$ additional integrators.

With nFOSM, stability is guaranteed in a similar manner. Again, n sliding manifolds and virtual controls are designed. However, no additional integrators are introduced and the convergence is ensured by a state feedback instead of a higher-order sliding-mode control. Both approaches guarantee stability subject to sufficiently smooth matched and unmatched disturbances.

Compensation of the unmatched disturbances for nHOSM is achieved using numerical derivatives. For the higher-order sliding-mode control laws (9.43d) and (9.44d), the numerical derivatives of the sliding manifolds (9.43a), (9.44a) are needed. Using these numerical derivatives for a transformation in a similar fashion as in (9.22) leads to a system similar to (9.29) with only matched disturbances which are compensated by the sliding-mode control.

Disturbance compensation in the nFOSM approach is achieved by estimating the disturbances in finite time. These estimations are obtained using a higher-order sliding-mode observer exploiting numerical differentiation as well. Then the estimations are incorporated into the control law (9.47b), (9.48b), (9.49b) to compensate the disturbance.

Note that both approaches, nHOSM and nFOSM, can be augmented to tackle systems with internal dynamics under the same conditions as for the FOSM-D approach in Sect. 9.3.2.

9.6 Case Studies

In this section, we investigate the capabilities of the various sliding-mode techniques toward the compensation of unmatched disturbances via some examples. In particular, we shall distinguish state-dependent uncertainties and time-varying disturbances. We consider example systems with stable and unstable or even no internal dynamics. All cases are constructed from a base system of third order that exhibits different relative degrees and internal dynamics depending on the chosen set of parameters:

$$\dot{x}_1 = -ax_1 + bx_2 + (1-b)x_3 + \phi_1(x_1, t) \tag{9.50a}$$

$$\dot{x}_2 = ax_2 + x_3 + \phi_2(x_1, x_2, t) \tag{9.50b}$$

$$\dot{x}_3 = u \tag{9.50c}$$

$$y = x_1 \, . \tag{9.50d}$$

With the parameters $a \in \{-1, 0, 1\}$ and $b \in \{0, 1\}$, we can describe three different system characteristics: for $a = 0$, $b = 1$ the system has relative degree 3 and no internal dynamics, for $a = -1, b = 0$ we have relative degree 2 and stable internal dynamics, and $a = 1, b = 0$ yields relative degree 2 and unstable internal dynamics.

Note that the disturbances ϕ_1 and ϕ_2 are unmatched and enter the system in lower triangular form, to ensure that they do not alter the relative degree. Since we are focussing on the effect of the unmatched disturbances, we ignore matched disturbances as they are fully compensated by all sliding-mode controllers. It is assumed that all disturbances are bounded with respect to time and state for global stability.

Note that the resulting systems for different parameter configurations can also be obtained by state transformation via a diffeomorphism yielding the output $y = h(x)$.

9.6.1 Transformations and Resulting Internal Dynamics

For the implementation of the control concepts, we use different state and input transformations. In the following, we analyze the system (9.50) for the case that nontrivial internal dynamics exist, i.e., the case $b = 0$:

$$\dot{x}_1 = -ax_1 + x_3 + \phi_1(x_1, t) \tag{9.51a}$$

$$\dot{x}_2 = ax_2 + x_3 + \phi_2(x_1, x_2, t) \tag{9.51b}$$

$$\dot{x}_3 = u \tag{9.51c}$$

$$y = x_1 \,. \tag{9.51d}$$

The state transformation (9.6), (9.7) with respect to y defines the states

$$\xi_1 = x_1 \tag{9.52a}$$

$$\xi_2 = -ax_1 + x_3 \tag{9.52b}$$

$$\eta_1 = x_2 \,. \tag{9.52c}$$

With the input transformation

$$u = -a^2 x_1 + ax_3 + w, \tag{9.53}$$

we obtain

$$\dot{\xi}_1 = \xi_2 + \phi_1(\xi_1, t) \tag{9.54a}$$

$$\dot{\xi}_2 = w - a\phi_1(\xi_1, t) \tag{9.54b}$$

$$\dot{\eta}_1 = a\eta_1 + a\xi_1 + \xi_2 + \phi_2(\xi_1, \eta_1, t) \,. \tag{9.54c}$$

Note that for $a = 1$ the internal dynamics η_1 are stable, whereas for $a = -1$ they are unstable.

The unmatched disturbance ϕ_2 has no effect on the external state ξ_1. The unmatched disturbance ϕ_1 is rendered matched in (9.54b) but remains unmatched in (9.54a).

If we allow for differentiation of x_1 (including disturbance), we can choose

$$\xi_1 = x_1 \tag{9.55a}$$
$$\xi_2 = \dot{x}_1 = -ax_1 + x_3 + \phi_1(x_1, t) \tag{9.55b}$$
$$\eta_1 = x_2 \,. \tag{9.55c}$$

With the input transformation (9.53), we obtain the dynamics

$$\dot{\xi}_1 = \xi_2 \tag{9.56a}$$
$$\dot{\xi}_2 = w - a\phi_1(\xi_1, t) + \dot{\phi}_1(\xi_1, t) \tag{9.56b}$$
$$\dot{\eta}_1 = a\eta_1 + a\xi_1 + \xi_2 - \phi_1(\xi_1, t) + \phi_2(\xi_1, \eta_1, t) \,. \tag{9.56c}$$

None of the original unmatched disturbances appears in (9.56a). The disturbance in (9.56b) is matched by the input w, and both unmatched disturbances ϕ_1 and ϕ_2 affect the internal dynamics (9.56c). In case of stable internal dynamics, i.e., $a < 0$, a sliding-mode control law depending on ξ_1 and ξ_2 only will compensate these matched disturbances. Note that the new matched disturbance $a\phi_1(\xi_1, t) + \dot{\phi}_1(\xi_1, t)$ involves the derivative of ϕ_1, which is required to be bounded. This has to be taken into account when choosing the gain of the sliding-mode controller.

In case of unstable internal dynamics, the control law for w has to depend on η_1 in order to stabilize the internal dynamics. Then, however, η_1 enters (9.56b). As a consequence, η_1 are no internal dynamics for the closed-loop system anymore and ϕ_1, ϕ_2 in (9.56c) are not rendered matched. Thus, these disturbances cannot be compensated by the sliding-mode techniques without integrator considered in this paper.

For the system with no internal dynamics, i.e., $r = 3$, we obtain similar findings without internal state η_1 of course. All controllers designed can be found in Appendix 4.

9.6.2 Simulation Results

We shall consider three cases of disturbance configurations. In the first case, the unmatched uncertainties ϕ_1, ϕ_2 are assumed to be time-invariant, but state-dependent. In the second case, the unmatched uncertainty ϕ_1 is time-invariant, while ϕ_2 may be a time-varying disturbance. In the third case, all disturbances are time-varying.

Table 9.2 summarizes the resulting nine cases and indicates which of the considered controllers are able to compensate the respective uncertainties and disturbances. Essentially, first-order sliding mode with differentiator (FOSM-D), nested higher-order sliding mode (nHOSM), and nested first-order sliding mode (nFOSM) are capable to compensate all disturbance configurations for stable or non-existing internal dynamics. Integral sliding mode (ISMC) can also compensate time-invariant

Table 9.2 Comparison of the ability of the considered SMC methods to compensate unmatched disturbances for various system classes

Class of unmatched uncertainties	Internal dynamics		
	Non-existing	Stable	Unstable
$\phi_1(x_1), \phi_2(x_1, x_2)$	FOSM-D, ISMC, nHOSM, nFOSM	FOSM-D, ISMC nHOSM, nFOSM	ISMC
$\phi_1(x_1), \phi_2(x_1, x_2, t)$	FOSM-D, ISMC-D nHOSM, nFOSM	FOSM-D, ISMC nHOSM, nFOSM	Non
$\phi_1(x_1, t), \phi_2(x_1, x_2, t)$	FOSM-D, ISMC-D nHOSM, nFOSM	FOSM-D, ISMC-D nHOSM, nFOSM	Non

unmatched uncertainties for unstable internal dynamics. Note, however, that ISMC needs the additional use of a differentiator (ISMC-D) for three lower left cases of time-varying disturbances.

We shall now present the simulation results for three cases: To the systems with no and stable internal dynamics, i.e., $b = 1$ and $a = -1, b = 0$, respectively, we apply the disturbances

$$\phi_1(x_1) = 0.2 + 0.05x_1^3, \quad \text{and} \quad \phi_2(t) = -0.3\sin\left(\tfrac{2\pi}{8}t\right).$$

As ϕ_1 is not bounded in state, stability is only guaranteed locally in this case. The resulting systems and all controllers can be found in Appendix 4.

Figure 9.1 presents the simulation results. The plots on the left show the system states for the system with no internal dynamics; the system with stable internal dynamics is depicted on the right.

For the system without internal dynamics the approaches using numerical derivatives, namely, FOSM-D (blue line), nHOSM (yellow), and nFOSM (purple), behave very similar. All three compensate both unmatched disturbances ϕ_1 and ϕ_2 on the state of interest x_1. The state x_2 is only disturbed by ϕ_1, while on x_3 the effect of $\dot{\phi}_1$ and ϕ_2 can be seen. The ISMC (red) does not use the numerical but the nominal derivatives of h. Thus, due to its dynamic behavior, the disturbance ϕ_2 cannot be completely compensated on x_1. However, due to the integrator the effect of ϕ_1 on x_1 is compensated for $t \to \infty$. The dynamics of x_2 and x_3 are influenced by ϕ_1 as well as ϕ_2.

For the system with stable internal dynamics (right column), it can be seen that all four proposed approaches are able to compensate the unmatched disturbances on x_1. The approaches FOSM-D (blue line), nHOSM (yellow), and nFOSM (purple) compensate ϕ_1 due to the numerical derivatives, while the ISMC approach uses the integrator. All approaches decouple the states x_1 and x_3 of the disturbance ϕ_2. The state x_2 is affected by ϕ_1 as well as ϕ_2 while x_3 is only disturbed by ϕ_1.

If we add the differentiator used in the other approaches to the ISMC scheme, the time-varying uncertainties can be compensated in any of the two cases. Figure 9.2 shows the closed-loop system with no internal dynamics (9.69) with ISMC-D control law using numerical derivatives (9.22)–(9.24) subject to the triangular and sinusoidal

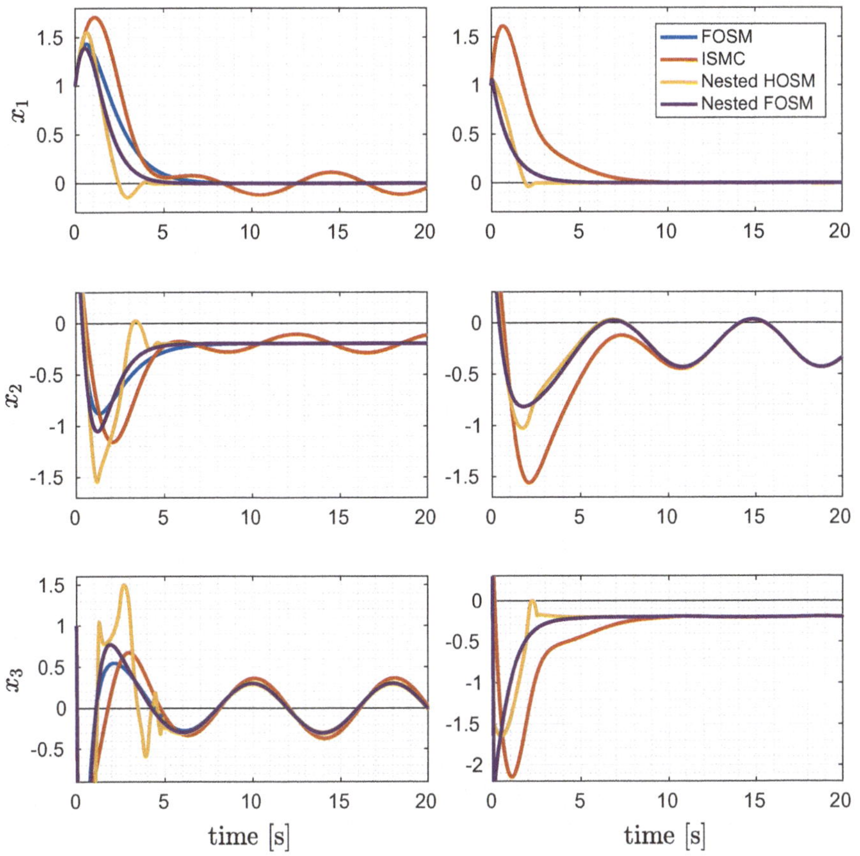

Fig. 9.1 Closed-loop state evolution of system (9.50) with no internal dynamics, i.e., $b = 1$ (left) and stable internal dynamics, i.e., $a = -1, b = 0$ (right)

disturbances

$$\phi_1(t) = \begin{cases} \frac{4}{3}t - 4k - 1, & t \in \left[3k, 3k + \frac{3}{2}\right], \\ -\frac{4}{3}t + 4k + 3, & t \in \left[3k + \frac{3}{2}, 3(k+1)\right], \quad k = 0, 1, 2, \ldots \end{cases}$$
$$\phi_2(t) = -0.3 \sin\left(\frac{2\pi}{8}t\right) .$$

Without loss of generality, we choose these disturbances time-dependent only, in order to illustrate their respective effects on each state. It can be seen that the ISMC-D control law compensates the unmatched disturbances on x_1. The effect of ϕ_1 is modulated on x_2, while x_3 is disturbed by $\dot{\phi}_1 + \phi_2$. The compensation on x_1 is achieved through the usage of the numerical derivatives of x_1. The integrator does not achieve any extra robustness in this scenario. Note that FOSM-D, nHOSM,

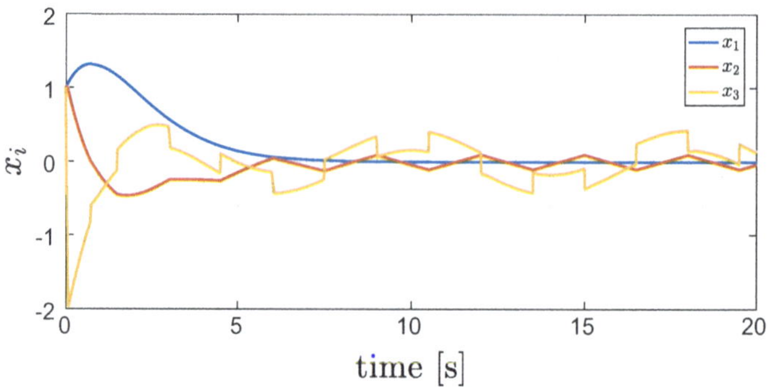

Fig. 9.2 Closed-loop state evolution of system (9.69) with no internal dynamics and ISMC-D control law (9.22)–(9.24) under time-varying disturbances

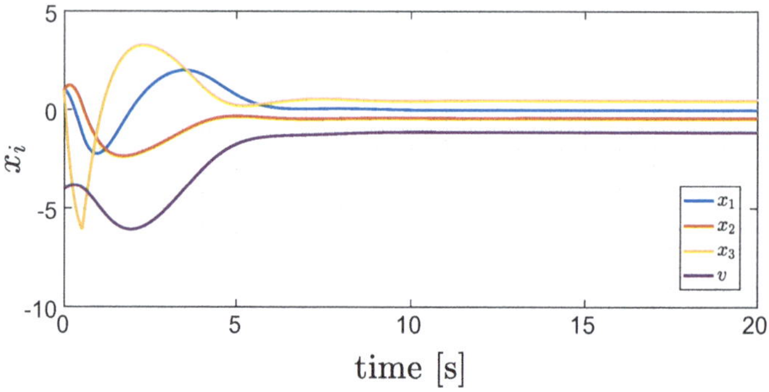

Fig. 9.3 Closed-loop state evolution of the system (9.76) with unstable internal dynamics and ISCM (9.78) subject to state-dependent uncertainties

and nFOSM achieve the same compensation in this scenario due to the usage of the numerical derivatives.

The state-dependent uncertainties on the system with unstable internal dynamics can only be compensated by ISMC. In Fig. 9.3, the closed-loop system (9.76) with ISMC (9.78) and uncertainties

$$\phi_1(x_1) = -0.5 + 0.05x_1^3 \quad \text{and} \quad \phi_2(x_1, x_2) = 0.3 + 0.1x_1^2 + \sin(x_2)$$

is shown. It can be seen that both uncertainties are compensated on x_1 for $t \to \infty$. The other states x_2, x_3, and v are perturbed by the uncertainties. The compensation of the uncertainties is due to the introduction of the additional integrator forcing an equilibrium with $x_1 = 0$.

9.7 Discussion and Concluding Remarks

The previous analysis of the various controller designs has shown that unmatched disturbances may be compensated for different classes of systems and disturbances. All controller approaches make a different use of the system structure and use various methods to compensate the disturbances. However, we can identify essentially three mechanisms that are responsible for the compensation: integration, differentiation, and decoupling.

To illustrate this, we consider the simplified system with stable internal dynamics:

$$\begin{aligned}\dot{x}_1 &= x_3 + \phi_1(x_1, x_2)\\ \dot{x}_2 &= -x_2 + \phi_2(x_1, x_2, t)\\ \dot{x}_3 &= u\,.\end{aligned}$$

Note that in contrast to the example of the previous section the uncertainty ϕ_1 may depend on x_2. However, this does not alter the relative degree of y with respect to u. For brevity of exposition, we consider ϕ_2 to be a time-dependent disturbance and ϕ_1 an uncertainty. However, it is not hard to generalize these findings to static ϕ_2 or time-dependent ϕ_1.

Choosing the ISMC approach, we introduce the integrator v with $\dot{v} = x_1$. Using the transformation $\xi_1 = v$, $\xi_2 = x_1$, $\xi_3 = x_3$ and $\eta = x_2$, leads to the system

$$\begin{aligned}\dot{\xi}_1 &= \xi_2\\ \dot{\xi}_2 &= \xi_3 + \phi_1(\xi_1, \eta)\\ \dot{\xi}_3 &= u\\ \dot{\eta} &= -\eta + \phi_2(x_1, x_2, t)\,.\end{aligned}$$

Suppose that ϕ_1 is independent of $x_2 = \eta$, then the output ξ_2 is decoupled from η and thus the disturbance ϕ_2 does not influence the closed-loop evolution of the output x_1. If the system is rendered stable by the control (independent of η), then the integrator state compensates the uncertainty ϕ_1. Otherwise, the integrator approach cannot compensate the time-varying unmatched disturbance ϕ_2. However, if ϕ_2 is time-invariant, the integrator compensates both unmatched uncertainties in any case.

Consider now a transformation with differentiator

$$\begin{aligned}\xi_1 &= x_1\\ \xi_2 &= x_3 + \phi_1(x_1, x_2)\\ \eta &= x_2\,.\end{aligned}$$

Then the system reads

Table 9.3 Comparison of mechanism responsible for compensation: Integrator (Int), differentiator in combination with SMC (Diff+SMC), and decoupling

Class of unmatched	Compensation of	
Uncertainties	ϕ_1	ϕ_2
$\phi_1(x_1, x_2)$, $\phi_2(x_1, x_2)$	Int or Diff+SMC	Int or Diff+SMC
$\phi_1(x_1)$, $\phi_2(x_1, x_2, t)$	Int or Diff+SMC	Decoupling
$\phi_1(x_1, x_2)$, $\phi_2(x_1, x_2, t)$	Diff+SMC	Diff+SMC
$\phi_1(x_1, t)$, $\phi_2(x_1, x_2, t)$	Diff+SMC	Decoupling
$\phi_1(x_1, x_2, t)$, $\phi_2(x_1, x_2, t)$	Diff+SMC	Diff+SMC

$$\dot{\xi}_1 = \xi_2 \tag{9.57a}$$

$$\dot{\xi}_2 = u + \frac{\partial \phi_1(\xi_1, \eta)}{\partial x_1}\xi_2 + \frac{\partial \phi_1(\xi_1, \eta)}{\partial x_2}\phi_2(x_1, x_2, t) \tag{9.57b}$$

$$\dot{\eta} = -\eta + \phi_2(x_1, x_2, t)\,. \tag{9.57c}$$

Inspecting equation (9.57b), we observe that the effects of both disturbances are rendered matched. Note, however, that $\frac{\partial}{\partial x_2}\phi_1 = 0$ implies also that $\frac{\partial}{\partial x_1}\phi_1$ is independent of $x_2 = \eta$. In such case, ξ_2 is independent of η and ϕ_2, and thus ϕ_2 is decoupled from the output x_1. Therefore, the control u only compensates the uncertainty injected by ϕ_1. The disturbance ϕ_2 is compensated via decoupling.

In summary, ϕ_1 is compensated by the effect that the transformation using the differentiator renders this uncertainty matched which can be compensated via a sliding-mode control law. The disturbance ϕ_2 is either decoupled or also rendered a matched disturbance, depending on the dependency of ϕ_1.

Note that the conditions for decoupling of ϕ_2, i.e., $\frac{\partial}{\partial x_2}\phi_1 = 0$, are equivalent for the ISMC and the approaches using differentiation.

The effects that compensate the respective unmatched disturbance for various disturbance configurations can be found in Table 9.3. For other combinations of disturbances not discussed above, similar reasoning yields the results.

All sliding-mode controllers considered in this contribution use one or more of the compensating mechanisms summarized in Table 9.3. The ISMC approach discussed in Sect. 9.4 uses integration and decoupling. This approach as well as the classical FOSM can be augmented by using a differentiator as discussed in Sects. 9.3.2 and 9.6, respectively. The nested approaches, nHOSM and nFOSM, use a differentiator and can be easily adapted to systems with internal dynamics and therefore make use of the decoupling effect also.

Certainly, resorting to a differentiator is the most powerful concept here. However, there are some costs involved: implementation of an adequate differentiator, numerical difficulties in calculation, and additional smoothness requirements for the disturbances. Furthermore, the sliding-mode controller gain has to be adjusted to cover also the bounds for the derivatives of the uncertainties.

Three approaches, FOSM-D, nHOSM, and nFOSM, yield the same stability and compensation properties as they all lead to a transformed system where the disturbance acts only on the internal dynamics or as a matched disturbance.

Acknowledgements The authors kindly acknowledge support by the European Union Horizon 2020 research and innovation program under Marie Skłodowska-Curie grant agreement No. 734832. Parts of Sect. 9.4 have been previously published © 2018 IEEE. Reprinted, with permission, from T. Posielek, K. Wulff, and J. Reger. "Disturbance decoupling using a novel approach to integral sliding-mode". In: Proceedings of the IEEE International Workshop on Variable Structure Systems, pages 420–426, 2018.

Appendix 1: Relative Degree Under Disturbance

A disturbance $\phi : \mathbb{R}^n \times \mathbb{R} \to \mathbb{R}^n$ may change the relative degree of the nominal system for a given output y as the following example illustrates. Consider the system

$$\begin{aligned} \dot{x}_1 &= x_2 + \phi_1(x, t) \\ \dot{x}_2 &= x_3 \\ \dot{x}_3 &= u \\ y &= x_1 \,. \end{aligned}$$

For $\phi_1 = 0$, the output y has relative degree 3, whereas for $\phi_1(x, t) = x_3$ the output y has relative degree 2.

In order to formulate a condition that ensures that the relative degree is not impaired by the disturbance ϕ, we introduce the following multi-index notation for the Lie derivative of the sum of two vector fields f and ϕ. Let

$$(L_f L_\phi)^{(\alpha,\beta)} := L_f^{\alpha_1} L_\phi^{\beta_1} \dots L_f^{\alpha_n} L_\phi^{\beta_n} \tag{9.58}$$

denote the composition of Lie derivatives according to the combination indicated by $\alpha_i, \beta_i \in \{0, 1\}$. Observe that the Lie derivatives L_f and L_ϕ do not commute in general. Therefore, we write for the nth Lie derivative of the vector field $f + \phi$

$$L_{f+\phi}^n = \sum_{|(\alpha,\beta)|=n} (L_f L_\phi)^{(\alpha,\beta)} \,, \tag{9.59}$$

where $\alpha, \beta \in \{0, 1\}^n$ and

$$|(\alpha, \beta)| := \sum_{i=1}^{n} \alpha_i + \beta_i \,. \tag{9.60}$$

The composition $(L_f L_\phi)^{(\alpha,\beta)}$ denotes permutations of Lie derivatives of length n.

Consider the system (9.1) with relative degree r, and then the definition (9.3) ensures $L_g L_f^k h(x) = 0$ and $L_g L_f^{r-1} h(x) \neq 0$ for $k \in \{0, \ldots, r-2\}$. In order to retain the relative degree, the condition

$$L_g L_{f+\phi}^k h(x) = 0 \tag{9.61a}$$

$$L_g L_{f+\phi}^{r-1} h(x) \neq 0 \tag{9.61b}$$

must hold. Using the proposed notation (9.58)–(9.60), we may write

$$L_g L_{f+\phi}^k h(x) = L_g L_f^k h(x) + \sum_{|(\alpha,\beta)|=k, \alpha \neq 1} L_g\big((L_f L_\phi)^{(\alpha,\beta)} h(x)\big) \quad \text{for} \quad k \in \mathbb{N}. \tag{9.62}$$

Thus, with (9.62) we can see that condition (9.61a) is equivalent to

$$\sum_{|(\alpha,\beta)|=k, \alpha \neq 1} L_g\big((L_f L_\phi)^{(\alpha,\beta)} h(x)\big) = 0 \quad \text{for} \quad k \in \{1, \ldots, r-2\}. \tag{9.63}$$

Further, we can see that with (9.3b) and (9.62), the condition

$$\sum_{|(\alpha,\beta)|=r-1, \alpha \neq 1} L_g\big((L_f L_\phi)^{(\alpha,\beta)} h(x)\big) = 0 \tag{9.64}$$

is sufficient but not necessary to ensure (9.61b). These two conditions are fulfilled if

$$L_g\big((L_f L_\phi)^{(\alpha,\beta)} h(x)\big) = 0, \qquad \alpha, \beta \in \{0,1\}^k, \; |(\alpha,\beta)| = k \tag{9.65a}$$

$$L_g\big((L_f L_\phi)^{(\alpha,\beta)} h(x)\big) = 0, \qquad \alpha, \beta \in \{0,1\}^{r-1}, \; |(\alpha,\beta)| = r-1 \tag{9.65b}$$

for $k \in \{0, \ldots, r-2\}$, $|(\alpha, \beta)| = k$ and $\alpha \neq 1$ hold. Finally, this can be expressed by

$$L_g\big((L_f L_\phi)^{(\alpha,\beta)} h(x)\big) = 0 \tag{9.66}$$

for $k \in \{0, \ldots, r-1\}$, $\alpha, \beta \in \{0,1\}^k$, $|(\alpha, \beta)| = k$ and $\beta \neq 0$. Note that the nominal term $L_g L_f^k h(x)$ is deliberately excluded in condition (9.66) for all $k \in \{0, \ldots, r-1\}$. But mind that for $k \in \{0, \ldots, r-2\}$ it is $L_g L_f^k h(x) = 0$, while $L_g L_f^{r-1} h(x) \neq 0$ by definition of the relative degree of the nominal system. The proposed condition (9.66) is sufficient (but not necessary) to guarantee that the relative degree is not impaired by the disturbance. However, there are a number of system classes that satisfy this condition, such as systems in nonlinear block controllable form.

Appendix 2: Sliding-Mode Differentiator

As introduced in [26] and used in [4], the sliding-mode differentiator for a function f of order r takes the form

$$\begin{aligned}
\dot{\nu}_0 &= -\lambda_0 \Lambda^{\frac{1}{r+1}} \Psi^{\frac{r}{r+1}}(\nu_0 - f(t)) + \nu_1 \\
\dot{\nu}_i &= -\lambda_i \Lambda^{\frac{1}{r+1-i}} \Psi^{\frac{r-i}{r-i+1}}(\nu_i - \dot{\nu}_{i-1}) + \nu_{i+1} \\
\dot{\nu}_r &= -\lambda_r \Lambda \Psi^0(\nu_r - \dot{\nu}_{r-1}),
\end{aligned}$$

where

$$\Psi^\beta(\zeta) = \left[|\zeta_1|^\beta \operatorname{sign}(\zeta_1) \;\dots\; |\zeta_n|^\beta \operatorname{sign}(\zeta_n) \right]^\top$$

and ν_i is the ith derivative of f.

Appendix 3: Quasi-continuous Higher-Order Sliding-Mode Controller

We will consider the quasi-continuous controller from [9] for a switching function with relative degree r as the benchmark HOSM. The control is given by

$$\begin{aligned}
u &= -\alpha \Psi_{r-1,r}(\sigma, \dot{\sigma}, \dots, \sigma^{r-1}), \\
\varphi_{0,r} &= \sigma, \qquad N_{0,r} = |\sigma|, \qquad \Psi_{0,r} = \frac{\varphi_{0,r}}{N_{0,r}} = \operatorname{sign}(\sigma), \\
\varphi_{i,r} &= \sigma^{(i)} + \beta_i N_{i-1,r}^{\frac{r-i}{r-i+1}} \Psi_{i-1,r} \\
N_{i,r} &= |\sigma^{(i)}| + \beta_i N_{i-1,r}^{\frac{r-i}{r-i+1}} \\
\Psi_{i,r}(\cdot) &= \frac{\varphi_{i,r}}{N_{i,r}}, \qquad i = 0, \dots, r-1,
\end{aligned}$$

where $\beta_1, \dots, \beta_{r-1} \in \mathbb{R}$. The first-order sliding-mode control law has the form

$$u = \Psi_{0,1}(\sigma) = -\alpha \operatorname{sign}(\sigma), \tag{9.67}$$

and the second-order control law takes the form

$$u = -\alpha \Psi_{1,2}(\sigma, \dot{\sigma}) = -\alpha \frac{\dot{\sigma} + |\sigma|^{\frac{1}{2}} \operatorname{sign}(\sigma)}{|\dot{\sigma}| + |\sigma|^{\frac{1}{2}}}. \tag{9.68}$$

Appendix 4: Controller Design for the Simulation Examples

In this section, we state the control laws implemented for case studies in Sect. 9.6.

A 4.1 Case: No Internal Dynamics

For $a = 0$ and $b = 1$, we obtain the system

$$\dot{x}_1 = x_2 + \phi_1(x, t) \tag{9.69a}$$

$$\dot{x}_2 = x_3 + \phi_2(x, t) \tag{9.69b}$$

$$\dot{x}_3 = u \tag{9.69c}$$

$$y = x_1 \tag{9.69d}$$

in controller normal form with y being an output with full relative degree $r = 3$.

FOSM and FOSM-D

Define the sliding manifold as in (9.22) by

$$\sigma(x) = x_1 + 2\dot{x}_1 + \ddot{x}_1 \,. \tag{9.70}$$

For FOSM-D $\dot{x}_1$ and $\ddot{x}_1$ are numerically calculated. For FOSM, we use $\dot{x}_1 = x_2$ and $\ddot{x}_1 = x_3$ and use the sliding manifold proposed in (9.12) with

$$\sigma(x) = x_1 + 2x_2 + x_3 \,. \tag{9.71}$$

Then, choose the control law $w = -\alpha \operatorname{sign}(\sigma(x))$ from (9.13)/(9.23) with $\alpha = 40$ to obtain the desired control law by using the input transformation (9.14)/(9.24)

$$u = -x_2 - 2x_3 + w \,. \tag{9.72}$$

ISMC and ISMC-D

Introduce the integrator v as in (9.31) with $\dot{v} = x_1$ with $v(0)$ so that the initial state lies on the sliding manifold. Then, define the sliding manifold

$$\sigma(v) = v + 3\dot{v} + 3\ddot{v} + \dddot{v} \,.$$

For ISMC-D, we use the numerical derivatives of v. For ISMC, we substitute $\dot{v} = x_1$, $\ddot{v} = x_2$, $\dddot{v} = x_3$ and obtain the switching function as in (9.32) with

$$\sigma(v, x) = v + 3x_1 + 3x_2 + x_3 \,.$$

Use the control law $w = -\alpha \operatorname{sign}(\sigma(v, x))$ with $\alpha = 40$ and use the input transformation

$$u = -x_1 - 3x_2 - 3x_3 + w$$

to obtain the control law as in (9.34).

nHOSM

Define the switching functions and virtual control laws as in (9.43)–(9.45)

$$\begin{aligned}
&\sigma_1(x) = x_1 && \sigma_2(x) = x_2 - w_1 && \sigma_3(x) = x_3 - w_2 \\
&\dot{u}_{1,1} = u_{1,2} && \dot{u}_{2,1} = -\alpha_2 \Psi_{2,3}(\sigma_2, \dot{\sigma}_2) && \\
&\dot{u}_{1,2} = -\alpha_1 \Psi_{2,3}(\sigma_1, \dot{\sigma}_1, \ddot{\sigma}_1) && && \\
&w_1 = u_{1,1} && w_2 = u_{2,1} && w_3 = -\alpha_3 \operatorname{sign}(\sigma_3)
\end{aligned}$$

with $\alpha_1 = 15$, $\alpha_2 = 13$, and $\alpha_3 = 40$. The derivatives $\dot{\sigma}_1$, $\ddot{\sigma}_1$, and $\dot{\sigma}_2$ are numerically calculated. Finally, use the first-order sliding-mode input transformation (9.46a)

$$u = -\alpha_2 \Psi_{2,3}(\sigma_2, \dot{\sigma}_2) + w_3 \,. \tag{9.73}$$

Note that we use an additional first-order sliding-mode transformation to make this control law more comparable to FOSM, FOSM-D, ISMC, and ISMC-D. This leads to a slightly different notation in comparison to (9.45).

nFOSM

Define the switching functions and virtual control laws as in (9.47)–(9.49)

$$\begin{aligned}
&\sigma_1(x) = x_1 && \sigma_2(x) = x_2 - w_1 && \sigma_3(x) = x_3 - w_2 \\
&w_1 = -\sigma_1 - \phi_1 && w_2 = -\sigma_2 + \phi_2 - x_2 - \dot{\phi}_1 - \phi_1 && w_3 = -\alpha \operatorname{sign}(\sigma_3)
\end{aligned}$$

with $\alpha = 40$. Use the first-order sliding-mode input transformation

$$u = -2x_3 + x_2 + w_3 \,.$$

Again, we have introduced the additional first-order sliding-mode transformation which changes the notation in comparison to (9.49) slightly.

A 4.2 Case: Stable Internal Dynamics

Consider the system (9.51) with $a = -1$

$$\dot{x}_1 = x_1 + x_3 + \phi_1(x_1, t) \tag{9.74a}$$
$$\dot{x}_2 = -x_2 + x_3 + \phi_2(x_1, x_2, t) \tag{9.74b}$$
$$\dot{x}_3 = u \tag{9.74c}$$
$$y = x_1 \,. \tag{9.74d}$$

This system has the internal dynamics

$$\dot{\eta}_1 = -\eta_1 + \xi_2 - \xi_1 \,. \tag{9.75}$$

It is easy to see that the zero dynamics are asymptotically stable at 0. Thus, it is sufficient to find a control law w for the system (9.54a), (9.54b) to stabilize the system (9.54a)–(9.54c).

FOSM and FOSM-D

For ISMC-D, use the sliding manifold

$$\sigma(x) = x_1 + \dot{x}_1$$

with numerical derivatives. For ISMC, use the substitution $\dot{x}_1 = x_1 + x_3$ to obtain the sliding manifold

$$\sigma(x) = 2x_1 + x_3 \,.$$

Then, choose $w = -\alpha \operatorname{sign}(\sigma)$ with $\alpha = 40$ to obtain the desired control law by using the input transformation

$$u = -2(x_1 + x_3) + w \,.$$

ISMC and ISMC-D

Introduce the integrator v with $\dot{v} = x_1$. For ISMC-D, define the sliding manifold

$$\sigma(v) = v + 2\dot{v} + \ddot{v} \,.$$

For ISMC, we use $\dot{v} = x_1$ and $\ddot{v} = x_1 + x_3$ to obtain

$$\sigma(v, x) = v + 2x_1 + x_1 + x_3$$

and choose $w = -\alpha \operatorname{sign}(\sigma)$ with $\alpha = 40$. Finally, use the input transformation

$$u = x_1 + 3(x_1 + x_3) + w\,.$$

nHOSM

Define

$$\begin{aligned} \sigma_1 &= \xi_1 & \sigma_2 &= \xi_2 - w_1 \\ \dot{u}_{1,1} &= -\alpha_1 \Psi_{1,2}(\sigma_1, \dot{\sigma}_1) & & \\ w_1 &= u_{1,1} & w_2 &= -\alpha_2 \operatorname{sign}(\sigma_2) \end{aligned}$$

with $\alpha_1 = 2$, $\alpha_2 = 40$. Use the first-order sliding-mode transformation to obtain

$$w = -\alpha_1 \Psi_{1,2}(\sigma_1, \dot{\sigma}_1) + w_2$$

and the final control law

$$u = -x_1 - x_3 + w\,.$$

Note that we use an additional first-order sliding-mode transformation to make this control law more comparable to FOSM, FOSM-D, ISMC, and ISMC-D. This leads to a slightly different notation in comparison to (9.45).

nFOSM

Define

$$\begin{aligned} \sigma_1 &= \xi_1 & \sigma_2 &= \xi_2 - w_1 \\ w_1 &= -\phi_1 - \xi_1\,. & w_2 &= -\alpha \operatorname{sign}(\sigma_2) \end{aligned}$$

with $\alpha = 40$. Use the first-order sliding-mode transformation

$$w = -x_1 + x_3 + w_2$$

and obtain the final control law

$$u = -x_1 - x_3 + w\,.$$

A 4.3 Case: Unstable Internal Dynamics

Consider the system (9.51) with $a = 1$

$$\dot{x}_1 = -x_1 + x_3 + \phi_1(x_1, t) \tag{9.76a}$$
$$\dot{x}_2 = x_2 + x_3 + \phi_2(x_1, x_2, t) \tag{9.76b}$$
$$\dot{x}_3 = u \tag{9.76c}$$
$$y = x_1 \,. \tag{9.76d}$$

This system has the internal dynamics

$$\dot{\eta}_1 = \eta_1 + \xi_2 - \xi_1 \,. \tag{9.77}$$

These internal dynamics are unstable. Thus, it is not sufficient anymore to find a control law w for the system (9.54a), (9.54b) to stabilize the system (9.54a)–(9.54c). This makes FOSM, FOSM-D, nHOSM, and nFOSM not straightforwardly applicable. Thus, we only consider ISMC and discuss its results.

ISMC

Introduce the integrator v with $\dot{v} = x_1$. Consider the new output

$$h_v(v, x) = 2v + x_1 - x_2$$

with full relative degree with respect to the input u. Then define the sliding manifold

$$\sigma(v, x) = h_v(v, x) + 3\frac{dh_v(v, x)}{dt} + 3\frac{d^2h_v(v, x)}{dt^2} + \frac{d^3h_v(v, x)}{dt^3} \,.$$

Substitute $\frac{dh_v(v,x)}{dt} = x_1 - x_2$, $\frac{d^2h_v(v,x)}{dt^2} = -x_1 - x_2$, and $\frac{d^3h_v(v,x)}{dt^3} = -x_1 - x_2 - 2x_3$ and define $w = -\alpha \operatorname{sign}(\sigma)$ with $\alpha = 40$. Finally, use the first-order sliding-mode input transformation

$$u = -\frac{8x_2 + 6x_3 - \alpha \operatorname{sign}(\sigma)}{2} \,. \tag{9.78}$$

References

1. Cao, W.-J., Jian-Xin, X.: Nonlinear integral-type sliding surface for both matched and unmatched uncertain systems. IEEE Trans. Autom. Control **49**(8), 1355–1360 (2004)
2. Castaños, F., Fridman, L.: Analysis and design of integral sliding manifolds for systems with unmatched perturbations. IEEE Trans. Autom. Control. **51**(5), 853–858 (2006)

3. Estrada, A., Fridman, L.: Integral HOSM semiglobal controller for finite-time exact compensation of unmatched perturbations. IEEE Trans. Autom. Control **55**(11), 2645–2649 (2010)
4. Ferreira de Loza, A., Puntab, E., Fridman, L., Bartolini, G., Delprat, S.: Nested backward compensation of unmatched perturbations via HOSM observation. J. Frankl. Inst. **351**(5), 2397–2410 (2014)
5. Rubagotti, M., Estrada, A., Castaños, F., Ferrara, A., Fridman, L.: Integral sliding mode control for nonlinear systems with matched and unmatched perturbations. IEEE Trans. Autom. Control **56**(11), 2699–2704 (2011)
6. Posielek, T., Wulff, K., Reger, J.: Disturbance decoupling using a novel approach to integral sliding-mode. In: Proceedings of the IEEE International Workshop on Variable Structure Systems, pp. 420–426 (2018)
7. Bin, Y., Li, K.Q., Feng, N.L.: Disturbance decoupling robust control of vehicle full speed cruise dynamic system. Sci. China Ser. E: Technol. Sci. **52**(12), 3545–3564 (2009)
8. Thomsen, S.C., Poulsen, N.K.: A disturbance decoupling nonlinear control law for variable speed wind turbines. In: Proceedings of the IEEE Mediterranean Conference on Control and Automation, pp. 1–6 (2007)
9. Levant, A.: Quasi-continuous high-order sliding-mode controllers. Proceedings of the IEEE Conference on Decision and Control **5**, 4605–4610 (2003)
10. Estrada, A., Fridman, L.: Exact compensation of unmatched perturbation via quasi-continuous HOSM. In: Proceedings of the IEEE Conference on Decision and Control, pp. 2202–2207 (2008)
11. Estrada, A., Fridman, L., Iriarte, R.: Combined backstepping and HOSM control design for a class of nonlinear MIMO systems. Int. J. Robust Nonlinear Control **27**(4), 566–581 (2016)
12. Dávila, J.: Exact tracking using backstepping control design and high-order sliding modes. IEEE Trans. Autom. Control **58**(8), 2077–2081 (2013)
13. Yang, J., Li, S., Jinya, S., Xinghuo, Y.: Continuous nonsingular terminal sliding mode control for systems with mismatched disturbances. Automatica **49**(7), 2287–2291 (2013)
14. Li, X., Zhou, J.: A sliding mode control design for mismatched uncertain systems based on states transformation scheme and chattering alleviating scheme. Trans. Inst. Meas. Control **40**(8), 2509–2516 (2018)
15. Ferreira de Loza, A., Cieslak, J., Henry, D., Zolghadri, A., Fridman, L.: Output tracking of systems subjected to perturbations and a class of actuator faults based on HOSM observation and identification. Automatica **59**, 200–205 (2015)
16. Smidt Gabbi, T., Abílio Gründling, H., Padilha Vieira, R.: Sliding mode current control based on disturbance observer applied to permanent magnet synchronous motor. In: Proceedings of the Brazilian Power Electronics Conference and Southern Power Electronics Conference, pp. 122–127 (2015)
17. Krstić, M., Kanellakopoulos, I., Kokotović, P.: Nonlinear and Adaptive Control Design. Wiley, New York (1995)
18. Isidori, A.: Nonlinear Control Systems. Springer Science & Business Media, Berlin (2013)
19. Shtessel, Y., Edwards, C., Fridman, L., Levant, A: Sliding Mode Control and Observation. Birkhäuser (2013)
20. Utkin, V., Shi, J.: Integral sliding mode in systems operating under uncertainty conditions. In: Proceedings of the Conference on Decision and Control, pp. 4591–4596 (1996)
21. Rubagotti, M., Estrada, A., Castaños, F., Ferrara, A., Fridman, L.: Optimal disturbance rejection via integral sliding mode control for uncertain systems in regular form. In: Proceedings of the IEEE International Workshop on Variable Structure Systems, pp. 78–82 (2010)
22. Kim, J.H., Gadsden, S.A., Wilkerson, S.A.: Adaptive integral sliding mode controller for longitudinal rotation control of a tilt-rotor aircraft. In: Proceedings of the IEEE Mediterranean Conference on Control and Automation, pp. 820–825 (2016)
23. Xu, J.-X., Guo, Z.-Q., Lee, T.H.: Design and implementation of integral sliding-mode control on an underactuated two-wheeled mobile robot. IEEE Trans. Ind. Electron. **61**(7), 3671–3681 (2014)

24. Louk'yanov, A., Utkin, V.: Methods of reducing equations of dynamic systems to regular form. Autom. Remote. Control **42**(4), 413–420 (1981)
25. Fridman, L., Estrada, A., de Loza, A.F.: Higher order sliding mode based accurate tracking of unmatched perturbed outputs. In: Bandyopadhyay, B., Sivaramakrishnan, J., Spurgeon, S. (eds.) Advances in Sliding Mode Control. Lecture Notes in Control and Information Sciences, vol. 440, pp. 117–144. Springer, Berlin (2013)
26. Levant, A.: Higher-order sliding modes, differentiation and output-feedback control. Int. J. Control **76**(9–10), 924–941 (2003)
27. Fridman, L., Dávila, J., Levant, A.: High-order sliding-mode observation for linear systems with unknown inputs. Nonlinear Anal.: Hybrid Syst. **5**, 189–205 (2011)

Part IV
Applications of VSS

Chapter 10
Grid-Connected Shunt Active LCL Control via Continuous SMC and HOSMC Techniques

Mohamad A. E. Alali, Yuri. B. Shtessel and Jean-Pierre Barbot

Abstract A LCL grid-connected three-phase and three-wire shunt active filter (SAF) is studied and controlled. It is known that SAFs generate distortive components caused by a high switching frequency of a voltage source inverter (VSI). In order to prevent spreading these distortive components to the grid, the LCL filter (usually controlled by a linear control feedback) is used; while a parasitic phase shift/lag between the reference and injected currents emerges and severely deteriorates the filtration quality. In order to overcome phase shift problem, while ensuring good and robust filtration performance of the SAF, linear and nonlinear controllers are studied. An improved linear RST controller (RST_{imp}) presents a good filtration solution in continuous time, but technological limitations prevent its discrete-time implementation. This work proves the advantages of sliding mode over the linear controllers by improving tracking accuracy over the entire bandwidth of harmonics. Specifically, an SMC with *sigmoid* approximation function, an SMC with artificial increase of relative degree (AIRD), and a continuous higher order sliding mode controller (C-HOSMC) are studied, in order to prevent a very high-frequency switching of control that can severely hurt the switching elements. The output of the continuous SMCs is pulse-width modulated (PWM) in order to provide a fixed given frequency of control switching, required for the VSI safe operation. A new modeling approach of nonlinear loads in a real textile factory is proposed, and the efficacy of the studied controllers for SAF/LCL filter, even under unbalanced conditions, is validated via simulations based on real measurements coming from power quality analyzers.

M. A. E. Alali (✉) · J.-P. Barbot
QUARTZ Laboratory, EA7393, ENSEA, 6 Avenue du Ponceau, 95014 Cergy-Pontoise, France
e-mail: mohamad-alaa-eddin.alali@ensea.fr

J.-P. Barbot
e-mail: barbot@ensea.fr

Yuri. B. Shtessel
Department of Electrical and Computer Engineering, The University of Alabama in Huntsville, Huntsville, AL 35899, USA
e-mail: shtessy@uah.edu

M. Steinberger et al. (eds.), *Variable-Structure Systems and Sliding-Mode Control*, Studies in Systems, Decision and Control 271,
https://doi.org/10.1007/978-3-030-36621-6_10

10.1 Introduction

Efficiency and sustainability of modern electric systems and machines significantly depends on the operating quality of power electronics. In this context, voltage source inverters (VSI) are used for energy conversion from a DC source to an AC output, either in a standalone mode or when connected to the utility grid.

A simple series inductor (first-order output filter) is often employed [1]. However, the attenuation of high-frequency components, due to a high switching frequency, appears not very effective. Furthermore, a high voltage drop is observed, and the high-inductance inductor, required in the design, becomes very bulky [2]. Commonly, a high-order output LCL filter is used instead of the conventional L-filter for smoothing the high-frequency output currents from a VSI [2]. Indeed, the higher-order LCL filter provides, in comparison with the first-order one, higher attenuation of the high-frequency components as well as the weight and size reduction of the components. A grid-connected VSI with LCL filter is commonly used for the renewable energy (photovoltaic and wind) systems [3]. In this case, the grid-connected VSIs have to inject only fundamental components to the grid.

Lately, shunt active filters became an alternative solution for compensating all current disturbances, including harmonic, unbalance and reactive currents, in electrical grids. In these systems, the current control loop of the shunt active compensator ensures a good compensating performance over the entire bandwidth harmonic frequencies, including the fundamental one. For this purpose, a PWM-VSI with an adaptive linear controller (such as RST) is proposed in [4, 5]. However, this controller, as well as any other linear controller, cannot guarantee good filtration quality. Indeed, although the tracking of the magnitude of identified perturbation currents is quite good, a phase shift between the reference and the injected currents that occurs in the current control loop deteriorates the filtration quality and limits the integration property of the LCL filters in the block of the shunt active filters [5].

Addressing the phase shift and robustness problems, a conventional structure of the VSI (a two-level system configuration) with LCL filter, controlled by adaptive nonlinear controllers, as well as by the sliding-mode control (SMC) was proposed in [6]. This technique provides a good attenuation of matched disturbances [7, 8]. However, the very high-frequency switching control signal may destroy the inverter and may cause a Zenon phenomenon.

It is worth noting that other phase shift compensation solutions were focused on achieving the least polluted load current. In this context, sinusoidal load current with unity power factor is to be provided for reactive power and/or harmonics canceling [9, 10]. However, these solutions involve additional costs and do not completely solve the problems caused by the polluting loads that are observed in industrial systems. The contribution of this chapter is as follows:

1. Exploring the continuous SMC algorithms with a consecutive pulse-width modulation (PWM) that guarantees a required switching frequency of the control function in the studied LCL system. In this case, no overheating of inverter transistors is expected. This control must retain the robustness of classical SMC and increases

reliability of the entire LCL system. In this work, two different continuous SMC algorithms are explored: the one is the SMC with a *sigmoid* approximation, and the other is the classical SMC with artificially increased relative degree (AIRD), followed by an integrator that provides continuity to the control function. The performances of the LCL system, controlled by the high-frequency switching SMC and the studied continuous SMCs with PWM, are compared.
2. Studying continuous higher-order sliding-mode algorithm (C-HOSM) followed by pulse-width modulation that ensures also a fixed switching frequency of the control function devoid of destructive very high switching frequency components.
3. A new modeling approach, when the simulations were performed based on the real measurements, is obtained by the power quality analyzer. This improves accuracy and fidelity of the simulation analysis.

10.2 Shunt Active Filter Structure and Control Problem Formulation

10.2.1 General Basic Structure

In the distribution electrical power systems, the main power quality problems are associated with voltage harmonics, unbalanced voltages, and reactive power. These perturbations are usually due to the unbalanced and inductive current consumption and nonlinear loads. Therefore, shunt active compensators are generally proposed to cancel the current perturbations and subsequently reduce the voltage perturbations of the same kind [4, 11].

A conventional structure of shunt active filter and its power environment, including a three-wire power network and a nonlinear load, is shown in Fig. 10.1. The shunt active filter is generally composed of a control part and a power part, including a PWM voltage source inverter (VSI), a DC capacitive storage system, and an output filter. The control part usually consists of the current perturbation identification block and the control loops for injecting currents into the grid and for regulating the direct voltage storage capacitor.

10.2.2 Transfer Function and State-Space Models of the LCL Filter

A PWM three-phase voltage source inverter (VSI) is used to generate the current to be injected to the network supply (see Fig. 10.1). The VSI is connected to the electrical network via a passive output filter. The output filter must be designed to filter out the components of the switching frequency, generated by the PWM-VSI.

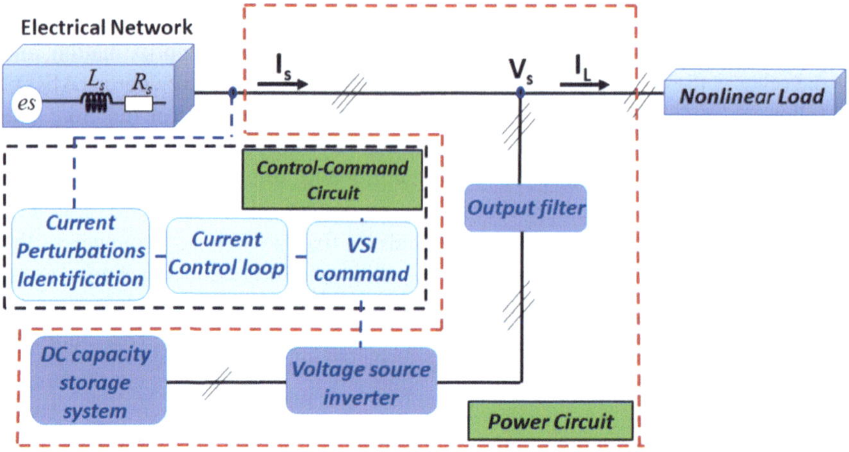

Fig. 10.1 Basic structure of shunt active filters

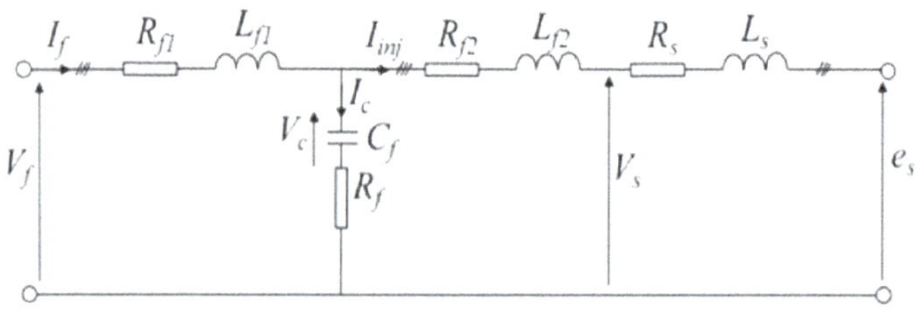

Fig. 10.2 A single-phase model of LCL output filter

A single-phase equivalent circuit of the LCL output filter is shown in Fig. 10.2. The LCL filter is modeled by the following Laplace domain equations [4]:

$$I_{inj} = \frac{B_1(s)}{A(s)} V_f(s) + \frac{B_2(s)}{A(s)} V_s(s) \tag{10.1}$$

with

$$\begin{cases} A(s) = a_1 s^3 + a_2 s^2 + a_3 s + a_4 \\ B_1(s) = b_{11} s + b_{12} \\ B_2(s) = -(b_{21} s^2 + b_{22} s + b_{23}) \end{cases}$$

and

$$\begin{aligned} a_1 &= (L_s + L_{f2}) L_{f1} C_f \\ a_2 &= (L_s + L_{f2}) R_{f1} C_f + (R_s + R_{f2}) L_{f1} C_f + (L_{f1} + L_s + L_{f2}) R_f C_f \\ a_3 &= L_s + L_{f2} + L_{f1} + (R_s + R_{f2}) R_{f1} C_f + (R_{f1} + R_s + R_{f2}) R_f C_f \\ a_4 &= R_{f1} + R_s + R_{f2} \end{aligned}$$

$$\begin{aligned}
b_{11} &= R_f C_f \\
b_{12} &= 1 \\
b_{21} &= L_{f1} C_f \\
b_{22} &= (R_f + R_{f1}) C_f \\
b_{23} &= 1,
\end{aligned}$$

where L_{f1} is the inverter side inductor, L_{f2} is the grid-side inductor, C_f is a capacitor with a series R_f damping resistor, R_{f1} and R_{f2} are the inductor resistances, L_s and R_s are grid inductor and resistor, respectively, $\frac{B_1(s)}{A(s)}$ is the transfer function model of the LCL filter (including the network impedance), $\frac{B_2(s)}{A)(s)}$ represents a disturbance model, $V_f(s)$ is the inverter output voltage, and $V_s(s)$ is the point of common coupling (PCC) network voltage, which is considered as a bounded disturbance. Currents I_f, I_c, and I_{inj} are the inverter output current, the capacitor current, and the grid-injected current, respectively; and V_c is the capacitor voltage. It is worth noting that the grid inductor and resistor are well below those of the output filter, and then the electromotive force of the secondary of the transformer $e_s(s)$ can be replaced by $V_s(s)$.

If we neglect all very small resistances (except the damping one), the simplified transfer function of the LCL filter will be

$$\frac{B_1(s)}{A(s)} = \frac{(R_f C_f)s + 1}{(L_{f1} L_{f2} C_f)s^3 + R_f C_f (L_{f1} + L_{f2})s^2 + (L_{f1} + L_{f2})s}.$$

Next, for industrial applications, R_f is to be zero, then the approximate cutoff frequency of the LCL filter becomes

$$f_{cp} = \frac{1}{2\pi \sqrt{\frac{L_{f1} L_{f2} C_f}{L_{f1} + L_{f2}}}}.$$

Based on the Laplace transform model (10.1), the state-space model of the single-phase LCL output filter is derived [12]:

$$\begin{aligned}
\frac{dv_c}{dt} &= \frac{i_f - i_{inj}}{C_f} \\
\frac{di_f}{dt} &= \frac{1}{L_{f1}} \left(v_f - v_c - R_f (i_f - i_{inj}) - R_{f1} i_f \right) \\
\frac{di_{inj}}{dt} &= \frac{1}{L_{f2}} \left(v_c + R_f (i_f - i_{inj}) - v_s - R_{f2} i_{inj} \right)
\end{aligned} \tag{10.2}$$

and presented in a shortcut vector–matrix format as

$$\dot{x} = Ax + Bv_f + Pv_s, \tag{10.3}$$

where

$$A = \begin{bmatrix} -\frac{R_{f1}+R_f}{L_{f1}} & \frac{R_f}{L_{f1}} & -\frac{1}{L_{f1}} \\ \frac{R_f}{L_{f2}} & -\frac{R_{f2}+R_f}{L_{f2}} & \frac{1}{L_{f2}} \\ \frac{1}{C_f} & -\frac{1}{C_f} & 0 \end{bmatrix}, \quad B = \begin{bmatrix} \frac{1}{L_{f1}} \\ 0 \\ 0 \end{bmatrix}, \quad P = \begin{bmatrix} 0 \\ -\frac{1}{L_{f2}} \\ 0 \end{bmatrix}, x = \begin{bmatrix} i_f \\ i_{inj} \\ v_c \end{bmatrix}.$$

Note that small letters represent time-dependent instantaneous variables, while capital letters represent their corresponding Laplace transform or RMS values.

10.2.3 Control Problem Formulation

The control objectives of the current control loop of the shunt active compensator are, among others, to ensure a high level of compensating quality over the entire bandwidth harmonic frequencies, including the fundamental one. Thus, the controller must ensure asymptotic convergence of the tracking error, i.e., the difference between the current reference command $i_{ref}(t)$ and the injection current $i_{inj}(t)$, to zero in amplitude and phase.

Therefore, the control design problem is in designing the feedback control so that

$$\lim_{t\to\infty}(i_{ref} - i_{inj}) = 0$$

in the presence of the bounded disturbance v_s: $|v_s| \leq V_{max}$, $V_{max} > 0$.

10.3 Linear Root Locus-Based Controller Design for Grid-Connected VSI with LCL Filter

10.3.1 Root Locus Controller

Linear controllers are generally used to guarantee good performance in compensation for reactive power, unbalance, etc. at the fundamental frequency. However, these controllers cause non-neglected phase shift between the identified and injected currents. This phase delay, which increases with the frequencies and the order of the controlled system, prevents the applicability of the SAF associated with the LCL output filter.

In this context, improved linear controller is first proposed to overcome this problem of phase tracking. Then, continuous nonlinear controllers are proposed as a more efficient and universal alternative to the improved linear controller.

The control strategy is based on detecting the current perturbations (the current reference I_{ref}) by means of an identification algorithm. Next, the VSI driven by the

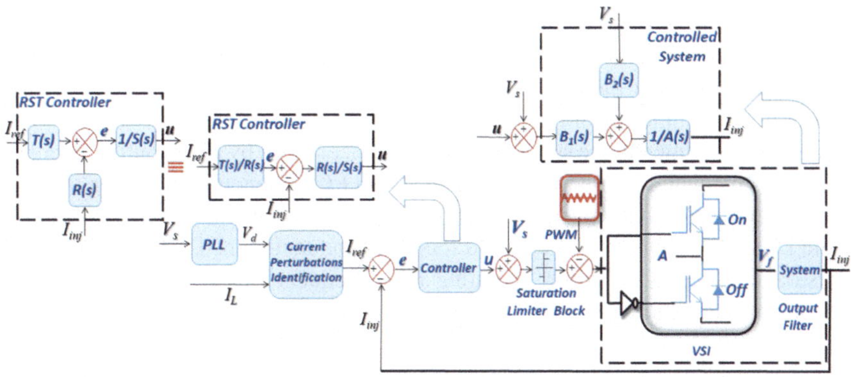

Fig. 10.3 General block scheme of current control algorithm

PWM control generates the current (I_{inj}) references (I_{ref}) to be followed and injected to the grid.

To provide asymptotic tracking $I_{inj} \to I_{ref}$, the control strategy includes the current control loop. The general block diagram of the current control system is shown in Fig. 10.3. Specifically, the LCL grid-connected PWM-VSI with a roots locus (RST) controller and an instantaneous power method for current perturbation identification were proposed in [5, 11]. A phase-locked loop (PLL) is used to extract the positive component of the network voltage (V_d), necessary for the good performance of the identification method [4, 5].

The voltage network V_s represents here external disturbance, whose effects are compensated by adding the same network voltage to the control signal $u = v_f$. This will prevent the fundamental current from passing from the network to the active filter [4].

Equation (10.1) with RST controller becomes [4, 5]:

$$I_{inj}(s) = \frac{T(s)B_1(s)}{S(s)A(s) + R(s)B_1(s)} I_{ref}(s) + \frac{S(s)B_2(s)}{S(s)A(s) + R(s)B_1(s)} V_s(s),$$

where $R(s)$, $T(s)$, and $S(s)$ are the controller polynomials.

The two parts of the previous equation represent the closed-loop transfer function in terms of tracking and disturbance rejection. The order of R(s) and S(s) is the same as the LCL filter: $\frac{B_1(s)}{A(s)}$; thus $R(s)$ and $S(s)$ are of third order [4]. In addition, the polynomial $S(s)$ is chosen so that the closed-loop transfer function between the disturbances and the output is stable and will be zero in steady state, because we set $S(0) = 0$ [4]. $T(s)$ can be given by a simple gain as $\frac{I_{ref}(s)}{I_{inj}(s)} = 1$ for all the bandwidth of the harmonics contained in the reference signal. Knowing that $\frac{R(s)}{S(s)}$ (Cf. Fig. 10.3) represents the transfer function of the controller, the common denominator $D(s) = S(s)A(s) + R(s)B_1(s)$ is called *arbitrary stability polynomial*; it contains the closed-loop poles. These poles are placed in a $(2 \times 45°)$ sector/cone, in order to ensure a 0.7 damping ratio. Finally, the closed-loop poles are placed so that they provide fast and accurate response tracking with a good disturbances rejection [4]. It is worth noting that the poles' values are limited by the cutoff frequency of the transfer function in terms of tracking.

It is worth noting that the stability of the system, ensured by RST controller, could be affected by changes in network voltage and in LCL elements (including the network impedance). On the basis that the grid inductance inductor is well below those of the LCL, the stability of the closed loop remains maintained even under severe industrial constraints including, operating at two frequencies (50 Hz and 60 Hz) with network power variations from 100 kVA to 2 MVA and grid voltage variation of 15% [4].

10.3.2 Phase Shift Effect Problems

The RST control algorithm, as well as other linear controllers, are generally used when the reference to be tracked is either a constant or a single low-frequency signal. In the single low-frequency case (i.e., unbalance or reactive power compensation), phase shift between reference (I_{ref}) and injected (I_{inj}) signals is acceptable. However, the phase shift is unacceptable when the reference signal includes multiple frequencies, as phase shift increases with frequency. Furthermore, higher order of controlled system (output filter here) implies larger phase shift. Figure 10.4 illustrates the phase shift effect for the structure presented in Fig. 10.3. In Fig. 10.4, it is simple to observe that the perturbed current (I_{Load}) is not correctly compensated for (I_{real}: compensation with phase shift), comparing with the ideal form (I_{ideal}: compensation without phase shift).

We can clearly see, on the last curves of Fig. 10.4, how much the phase shift problem can degrade the filtration quality of the shunt active filter. This problem is not related only to the RST controller, it is also inherent to all known linear controllers such as *PI*, *PID*, H_2, H_∞, etc. The application of this type of "linear" controller, as it is presented in the literature, is therefore not a solution, and an improved RST method is proposed.

10.3.3 Improved RST Control Method (RST_{imp})

To overcome the phase shift effect and achieve good filtration performance, a control method based on an advanced linear control mode is proposed; this method treats particularly the phase shift problem [5]. This advanced control method is based on the same principle used in a classic RST where zeros have been placed in the transfer function of the closed loop. These zeros are placed to reduce gain and phase error. Error minimization $e(s)$ is determined by the equation from Fig. 10.3:

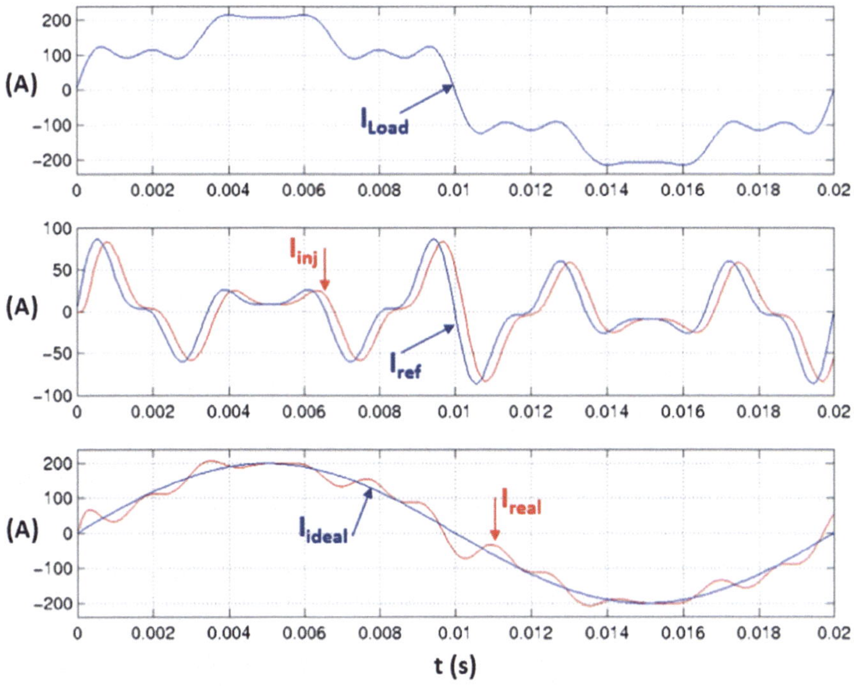

Fig. 10.4 Phase lag effects on compensation quality

$$e(s) = I_{inj}(s) - I_{ref}(s) = \left(\frac{T(s)B_1(s)}{D(s)} - 1\right) I_{ref}(s) \tag{10.4}$$

with $D(s) = S(s)A(s) + R(s)B_1(s)$ is called *arbitrary stability polynomial*. From Eq. (10.4), we can obtain the transfer function as given by

$$\frac{e(s)}{I_{ref}(s)} = \frac{T(s)B_1(s) - D(s)}{D(s)} I_{ref}(s). \tag{10.5}$$

To reduce the error $e(s)$ to zero, the numerator of equation (10.5) must tend to zero. Since $R(s)$ and $S(s)$ are already defined, only $T(s)$ can reduce the error $e(s)$. Moreover, the order of $T(s)$ must be chosen so that the transfer function $\frac{T(s)}{R(s)}$ (cf. Fig. 10.3) is causal ($degree(T(s)) \leq degree(R(s))$), as given in Eq. (10.6). The system, defined by a classic RST method, stay minimum phase, if the $T(s)$ polynomial is stable:

$$T(s) = t_1 s^n + t_2 s^{n-1} + ... + t_n s + t_{n+1}. \tag{10.6}$$

By using Eq. (10.6) and replacing $s = j\omega$ with ω the angular frequency, the numerator of Eq. (10.5) becomes

$$T(j\omega)B_1(j\omega) - D(j\omega) = R_e(\omega) + jI_m(\omega). \qquad (10.7)$$

To make (10.7) tend to zero, both the real $R_e(\omega)$ and imaginary $I_m(\omega)$ parts must tend to zero. This constraint means that there are two equations for each frequency of a placed zero.

In this study, $T(s)$ is a third-order polynomial. Indeed, since the order of $R(s)$ is three, a third-order $T(s)$ polynomial is the maximum order satisfying the causality of the transfer function $\frac{T(s)}{R(s)}$ (cf. Fig. 10.3). Taking account of the unknown parameters in this case (t_1, t_2, t_3, t_4), only two frequencies have to be chosen. Zeros should be placed near the slowest poles of the closed-loop transfer function.

10.3.4 Control Method Synthesis

Figure 10.5 and Table 10.1 compare current reference tracking of the RST and RST_{imp} control methods. The gain and the phase of the current closed-loop $\frac{I_{inj}(s)}{I_{ref}(s)}$ transfer function are given in Table 10.1, for multiples of the fundamental frequency, from 50 Hz to (31×50) Hz. In both control methods, gain remains at 0 dB for all frequencies, as shown in Table 10.1.

The RST control method (as the other linear controllers) ensures a 0 dB gain (zero gain tracking error) for all frequency bands. Besides, the RST control method is only used to compensate for the unbalanced current or the reactive current that occurs at

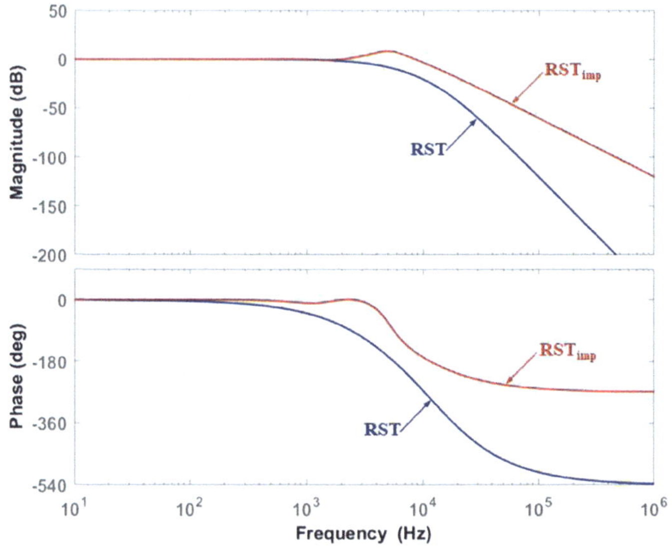

Fig. 10.5 Current closed-loop transfer function with RST and RST$_{inp}$

Table 10.1 Gain and phase of current closed-loop transfer function with RST and RST_{inp} controllers

F (Hz)		50	250	350	550	650	850	950	1150	1250	1450	1550
RST	G	1	1	1	1	0.99	0.98	0.98	0.97	0.96	0.95	0.94
	$PH°$	−1	−8	−11	−18	−21	−27	−30	−37	−40	−46	−49
RST_{imp}	G	1	1	1	1.1	1.06	1.03	0.98	0.93	0.9	0.89	0.9
	$PH°$	−0.01	−0.54	−1.4	−4	−6	−10	−11	−10	−10	−7	−5

the fundamental frequency 50 Hz. Indeed, at this frequency, a phase lag of (−1°) is negligible. Beyond this, phase lag is no longer negligible and the shunt active filter cannot compensate for current harmonics.

The improved RST control method (RST_{imp}) works for phase shift over the entire bandwidth. In this method, little overshoot occurred at the cutoff frequency of the closed loop, as shown in the gain diagram of Fig. 10.5. This frequency is high enough so that there is no risk of having a harmonic current component. Furthermore, a little dip between 850 and 2000 Hz can be observed in the phase diagram of Fig. 10.5. This little dip is caused by the order of the $T(s)$ polynomial, which is 3 in this study and insufficient to ensure zero phase shift. Indeed, the polynomial degree of $T(s)$ is limited by the causality of the transfer function $\frac{T(s)}{R(s)}$ (cf. Fig. 10.3).

10.3.5 Controller Feasibility Study RST_{imp} in Discrete Time

Given that RST controllers in this research, modeling, design, and simulation have been made in continuous-time mode, and for a hardware discrete-time realization, all the models and designs should be made in discrete time. Then, the controller has new constraints such as Nyquist frequency, sampling frequency, delay time, etc.

The general scheme of the injected current control loop system of the active filter, as it will be implemented, is given in Fig. 10.6. The term (z^{-1}) of the control loop represents the delay of a sampling period, due to the digitization/discretization and reconstruction of the continuous signal and to the computation time by the digital processor. In addition, the calculations are made during a first sampling period and can only be used at the next sampling period; this justifies the fact that sample frequency is equal to switching frequency in this case. The gains, in this scheme, represent the reduction rate of the implemented values, compared to those real of the electrical network, with

- G_{Iinj} the measurement gain of the injected current,
- G_{VI} the gain resulting from measurements of the currents and voltages of the electrical network,

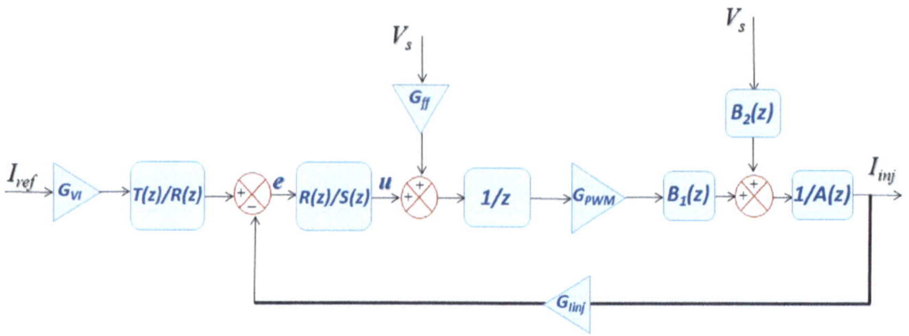

Fig. 10.6 Implementation scheme of the current control loop

- G_{ff} the measurement gain of the network voltage added to prevent the drawing/calling of a fundamental current from the network by the LCL filter, and
- G_{PWM} the gain of the pulse-width modulator.

From Fig. 10.6, we find that the transfer function of the controlled system, performed in s-plane in Eq. (10.1), becomes

$$\frac{B_{tot-d}(z)}{A_{tot-d}(z)} = \frac{G_{PWM}G_{Iinj}B_1(z)}{zA(z)} \tag{10.8}$$

with $\frac{B_1(z)}{A(z)}$ is the transfer function model of the LCL filter (including the network impedance), converted to discrete time using zero-order hold (Zoh). It is worth noting that the controlled system is of order four in discrete time, while it is of third order in continuous time. Therefore, in order to obtain the controller (RST_{imp}) polynomials in discrete time, we have to recalculate the RST_{imp} in continuous time domain, taking into account the delay z^{-1}. In this case, the overall controlled system in continuous time becomes

$$\frac{B_{tot}(s)}{A_{tot}(s)} = \frac{B_1(s)}{(\tau s + 1)A(s)} \tag{10.9}$$

with the time constant τ is equal to the sample time T_s applied in discrete-time scheme; it represents the delay time of the discrete-time model.

Following the same procedures of RST_{imp} polynomials design, described previously in Sects. 10.3.1 and 10.3.3, we have established the expression of the polynomials $R_{tot}(s)$, $S_{tot}(s)$, and $T_{tot}(s)$ of the controller. In this case, the order of $R_{tot}(s)$ and $R_{tot}(s)$ is the same as this of the overall controlled system presented in Eq. (10.9): $\frac{B_{tot}(s)}{A_{tot}(s)}$; thus $R_{tot}(s)$ and $S_{tot}(s)$ are of fourth order [5]. Then, $T_{tot}(s)$, presented in Eq. (10.10), is a fourth-order polynomial, which is the maximum order satisfying the causality of the transfer function $\frac{T_{tot}(s)}{R_{tot}(s)}$.

$$T_{tot}(s) = t_1 s^4 + t_2 s^3 + t_3 s^2 + t_4 s + t_5. \tag{10.10}$$

Taking account of the unknown parameters in this case (t_1, t_2, t_3, t_4, t_5), only three frequencies have to be chosen. Zeros should be placed near the slowest poles of the closed-loop transfer function $\frac{I_{inj}(s)}{I_{ref}(s)}$.

In this context, two methodologies of designing discrete-time controller can be employed. The first one is based on the discretization of the controllers, already designed in continuous time, by using the zero-order hold (Zoh) discretization method and bilinear (Tustin) method for the controlled system and the controller, respectively. The second methodology directly proceeds to design the controllers in z space.

10.3.6 Synthesis of Controller Feasibility RST_{imp} in Discrete Time

In the two methodologies' design, the conversion from continuous to discrete time, add two major constraints to controller design with respect to those treated in continuous time [5]. The first one is an additional degree added to the controlled system, and caused by the delay time (z^{-1}), while the second constraint is due to the Nyquist frequency which is equal to half of the sampling frequency f_s ($f_N = \frac{f_s}{2}$) [4]. It is worth noting that the two constraints influence the phase shift between I_{ref} and I_{inj}, in closed loop. Indeed, the phase shift increases with the order of controlled system and consequently with the order of the closed-loop transfer function in terms of tracking. Furthermore, the limitation by Nyquist frequency is very restrictive regarding the design principle of the improved controller (RST_{imp}) in continuous time. Indeed, a limited/halved bandwidth, by Nyquist frequency, does not allow sufficiently raise the phase curve, without influencing the reference tracking in terms of gain [4]. It is worth noting that in continuous time, the cutoff frequency/bandwidth of the transfer function, in terms of tracking, is limited by the switching frequency, which is equal, in this case, to the entire and not to half of the sampling frequency f_s.

In order to ensure a perfect conversion/design of the RST_{imp} controller in discrete time, while preserving robustness in stability and performances of the continuous time, we have to satisfy Shannon theorem. This implies increasing the sample frequency and consequently the switching frequency 10 times the current frequency (16 kHz). Such an increase is currently constrained by the technological limitation of high power industrial applications.

Despite the improved RST_{imp} presents, as a linear controller, a viable solution to circumvent the phase shift problem and, consequently to ensure a high-level harmonic filtration quality of the SAF, the ultimate solution can be obtained using the sliding-mode control (SMC) techniques that is inherently robust/insensitive to matched bounded disturbances.

10.4 Continuous SMC and HOSMC for Controlling Grid-Connected VSI with LCL Filter

10.4.1 Motivations

In most cases, the control laws designed for grid-connected VSI with LCL filter were originally established for the renewable energy (photovoltaic and wind) systems [13]. Usually, linear controllers are proposed for this kind of power converters (which have to inject only fundamental components into the grid), such as proportional–integral (PI) controller [14], proportional–resonant (PR) control strategies [15], repetitive control strategies [16], deadbeat control strategies [17], and so on. However, in the case of shunt active filter, the VSI has to inject, into the grid, both fundamental (reactive and unbalance) and harmonic components, which implies full control of a very large frequency bandwidth. Unfortunately, as was previously stated (Cf. Sect. 10.3.2), phase shift problem limits the possibility of applying linear controllers in the case of harmonic filtration.

In this work, it is proposed to employ SMC that ensures a desired dynamic response, strong robustness/insensitivity to matched bounded disturbances, and good regulation properties in a wide range of operating conditions [7, 8]. The SMC provides finite-time convergence to a sliding surface, which, usually, is a linear combination of the system's state variables and the generated references. It is a well-known fact that the ideal SMC generates the infinity (very high) frequency switching control that is needed to provide finite-time convergence to the sliding surface and keep the states there for all consecutive time in the presence of bounded disturbances. Physically, the VSI is run by the switching function of limited/fixed frequency [12], and the very high switching frequency, if happens, could lead to overheating and destruction of the inverter.

Therefore, in this work, the continuous SMCs are proposed to be employed. In order to prevent the generation of variable high switching frequency components, the use of a "sign" function approximated by a "*sigmoid*" function [18] and AIRD method [18, 19], processed by a PWM, are first proposed. In addition, continuous higher-order sliding-mode controller C-HOMC algorithm whose performance in a concert with PWM guarantees a given fixed switching frequency is proposed to be studied and compared to the other proposed SM controllers. Finally, the sliding variable for the SMC is chosen carefully, taking into account relative degree of the third-order output LCL filter in (10.2) and (10.3).

10.4.2 SMC Design

Considering the system (10.3) with $y := i_{inj}$ as the output, $u = v_f$ as the control input, and $w = v_s$ as the disturbances and $R_f = 0$. Then, Eq. (10.3) becomes

$$\dot{x} = Ax + Bu + Pw, \quad y = Cx, \quad C = [0\,1\,0], \tag{10.11}$$

where

$$A = \begin{bmatrix} -\frac{R_{f1}+R_f}{L_{f1}} & 0 & -\frac{1}{L_{f1}} \\ 0 & -\frac{R_{f2}+R_f}{L_{f2}} & \frac{1}{L_{f2}} \\ \frac{1}{C_f} & -\frac{1}{C_f} & 0 \end{bmatrix}, \quad B = \begin{bmatrix} \frac{1}{L_{f1}} \\ 0 \\ 0 \end{bmatrix}, \quad P = \begin{bmatrix} 0 \\ -\frac{1}{L_{f2}} \\ 0 \end{bmatrix}.$$

It is clear that the input–output $\frac{y}{u}$ dynamics of system (10.11) is of relative degree $r = 3$, since $CB = CAB = 0$ and $CA^2B = \frac{1}{L_{f1}L_{f2}C_f}$, then a sliding variable is chosen in the form [18]:

$$\sigma = K_0 e + K_1 \dot{e} + K_2 \ddot{e}, \tag{10.12}$$

where $e = i_{inj} - i_{ref}$. Apparently, the sliding variable (10.12) dynamics is of relative degree $r = 1$. Note that the derivatives $\dot{e}, \ddot{e}$ are obtained by using higher-order sliding-mode differentiator [20]. The positive coefficients K_0, K_1, and K_2 are selected to make the system (10.11) exponentially stable with the desired convergence rate in the sliding mode defined by $\sigma = 0$.

In order to design SMC that drives $\sigma \to 0$ in finite time and keeps the states of the system (10.11) in the sliding surface $\sigma = 0$ for all consecutive time, the sliding variable dynamics are derived

$$\dot{\sigma} = F\left(\frac{di_{ref}}{dt}, \frac{d^2 i_{ref}}{dt^2}, v_s, \frac{dv_s}{dt}, \frac{d^2 v_s}{dt^2}, i_{inj}, i_f, v_c\right) + \frac{K_2}{L_{f1}L_{f2}C_f} u, \tag{10.13}$$

where

$$\begin{aligned} & F\left(\tfrac{di_{ref}}{dt}, \tfrac{d^2 i_{ref}}{dt^2}, v_s, \tfrac{dv_s}{dt}, \tfrac{d^2 v_s}{dt^2}, i_{inj}, i_f, v_c\right) = K_0 \dot{e} + K_1 \ddot{e} + K_2 \left\{ \tfrac{d^3 i_{ref}}{dt^3} + \right. \\ & \left(-\left[\left(\tfrac{R_{f2}}{L_{f2}}\right)^2 - \tfrac{1}{L_{f2}C_f}\right]\tfrac{R_{f2}}{L_{f2}} + \tfrac{R_{f2}}{L_{f2}^2 C_f}\right) i_{inj} - \left(\tfrac{R_{f2}}{L_{f2}^2 C_f} + \tfrac{R_{f1}}{L_{f1}L_{f2}C_f}\right) i_f + \\ & \left(\left[\left(\tfrac{R_{f2}}{L_{f2}}\right)^2 \tfrac{1}{L_{f2}} - \tfrac{1}{L_{f2}^2 C_f}\right] \tfrac{1}{L_{f1}L_{f2}C_f}\right) v_c - \left(\left[\left(\tfrac{R_{f2}}{L_{f2}}\right)^2 - \tfrac{1}{L_{f2}C_f}\right]\tfrac{1}{L_{f2}}\right) v_s + \\ & \left. \tfrac{R_{f2}}{L_{f2}}\tfrac{dv_s}{dt} - \tfrac{1}{L_{f2}}\tfrac{d^2 v_s}{dt^2} \right\}. \end{aligned} \tag{10.14}$$

It is assumed that

- the disturbance v_s and its two first derivatives $\frac{dv_s}{dt}$, $\frac{d^2 v_s}{dt^2}$, and $\frac{d^3 i_{ref}}{dt^3}$ are bounded,
- the first two derivatives of the output tracking error $\dot{e}$, $\ddot{e}$ and the variables i_{inj}, i_f, v_c are bounded in a reasonable state domain that includes the operation point.

Then there exists $\eta > 0$ such that

$$\left| F\left(\frac{di_{ref}}{dt}, \frac{d^2 i_{ref}}{dt^2}, \frac{d^3 i_{ref}}{dt^3}, v_s, \frac{dv_s}{dt}, \frac{d^2 v_s}{dt^2}, i_{inj}, i_f, v_c\right)\right| \leq \eta$$

at least locally. The sliding-mode existence condition [8] $\sigma\dot{\sigma} < 0$ can be easily achieved by the SMC

$$u := v_f = -\lambda sign(\sigma), \tag{10.15}$$

where λ is fixed and defined by the saturation limiter block to $420V$ (see Fig. 10.3). Then the action of the control on is proportional to $\frac{K_2}{L_{f1}L_{f2}C_f}u$, which explains the choice of the K_0, K_1, and K_2 in (10.14).

Note that the stability of the system (10.11) motion in the sliding mode ($\sigma = 0$ in (10.12)) in the presence of the bounded disturbance w is guaranteed by

- the SMC (10.15) with the gain $\lambda > 0$ that meets the sliding-mode existence condition $\sigma\dot{\sigma} < 0$ and
- by an adequate choice of the coefficients $K_0, K_1, K_2 > 0$ of sliding variable (10.12) so that $\sigma = 0$ yields an asymptotically stable dynamics.

Next, in order to achieve the fixed switching frequency of the control (10.15),

1. the "*sign*" function in control (10.15) is approximated by the "*sigmoid*" function $sign(\sigma) \geq \frac{\sigma}{|\sigma|+\varepsilon}$. It is worth noting that ε should be chosen to ensure a good compromise between conserving the robustness of the SMC and avoiding the generation of very high switching frequency.
2. a consequent processing of the control (10.15) with the *sigmoid* approximation by the PWM block with a fixed frequency is executed.

10.4.3 SMC with an Artificial Increase of Relative Degree

The artificial increase of relative degree and designing SMC in terms of the control derivative yields the continuous robust SMC control function. In order to do it, we extend the system (10.11) as follows:

$$x_4 = u, \quad \dot{x}_4 = v$$

with v as the new SMC command. Then, the system (10.11) becomes

$$\dot{x} = A_e x + B_e v + P_e w, \quad y = C_e x, \quad C_e = [0\,1\,0\,0], \tag{10.16}$$

where

$$A_e = \begin{bmatrix} -\frac{R_{f1}}{L_{f1}} & 0 & -\frac{1}{L_{f1}} & \frac{1}{L_{f1}} \\ 0 & -\frac{R_{f2}}{L_{f2}} & \frac{1}{L_{f2}} & 0 \\ \frac{1}{C_f} & -\frac{1}{C_f} & 0 & 0 \\ 0 & 0 & 0 & 0 \end{bmatrix}, \quad B_e = \begin{bmatrix} 0 \\ 0 \\ 0 \\ 1 \end{bmatrix}, \quad P_e = \begin{bmatrix} 0 \\ -\frac{1}{L_{f2}} \\ 0 \\ 0 \end{bmatrix}.$$

Since relative degree $r = 4$ in system (10.16), the following sliding variable is chosen:

$$\sigma_e = K_0 e + K_1 \dot{e} + K_2 \ddot{e} + K_3 e^{(3)} \tag{10.17}$$

with $e = i_{inj} - i_{ref}$ and $e^{(3)} := \frac{d^3 e}{dt^3}$. The time derivative of the sliding variable (10.17) is obtained

$$\dot{\sigma}_e = K_0 \dot{e} + K_1 \ddot{e} + K_2 e^{(3)} + K_3 F\left(i_f, i_{inj}, v_c, \frac{d^4 i_{ref}}{dt^4}\right) + K_3 \frac{1}{L_{f1} L_{f2} C_f} v \tag{10.18}$$

with

$$F = -\frac{d^4 i_{ref}}{dt^4}\left(\frac{R_{f1}}{L_{f1}L_{f2}C_f} + \frac{R_{f2}}{L_{f2}^2 C_f}\right)\frac{di_f}{dt} + \left[-\left(\left(\frac{R_{f2}}{L_{f2}}\right)^2 \frac{1}{L_{f2}C_f}\right)\frac{R_{f2}}{L_{f2}} + \frac{R_{f2}}{L_{f2}^2 C_f}\right]\frac{di_{inj}}{dt} - \left[\left(\left(\frac{R_{f2}}{L_{f2}}\right)^2 - \frac{1}{(L_{f2}C_f)}\right)\frac{1}{L_{f2}} - \frac{1}{L_{f1}L_{f2}C_f}\right]\frac{dv_c}{dt} - \left[\left(\left(\frac{R_{f2}}{L_{f2}}\right)^2 - \frac{1}{(L_{f2}C_f)}\right)\frac{1}{L_{f2}}\right]\frac{dv_s}{dt} + \frac{R_{f2}}{L_{f2}^2}\frac{d^2 v_s}{dt^2} - \frac{1}{L_{f2}}\frac{d^3 v_s}{dt^3}. \tag{10.19}$$

It is worth noting that the derivatives $\dot{e}$, $\ddot{e}$, and $e^{(3)}$ are obtained by using a higher-order sliding-mode differentiator [20]. The positive coefficients K_0, K_1, K_2, and K_3 are selected in this case also to make the systems' (10.16), (10.17) dynamics exponentially stable with the desired convergence rate in the sliding mode $\sigma_e = 0$.

We assume that, as in the previous case, all disturbances and their derivatives, all the derivatives of the tracking errors, and all state variables of the system are bounded in a reasonable state domain that includes the operation point.

Then there exists $\eta > 0$ that $|\dot{\sigma}_e|_{v=0} \leq \eta$ at least locally. The sliding-mode existence condition [8] $\sigma_e \dot{\sigma}_e < 0$ for $\sigma_e \neq 0$ can be easily achieved by the SMC

$$v := -\lambda sign(\sigma_e), \tag{10.20}$$

where the value $\lambda = 10^7$ is tuned during the simulations. Finally, the continuous control function $u = \int vdt$ is generated and limited by the saturation block to 420 V (see Fig. 10.3). This is reflected in the scheme (presented in Fig. 10.3) by an integrator at the output of the SMC. Note that $\dot{\sigma}_e$ is proportional to $\frac{K_3}{L_{f1}L_{f2}C_f} v$, which explains the choice of the K_0, K_1, K_2, and K_3 in (10.20).

10.4.4 C-HOSMC Design

Note that the continuous higher-order sliding-mode controller C-HOSMC algorithm [20, 21] can handle systems with arbitrary relative degree. In our case, the sliding variable is selected as

$$\sigma = \dot{e} + ce \tag{10.21}$$

with relative degree $r = 2$ in accordance with Eq. (10.11).

The rationale behind such selection is as follows. In a noisy measurement environment, the sliding variable σ will converge to a domain whose size is proportional to the amplitude of the noise $w(t)$ of the σ measurement. Therefore, in the real sliding mode $\sigma = \dot{e} + ce = w(t)$, and the noise effect on the tracking error e is be attenuated due to a low-pass filtering with a cutoff frequency equal to c (in our case $c = 10^4$).

The sliding variable σ input–output dynamics are derived in accordance with Eqs. (10.11), (10.21)

$$\ddot{\sigma} = v + f(x, t), \tag{10.22}$$

where $v = \frac{1}{(L_s + L_{f2})L_{f1}C_f}u$, with u is the controller output (see Fig. 10.3) and $f(x, t)$, representing the other terms in Eq. (10.22), is derived in accordance with Eq. (10.11); the derivative of the term $f(x, t)$ is assumed bounded at least locally, i.e., $|\dot{f}(x, t)| \leq L$, $L > 0$, and is treated as the bounded disturbance in (10.22).

In accordance with the C-HOSMC algorithm [20, 21], the control law is designed as

$$v = v_1 - v_2, \tag{10.23}$$

where

$$v_1 = -\gamma_1 \lceil \sigma \rfloor^{\alpha_1} - \gamma_2 \lceil \dot{\sigma} \rfloor^{\alpha_2} \tag{10.24}$$

with

- $\gamma_1, \gamma_2 > 0$ are selected to make the polynomial $s^2 + \gamma_2 s + \gamma_1$ Hurwitz with desired root placement,
- (α_1, α_2) are to be computed as $\alpha_1 = \frac{\alpha}{2-\alpha}$, $\alpha_2 = \alpha$, $\alpha \in (0, 1)$; if $\alpha = \frac{1}{2}$ is selected, then $\alpha_1 = \frac{1}{3}$, $\alpha_2 = \frac{1}{2}$, and v_2 chosen for disturbances rejection purposes as

$$v_2 = -\omega, \tag{10.25}$$

where

$$\omega = -\lambda \lceil \varsigma \rfloor^{\frac{1}{2}} - \beta \int_0^t sign(\varsigma) d\tau \quad \text{and} \quad \varsigma = \dot{\sigma} - \int_0^t v_1 d\tau. \tag{10.26}$$

It is proven in [20, 21] that the control law in (10.23)–(10.26)

- drives σ, $\dot{\sigma}$ to zero in finite time,
- is continuous, and

- therefore is a second-order continuous sliding-mode control (C-HOSMC). Note that the control continuity and robustness are achieved here without artificially increasing relative degree.

Computation of $\gamma_1,\ \gamma_2$: They are computed [20, 22] as coefficients of a second-order polynomial; the eigenvalues of this polynomial are chosen to provide a given transient response while being limited by the VSI switching frequency, enforced by the PWM carrier signal.
Computation of $\lambda,\ \beta$: They are chosen [20] as follows:

$$\lambda = 1.5\sqrt{L}, \quad \beta = 1.1L; \tag{10.27}$$

in our case, L is set to 3×10^{18}. Finally, the command u applied to the inverters switches is given

$$u = a_1 v$$

with $a_1 = (L_s + L_{f2})L_{f1}C_f$ of the transfer function $\frac{B_1(s)}{A(s)}$ of Eq. (10.1).

10.5 Simulation Results

The performances of the active filter controlled by two continuous SMC and one C-HOSMC algorithms will be validated in two simulation setups. The first one is based on a conventional nonlinear load, and the second one will be based on a practical commercial case study.

For the first simulation setup, the network consists of a power transformer 20/0.4 kV, 1 MVA, $ucc = 6\%$, powered by a network whose short-circuit power is $S_{sc} = 500$ MVA. The nonlinear load is a conventional 100 kVA six-pulse diode rectifier with the $R//C$ load.

The harmonics filtration performance of the shunt active filter is tested in order to eliminate the harmonics generated by the nonlinear load using MATLAB, Simulink, and Simscap-Sim_Power_System. In this simulation, the shunt active filter consists of the LCL filter, connected to the grid via the PWM-VSI, while the injected current is controlled, first, by SMC, then by SMC with $sigmoid$ function approximation, then by SMC-AIRD, and finally by C-HOSMC algorithms, with all controls processed by the PWM with 16 kHz switching frequency. The system rating values, considered in the general structure, are given in Table 10.2. The LCL parameters were chosen to impose 2000 Hz as cutoff frequency; this will ensure a −50 dB rejection rate of the switching frequency components. The choice of the inductances (L_{f1}, L_{f2}) and consequently of the capacitance C_f, is constrained by an industrial applicability and economic conditions. Besides, the storage parameter values ($C,\ V_{dc}$) are selected to provide the given ripple ratio of the voltage of the DC bus, and also the dynamic performance of the active filter [4].

Table 10.2 Electrical network characteristics

Electrical network	
E, Ssc, Rsc, Lsc	20 kV, 500 MVA, Rsc = 0.253, Lsc = 0.0024 H
Power transformer	
S_n, ΔP_{Cu}, ΔPo, u_{cc}, I_o	1 MVA, 10.5 kW, 1.7 kW, 6%, 1.3%
Shunt active filter	
Output LCL filter	$L_{f1} = 90\,\mu H$, $R_{fl} = 5\,m\Omega$
	$L_{f2} = 100\,\mu H$, $R_{fl} = 5\,m\Omega$
	$C_f = 130\,\mu F$,
Storage capacity	$C = 0.6\,\mu F$, $V_{dc} = 840\,V$
LCL cutoff frequency	2000 Hz
Switching PWM frequency	16 kHz

10.5.1 Simulink Simulation Results

The simulations with SMC, SMC-*sigmoid* function approximation, SMC-AIRD, and then with C-HOSMC are performed using, at the first step, a simple Simulink block scheme. In this scheme, current harmonics are modeled by harmonic current sources of ranks 5, 7, 11, 13, 17, and 19, with respect to the practical case study in the second simulation setup. The classical *sign* function is approximated by a hysteresis function, with very small band (10^{-16}), while the *sigmoid* function will be the following: $sign(\sigma) = \frac{\sigma}{|\sigma|+\varepsilon}$ with $\varepsilon = 2 \times 10^8$; note that the absolute value of σ is almost 2×10^9. Consequently, in normalized value, ε will be equal to 0.1.

Simulation plots of I_{ref}, I_{inj}, and the control function u are shown in Fig. 10.7. It can be observed that the very accurate tracking is achieved by the high-frequency switching SMC, AIRD continuous controller, and C-HOSMC controller, while the tracking accuracy is slightly degraded for continuous SMC with *sigmoid* function approximation that is expected.

10.5.2 Simscap-Sim_Power_System Simulation Results

In this simulation setup, the simulation is made using the complete model of the Simscap-Sim_Power_System. The active filter is activated after 5 operating periods (0.1 s); it operates, first, during 3 periods (to 0.16 s), with SMC. Next, it continues, during 3 periods (up to 0.22 s) controlled by the continuous SMC with *sigmoid* function approximation, and, finally, after 0.22 s to the end of the simulation (0.28 s), the active filter is driven by the continuous SMC-AIRD control. Another simulation is performed within the same environment as the previous case, for a SAF operating with C-HOSMC controller. The SAF is deactivated in this case after five periods (up to 0.16 s).

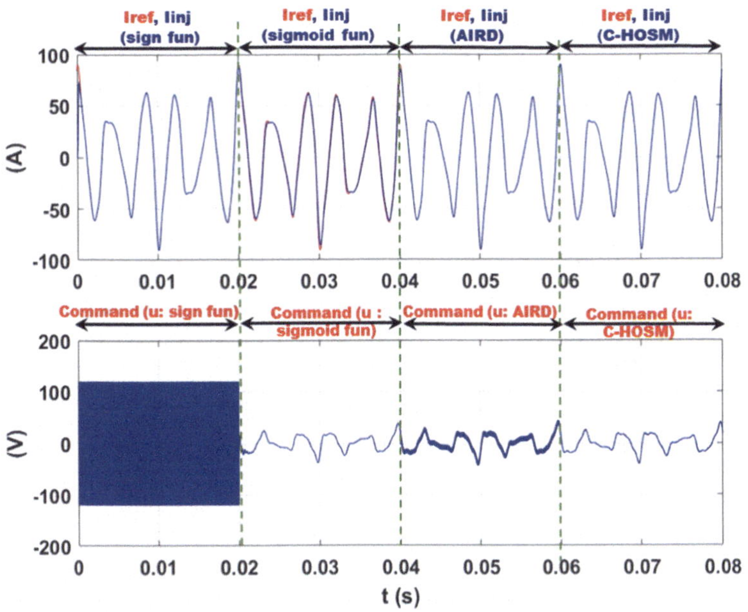

Fig. 10.7 Tracking and controls shapes for SMCs: sign, *sigmoid* functions, AIRD, and C-HOSMC algorithms

10.5.2.1 SMC Algorithms

Figure 10.8 shows simulation results in the time domain of the shunt active filter, for the three-phase network currents (I_s) before and after compensation, the identified harmonic current I_{ref1} and the injected current to the network I_{inj1}, for phase1, superimposed and the total harmonic distortion of current ($THD - I_s$) at network side, before and after compensation.

From Fig. 10.8, one can observe that, despite the presence of the LCL filter, there is no phase lag between the current reference and the injected current for high-frequency switching SMC and the continuous AIRD control, while the continuous SMC with the *sigmoid* function approximation causes a very small phase shift (see the zoomed plot presented in Fig. 10.8). Thus, the source line current pattern is sinusoidal for classical SMC, continuous AIRD-SMC, and SMC with the *sigmoid* approximation without any presence of high-frequency components at the network side, due to the presence of LCL filter.

This is shown in Fig. 10.8 by a steep decrease of current *THD* ($THD - I_s$) of the phase 1 at the network side from 22.4% before compensation to 0.9, 1.5, and 1.15% after compensation, for SMC, SMC-*sigmoid*, and AIRD-SMC algorithms, respectively.

The same observation can be noticed for the network voltage, in Fig. 10.9. Indeed, a significant decrease of voltage *THD* ($THD - V_s$) at the grid side, from about 4.5% before filtration to 0.7, 0.93, and 0.75% considering classical SMC, SMC-*sigmoid*, and AIRD- SMC algorithms, respectively, could be observed.

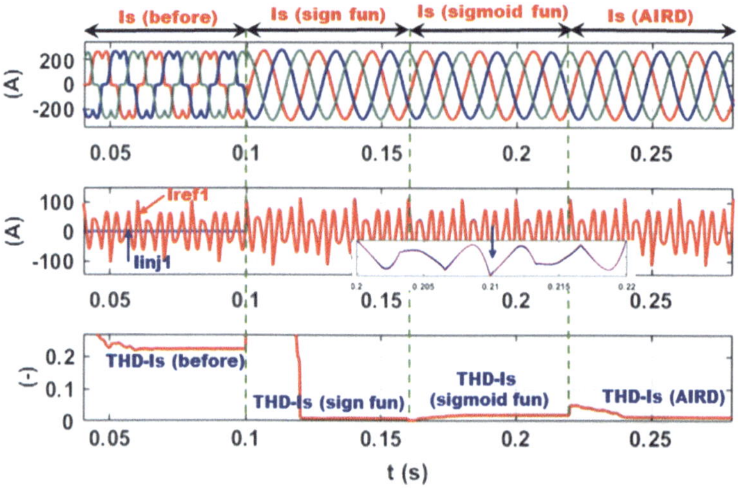

Fig. 10.8 Current filtration using time- and frequency-domain SM controllers

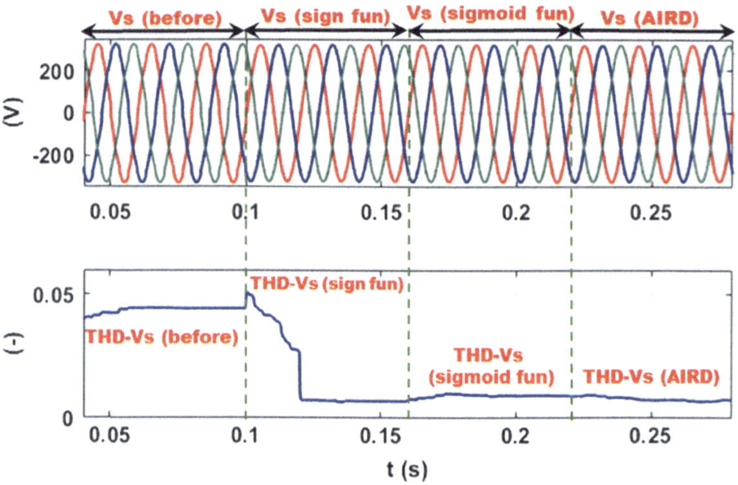

Fig. 10.9 Voltage filtration using time- and frequency-domain SM controllers

It is worth noting that $\varepsilon = 2 \times 10^8$ yelled SMC-*sigmoid* controller to maintain, as much as possible, the robustness of the SMC, while avoiding the generation of very high switching frequency.

The very good filtration performance of the active filter demonstrates that the two continuous control methods (i.e., SMC with *sigmoid* approximation and AIRD algorithm) represent very good alternatives to the discontinuous SMC method. This results in a much lower *THD* voltage than the 1.6%, required by Electricité de France (EDF) recommendations [4]. It is worth noting that in industrial zones, recommendations (e.g., IEEE STD 519-2014) limit *THD* voltage to 5–8%.

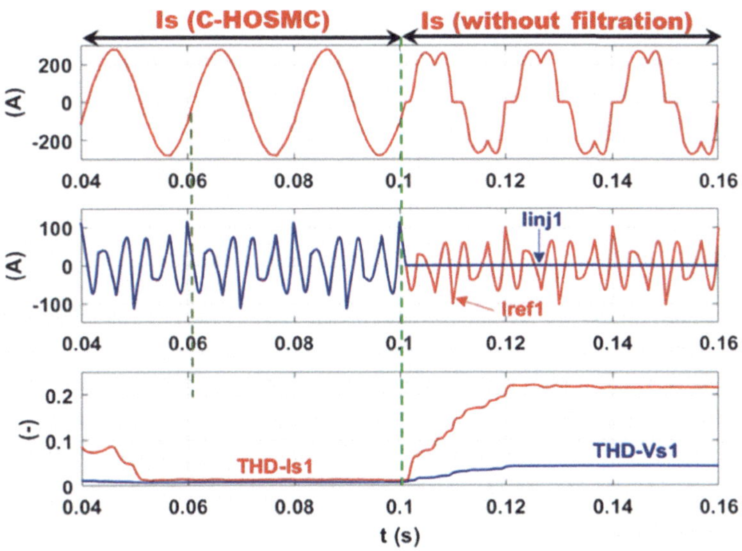

Fig. 10.10 Current and voltage filtration using time- and frequency-domain C-HOSM controller

10.5.2.2 C-HOSMC Controller

In this simulation case, the active filter, operating with C-HOSM controller, is, unlike the previous simulation case, deactivated after 0.1 s, in order to estimate its behavior (the response time, overshoot, oscillation, etc.) in this critical case. From Fig. 10.10, an almost sinusoidal source line current on the network side with filtration can be observed. This is affirmed, in same Fig. 10.10, by a significant decrease of current *THD* ($THD - I_s$) on the network side to 1.32% with filtration. This is accompanied by a steep decrease of voltage *THD* ($THD - V_s$) at the grid side after filtration to 0.9% as shown in Fig. 10.10.

It is worth noting that this controller also ensures continuous command signal with current and voltage *THDs* satisfying both EDF and IEEE recommendations.

10.5.2.3 Simulation Results' Summary

Table 10.3 presents a summary report of all the simulation results. From this table and based on all the simulation results, it can be noticed that all SMC-*sigmoid* function approximation, SMC-AIRD, and C-HOSM controllers ensure continuous command signal with very good filtration quality of the SAF.

On the other hand, and although the SMC-*sigmoid* function approximation controller provides continuous control signal with good filtration quality, it does not generally preserve the robustness provided by the SMC, especially in terms of disturbances rejection [18]. In this research, ε has been chosen to retain as much as

Table 10.3 Simulation results' summary

Control algorithm	SMC-sign (%)	SMC-sigmoid (%)	SMC-AIRD (%)	C-HOSM
Without filtration: THD-Is =22.4%, THD-Vs = 4.5%				
THD-Is	0.9	1.5	1.15	1.32
THD-Vs	0.7	0.93	0.75	0.9%

possible the robustness of the SMC, while avoiding the generation of very high switching frequency.

In the same context, the advantage of both SMC-AIRD and C-HOSM controllers is that they guarantee a robust performance, especially in terms of disturbances rejection [19–22]. It is worth noting that the third-order manifold SMC-AIRD controller is more difficult to implement compared to the second-order C-HOSM controller, due to the additional derivative which is difficult to implement in the presence of harmonics.

10.6 Practical Case Study

This case study deals with a real case, i.e., the textile factory of Melaiece, located in the industrial zone of Sheikh Najjar, in Aleppo, Syria.

10.6.1 General Description of the Industrial Site

This is a yarn-producing factory, working at a circadian rhythm (two circadian per day). The yarn manufacturing is based on four stages: the opening phase of cotton, the stage of carding or combing (it breaks up locks and unorganized clumps of fiber), the stage of throwing (for producing a continuous thread), and the spinning stage.

10.6.2 Electrical Data/Description of the Site

This industrial site is powered by a half-ring of two lines of medium voltage of 20 kV, coming from the Sheikh Najjar's substation transformer of 66/20 kV. These two lines provide power to the transformer of the site (20/0.4 kV of 1 MVA) that powers the factory production line, via underground cables of two single-core ($3 \times 300\,\text{mm}^2$), with 0.4 kV, as shown in Fig. 10.11.

The production line contains 56 asynchronous motors, 40 of them being driven by inverters and the nominal power of each one varying between 0.25 and 75 kW. It is worth noting that the asynchronous-driven motors represent nonlinear loads, where each asynchronous machine is supplied via a combination of three-phase rectifier followed by three-phase voltage inverter.

The transformer load factor is only 42%, because the factory was originally designed to operate with two production lines, but currently, there is only one line left.

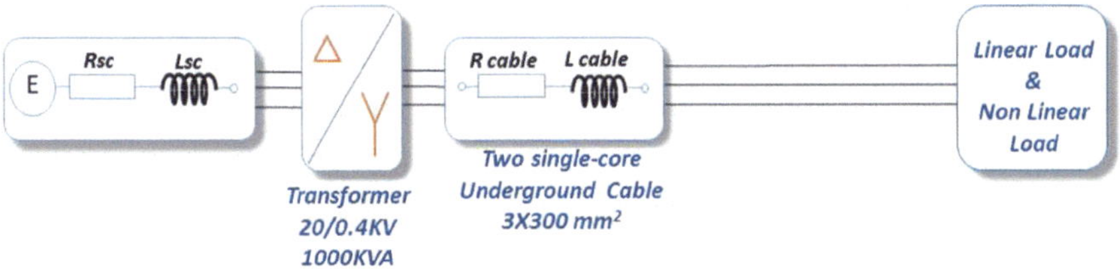

Fig. 10.11 Electrical network of the studied site

Fig. 10.12 Electrical measurements using power quality analyzer

The factory power factor is 0.82 without reactive power compensation and 0.98 with automatic power factor correction capacitors. All loads could be switched-off or activated via the main distribution panel.

In order to identify electrical problems deteriorating the production quality, we carried out many measurements during one week, using several devices of power quality analyzers/power analyzer, as shown in Fig. 10.12.

Finally, we present (see Fig. 10.13 from "*a*" till "*i*") the resulting measurement plots, for the most significant period from 10/03/2009 : 19h:10 to 11/03/2009 : 16h:55.

Figure 10.13 shows the plots for the three phases ($Ph1$, $Ph2$, and $Ph3$), (on a) the total load currents (I_{L123} in A), (on b) the phase network voltages (V_{s123} in V), (on c) the power factors (PF_{123}), (on d) the total harmonic distortion of the load current (THD_{i123} in %), (on e) the total harmonic distortion of the network voltage (THD_{V123} in %), and (on $f - i$) the individual harmonic distortion (in %) of the load current, respectively, for the 5th, 7th, 11th, and 13th components. It is worth noting that the measurements were made up to the 50th component. The details are presented in the next subsection.

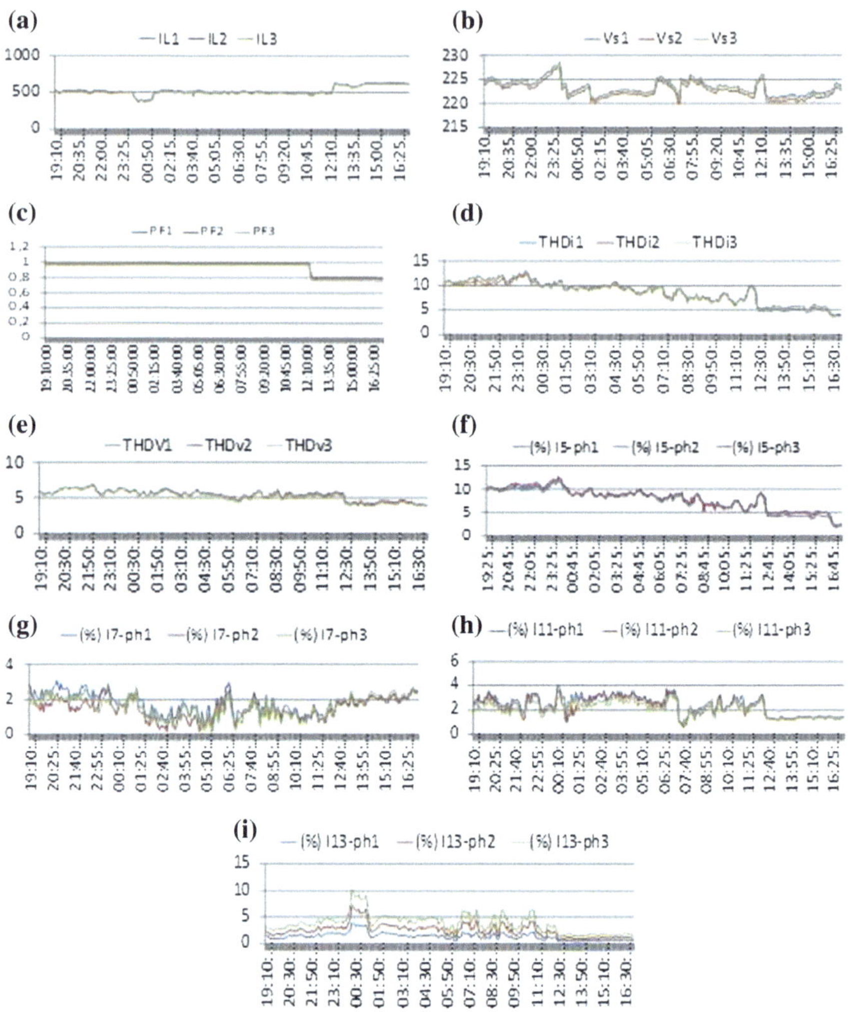

Fig. 10.13 Measurement curves using power quality analyzer

10.6.3 Modeling of the Industrial Site Network

The factory network consists of a power transformer of 20/0.4 kV, 1 MVA, $ucc = 6\%$, supplied by a network whose short-circuit power is $Ssc = 500\,\text{MVA}$. The cable between the transformer and the main distribution panel is two single-core $3 \times 300\,\text{mm}^2$ with 20 m long. The active and reactive power consumptions are modeled by an inductive load. It is worth noting that harmonic loads are not available in MATLAB toolbox libraries; the only ones that MATLAB offers are rectifiers and inverters that drive motors, charge batteries, etc. Therefore, it becomes almost impos-

Table 10.4 Catalogue characteristics of the electrical network

Electrical network	
E, Ssc, Rsc, Lsc	20 kV, 500 MVA, 0.253 Ω, Ls = 0.002425 H
Power transformer	
S_n, ΔPc_u, Δ Po, u_{cc}, I_o	1 MVA, 10.5 kW, 1.7 kW, 6%, 1.3%
Electrical cable	
3 × 300 mm^2, Len = 20 m	Xcable = 0.16 Ω/km, Rcable = 0.059 Ω/km

sible to model and to simulate an industrial network, containing 40 driven motors like as in the case study. Furthermore, in order to carry out simultaneous measurements for all nonlinear loads on an industrial site, we should use as much measurement devices (power analyzers) as driven loads (40 power analyzers in our case).

In this work, it is proposed replacing the nonlinear loads (asynchronous-driven motors) models by their individual harmonics currents, measured on the main distribution panel side [23, 24]. These individual harmonics currents will be modeled by sources of current harmonic with parallel resistors. This modeling approach is very easy to implement for the network of an arbitrary size, regardless of the types of the load and command. Furthermore, measurements can be carried out by only one or few measuring devices.

Finally, the system rating values of the industrial network, illustrated in Fig. 10.13, are given in Table 10.4.

10.6.4 Modeling Validation

Based on the measurement (Fig. 10.13), we pick up the time instant (22h: 05 : 00 in our case) when the total harmonic distortion of voltage reaches its maximum value. The results are presented in Table 10.5.

The simulation results of the factory network are performed by MATLAB, Simulink, Simscap-Sim_Power_System code, and shown in Fig. 10.14. The electrical network, presented in Fig. 10.11, is modeled on the basis of the previous measurements (shown in Fig. 10.13 and Table 10.5) and of the network components characteristics (see Table 10.4). In Fig. 10.14, the following simulation plots (in time and frequency domains) are shown: phase voltages ($Vs1 = 223.9$ V, $Vs2 = 223.07$ V, $Vs3 = 223.68$ V), load currents ($IL1 = 520.7841$ A, $IL2$=518.38 A, $IL3$=507.13 A), power factors ($PF1 = 0.967$, $PF2 = 0.978$, $PF3 = 0.959$), $THD - V$ ($THD - Vs1 = 6.46\%$, $THD - Vs2 = 5.9\%$, $THD - Vs3 = 6.65\%$), and $THD - I$($THD - I1 = 11.65\%$, $THD - I2 = 10.76\%$, $THD - I3 = 11.95\%$). From the same figures, individual harmonic distortion of the load currents are $I5 - ph1, 2, 3 =$ (11.02%, 10.42%, 11.51%), $I7 - ph1, 2, 3 =$ (2.72%, 1.78%, 2.36%), $I11 - ph1, 2, 3 =$ (1.71%, 1.27%, 1.36%), $I13 - ph1, 2, 3 =$ (1.5%, 1.08%, 1.17%), $I17 - ph1, 2, 3 =$ (1.13%, 0.61%, 1.17%), and $I19 - ph1, 2, 3 =$ (0.94%, 0%, 0.72%).

Table 10.5 Electrical network quantities

	DATE 03/10/2009 Ph1	Time 22:05:00 Ph2	Hz 49.98 Ph3
V_S (V)	224.5	223.8	224.2
I_L (A)	522.19	518.3	507.1
P_{Load} (W)	114379.2	114054.2	110071.88
Q_{Load} (VAR)	18812.18	15860.68	24532.81
S_{Load} (VA)	116769.1	115968.4	113611.6
PF	0.97	098	0.96
THD-V_S (%)	6.7	6.2	6.8
THD-I_L (%)	11.6	10.7	12
Ih5 (%)	11	10.4	11.5
Ih7 (%)	2.8	1.7	2.4
Ih11 (%)	1.7	1.2	1.4
Ih13 (%)	1.4	0.9	1.1
Ih17 (%)	1	0.5	1.1
Ih19 (%)	0.9	0	0.7

with P_{Load}, Q_{Load}, S_{Load} are respectively active, reactive and apparent power measured on the main distribution side

By comparing simulation and measurement results (presented on Fig. 10.13 and Table 10.5), it can be observed that all quantities are almost identical with maximum error of utmost 3%. This result validates the proposed models, in particular, the proposed nonlinear loads model. Therefore, all the subsequent simulations results, including the case of integrating the shunt active filter, are credible.

10.6.5 Simscap-Sim_Power_System Simulation of the SAF Within the Industrial Site

10.6.5.1 SMC Algorithms

In the second simulation setup, the active filter also operates at the same sequences as in the first simulation setup.

Figure 10.15 shows simulation results for the shunt active filter, the three-phase network currents (I_s), and the total harmonic distortion of current ($THD - I_s$) at network side, before and after compensation.

One can observe that, despite the unbalanced current and voltage of the site network as well as the presence of the *LCL* filter, all SM controllers ensure a very accurate tracking and, consequently, a sinusoidal shape of the network current after filtration. This can be observed at Fig. 10.15, by neglected current THD of the phases 1 , 2, 3 after compensation (0.4%, 0.8, % 0.5%: the biggest THDs, respectively, for classical SMC, SMC-*sigmoid* functions, and the AIRD control algorithm); the cur-

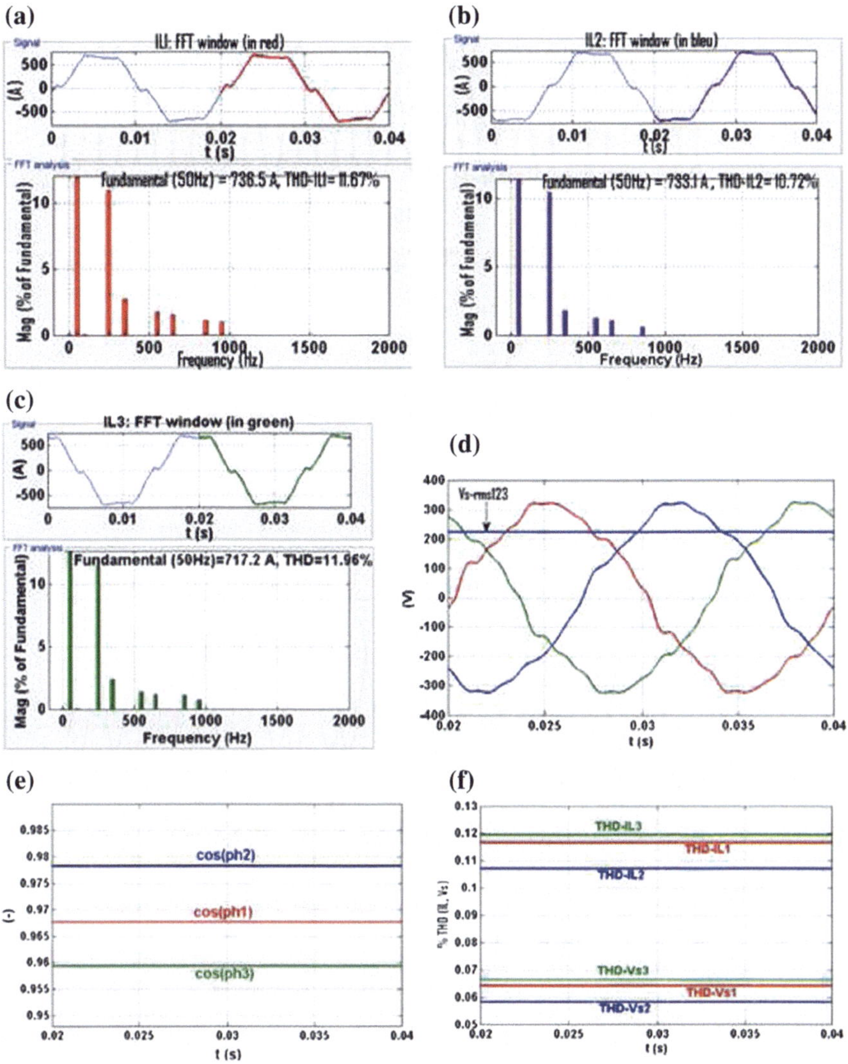

Fig. 10.14 Simulation results of the studied network

rent *THDs* ($THD - I_s$) before compensation are, respectively, 11.6, 10.7, and 12% for the three phases.

The significant decrease in the current *THD* leads, as shown in Fig. 10.16, to a steep decrease of voltage *THD* ($THD - V_s$) at *PCC*, from about 6.7, 6.2, and 6.8% before filtration to 0.28, 0.5, and 0.3%: the biggest *THDs*, respectively, for the three tested controllers, after filtration.

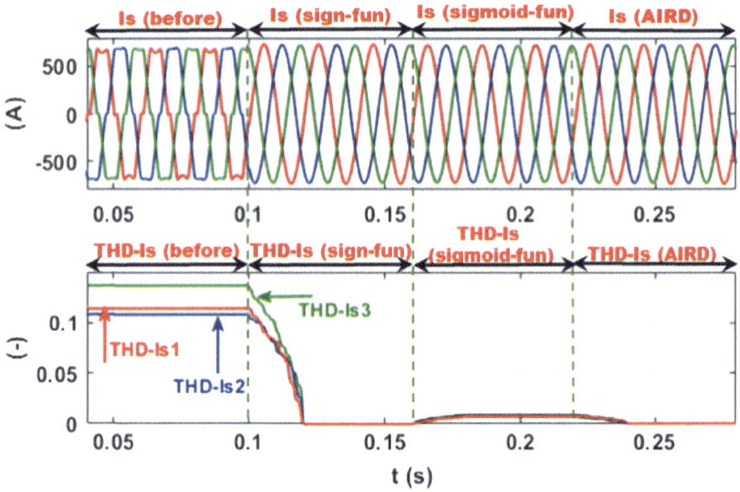

Fig. 10.15 Current filtration using time- and frequency-domain SM controllers

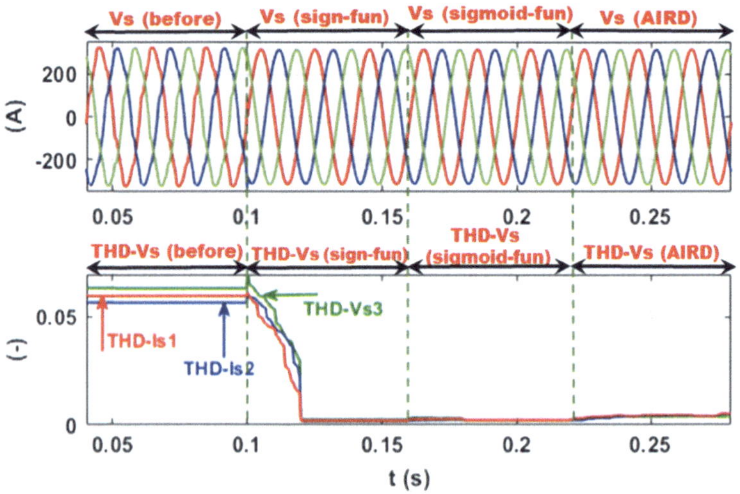

Fig. 10.16 Voltage filtration using time- and frequency-domain SM controllers

Once again and for a real disturbed network, the choice of $\varepsilon = 2 \times 10^8$ allowed SMC-*sigmoid* controller to almost preserve the robustness of the SMC, while also avoiding the generation of very high switching frequency.

These results prove, once again, that despite the unbalance and the significant harmonic distortion of the network voltage, the active filter ensures, thanks to SMC with *sigmoid* approximation and AIRD algorithms, a *THD* voltage well below the most constrained recommendation standards (the 1.6% of EDF norm).

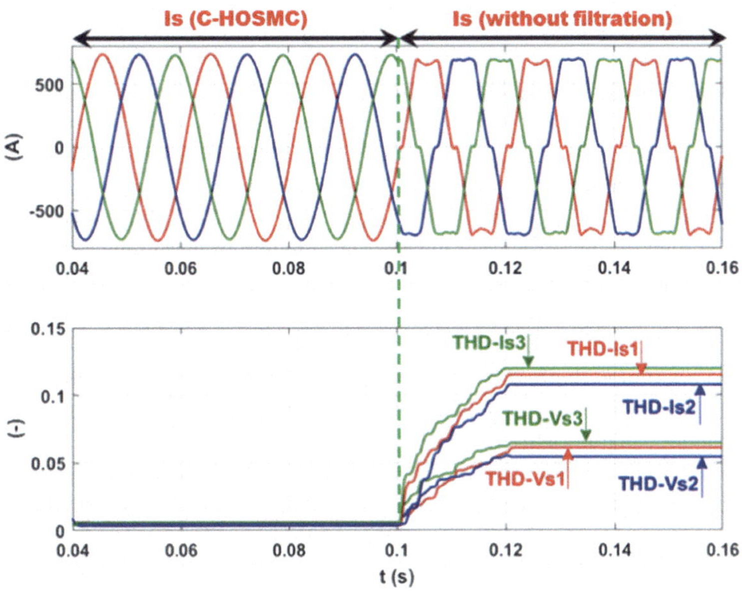

Fig. 10.17 Current and voltage filtration using time- and frequency-domain C-HOSM controller

10.6.5.2 C-HOSMC Algorithm

In this simulation setup, the active filter, operating with C-HOSM controller, is deactivated after five periods (see Sect. 10.5.2.2, within the same simulation characteristics as the previous cases. Figure 10.17 presents simulation results for the SAF, the three-phase network currents (I_s), and the total harmonic distortion of both current ($THD - I_s$) and voltage ($THD - V_s$) at network side, before and after filtration.

From this figure, an almost sinusoidal shape source line current on the network side witness a good filtration performance of the SAF. This observation is validated, in Fig. 10.17, by a significant decrease of current THD ($THD - I_s$) of the phases $1,\ 2,\ 3$ on the network side to 0.69%, 0.46%, and 0.54%, respectively. This steep decrease in the current THD leads to a significant decrease of voltage THD ($THD - V_s$) at PCC to $0.4,\ 0.34$, and 0.42% as shown in Fig. 10.17.

10.7 Conclusion

In this chapter, the continuous sliding-mode control algorithms, including SMC with *sigmoid* function approximation, SMC-AIRD, and continuous higher-order sliding mode (C-HOSMC) for controlling the LCL grid-connected three-phase and three-wire shunt active filter (SAF), are studied.

It is known that a linearly controlled LCL filter exhibits a phase lag between identified current reference and the injected one, preventing a good harmonic filtration.

The linear improved root locus controller (RST_{imp}) in continuous- but especially in discrete-time implementation reduces the phase lag to some extent, while the remaining phase lag depends on both switching/sample frequency and the order of the controlled system.

This work proves the advantages of sliding mode over the linear controllers by improving tracking accuracy over the entire bandwidth of harmonics. First, sliding-mode controller (SMC) was tested. Then, in order to prevent a very high-frequency switching effects of the SMC, both the SMC with the *sigmoid* approximation and the SMC with artificial increase of relative degree (AIRD) followed by an integrator were studied. A continuous higher/second-order sliding-mode controller (C-HOSMC) was studied as well. Moreover, a PWM technique was used to allow VSI operating at the fixed frequency that is easily filtered by the predesigned LCL filter.

A robust high-accuracy tracking performance of the considered electric power system driven by the proposed control algorithms was demonstrated via the simulations, carried out by MATLAB, Simulink, and Simscap-Sim_Power_System code, for the tutorial example and the practical case study. Indeed, it was observed that the phase lag between the current reference and the injected current has disappeared for the SMC, continuous AIRD, and C-HOSMC algorithms, while continuous SMC with the *sigmoid* function approximation yielded the very good performance with a small (negligible) phase shift. Thus, the source line current pattern (and consequently the PCC network voltage shape) remains sinusoidal for all controllers processed by the PWM. The adversary effects of discontinuous very high-frequency switching command, generated by SMC, were accommodated by using PWM-SMC with *sigmoid* function approximation, SMC-AIRD, and C-HOSM algorithm.

In the case study, the unbalanced electrical network that contains 56 asynchronous motors (between 0.25 and 75 kW), among which 40 are driven by inverters, is considered. This industrial textile site was modeled, in this work, using the real measurements carried out using power quality analyzer device. This work proposed replacing the complicated load-driven models by their individual harmonic current models. The use of this novel modeling approach facilitates the study of an industrial site, with the arbitrary size of the network, without taking care of the nonlinear load types and the command signals. The proposed modeling approach reduces a complexity of simulating the real models. Furthermore, in accordance with this modeling approach, the real measurements could be carried out by only one or few measuring devices. The obtained results have validated the proposed approach of modeling with almost 97% accuracy. Besides, filtration results have proved the efficacy of the all proposed control methods, even under unbalanced condition of the network currents and voltages.

It is worth noting that the SMC-*sigmoid* function approximation controller has the drawback/disadvantage of not preserving the robustness provided by the SMC, especially in terms of disturbances rejection. In this research, ε of the *sigmoid* has been chosen to yelled SMC-*sigmoid* controller to preserve, as much as possible, the robustness of the SMC, without generation of very high switching frequency.

On the other hand, although both SMC-AIRD and C-HOSM controllers offer a robust performance, especially in terms of disturbances rejection, the SMC-AIRD controller is harder to implement than the C-HOSM controller, because of its additional derivative, complicated to realize in the presence of harmonics.

References

1. Stefanutti, W., Mattavelli, P.: Fully digital hysteresis modulation with switching-time prediction. IEEE Trans. Ind. Appl. **42**, 763–769 (2006)
2. Bouchafaa, F., Beriber, D., Boucherit, M.S.: Modeling and control of a gird connected PV generation system. In: 18th Mediterranean Conference on Control and Automation (MED), Marrakech, Morocco, pp. 315–320 (2010)
3. Wang, F., Duarte, J.L., Hendrix, M.A.M., Ribeiro, P.F.: Modeling and analysis of grid harmonic distortion impact of aggregated DG inverters. IEEE Trans. Power Electron. **26**(3), 786–797 (2011)
4. Alali, M.A.E.: Contribution à l'Etude des Compensateurs Actifs des réseaux Electriques Basse Tension. PhD thesis, Louis Pasteur University, Srasbourg I, Strasbourg, France (2002)
5. Alali, M.A.E., Chapuis, Y.A., Saadate, S., Braun, F.: Advanced common control method for shunt and series active compensators used in power quality improvement. IEE Proc.-Electr. Power Appl. **151**(6), 658–665 (2004)
6. Cruz-Zavala, E., Moreno, J. A.: Lyapunov approach to higher order sliding mode design. In: Fridman, L., Barbot, J-P., Plestan, F. (eds.) Recent Trends in Sliding Mode Control, 1st edn., pp. 03–28. IET Digital Library (2016)
7. Jung, S.L., Tzou, Y.Y.: Discrete sliding-mode control of a PWM inverter for sinusoidal output waveform synthesis with optimal sliding curve. IEEE Trans. Power Electron. **11**(4), 567–577 (1996)
8. Utkin, V.I.: Sliding Modes in Control and Optimization, 1st edn. Springer, Berlin (1992)
9. Shtessel, Y., Baev, S., Biglari, H.: Unity power factor control in three phase AC/DC boost converter using sliding modes. IEEE Trans. Ind. Electron. **55**(11), 3874–3882 (2008)
10. Doria-Cerezo, A., Biel, D., Fossas, E.: Sliding mode control of LCL full-bridge rectifier. In: Fridman, L., Barbot, J-P., Plestan, F. (eds.) Recent Trends in Sliding Mode Control, 1st edn., pp. 361–405 IET Digital Library (2016)
11. Akagi, H.: New trends in active filters for power conditioning. IEEE Trans. Ind. Appl. **32**(6), 1312–1322 (1996)
12. Aali, M.A.E., Barbot, J.-P.: A first order sliding mode controller for grid connected shunt active filter with a LCL filter. In: 20th World Congress of the International Federation of Automatic Control (IFAC), Toulouse, France, pp. 377–382 (2017)
13. Wang, F., Duarte, J. L., Hendrix, M.A.M., Ribeiro, P.F.: Modeling and analysis of grid harmonic distortion impact of aggregated DG inverters. IEEE Trans. Power Electron. **26**(3), 786–797 (2010)
14. Dannehl, J., Wessels, C., Fuchs, F.W.: Limitations of voltage-oriented PI current control of grid-connected PWM rectifiers with LCL filters. IEEE Trans. Ind. Electron. **56**(2), 380–388 (2009)
15. Shen, G., Zhu, X., Zhang, J., Xu, D.: A new feedback method for PR current control of LCL-filter-based grid-connected inverter. IEEE Trans. Ind. Electron. **57**(6), 2033–2041 (2010)
16. Mattavelli, P., Marafão, F.P.: Repetitive-based control for selective harmonic compensation in active power filters. IEEE Trans. Ind. Electron. **51**(5), 1018–1024 (2004)
17. Kawabata, T., Miyashita, T., Yamamoto, Y.: Dead beat control of three phase PWM inverter. IEEE Trans. Ind. Electron. **5**(1), 21–28 (1990)

18. Aali, M.A.E., Shtessel, Y., Barbot, J.-P.: Control of grid-connected shunt active/LCL filter: continuous sliding mode control approach. In: 2018 15th International Workshop on Variable Structure Systems (VSS), Graz, Austria, pp. 319–324 (2018)
19. Aali, M.A.E., Shtessel, Y., Barbot, J.-P.: Continuous SM controller with artificial increase of relative degree for grid-connected shunt active/LCL filter. In: 2018 IEEE 18th International Power Electronics and Motion Control Conference (IEEE-PEMC), Budapest, Hungry, pp. 849–899 (2018)
20. Shtessel, Y., Edwards, C., Fridman, L., Levant, A.: Sliding Mode Control and Observation, 1st edn. Birkhäuser, Basel (2014)
21. Edwards, C., Shtessel, Y.: Adaptive continuous higher order sliding mode control. Automatica **65**, 183–190 (2016)
22. Bhat, S., Bernstein, D.S.: Geometric homogeneity with applications to finite-time stability. Math. Control Signals Syst. **17**, 101–127 (2005)
23. Karania, N., Alali, M.A.E.: Nonlinear hysteresis control for parallel active filters (optimization, comparative study with PWM and case study in industrial site). R. J. of Aleppo University, Engineering Science Series, vol. 84 (2010) (In Arabic)
24. Aali, M.A.E., Shtessel, Y., Barbot, J.-P.: Grid-connected shunt active LCL control via continuous sliding modes. IEEE/ASME Trans. Mechatron. **24**(2), 729–740 (2019)

Chapter 11
On the Robust Distributed Secondary Control of Islanded Inverter-Based Microgrids

Alessandro Pilloni, Milad Gholami, Alessandro Pisano and Elio Usai

Abstract In this chapter, some recent results of applications of sliding-mode control strategies for solving the AC microgrid secondary restoration control problem are discussed. The control problem is formulated in a distributed way in accordance with the leader–follower consensus paradigm thus avoiding centralized decision-making. The control schemes are robust in the sense that the voltage and frequency restoration of an inverter-based islanded microgrid is achieved under severe uncertainties affecting the grid and load dynamical models and parameters. In this chapter, two distributed sliding-mode-based control strategies are discussed. The first approach performs the task in finite time exploiting instantaneous communications among distributed generators, whereas the second one attains the goal with an exponential convergence rate by taking into account delayed communications among generators. Both the proposed strategies yield continuous control actions, with discontinuous time derivative, that can safely be pulse-width modulated by a fixed-frequency carrier, as required to not hurt the switching power artifacts. The performance of the proposed schemes is analyzed by means of Lyapunov tools and verified by means of numerical simulations taken in different operative scenarios.

A. Pilloni (✉) · M. Gholami · A. Pisano · E. Usai
Department of Electrical and Electronic Engineering (DIEE), University of Cagliari, Piazza d'Armi, Cagliari, Italy
e-mail: alessandro.pilloni@diee.unica.it

M. Gholami
e-mail: milad.gholami@diee.unica.it

A. Pisano
e-mail: pisano@diee.unica.it

E. Usai
e-mail: eusai@diee.unica.it

M. Steinberger et al. (eds.), *Variable-Structure Systems and Sliding-Mode Control*, Studies in Systems, Decision and Control 271,
https://doi.org/10.1007/978-3-030-36621-6_11

11.1 Introduction

Microgrids (MGs) are localized grouping of heterogeneous distributed generators (DGs), storage systems, and loads operating either in autonomous islanded mode or connected to the main grid. Fundamental task of control systems in a MG is that of keeping the voltage and frequency values to certain setpoints while properly dispatching the injected power among the DGs. The control of a MG has been recently standardized into a nested hierarchical cascade architecture [1] including:

- a decentralized "primary control" (PC) loop that enforces the MG's stability while establishing the power sharing among DGs, as well as perform plug-and-play functionalities;
- a centralized "secondary control" (SC) loop that compensates for the frequency and voltage deviations introduced by the PCs due to the presence of the loads;
- a "tertiary control" (TC) loop that optimizes the power flows, respectively, among DGs during the islanded operations, and from the MG to the main grid, and vice versa, during the grid-tied mode in accordance with several optimality criteria taking into account, e.g., the power demand, the energy cost variations, the proper use of the storage units, etc.

This chapter focuses on the distributed cooperative restoration of the DG's frequencies and voltages at the SC layer of an inverter-based islanded MG. As mentioned, traditional implementations of SCs are of centralized nature [1–6]. However, the current trend is to discourage centralized strategies [7] which bring disadvantages such as latency and delays due to all-to-one communication; the need of costly central computing and communication units; limited scalability and reliability of the power system due to single-point failures.

In the literature also decentralized approaches have been proposed. Worth to mention, for instance [5, 8], where second-order sliding-mode control strategies are proposed, respectively, to regulate voltages while compensating the effects of load variations, nonlinearities, and model uncertainties, and for the robust load frequency control and economic dispatch of power in partitioned power networks. On the other hand, although decentralized approached works satisfactorily in most of today's power systems, in accordance with [9], strategies that employ only local information may become unfeasible in future power system developments, for which, due to the large penetration of renewable power generation which increases power fluctuations, more flexibility in the control system is needed.

To overcome these limitations, multi-agent consensus-based controllers have been proposed to take advantage of their inherent scalability and flexibility features [10], in order to address the frequency regulation while minimizing generation costs (economic dispatch) [11, 12] or the SC restoration problems [13–19]. Additionally, multi-agent-based distributed approaches can more easily deal with packet loss and/or latency in communication as compared to the centralized solutions [20].

In [13], the secondary distributed voltage and frequency restoration task is addressed. However, due to the requirement of all-to-all communication among DGs, its communication overhead was significantly greater than that of the centralized strategies [3]. Furthermore, no formal stability analysis was presented. Among the existing investigations, references [14–19] appear to be the more closely related to the present research. An overview of the main features of these existing proposals as well as the main improvements provided by the results presented in this chapter are explained hereinafter, separately for the frequency and the voltage restoration problems.

Review of Frequency Secondary Controllers

Based on a distributed averaging proportional integral (DA-PI) scheme, [14] first proposed the use of the consensus paradigm to restore the frequencies in an islanded MG modeled in terms of coupled Kuramoto oscillators. A consensus-based distributed tracking (DT) approach, is proposed in [15]. Both papers [14, 15] also implement active power sharing functionalities, but the corresponding approached only possess local exponential stability properties. Frequency ω_{ref} must be constant and globally known to all DGs. The approach in [15] allows to arbitrarily modify the steady-state frequency value of the MG by only acting on a particular virtual DG, referred to as "leader", which directly communicates only with a small portion of DGs. This feature is particularly useful during islanded operation when more active power is required [21], or to seamlessly transfer the MG from islanded to grid-connected mode [22].

Review of Voltage Secondary Controllers

DA-PI and DT solutions have also been employed in the SC layer for voltage restoration purposes. The DA-PI approach in [16] is only capable of providing a tuneable compromise between the adverse tasks of voltage restoration and reactive power sharing accuracy to arbitrarily affect the restoration voltages. Other solutions can be found in [23–25]. restoration accuracy. DT schemes such as those in [15, 17–19] focus on the exact voltage restoration problem only, disregarding the reactive power sharing issue.

A commonly used approach is to convert the voltage restoration problem into a linear DT consensus problem by using feedback linearization techniques. Then, after linearization, voltage restoration can be achieved by using different DT consensus strategies, such as power fractional finite-time consensus control [15], linear proportional–derivative-based consensus [17], and sliding-mode (SM)-based adaptive neural networks [19]. Among them, it is worth to mention that only [19] considers the presence of model uncertainties.

Feedback linearization techniques yield the underlying requirement of a perfectly known MG mathematical model which is rather unrealistic in practical power system scenarios, due to the presence of unmodelled dynamics, parameter uncertainties, abrupt modifications of the power demand, and the presence of nonsmooth nonlinearities introduced by the PWM-based power converters, thus making the performance of the control system prone to these uncertainties. Furthermore, feedback linearization may also yield numerical problems (e.g., due to the online computation of nonlinear coordinate transformations or high-order Lie derivatives) that can compromise the effectiveness of the whole control system. For these reasons, in this work are discussed some robust distributed secondary control strategies which design is completely model-free and robust against system uncertainties.

It is also worth to remark that the efficacy of distributed controllers strongly depend on the efficiency of the communication infrastructure over which the distributed control is built. In spite of this, to the best of our knowledge only [26] deals with the nonideal communications among generators considering packet loss and latency. However, [26] provides only local stability features since the analysis considers a simplified linear MG model.

Statement of Contributions

Inspired by the most recent results in the area of SM-based DT consensus [27–29], and time-delay DT consensus control [20], in the remainder of the chapter we review two recently proposed robust distributed control architectures for designing the SC layer of an islanded inverter-based MG.

The first proposed control system exploits ideal, i.e., delay-free, communication among DGs, whereas the second architecture considers the most realistic scenario in which the communication among DGs can potentially be delayed due to packet losses. Both the proposed schemes are based on the leader–follower (i.e., DT) consensus paradigm.

The first one consists of a nonlinear sliding-mode-based consensus protocol that will be referred to as finite-time converging (FTC) controller since it is able to provide for the finite-time restoration of the DG voltage and frequency profiles.

The second one is the combination between an integral sliding-mode controller and a linear consensus scheme with adaptive weights. It can deal with delayed measurements exchanged between agents and provides an exponentially converging voltage and frequency restoration. The second scheme will be referred to as adaptive integral sliding-mode (A-ISM) controller. Both proposals are robust against model/parameter uncertainties and they dispense with the need to measure several grid variables as compared, e.g., with [15, 17–19].

CT-C, that was exponential in the quoted references [14, 15] whereas the proposed frequency SC CT-C provides finite-time convergence properties. On the other hand, at the price of achieving the SC task with an exponential rate, the SC DT-C may consider multiple time-varying delayed communication among DGs.

The stability properties of the proposed control system architectures are studied by means of Lyapunov arguments by considering a nonlinear modelization of the MG system. In particular, the finite-time stability of the proposed FTC architecture, as well as the exponential stability for the proposed A-ISM scheme are, respectively, demonstrated, for both the frequency and voltage secondary controllers. Worth also to mention that these results are obtained without relying on local linearization arguments, cfr. [15, 17–19, 26].

The analysis also confirms that the active power sharing constraints are enforced with the same convergence rate, respectively, in finite time for the FTC, and with an exponential rate for the A-ISM scheme.

It is worth to remark that both the proposed SC architectures are SM-based; thus, they are inherently non-smooth. However, since the all the proposed schemes are designed by relying on ad hoc input dynamic extension techniques where the discontinuous sign functions only appear in the time derivatives of the actual control inputs, it results that the control actions that can safely be pulse-width modulated with a fixed-frequency carrier, as required to not hurt the switching power artifacts (cfr. with [5]).

Simulations will be carried out with reference to the considered MG model by also adding measurement noise, grid faults, and parameter variations. Additional simulations are also presented which refer to a detailed characterization of the MG that also includes a realistic inverter-based three-phase microgrid modelization with PWM modules [30].

Chapter Organization

The chapter is organized as follows: Sect. 11.2 presents the nonlinear modeling of a MG. Section 11.2.1 outlines the problem statement. In Sect. 11.3, the proposed FTC control architecture is presented and its performance are evaluated by means of Lyapunov tools. Similarly, Sect. 11.4 discusses the proposed A-ISM control architecture. The effectiveness of the proposed controllers are verified in Sect. 11.5 by computer simulations involving different MG modelizations and use-cases. Finally, in Sect. 11.4 some concluding remarks are given.

Nomenclature

The adopted nomenclature of an inverter-based MG consisting of several interconnected DGs is defined hereinafter.

- ω_i: The output frequency of the ith DG.
- δ_i: The phase angle droop of the ith DG.
- $\omega_{ni}, \upsilon_{ni}$: Frequency and voltage droop-power's setpoints.
- υ_i The output voltage of the ith DG.
- k_{P_i}, k_{Q_i}: The frequency and voltage droop gains.
- τ_{P_i}, τ_{Q_i}: Time constants of the two stable filters.
- k_{v_i}: The voltage control gain.
- ω_0, υ_0: Desired values of frequency and voltage.
- P_i, Q_i: The active and reactive power outputs of the ith DG.
- P_i^m, Q_i^m: The measured active and reactive powers.

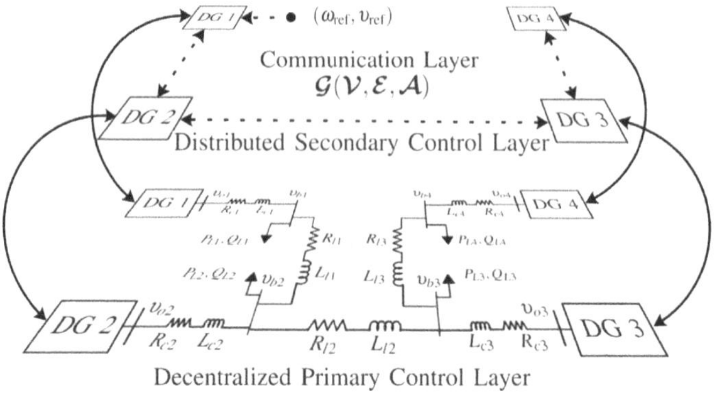

Fig. 11.1 An islanded MG with four DGs and four loads equipped with a leader–follower SC architecture

11.2 Microgrid Modeling

A MG is a geographically distributed power system consisting of DGs and loads physically connected by power lines. The DG control units exchange information for monitoring and control purposes over a communication infrastructure, as shown, for instance, in Fig. 11.1.

Distributed Generator Model

DG modeling includes a prime dc source, a 3-ph dc/ac power converter, an LCL filter, a RL output connector as shown in Fig. 11.2. A more detailed DG's modelization can be found in [30]. Here, as done in [15, 31], we refer to the next simplified representation

$$\dot{\delta}_i = \omega_i = \omega_{n_i} - k_{P_i} \cdot P_i^m \tag{11.1}$$

$$k_{v_i} \cdot \dot{v}_i = -v_i + v_{n_i} - k_{Q_i} \cdot Q_i^m, \tag{11.2}$$

where δ_i and v_i are the voltage phase angle and magnitude of the ith DG, including the power droop terms $k_{P_i} \cdot P_i^m$ and $k_{Q_i} \cdot Q_i^m$. $k_{P_i} \in \mathbb{R}^+$ and $k_{Q_i} \in \mathbb{R}^+$ are the strictly positive droop coefficients, selected to meet the MG's active and reactive power sharing specifications among DGs, respectively. $k_{v_i} \in \mathbb{R}^+$ is the voltage control gain. ω_{n_i} and $v_{n_i} \in \mathbb{R}$ are, respectively, the frequency and the voltage SC action which play the role of references for the PC layer. If the SC is inactive, their values correspond to the nominal setpoints $\omega_{n_i} = \omega_0 = 2\pi \cdot 50$ Hz and $v_{n_i} = v_0 = 220\text{V}_{\text{RMS}} \equiv 310\text{V}_{\text{ph}-0}$. Let τ_{P_i} and $\tau_{Q_i} \in \mathbb{R}^+$ be the time constants of two low-pass filters, P_i^m and Q_i^m denote the filtered measurements of the instantaneous active and reactive power flows P_i and Q_i such that

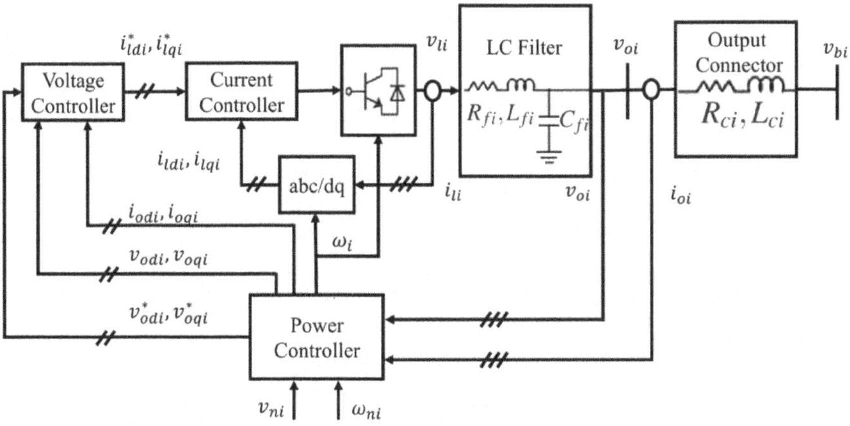

Fig. 11.2 Primary control block diagram of an inverter-based DG

$$\tau_{P_i} \cdot \dot{P}_i^m = -P_i^m + P_i \tag{11.3}$$

$$\tau_{Q_i} \cdot \dot{Q}_i^m = -Q_i^m + Q_i. \tag{11.4}$$

Before introducing the electrical and communication network modelization of a MG for SC purposes, some preliminary notions on graph theory are provided.

Preliminaries on graph theory: A directed graph (or *di-graph*) $\mathcal{G}_{\mathcal{N}}(\mathcal{V}, \mathcal{E}, \mathcal{A})$ is a mathematical tool to describe pairwise mutual interactions between objects, usually referred to as *agents*. $\mathcal{V} = \{1, \ldots, N\}$ denotes the agents' set. $\mathcal{E} \subseteq \{\mathcal{V} \times \mathcal{V}\}$ is the edges' set. $\mathcal{A} = [a_{ij}] \in \mathbb{R}^{N \times N}$ is the adjacency matrix of $\mathcal{G}$, with weight $a_{ij} = 1$ if i communicates with j $(i, j) \subseteq \mathcal{E}$, $a_{ij} = 0$ otherwise. $\mathcal{N}_i = \{j \in \mathcal{V} : (i, j) \in \mathcal{E}\}$ is the neighbors' set of i. A *directed path* is a sequence of agents and edges in $\mathcal{G}_{\mathcal{N}}$ with both endpoints of an edge appearing adjacent to it in the sequence. An agent in $\mathcal{G}_{\mathcal{N}}$ is *globally reachable* if it can be reached from any other agent by traversing a directed path. Information on $\mathcal{G}_N$ is also encoded by the *Laplacian matrix* $\mathcal{L} = [\ell_{ij}] \in \mathbb{R}^{N \times N}$, whose entries are

$$\ell_{ij} = \begin{cases} |\mathcal{N}_i|, & \text{if } i = j, \\ -1, & \text{if } (i, j) \in \varepsilon \text{ and } i \neq j, \\ 0, & otherwise. \end{cases} \tag{11.5}$$

Let $\mathbf{1}_N = [1, \ldots, 1]' \in \mathbb{R}^N$ and $\mathbf{0}_N = [0, \ldots, 0]' \in \mathbb{R}^N$, if $\mathcal{G}_N$ is undirected then $\mathcal{L}$ is a symmetric, positive semi-definite matrix with a simple zero eigenvalue, and it is such that $\mathcal{L}\mathbf{1}_N = \mathbf{0}_N$. For later use, an instrumental lemma is further presented.

Lemma 11.1 *Let us consider the Laplacian Matrix $\mathcal{L}$ associated to a connected undirected graph $\mathcal{G}_N(\mathcal{V}, \mathcal{E}, \mathcal{A})$ with N nodes. Let $G = \text{diag}([g_i])$ be a diagonal matrix with all diagonal elements being nonnegative and at least one of them being strictly positive. Then, $G + \mathcal{L}$ is a positive definite matrix.* ■

Proof Let λ_i be the ordered eigenvalues of $\mathcal{L}$, such that $0 = \lambda_1 < \lambda_2 \leqslant \cdots \leqslant \lambda_N$, and let $\zeta_i \in \mathbb{R}^N$ be their corresponding eigenvectors, where $\zeta_1 = \mathbf{1}_N$. Since the eigenbasis of $\mathcal{L}$ constitutes a basis in $\mathbb{R}^n$, it yields that any nonzero vector $z = [z_i] \in \mathbb{R}^N$, $i = 1 \ldots, n$ can be expressed as a linear combination of the eigenvectors ζ_i as $z = \sum_{i=1}^{n} c_i \zeta_i$, for some constants $c_i \in \mathbb{R}$. By standard manipulation, it results

$$z'(G + \mathcal{L})z = \sum_{i=1}^{n} g_i z_i^2 + \sum_{i=1}^{n} \lambda_i c_i^2 \|\zeta_i\|_2^2 \geqslant \sum_{i=1}^{n} g_i z_i^2 + \sum_{i=2}^{n} \lambda_i c_i^2 \|\zeta_i\|_2^2.$$

Since there exists at least one $g_j > 0$, for some j, then it yields that

$$\zeta_1' G \zeta_1 = \mathbf{1}_N' G \mathbf{1}_N \geq g_j.$$

Therefore, in either the case when $c_2 = \cdots = c_n = 0$, (so $c_1 \neq 0$, since z is a nonzero vector), we always have that

$$\sum_{i=1}^{n} g_i z_i^2 + \sum_{i=2}^{n} \lambda_i c_i^2 \|\zeta_i\|_2^2 \succ 0, \quad \forall z \neq 0.$$

The claimed result is thus proved. ■

Communication network model: Let $\mathcal{V} = \{1, ..., N\}$ be the set of DGs in the MG, the communication topology among DGs can be modeled as a di-graph $\mathcal{G}_N^c = (\mathcal{V}, \mathcal{E}^c, \mathcal{A}^c)$. Edges have real positive weights $\alpha_{ij} \in \mathbb{R}^+$ such that $\alpha_{ij} = 1$ if $(i, j) \in \mathcal{E}^c$, $\alpha_{ij} = 0$ otherwise. $\mathcal{A}^c = [\alpha_{ij}] \in \mathbb{R}^{N \times N}$. $\mathcal{N}_i^c = \{j | (i, j) \in \mathcal{E}^c\}$. Finally, the *Laplacian matrix* associated with $\mathcal{G}_N^c$ is denoted by $\mathcal{L}^c = [\ell_{ij}] \in \mathbb{R}^{N \times N}$, which entries satisfy (11.5). Let agent "0" be an additional virtual object in the augmented graph $\mathcal{G}_{N+1}^c$, in the remainder "0" is considered as the virtual leader for the SCs. It knows and provides the frequency and voltage setpoints ω_0 and $v_0 \in \mathbb{R}$ only to a subset of DGs of $\mathcal{V}$. Node 0 is assumed to be globally reachable on $\mathcal{G}_{N+1}^c$.

Electrical network model: The power lines connecting DGs and loads can be modeled in accordance with the power flow equation by a di-graph $\mathcal{G}_N^e(\mathcal{V}, \mathcal{E}^e, \mathcal{A}^e)$ with complex weights $Y_{ik} = G_{ik} + \imath \cdot B_{ik} \in \mathbb{C}$. G_{ik} and B_{ik} denote the line conductance and susceptance between DG "i" and "k". $B_{ik} > 0$ is positive if inductive. $\imath$ is the imaginary unit. If no connection exists between DG "i" and "k" $Y_{ik} = 0$. $\mathcal{N}_i^e = \{k : k \in \mathcal{V}, k \neq i, Y_{ik} \neq 0\}$ is the electrical neighbors' set of DG "i". We also define $G_{ii} = \sum_{k \in \mathcal{N}_i^e} G_{ik}$ and $B_{ii} = \sum_{k \in \mathcal{N}_i^e} B_{ik}$. Consider the DG's model (11.1)–(11.4), the electrical coupling among DGs can be expressed in terms of the power flowing throughout $\mathcal{G}_N^e$. Under the assumption that the inverter's output admittance is purely inductive and dominates any resistive effect [23–31], $Y_{ik} \approx \iota \cdot B_{ik}\ \forall\, i, k \in \mathcal{V}$, then the injected powers at each bus are

$$\hat{P}_i = \sum_{k \in \mathcal{N}_i^e} v_i v_k B_{ik} \sin(\delta_i - \delta_k) \tag{11.6}$$

$$\hat{Q}_i = v_i^2 B_{ii} + \sum_{k \in \mathcal{N}_i^e} v_i v_k B_{ik} \cos(\delta_i - \delta_k). \tag{11.7}$$

Furthermore, in order to consider also the presence of local loads (P_{Li}, Q_{Li}) connected at the DG's output port, we further define (P_i, Q_i) as

$$P_i = P_{Li} + \hat{P}_i \tag{11.8}$$

$$Q_i = Q_{Li} + \hat{Q}_i, \tag{11.9}$$

where $P_{Li} = P_{1i} v_i^2 + P_{2i} v_i + P_{3i}$, and $Q_{Li} = Q_{1i} v_i^2 + Q_{2i} v_i + Q_{3i}$ shows the power flows load behavior under varying voltage conditions in accordance with the ZIP load power flow modelization. There, the pairs of parameters (P_{ki}, Q_{ki}), with $k = 1, 2, 3$, describe the active and reactive power flows absorbed by the ith load at, resp., constant impedance (Z) that are (P_{1i}, Q_{1i}), constant current (I) that are (P_{2i}, Q_{2i}), and constant power (P) that are (P_{3i}, Q_{3i}). Worth also to remark that the ZIP load modelization is widely accepted in power engendering, see, for instance [15, 32, 33]. That's because it completely describe the power flows absorbed by an electrical load throughout its operation by means of only six parameters, while meeting the constraints of the power system load flow analysis paradigm [34]. It is worth also to mention that in the literature, there exist simple automated measurement-based identification procedures for the identification of the ZIP load parameters, see [35, 36].

Combining (11.1) to (11.9), the overall dynamical MG model is obtained. Following [30], the next assumption is made.

Assumption 11.1 Consider a MG described by (11.1)–(11.9). We assume the active and reactive powers (11.8), (11.9) to be bounded-in-magnitude according to $|P_i| \le \Pi^P$, $|Q_i| \le \Pi^Q$, $\forall\, i \in \mathcal{V}$, $\Pi^P, \Pi^Q \in \mathbb{R}^+$. Due to (11.3), (11.4), it follows that the time derivatives of the measured powers are uniformly bounded as well. ■

Remark 11.1 Assumption 11.1 is sensible due to the physical limitation in the local generated/demanded power and the stabilizing features of the PCs which keep these quantities within prespecified ranges [18, 30, 31]. ■

Let $\tilde{w}_i(t)$ and $w_i(t)$ be equal to the right-hand sides of, resp., (11.1) and (11.2) except for the control terms ω_{n_i} and v_{n_i}, i.e.,

$$\tilde{w}_i(t) = -k_{Pi} \cdot P_i^m, \quad w_i(t) = -v_i - k_{Qi} \cdot Q_i^m. \tag{11.10}$$

By (11.1)–(11.9) one derives that

$$\dot{\tilde{w}}_i = \tilde{h}_i(v_i, P_i^m, P_i), \quad \dot{w}_i = h_i(v_i, Q_i^m, Q_i), \tag{11.11}$$

where $\tilde{h}_i$ and h_i are nonlinear functions of the respective arguments, whose exact form can be derived after straightforward computations. According to Assumption 11.1, $\dot{\tilde{w}}_i$ and $\dot{w}_i$ meet boundedness restrictions as follows:

$$\exists \;\; \Gamma \in \mathbb{R}^+ \;\; : \;\; |\dot{\tilde{w}}_i(t)| \leq \Gamma \;\; \forall\, i \in \mathcal{V} \tag{11.12}$$

$$\exists \;\; \Pi \in \mathbb{R}^+ \;\; : \;\; |\dot{w}_i(t)| \leq \Pi \;\; \forall\, i \in \mathcal{V}. \tag{11.13}$$

By differentiating (11.1), (11.2), and by virtue of (11.6), (11.7), the next augmented frequency and voltage dynamics arise

$$\dot{\omega}_i = \dot{\omega}_{n_i} + \dot{\tilde{w}}_i(t) \tag{11.14}$$

$$\begin{bmatrix} \dot{v}_i \\ \ddot{v}_i \end{bmatrix} = A \cdot \begin{bmatrix} v_i \\ \dot{v}_i \end{bmatrix} + B_i \cdot \dot{v}_{n_i} + \begin{bmatrix} 0 \\ \frac{1}{k_{v_i}} \dot{w}_i(t) \end{bmatrix} \tag{11.15}$$

with

$$A = \begin{bmatrix} 0 & 1 \\ 0 & 0 \end{bmatrix}, \quad B_i = \begin{bmatrix} 0 \\ \frac{1}{k_{v_i}} \end{bmatrix}, \quad \forall\, i \in \mathcal{V}.$$

In the remainder of the chapter, and according to the input dynamic extension principle, the time derivatives $\dot{\omega}_{n_i}$ and $\dot{v}_{n_i}$ will be used as control signals. By doing this, we shall be able to define such frequency and voltage SC control signals in a discontinuous manner and then the actual control signals ω_{n_i} and v_{n_i} will be continuous functions of time.

11.2.1 Secondary Control Objectives for Islanded Microgrids

In the absence of a SC layer, all the DG's frequencies and output voltages deviate from their reference values. Following [14, 15], it is known that relation (11.1) along with the constraints $k_{P_i} P_i^m = k_{P_k} P_k^m \; \forall\, i, k \in \mathcal{V}$, enforces a steady-state ("ss") frequency synchronization condition depending on the active power flowing in the MG and on the droop coefficients, such that,

$$\lim_{t \to \infty} \omega_i(t) = \omega_{i,ss} = \omega_0 - \frac{\sum_{k=1}^{n} P_k^m}{\sum_{k=1}^{n} \frac{1}{k_{P_k}}} \quad \forall \;\; i = 1, \ldots, n. \tag{11.16}$$

Similarly, due to (11.2), the steady-stage voltages deviate from v_0. It follows that the two main tasks of the SC layer are:

1. Restore the frequencies and voltages of each DG to their reference values, i.e.,

$$\omega_{i,ss} = \omega_0 \quad \forall \ i \in \mathcal{V} \tag{11.17}$$

$$v_{i,ss} = v_0 \quad \forall \ i \in \mathcal{V}. \tag{11.18}$$

2. Guarantee the active power sharing ratio, i.e.,

$$\frac{P^m_{i,ss}}{P^m_{k,ss}} = \frac{k_{P_k}}{k_{P_i}} \quad \forall \ i, k \in \mathcal{V}. \tag{11.19}$$

With reference to the voltage SC problem, and in accordance with the related literature on DT-based SC, here we focus our attention only to the voltage restoration problem (11.18), disregarding the reactive power sharing issue, as, for instance, in [15, 17–19, 30]. Further investigation devoted to consider both the voltage restoration problem, and the reactive power sharing issue is left for future research activities.

11.3 Robust Distributed Secondary Control with Finite-Time Convergence

Strategies ranging from centralized to completely decentralized have been suggested to achieve the aforementioned SC purpose. However, centralized approaches conflict with the MG paradigm of autonomous management [16]. On the other hand, decentralized strategies appear to be unfeasible by using only local information [9]. As such, the communication between DGs has been identified as the key ingredient in achieving these goals while avoiding a centralized architecture. In accordance with the DT paradigm, and similar to [15, 17–19], we assume that at least one DG knows the voltage and frequency setpoints established by the TC. We also assume that the DGs may only communicate according to the communication topological graph $\mathcal{G}_N^c$ described in Sect. 11.2.

11.3.1 Voltage Secondary Controller Design

The secondary voltage controller has the task of synchronizing the DG's output voltages v_i to the desired setpoint v_0 by designing appropriate control inputs v_{ni} in (11.2). We propose a novel second-order SM distributed control protocol that guarantees the finite-time attainment of (11.18). The proposed controller is

$$\dot{v}_{ni} = -\varsigma_1 \cdot \text{sign}\left(\sum_{j\in\mathcal{N}_i^c} (v_i - v_j) + g_i\,(v_i - v_0)\right)$$
$$-\varsigma_2 \cdot \text{sign}\left(\sum_{j\in\mathcal{N}_i^c} (\dot{v}_i - \dot{v}_j) + g_i\,(\dot{v}_i - \dot{v}_0)\right), \tag{11.20}$$

where v_{ni} is the primary reference supplied by the secondary controller (11.20), ς_1, $\varsigma_2 \in \mathbb{R}^+$ are the control gains, and $g_i \in \mathbb{R}^+$ is the pinning gain, assumed to be equal to 1 if the ith DG has a direct access to the reference voltage, otherwise $g_i = 0$. Finally, the sign($\mathfrak{S}$) operator denotes the set-valued sign function

$$\text{sign}(\mathfrak{S}) = \begin{cases} 1 & if\ \mathfrak{S} > 0 \\ [-1, 1] & if\ \mathfrak{S} = 0. \\ -1 & if\ \mathfrak{S} < 0 \end{cases} \tag{11.21}$$

Remark 11.2 Notice that the derivative of the DG's voltages $\dot{v}_i$ in the voltage SC (11.20) are unknown and are not available from measurements. However, following [37], if the relative degree of the plant is known and constant, then robust exact differentiators with finite-time convergence properties can be employed to provide the full output-feedback control of any output variable of an uncertain dynamic system. Since the voltage dynamic (11.15) met this condition, and thanks to the finite-time convergence properties of standard high-order sliding-mode differentiator [38, 39], in the remainder of the paper we will refer to the quantities $\dot{v}_i$ as they were known, although they are estimated in practice by means of Levant differentiators. The Levant differentiation scheme is given below

$$\begin{aligned} \dot{\hat{v}}_i(t) &= -c_1 \cdot \left|\hat{e}_{v,i}(t)\right|^{\frac{1}{2}} \cdot \text{sign}(\hat{e}_{v,i}(t)) + \hat{z}_i(t) \\ \dot{\hat{z}}_i(t) &= -c_2 \cdot \text{sign}(\hat{e}_{v,i}(t)), \end{aligned} \tag{11.22}$$

where $\dot{\hat{v}}_i(t) \equiv \dot{v}_i(t)$, $\hat{e}_{v,i} = \hat{v}_i - \int \dot{\hat{v}}_i(t)dt$ is the sliding manifold of the differentiator and $c_1 > 1.5\bar{C}^{\frac{1}{2}}$, $c_2 > 1.1\bar{C}$, and $\bar{C}$ are the constant gains of the differentiator with $\bar{C}$ to be chosen large enough [38]. ■

Here, the following assumption is made.

Assumption 11.2 The reference voltage $v_0(t)$ is such that $|\ddot{v}_0(t)| \leq \Omega$ for some nonnegative constant Ω and for all $t \geq 0$. ■

Before presenting the main result of this section and because of the proof of the convergence property of the proposed voltage secondary controllers (11.20) exploits properties of homogeneous systems, we first introduce the definition of *local homogeneity* and the *Quasi-homogeneity Principle*, both taken from [40], upon which our reasoning will rely on.

Definition 11.1 A piecewise continuous vector field

$$f(t) = [f_1(x), \ldots, f_n(x)]' \in \mathbb{R}^n \quad \text{with } x \in \mathbb{R}^n, \quad f_i : \mathbb{R}^n \mapsto \mathbb{R}, \quad i = 1, 2, \ldots, n \tag{11.23}$$

is called locally homogeneous of degree $q \in \mathbb{R}$ with respect to the dilation vector $r = [r_1, \ldots, r_n]'$, $r_i > 0$, if for all $\varepsilon > 0$

$$f_i\left(\varepsilon^{r_1} x_1, \ldots, \varepsilon^{r_n} x_n, \varepsilon^{-q} t\right) = \varepsilon^{q+r_i} f_i(x, t), \quad i = 1, \ldots, n. \tag{11.24}$$

Theorem 11.1 *Quasi-homogeneity Principle. The uncertain system*

$$\dot{\theta} = \Phi(\theta) + \Psi(t), \quad \theta = [\theta_i]' \in \mathbb{R}^n, \tag{11.25}$$

where $\Psi(t) = [\psi_1, \ldots, \psi_n]'$ is an uncertain vector, is globally equi-uniformly finite-time stable if the following conditions are satisfied:

1. *the right-hand side of* (11.25) *consists of a locally homogeneous piecewise continuous function $\Phi(\theta)$ of degree $q < 0$ with respect to dilation $r = [r_1, \ldots, r_n]'$ and a piecewise continuous function $\Psi(t)$ which components ψ_i, $i = 1, \ldots, n$ are locally uniformly bounded by constants $H_i \geq 0$, such that $|\psi_i(t)| \leq H_i$, within a homogeneity ball;*
2. *$H_i = 0$ whenever $q + r_i > 0$;*
3. *the uncertain system* (11.25) *is globally equi-uniformly asymptotically stable around the origin.* ■

We are now in position to present the main result of this section.

Theorem 11.2 ([41]) *Consider a MG of N DGs communicating over a connected undirected communication network $\mathcal{G}^c$, which voltage dynamics satisfies* (11.13) *and* (11.15). *Let Assumptions 11.1 and 11.2 be in force. If at least one DG has access to v_0, then the continuous consensus-based controller with discontinuous time derivative* (11.20) *with*

$$\varsigma_1 > \varsigma_2 + \Omega + \Pi, \quad \varsigma_2 > \Omega + \Pi \tag{11.26}$$

ensures the finite-time attainment of the voltage restoration condition (11.18). ■

Proof Let the auxiliary error variables be defined as

$$\sigma_{v1} = y_1 - \mathbf{1}_N v_0, \quad \sigma_{v2} = y_2 - \mathbf{1}_N \dot{v}_0 \tag{11.27}$$

with $y_1 = [v_i] \in \mathbb{R}^N$ and $y_2 = [\dot{v}_i] \in \mathbb{R}^N$. By differentiating (11.27) along the trajectories of (11.11), (11.15), and (11.20), and due to the zero row sum property $\mathcal{L}^c \mathbf{1}_N = \mathbf{0}_N$ of the Laplacian matrix, the resulting collective error output voltage dynamics are

$$\begin{aligned}\dot{\sigma}_{v1} &= \sigma_{v2} \\ \dot{\sigma}_{v2} &= \tfrac{1}{k_{v,i}}\left(\dot{w} - \mathbf{1}_N \ddot{v}_0 - \varsigma_1 \cdot \mathrm{Sign}((\mathcal{L}^c + G)\sigma_{v1}) - \varsigma_2 \cdot \mathrm{Sign}((\mathcal{L}^c + G)\sigma_{v2})\right),\end{aligned} \tag{11.28}$$

where $\mathrm{Sign}(\mathfrak{S}) = [\mathrm{sign}(\mathfrak{S}_i)]$ is the column wise sign operator in accordance with (11.21), $w = [w_i] \in \mathbb{R}^N$ and $\dot{w}_i$ satisfies (11.13). By invoking Assumptions 11.1 and 11.2, it follows that $\|\dot{w} - \mathbf{1}_N \otimes \ddot{v}_0\|_\infty \leq \Omega + \Pi$.

Now define $M = G + \mathcal{L}^c \in \mathbb{R}^{N\times N}$. Since by assumption at least one DG has a direct access to the reference voltage, one concludes from Lemma 11.1 that M is symmetric and positive definite. It follows that we can candidate

$$V_v(t) = \frac{\varsigma_1}{k_{v,i}} \cdot \|M\sigma_{v1}\|_1 + \frac{1}{2} \cdot \sigma_{v2}^T M \sigma_{v2} \succ 0 \tag{11.29}$$

as a Lyapunov function. Lengthy but straightforward computations, similar to those made in [42], yields

$$\dot{V}_v(t) \leq -\frac{1}{k_{v,i}}(\varsigma_2 - \Omega - \Pi) \cdot \|M\sigma_{v2}\|_1 \preceq 0 \quad \text{if} \quad \varsigma_{v2} > \Omega + \Pi. \tag{11.30}$$

By virtue of (11.30), the equi-uniform stability of the error dynamic (11.28) is established. From (11.30) follows that whenever system (11.28) is initialized in an arbitrary vicinity of the origin, its solution there remain confined, within the invariant set

$$\mathrm{d}_R = \left\{(\sigma_{v1}, \sigma_{v2}) \in \mathbb{R}^{2n} : V_v(\sigma_{v1}, \sigma_{v2}) \leq R\right\}, \quad R > V_v(0).$$

Let $\lambda_{min}(M)$ be the smallest eigenvalue of M, and $\kappa_R > 0$ be a sufficiently small positive constant such that $\kappa_R < \min\{2\varsigma_1^2/R, \lambda_{min}(M), \sqrt{\lambda_{min}(M)/2R} \cdot (\varsigma_2 - \Omega - \Pi)\}$, following the same arguments of [42] it results that the augmented Lyapunov function $W_v(t) = V_v(t) + \kappa_R \cdot \sigma_{v1}^T M \sigma_{v2}$ satisfies

$$W_v(t) \geq \underline{c}_R \cdot (\|M\sigma_{v1}\|_1 + \|M\sigma_{v2}\|_1) \succ 0 \tag{11.31}$$

with $\underline{c}_R = \min\{\varsigma_1 - \kappa_R R/(2\varsigma_1), (\lambda_{min}(M) - \kappa_R)/2\}$. Lengthy but straightforward computations yield

$$W_v(t) \leq \bar{c}_R \cdot (\|M\sigma_{v1}\|_1 + \|M\sigma_{v2}\|_1) \tag{11.32}$$

with $\bar{c}_R = \max\{\varsigma_1 + \kappa_R\sqrt{2R/\lambda_{min}(M)}, \sqrt{R/(2\lambda_{min}(M))}\}$. Finally, after differentiation of (11.31) along the system trajectories described by (11.28), the next estimation is shown to be in force

$$\frac{d}{dt} W_v(t) \leq -\left(\frac{c_R}{\bar{c}_R}\right) W_v(t) \prec 0 \tag{11.33}$$

with $c_R = \min\{\kappa_R(\varsigma_1 - \varsigma_2 - \Omega), \varsigma_2 - \Omega - \kappa_R\sqrt{2R/\lambda_{min}(M)}\} > 0$. From (11.33), the augmented Lyapunov function $W_v(t)$ globally exponentially decays to zero.

Finally, to proof the finite-time stability of the error dynamics (11.28), let us first rewrite the uncertain error dynamics (11.28) in the form (11.25), where

$$\Psi(t) = \begin{bmatrix} \mathbf{0}_N \\ \frac{1}{k_{v,i}}(\dot{w} - \mathbf{1}_N \ddot{v}_0) \end{bmatrix} = \begin{bmatrix} \psi_1 \\ \vdots \\ \psi_{2N} \end{bmatrix} \in \mathbb{R}^{2N} \tag{11.34}$$

is an uncertain vector of uniformly bounded functions, according to

$$|\psi_i(t)| \leq H_i, \; H_i = \begin{cases} 0, & i = 1, \ldots, N \\ \frac{\Omega+\Pi}{k_{vi}}, & i = N+1, \ldots, 2N \end{cases} \tag{11.35}$$

and

$$\Phi(\theta) = \begin{bmatrix} \sigma_{v2} \\ \frac{\varsigma_1}{k_{vi}}\mathrm{Sign}((\mathcal{L}^c + G)\sigma_{v1}) - \frac{\varsigma_2}{k_{vi}} \cdot \mathrm{Sign}((\mathcal{L}^c + G)\sigma_{v2}) \end{bmatrix} = \begin{bmatrix} \phi_1 \\ \vdots \\ \phi_{2N} \end{bmatrix} \in \mathbb{R}^{2N} \tag{11.36}$$

is a vector of piecewise continuous functions. In accordance with the Definition 11.1, we observe that $\Phi(\theta)$ is locally homogeneous of degree $q = -1$ with respect to the dilation vector $r = [r_i] \in \mathbb{R}^{2N}$ taking the form

$$r_i = \begin{cases} r_i = 2, \text{ with } i = 1, \ldots, N \\ r_i = 1, \text{ with } i = N+1, \ldots, 2N \end{cases}. \tag{11.37}$$

Then, since system (11.25) is globally uniformly exponentially stable, and the following conditions hold:

$$\begin{array}{lll} q + r_i > 0 \; \forall \; i & : & H_i = 0 \Rightarrow i = 1, \ldots, N \\ q + r_i \leq 0 \; \forall \; i & : & H_i \neq 0 \Rightarrow i = N+1, \ldots, 2N \end{array} \tag{11.38}$$

all the conditions of Theorem 11.1 are satisfied, and the proposed voltage SC local interaction rule (11.20) guarantees the finite-time stability of the error dynamics (11.28). As a consequence, the error variables (11.27) go to zero in finite time as well, while guaranteeing the voltage restoration control aims (11.18). This concludes the proof. ■

11.3.2 Frequency Secondary Controller Design

Thanks to the droop characteristic (11.1), and according to (11.16), the condition to achieve the frequency restoration (11.17) while preserving desired power sharing capability (11.19) is

$$\omega_{ni,ss}/\omega_{nk,ss} = 1 \;\forall\; i,\; k = 1, \ldots, n. \tag{11.39}$$

In fact, except from special cases, see, for instance, [9], achieving the frequency synchronization without guaranteeing (11.39) destroys the power sharing property established by the PC [9, 15, 16]. Achieving condition (11.17) subject to (11.39) is a more involved problem that cannot be solved by using standard consensus-based synchronization algorithms. Thus motivated, we propose the following novel sliding-mode-based frequency restoration control strategy:

$$\omega_{ni} = \tilde{\omega}_i + \bar{\omega} \tag{11.40}$$

$$\dot{\tilde{\omega}}_i = -\alpha \cdot \text{sign}\left(\sum_{j \in \mathcal{N}_i^c} \left(\tilde{\omega}_i - \tilde{\omega}_j\right) + \sum_{j \in \mathcal{N}_i^c} \left(\omega_i - \omega_j\right) + g_i \left(\omega_i - \omega_0\right) \right), \tag{11.41}$$

where $\bar{\omega}$ is the rated nominal MG's frequency, $\alpha \in \mathbb{R}^+$ is the protocol gain, $\tilde{\omega}_i$ is a local auxiliary control variable, $g_i \in \{0, 1\}$ is the pinning gain, assumed to be equal to 1 for those DGs having direct access to the desired reference frequency ω_0 and zero otherwise.

From (11.1), and (11.40), (11.41), a compact representation is straightforwardly derived as

$$\dot{\delta} = \omega = \tilde{\omega} + \mathbf{1}_N \bar{\omega} - K_P \cdot P^m \tag{11.42}$$

$$\dot{\tilde{\omega}} = -\alpha \cdot \text{Sign}\left(\mathcal{L}^c \left(\omega + \tilde{\omega}\right) + G\left(\omega - \mathbf{1}_N \omega_0\right)\right), \tag{11.43}$$

where $\tilde{\omega} = [\tilde{\omega}_i]$, $\omega = [\omega_i]$, $P^m = [P_i^m]$ are vectors in $\mathbb{R}^N$, and $K_P = \text{diag}([k_{P_i}])$, $G = \text{diag}([g_i])$ are diagonal matrices in $\mathbb{R}^{N \times N}$ with entries $k_{P_i} > 0$ and $g_i \geqslant 0$. $\mathcal{L}^c$ is the Laplacian matrix associated with $\mathcal{G}_N^c$.

Theorem 11.3 *Consider a MG of N DGs communicating over a connected undirected communication network $\mathcal{G}^c$, which frequency dynamics satisfies* (11.12) *and* (11.14). *Let $\mathcal{L}^c \succeq 0$ be the Laplacian matrix associated with $\mathcal{G}_N$. Let Assumption 11.1 be in force and let $\lambda_{max}(M)$ and $\lambda_{min}(M)$ be the maximum and minimum eigenvalues of matrix $M = G + \mathcal{L}^c \succ 0$. If at least one DG has access to ω_{ref}, then the frequency controller* (11.40), (11.41) *with*

$$\alpha > \frac{\lambda_{max}(M)}{\lambda_{min}(M)} \cdot \Gamma \tag{11.44}$$

ensures in finite time the restoration condition (11.17), *while preserving the power sharing accuracy condition* (11.19). ■

Proof Define the argument of the Sign(·) function in (11.43), as follows:

$$\sigma_\omega = [\sigma_{\omega,i}] = \mathcal{L}^c \left(\omega + \tilde{\omega}\right) + G\left(\omega - \mathbf{1}_N \omega_0\right). \tag{11.45}$$

It is straightforward to show that the constraint $\sigma_\omega = 0$ implies the achievement of both the control objectives (11.17) and (11.19). Thus, to prove Theorem 11.3, we must simply show the finite-time decaying to zero of σ_ω. Let us first differentiate (11.45) along the trajectory of the closed-loop dynamics (11.42), (11.43)

$$\begin{aligned}\dot{\sigma}_\omega &= (G + \mathcal{L}^c) \cdot \dot{\omega} + \mathcal{L}^c \cdot \dot{\tilde{\omega}} \\ &= (G + \mathcal{L}^c) \cdot \left(-\alpha \cdot \text{Sign}(\sigma_\omega) + \dot{\tilde{w}}\right) - \alpha \mathcal{L} \cdot \text{Sign}(\sigma_\omega),\end{aligned} \tag{11.46}$$

where, in accordance with (11.10), $\dot{\tilde{w}} = K_P \cdot \dot{P}^m$. Let V_ω be a candidate Lyapunov function such that

$$V_\omega(t) = \|\sigma_\omega(t)\|_1 = \sum_{i=1}^{n} \left|\sigma_{\omega,i}(t)\right|. \tag{11.47}$$

By differentiating $V_\omega(t)$ along the trajectories of (11.46), thanks to positive semidefiniteness of $\mathcal{L}^c \succeq 0$ and thanks to Lemma 11.1 and (11.12), it yields

$$\begin{aligned}\dot{V}_\omega(t) &= \text{Sign}(\sigma_\omega)'\dot{\sigma}_\omega \\ &= \text{Sign}(\sigma_\omega)' \left(G + \mathcal{L}^c) \cdot \left(-\alpha \cdot \text{Sign}(\sigma_\omega) + \dot{\tilde{w}}\right) - \alpha \mathcal{L} \cdot \text{Sign}(\sigma_\omega)\right) \\ &\leqslant \text{Sign}(\sigma_\omega)'(G + \mathcal{L}^c) \cdot \left(-\alpha \cdot \text{Sign}(\sigma_\omega) + \dot{\tilde{w}}\right) \\ &\leqslant - \text{Sign}(\sigma_\omega)' \left(\alpha \lambda_{min}(M) \cdot \text{Sign}(\sigma_\omega) - \lambda_{max}(M) \Gamma\right).\end{aligned} \tag{11.48}$$

From (11.21), (11.47), it follows that $V_\omega \neq 0$ implies that there exist at least one $\sigma_{\omega,i} \neq 0$. Since $\alpha > \lambda_{max}(M)\Gamma/\lambda_{min}(M)$, it follows that whenever $V_\omega(t) \neq 0$, (11.48) can be further estimated by

$$\begin{aligned}\dot{V}_\omega(t) &\leqslant - \sum_{i=1}^{n} \left(\alpha \lambda_{min}(M)\, \text{sign}(\sigma_{\omega,i})^2 - \lambda_{max}(M) \Gamma\, \text{sign}(\sigma_{\omega,i})\right) \\ &= - \sum_{\forall\, i\,:\, \sigma_{\omega,i} \neq 0} \left(\alpha \lambda_{min}(M) - \lambda_{max}(M) \Gamma\right) \leqslant -\rho \prec 0,\end{aligned} \tag{11.49}$$

where ρ is a strictly positive constant that defines the minimum decaying rate of V_ω, when $V_\omega(t)$ is nonzero. On the contrary, if $\sigma_{\omega,i} = 0\ \forall\, i \in \mathcal{V}$, then $V_\omega = 0$. Furthermore, from (11.49) σ_ω goes to zero in finite time, while preserving (11.17) and (11.19). ■

Remark 11.3 Some comments on the feasibility of the tuning rule (11.44) are given. In the right-hand side of such relation, parameters $\lambda_{max}(M)$, $\lambda_{min}(M)$ and Γ could be not obviously estimated. If $\mathcal{G}^c$ is fixed and known, then $\lambda_{max}(M)$ and $\lambda_{min}(M)$ are known as well. Otherwise, decentralized approaches are available to estimate them, see, for instance, [43]. Concerning Γ in (11.12), relation (11.3) gives a conservative upper bound. However, one can observe that all signals in the right-hand side of (11.3) are locally available for measurements; therefore, distributed estimation strategies based on max-consensus [44] could be used to obtain less conservative estimations of this bound. Clearly, these strategies would be built on the same communication infrastructure of the SC layer. ■

11.4 Robust Distributed Secondary Control with Time-Delayed Communications

Here, similar to the previous section, and in accordance with [15–36, 40, 45–48], we still assume that at least one DG may receive the information on the SC setpoints (ω_0, v_0) dispatched by the virtual leader (referred to as node "0"). Thus, node zero is globally reachable on $\mathcal{G}^c_{N+1}$. Furthermore, we assume $\mathcal{G}^c_N$ to be a directed connected graph. It is worth to remark that in [16, 30] as well as in the results discussed in the previous section undirected communications were needed, and moreover, in contrast with [15, 17–19, 30], but similar to [26], the communications among agents are assumed to be potentially delayed due to packet loss and/or latency in communications.

11.4.1 Voltage Secondary Controller Design

In order to solve the voltage SC problem (11.18) in the presence of communication delays among DGs we propose the following strategy:

$$\dot{v}_{n_i} = \tilde{u}_{n_i} + \tilde{u}_{d_i} \tag{11.50}$$

$$\tilde{u}_{n_i} = -\sum_{j\in\mathcal{N}_i^c} \alpha_{ij}\tilde{k}_{ij}(t)\begin{bmatrix} v_i(t-\tau_{ij}) - v_j(t-\tau_{ij}) \\ \dot{v}_i(t-\tau_{ij}) - \dot{v}_j(t-\tau_{ij}) \end{bmatrix} \tag{11.51}$$

$$\tilde{u}_{d_i} = -\tilde{m}_i \cdot sign(s_i), \tag{11.52}$$

where the sliding variable s_i in (11.52) is defined as follows:

$$s_i = c^T \begin{bmatrix} v_i \\ \dot{v}_i \end{bmatrix} + z_i, \quad c = \begin{bmatrix} 1 \\ 1 \end{bmatrix} \tag{11.53}$$

$$\dot{z}_i = -c^T \left[A \begin{bmatrix} v_i \\ \dot{v}_i \end{bmatrix} - B_i \sum_{j\in\mathcal{N}_i^c} \alpha_{ij}\tilde{k}_{ij}(t) \begin{bmatrix} v_i(t-\tau_{ij}) - v_j(t-\tau_{ij}) \\ \dot{v}_i(t-\tau_{ij}) - \dot{v}_j(t-\tau_{ij}) \end{bmatrix} \right]$$

$$z_i(0) = -c^T \begin{bmatrix} v_i(0) \\ \dot{v}_i(0) \end{bmatrix}. \tag{11.54}$$

The term $\tau_{ij}(t)$ in (11.51) is the time-varying communication delay between the "i"th and "j"th DG over the communication graph $\mathcal{G}_N^c$, whereas $\tilde{k}_{ij}(t)$=[$\tilde{k}_{ij,1}(t)$, $\tilde{k}_{ij,2}(t)$] $\in \mathbb{R}^{1\times 2}$ denotes a vector of adaptive gains which entries are updated according to

$$\begin{aligned} \dot{\tilde{k}}_{ij,1}(t) &= \tilde{\zeta}_{ij,1} \cdot |v_i(t-\tau_{ij}) - v_j(t-\tau_{ij})|^2 \\ \dot{\tilde{k}}_{ij,2}(t) &= \tilde{\zeta}_{ij,2} \cdot |\dot{v}_i(t-\tau_{ij}) - \dot{v}_j(t-\tau_{ij})|^2 \end{aligned} \tag{11.55}$$

with $\tilde{\zeta}_{ij,1}$ and $\tilde{\zeta}_{ij,2} \in \mathbb{R}^+$ and $\tilde{k}_{ij,1}(0)$ and $\tilde{k}_{ij,2}(0)$>0. vector $\tau_l(t) = [\tau_{l,1}(t), \ldots, \tau_{l,q}(t)]$. Let m be the number of communication links between DGs in $\mathcal{G}_N^c$. Let us stack the corresponding communication delays $\tau_{ij}(t)$ in the m-dimensional vector $\sigma_g(t) = [\sigma_{g,1}(t), \ldots, \sigma_{g,m}(t)]$. To exploit a more compact notation, in the remainder we split the delays τ_{ij} affecting the communication links as elements of two different sets. We indicate the delays affecting the communication of between each DG and the leader (node 0) as elements of the set $\tau_l(t) \in \{\tau_{i0}(t) : i = 1, \ldots, N\}$ for $l = 1, \ldots, q$ with $q \leq N$. Analogously, delays affecting the communication among the N DGs are referred as elements of the following delay set: $\sigma_g(t) \in \left\{\tau_{ij}(t) : i, j = 1, \ldots, N, i \neq j\right\}$ for $g = 1, \ldots, m$ with $m \leq N(N-1)$. Note that m and q are equal to their maximum value if the underlying network topology is a directed complete graph and all time delays are different. The following assumption is now made.

Assumption 11.3 Let a priori known bounds $\tau_l^\star, \sigma_g^\star, d_l, \overline{d}_g \in \mathbb{R}^+$ exist such that the elements of the sets $\tau_l(t)$ and $\sigma_g(t)$ satisfies

$$\begin{aligned} &\tau_{i0}(t) \in \left[0, \tau_l^\star\right), \quad |\dot{\tau}_{i0}(t)| \leq d_l < 1, \quad \forall\, t \geq 0,\ \forall\, \tau_{i0} \in \tau_l,\ l = 1, 2, \ldots, q \\ &\tau_{ij}(t) \in \left[0, \sigma_p^\star\right), \quad |\dot{\tau}_{ij}(t)| \leq \overline{d}_g < 1, \quad \forall\, t \geq 0,\ \forall\, \tau_{ij} \in \sigma_g,\ g = 1, 2, \ldots, m. \end{aligned} \tag{11.56}$$

■

Theorem 11.4 ([49]) *Consider the voltage dynamics* (11.15) *under the voltage restoration SC* (11.50)–(11.54). *Let Assumptions 11.1 and 11.3 be satisfied and let there exist* $\tilde{m} > \Pi$, *symmetric positive definite matrices* $P, R_g, Q_g, W_g, Q_l, W_l \in \mathbb{R}^{2N\times 2N}$, *and a positive scalar* η, *such that the following LMIs are feasible:*

$$-\frac{1}{\eta}\sum_{g=1}^{m} \sigma_g^\star R_g + A_0^T H_1 A_0 < 0 \tag{11.57}$$

$$A_l^T H_1 A_l - \frac{1}{\eta} Q_l (1 - d_l) < 0 \tag{11.58}$$

$$\hat{A}_g^T H_1 \hat{A}_g - \frac{1}{\eta} Q_g (1 - \overline{d}_g) < 0 \tag{11.59}$$

$$P\Phi^T + P\Phi + \sum_{g=1}^{m} \sigma_g^\star R_g + \sum_{g=1}^{m} Q_g + \sum_{l=1}^{q} Q_l < 0 \tag{11.60}$$

$$H_1 = \sum_{g=1}^{m} \sigma_g^\star W_g + \sum_{l=1}^{q} \tau_l^\star W_l \tag{11.61}$$

$$A_0 = \text{diag}\left([A, \dots, A]\right) \in \mathbb{R}^{2N\times 2N} \tag{11.62}$$

$$A_l(t) = \text{diag}\left(A_{10}, \dots, A_{N0}\right) \ , \ A_{i0} = -\alpha_{i0} \cdot \hat{k}_{i0}(t) \tag{11.63}$$

and the matrix $\hat{A}_g(t) = [\hat{A}_{g(r,y)}(t)]$ *has entries such that*

$$\begin{cases} \hat{A}_{g(r,r)}(t) = \hat{A}_{ij}(t) & \text{if } i \neq j,\ \sigma_g(\cdot) = \tau_{ij}(\cdot),\ r = y = i \\ \hat{A}_{g(r,y)}(t) = -\hat{A}_{ij}(t) & \text{if } i \neq j,\ \sigma_g(\cdot) = \tau_{ij}(\cdot),\ r = i, y = j \\ \hat{A}_{g(r,y)}(t) = 0 & \text{otherwise} \end{cases} \tag{11.64}$$

with

$$\hat{A}_{ij}(t) = -\alpha_{ij} \cdot \hat{k}_{ij,1}(t) \in \mathbb{R} \tag{11.65}$$

being $r, y = 1, 2, \dots, N$, *and*

$$\Phi(t) = \sum_{l=1}^{q} A_l(t) + \sum_{g=1}^{m} \hat{A}_g(t). \tag{11.66}$$

Then, condition (11.18) *is verified, and the adaptive gains in* (11.55) *asymptotically converge to positive constants* $\tilde{k}^\star_{ij,1}$ *and* $\tilde{k}^\star_{ij,2}$.

Proof By substituting (11.50)–(11.54) into (11.15), it yields

$$\begin{bmatrix} \dot{v}_i \\ \ddot{v}_i \end{bmatrix} = A \begin{bmatrix} v_i \\ \dot{v}_i \end{bmatrix} + \begin{bmatrix} 0 \\ \frac{1}{k_{v_i}} \end{bmatrix} + \left(\tilde{u}_{n_i} - \tilde{m}_i \cdot sign(s_i) + \dot{w}_i(t)\right) \tag{11.67}$$

$$\dot{z}_i = -\dot{v}_i - \frac{1}{k_{v_i}} \tilde{u}_{n_i} \tag{11.68}$$

$$\dot{s}_i = \frac{1}{k_{v_i}} (\dot{w}_i(t) - \tilde{m}_i \cdot sign(s_i)). \tag{11.69}$$

Let us consider the following candidate Lyapunov function:

$$V(t) = \frac{1}{2} \sum_{i=1}^{N} s_i^2 \tag{11.70}$$

which, in accordance with (11.54), satisfies $V(0) = 0$. By differentiating it along the trajectories of (11.1)–(11.9) and considering (11.69), yields

$$\dot{V}(t) = \sum_{i=1}^{N} s_i(t) \cdot \dot{s}_i(t) = \sum_{i=1}^{N} \frac{1}{k_{v_i}}(s_i(t) \cdot \dot{w}_i(t) - \tilde{m}_i \cdot |s_i|). \tag{11.71}$$

Then, by invoking the Hölder's inequality and thanks to (11.13), it follows that

$$\dot{V}(t) \leq -\sum_{i=1}^{N} \frac{(\tilde{m}_i - \Pi)}{k_{v_i}} \cdot |s_i(t)| < 0 \quad \forall\, i, \quad \tilde{m}_i > \Pi. \tag{11.72}$$

From (11.72), one thus concludes that $V(t) = 0 \;\forall\; t \geq 0$. It also follows that $s_i = \dot{s}_i = 0 \,\forall\, t \geq 0$. Thus, in accordance with equivalent control paradigm, by letting $\dot{s}_i = 0$ in (11.69), one has

$$sign(s_i) \equiv \frac{1}{\tilde{m}_i} \dot{w}_i(t). \tag{11.73}$$

Thus, by substituting (11.73) and (11.51) into equation (11.67) it yields

$$\begin{bmatrix} \dot{v}_i \\ \ddot{v}_i \end{bmatrix} = A \cdot \begin{bmatrix} v_i \\ \dot{v}_i \end{bmatrix} - B_i \sum_{j=0}^{n} \alpha_{ij} \tilde{k}_{ij}(t) \begin{bmatrix} v_i(t-\tau_{ij}) - v_j(t-\tau_{ij}) \\ \dot{v}_i(t-\tau_{ij}) - \dot{v}_j(t-\tau_{ij}) \end{bmatrix}. \tag{11.74}$$

From now on, since the closed-loop dynamic (11.74) is same as that considered in the proof of Theorem 1 of [20], by exploiting the same arguments and the same Lyapunov analysis of [20], it can be shown that the delay system (11.74), under the adaptation law (11.51), asymptotically achieve the restoration condition (11.18). Furthermore, it results that the adaptive gains of (11.51) also converge to same to constant quantities according to

$$\lim_{t\to\infty} \tilde{k}_{ij}(t) = \tilde{k}^{\star}_{ij}. \tag{11.75}$$

This concludes the proof. ■

11.4.2 Frequency Secondary Controller Design

Here, in order to solve the frequency SC problem (11.17), while preserving the constraint (11.39), and by taking into account delayed communications among DGs, we propose the following frequency SC strategy:

$$\dot{\omega}_{n_i} = \dot{u}_{n_i} + \dot{u}_{d_i} \tag{11.76}$$

$$\begin{aligned}\dot{u}_{n_i} =& -\sum_{j\in\mathcal{N}_i^c} \alpha_{ij}\cdot \hat{k}_{ij,1}(t)(\omega_i(t-\tau_{ij}(t)) - \omega_j(t-\tau_{ij}(t)))\\ &-\sum_{j\in\mathcal{N}_i^c} \alpha_{ij}\cdot \hat{k}_{ij,2}(t)(u_{n_i}(t-\tau_{ij}(t)) - u_{n_j}(t-\tau_{ij}(t)))\end{aligned} \tag{11.77}$$

$$\dot{u}_{d_i} = -\hat{m}_i \cdot sign(s_i), \tag{11.78}$$

where the sliding variable s_i is defined as follows:

$$s_i(t) = \omega_i(t) + z_i(t)\ ,\ \dot{z}_i(t) = -\dot{u}_{n_i}(t)\ ,\ z_i(0) = -\omega_i(0). \tag{11.79}$$

Moreover, let $\hat{\zeta}_{ij,1}, \hat{\zeta}_{ij,2} \in \mathbb{R}^+$ and $\hat{k}_{ij,1}(0), \hat{k}_{ij,2}(0) > 0$, $\hat{k}_{ij,1}(t), \hat{k}_{ij,2}(t) \in \mathbb{R}^+$ be adaptive gains whose update rules are

$$\dot{\hat{k}}_{ij,1}(t) = \hat{\zeta}_{ij,1}\cdot |\omega_i(t-\tau_{ij}(t)) - \omega_j(t-\tau_{ij}(t))|^2 \tag{11.80}$$

$$\dot{\hat{k}}_{ij,2}(t) = \hat{\zeta}_{ij,2}\cdot |u_{n_i}(t-\tau_{ij}(t)) - u_{n_j}(t-\tau_{ij}(t))|^2. \tag{11.81}$$

Similar to previous subsection, the delays $\tau_{ij}(t)$ satisfy Assumption 11.3.

Theorem 11.5 *Consider system* (11.14) *under the action of the adaptive law* (11.76)–(11.81). *Let Assumptions 11.1 and 11.3 be satisfied and let there exist* $\hat{m}_i > \Gamma$, *symmetric positive definite matrices* Q_l, Q_g, W_l, W_g, R_g, P, P_w, Q_w, R_w, $W_w \in \mathbb{R}^{N\times N}$, *and a positive scalar* η, *such that the following LMIs:*

$$A_l(t)^T H_1 A_l(t) + \tilde{P}^T A_l(t)^T M_1 \tilde{P} A_l(t) - \frac{1}{\eta} Q_l(1-d_l) < 0 \tag{11.82}$$

$$\hat{A}_g(t)^T H_1 \hat{A}_g(t) + \tilde{P}^T \hat{A}_g(t)^T M_1 \tilde{P} \hat{A}_g(t) - \frac{1}{\eta} Q_l(1-\overline{d}_g) < 0 \tag{11.83}$$

$$\tilde{A}_g(t)^T H_1(t) \tilde{A}_g(t) + \tilde{A}_g(t)^T M_1(t) \tilde{A}_g(t) - \frac{1}{\eta} Q_w(1-\overline{d}_g) < 0 \tag{11.84}$$

$$P\Phi^T + P\Phi + \sum_{g=1}^{m} \sigma_g^\star R_g + \sum_{g=1}^{m} Q_g + \sum_{l=1}^{q} Q_l < 0 \tag{11.85}$$

$$\sum_{g=1}^{m} P_w \tilde{A}_g^T + \sum_{g=1}^{m} P_w \tilde{A}_g + \sum_{g=1}^{m} Q_w + \sum_{g=l}^{m} R_w < 0 \tag{11.86}$$

are feasible, where

$$H_1 = \sum_{g=1}^{m} \sigma_g^\star W_g + \sum_{l=1}^{q} \tau_l^\star W_l \ , \ M_1 = \sum_{g=1}^{m} \sigma_g^\star W_w$$
$$\tilde{P} = \text{diag}\,(\mathbf{1}_N) - \tfrac{1}{N} \cdot \mathbf{1}_N \mathbf{1}_N^T \in \mathbb{R}^{N\times N} \tag{11.87}$$

$$A_l(t) = \text{diag}\,(A_{10}, \ldots, A_{N0}) \ , \ A_{i0} = -\alpha_{i0} \cdot \hat{k}_{i0}(t) \tag{11.88}$$

and the matrices $\hat{A}_g(t) = [\hat{A}_{g(r,y)}(t)]$ *and* $\tilde{A}_g(t) = [\tilde{A}_{g(r,y)}(t)] \in \mathbb{R}^{N\times N}$ *have entries such that*

$$\begin{cases} \tilde{A}_{g(r,r)}(t) = \tilde{A}_{ij}(t) & \text{if } i \neq j,\ \sigma_g(\cdot) = \tau_{ij}(\cdot),\ r = y = i \\ \tilde{A}_{g(r,y)}(t) = -\tilde{A}_{ij}(t) & \text{if } i \neq j,\ \sigma_g(\cdot) = \tau_{ij}(\cdot),\ r = i,\ y = j \\ \tilde{A}_{g(r,y)}(t) = 0 & \text{otherwise} \end{cases} \tag{11.89}$$

$$\begin{cases} \hat{A}_{g(r,r)}(t) = \hat{A}_{ij}(t) & \text{if } i \neq j,\ \sigma_g(\cdot) = \tau_{ij}(\cdot),\ r = y = i \\ \hat{A}_{g(r,y)}(t) = -\hat{A}_{ij}(t) & \text{if } i \neq j,\ \sigma_g(\cdot) = \tau_{ij}(\cdot),\ r = i,\ y = j \\ \hat{A}_{g(r,y)}(t) = 0 & \text{otherwise} \end{cases} \tag{11.90}$$

with

$$\hat{A}_{ij}(t) = -\alpha_{ij} \cdot \hat{k}_{ij,1}(t) \ , \ \tilde{A}_{ij}(t) = -\alpha_{ij} \cdot \hat{k}_{ij,2}(t) \in \mathbb{R} \tag{11.91}$$

being $r, y = 1, 2, \ldots, N$, *and*

$$\Phi(t) = \sum_{l=1}^{q} A_l(t) + \sum_{g=1}^{m} \hat{A}_g(t). \tag{11.92}$$

Then, conditions (11.17) *and* (11.19) *are verified, and the adaptive gains* (11.80) *and* (11.81) *asymptotically converge to positive constants* $\hat{k}_{ij,1} \to \hat{k}_{ij,1}^\star$ *and* $\hat{k}_{ij,2} \to \hat{k}_{ij,2}^\star$.

Proof Let us substitute (11.76)–(11.78) into (11.14) as

$$\dot{\omega}_i = \dot{u}_{n_i} - \hat{m}_i \cdot sign(s_i) + \dot{\tilde{w}}(t) \tag{11.93}$$
$$\dot{s}_i = \dot{\tilde{w}}(t) - \hat{m}_i \cdot sign(s_i). \tag{11.94}$$

Now, by using the Lyapunov function (11.70), and thank to (11.12) it results that

$$\dot{V}(t) \leq -\sum_{i=1}^{N} (\hat{m}_i - \Gamma) \cdot |s_i(t)| < 0, \quad \forall\, \hat{m}_i > \Gamma, \quad \forall\, t \geq 0.$$

One thus concludes that the condition $s_i = \dot{s}_i = 0$ is invariant along the trajectories of system (11.10), (11.11), (11.14), since the initial instant of time $t = 0$.

Then, according to the Utkin's equivalent control principle, we can substitute $\hat{m}_i \cdot sign(s_i) \equiv \tilde{\dot{w}}_i(t)$ (which is derived from $\dot{s}_i = 0$) in (11.93) as

$$\begin{aligned}\dot{\omega}_i = &- \textstyle\sum_{j=0}^{N} \alpha_{ij} \cdot \hat{k}_{ij,1}(t)(\omega_i(t - \tau_{ij}(t)) - \omega_j(t - \tau_{ij}(t)))\\ &- \textstyle\sum_{j=1}^{N} \alpha_{ij} \cdot \hat{k}_{ij,2}(t)(u_{n_i}(t - \tau_{ij}(t)) - u_{n_j}(t - \tau_{ij}(t))).\end{aligned} \tag{11.95}$$

Let's define errors between the ith and jth agent's frequencies to the leader's setpoint ω_0 as

$$e_i(t) = \omega_i(t) - \omega_0, \quad e_j(t) = \omega_j(t) - \omega_0 \tag{11.96}$$

after some algebraic manipulations, one derives from (11.95) that

$$\begin{aligned}\dot{e}_i(t) = &-\alpha_{i0} \cdot k_{i0}(t)e_i(t - \tau_{i0}(t))\\ &- \textstyle\sum_{j=1}^{N} \alpha_{ij} \cdot \hat{k}_{ij,1}(t)(e_i(t - \tau_{ij}(t)) - e_j(t - \tau_{ij}(t)))\\ &- \textstyle\sum_{j=1}^{N} \alpha_{ij} \cdot \hat{k}_{ij,2}(t)(u_{n_i}(t - \tau_{ij}(t)) - u_{n_j}(t - \tau_{ij}(t))).\end{aligned}$$

Then, by substituting (11.88) and (11.91), it results

$$\begin{aligned}\dot{e}_i(t) = &A_{i0}(t)e_i(t - \tau_{i0}(t)))\\ &+ \textstyle\sum_{j=1}^{N} \hat{A}_{ij}(t)e_i(t - \tau_{ij}(t)) - e_j(t - \tau_{ij}(t)))\\ &+ \textstyle\sum_{j=1}^{N} \tilde{A}_{ij}(t)(u_{n_i}(t - \tau_{ij}(t)) - u_{n_j}(t - \tau_{ij}(t))).\end{aligned} \tag{11.97}$$

By (11.87), let us also define the disagreement vector of the control actions as $\varepsilon(t) = \tilde{P} \cdot [\omega_{n_1}(t), \ldots, \omega_{n_N}(t)]'$. Since $s_i = \dot{s}_i = 0$, then it follows that ask for $\omega_{ni} = \omega_{nj}$ as in (11.39) is the same to ask for $u_{n_i} = u_{n_j}$ then, $\varepsilon(t) \equiv \tilde{P} \cdot [u_{n_1}(t), \ldots, u_{n_N}(t)]'$, which is zero element-wise if (11.39) is satisfied. To describe the multi-agent dynamics we define the error state vector as

$$\begin{aligned}\tilde{x}(t) &= [\,e_1(t)\; e_2(t)\; \cdots\; e_N(t)]^T\\ \varepsilon(t) &= [\,\varepsilon_1(t)\; \varepsilon_2(t)\; \cdots\; \varepsilon_N(t)]^T.\end{aligned} \tag{11.98}$$

It is worth noting that both the matrices $\hat{A}_g(t)$ and $\tilde{A}_g(t)$ satisfy the following conditions:

$$\hat{A}_g\varepsilon_i = \hat{A}_g u_{n_i}, \quad \tilde{A}_g\varepsilon_i = \tilde{A}_g u_{n_i} \quad i = 1, \cdots, N. \tag{11.99}$$

According to (11.87)–(11.89), and (11.98), (11.99), the multi-agent closed-loop dynamics can be written as

$$\begin{aligned}\dot{\tilde{x}}(t) = &\textstyle\sum_{l=1}^{q} A_l(t)\tilde{x}(t - \tau_l(t)) + \sum_{g=1}^{m} \hat{A}_g(t)\tilde{x}(t - \sigma_g(t))\\ &+ \textstyle\sum_{g=1}^{m} \tilde{A}_g(t)\varepsilon(t - \sigma_g(t))\end{aligned} \tag{11.100}$$

$$\dot{\varepsilon}(t) = \sum_{l=1}^{q} \tilde{P} A_l(t)\tilde{x}(t - \tau_l(t)) + \sum_{g=1}^{m} \tilde{P} \hat{A}_g(t)\tilde{x}(t - \sigma_g(t)) + \sum_{g=1}^{m} \tilde{A}_g(t)\varepsilon(t - \sigma_g(t)). \tag{11.101}$$

Then, with a similar treatment as that presented in [20], it can be proved that the following candidate positive definite functional

$$\begin{aligned}
\overline{V}(t) &= \tilde{x}(t)^T P\tilde{x}(t) + \sum_{l=1}^{q} \int_{t-\tau_l(t)}^{t} \tilde{x}(s)^T Q_l \tilde{x}(s)ds + \sum_{g=1}^{m} \int_{t-\sigma_g(t)}^{t} \tilde{x}(s)^T Q_g \tilde{x}(s)ds \\
&+ \eta \sum_{l=1}^{q} \int_{-\tau_l^\star}^{0} \int_{t+\theta}^{t} \dot{\tilde{x}}(s)^T W_l \dot{\tilde{x}}(s)dsd\theta + \eta \sum_{g=1}^{m} \int_{-\sigma_g^\star}^{0} \int_{t+\theta}^{t} \dot{\tilde{x}}(s)^T W_g \dot{\tilde{x}}(s)dsd\theta \\
&+ \sum_{i=1}^{N} \sum_{j=0}^{N} \frac{1}{2}(\hat{k}_{ij,1}^\star - \hat{k}_{ij,1}(t))^T (\hat{k}_{ij,1}^\star - \hat{k}_{ij,1}(t)) + \varepsilon(t)^T P_w \varepsilon(t) \\
&+ \sum_{g=1}^{m} \int_{t-\sigma_g(t)}^{t} \varepsilon(s)^T Q_w \varepsilon(s)ds + \eta \sum_{g=1}^{m} \int_{-\sigma_g^\star}^{0} \int_{t+\theta}^{t} \dot{\varepsilon}(s)^T W_w \dot{\varepsilon}(s)dsd\theta \\
&+ \sum_{i=1}^{N} \sum_{j=0}^{N} \frac{1}{2}(\hat{k}_{ij,2}^\star - \hat{k}_{ij,2}(t))^T (\hat{k}_{ij,2}^\star - \hat{k}_{ij,2}(t))
\end{aligned} \tag{11.102}$$

is a Lyapunov function for (11.100), (11.101) if the set of LMI (11.82)–(11.86) is satisfied. Thus, it yields that the error variables (11.98) go to zero with an exponential rate thus yielding the frequency SC objectives (11.17) and (11.19). Further details can be found in [50]. This concludes the proof. ■

11.5 Simulations and Discussion of Results

The two proposed SC architectures, respectively, the SC FTC scheme presented in Sect. 11.3 and the SC A-ISM scheme discussed in Sect. 11.4 are now tested by means of numerical simulations carried out in the MATLAB®/Simulink® environment. In particular, two different MG's modelizations are taken into account for validating the effectiveness of the proposed FTC and A-ISM control strategies.

The first modelization is based on the mathematical model (11.1)–(11.9); in the remainder we will refer to this model as "load-flow MG model". On the other hand, the proposed controllers will also be verified by means of a more realistic model developed by means of the Simscape™ Power Systems™ MATLAB®/Simulink® toolbox. This detailed model includes IGBT DC/AC power converters with PWM modules, corrupted noisy measurements, and grid faults. In the remainder we will refer to this model as "SimPowerSys MG model".

Table 11.1 Specification of the load-flow microgrid model described in Sect. 11.2

	DG1		DG2		DG3		DG4	
Model	τ_{P_1}	0.016	τ_{P_2}	0.016	τ_{P_3}	0.016	τ_{P_4}	0.016
	τ_{Q_1}	0.016	τ_{Q_2}	0.016	τ_{Q_3}	0.016	τ_{Q_4}	0.016
	k_{P_1}	$6e^{-5}$	k_{P_2}	$3e^{-5}$	k_{P_3}	$2e^{-5}$	k_{P_4}	$1.5e^{-5}$
	k_{Q_1}	$4.2e^{-4}$	k_{Q_2}	$4.2e^{-4}$	k_{Q_3}	$4.2e^{-4}$	k_{Q_4}	$4.2e^{-4}$
	k_{v_1}	$1e^{-2}$	k_{v_2}	$1e^{-2}$	k_{v_3}	$1e^{-2}$	k_{v_4}	$1e^{-2}$
Load	P_{1_1}	0.01	P_{1_2}	0.01	P_{1_3}	0.01	P_{1_4}	0.01
	P_{2_1}	1	P_{2_2}	2	P_{2_3}	3	P_{2_4}	4
	P_{3_1}	$1e^4$	P_{3_2}	$1e^4$	P_{3_3}	$1e^4$	P_{3_4}	$1e^4$
	Q_{1_1}	0.01	Q_{1_2}	0.01	Q_{1_3}	0.01	Q_{1_4}	0.01
	Q_{2_1}	1	Q_{2_2}	2	Q_{2_3}	3	Q_{2_4}	4
	Q_{3_1}	$1e^4$	Q_{3_2}	$1e^4$	Q_{3_3}	$1e^4$	Q_{3_4}	$1e^4$
Line	$B_{12} = 10\,\Omega^{-1},\ B_{23} = 10.67\,\Omega^{-1},\ B_{34} = 9.82\,\Omega^{-1}$							

11.5.1 MG Model Parameters and Details

Load-Flow Microgrid Model

The set of parameters for this MG modelization, expressed by (11.1)–(11.15), are listed in Table 11.1. Simulations were performed by using the Runge–Kutta fixed-step solver with sampling time $T_s = 500\,\mu$s.

The control gains of the FTC controllers (11.20) and (11.40), (11.41) have been set as

$$\varsigma_1 = 9\,,\ \varsigma_2 = 5\,,\ \alpha = 20. \tag{11.103}$$

The tuning parameters of the A-ISM control systems (11.50)–(11.55) and (11.76)–(11.81) are set as

$$\begin{array}{l} \tilde{k}_{ij}(0) = [1.5, 1.5],\ \ \hat{k}_{ij,1}(0) = 10,\ \ \hat{k}_{ij,2}(0) = 2, \\ \hat{\zeta}_{ij,1} = \tilde{\zeta}_{ij,1} = \hat{\zeta}_{ij,2} = \tilde{\zeta}_{ij,2} = 0.5,\ \ \hat{m}_i = 0.1,\ \ \tilde{m}_i = 0.01, \end{array} \quad \forall\,(i,j) \in \mathcal{E}^c_{N+1}. \tag{11.104}$$

Notice that DG 1 is the only one that knows the reference voltage v_0 and frequency ω_0, i.e., $g_1 = 1$ and $g_2 = g_3 = g_4 = 0$ [15, 19]. Finally, limited to the tests involving the A-ISM scheme, and in accordance with Assumption 11.3, the time derivatives of the communication delays between DGs, $\dot{\tau}_{ij} \in \mathbb{R}^+$, were modeled as random variables with a uniform distribution in the range $|\dot{\tau}_{ij}| \le d_l = \overline{d}_g = 1$ and conditions $\tau_{ij} \in [0, 0.1]$ s are enforced by means of limiters [20].

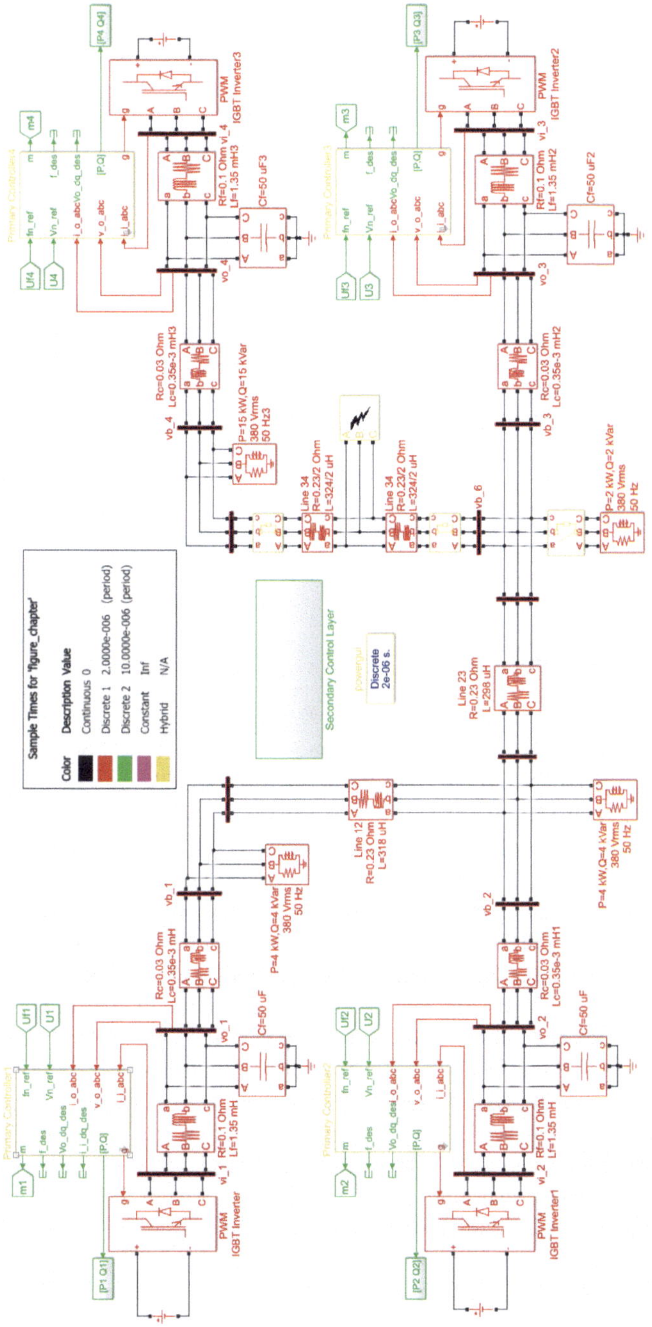

Fig. 11.3 MATLAB®/Simulink® diagram of the three-phase SimPowerSys MG model under test

Table 11.2 Specification of the 3-ph SymPowerSys microgrid model

DG's parameters		DG 1		DG 2	DG 3		DG 4	
Droop control	k_{Pi}	10×10^{-5}		6×10^{-5}	4×10^{-5}		3×10^{-5}	
	k_{Qi}	1×10^{-2}		1×10^{-2}	1×10^{-2}		1×10^{-2}	
Voltage control	k_{pv}	0.4		0.4	0.4		0.4	
	k_{iv}	500		500	500		500	
	k_{fv}	0.5		0.5	0.5		0.5	
Current control	k_{pc}	0.4		0.4	0.4		0.4	
	k_{ic}	700		700	700		700	
LC Filter [Ω], [mH], [μF]	R_f	0.1		0.1	0.1		0.1	
	L_f	1.35		1.35	1.35		1.35	
	C_f	50		50	50		50	
Connector [Ω], [mH]	R_c	0.03		0.03	0.03		0.03	
	L_c	0.35		0.35	0.35		0.35	
Lines [Ω], [μH]	Line 1		Line 2		Line 3			
	R_{l1}	0.23	R_{l2}	0.23	R_{l3}	0.23		
	L_{l1}	318	L_{l2}	324	L_{l3}	324		
Loads [kW], [kVar]	Load 1		Load 2		Load 3		Load 4	
	P_{L1}	3	P_{L2}	3	P_{L3}	2	P_{L4}	3
	Q_{L1}	1.5	Q_{L2}	1.5	Q_{L3}	1.3	Q_{L4}	1.5

SimPowerSys MG Model

Figure 11.3 shows the MATLAB®/Simulink® block diagram of the considered MG, using detailed electronic components of the Simscape™ Power Systems™ component library. Each DGs modelization includes a three-phase IGBT bridge with 10 kVA of rated power provided by a 800 V dc source. PWM Generators with 2 kHz carrier are used to control the switching devices. According to Fig. 11.2, and [30, Section II], the PC consists of three nested loops that control, respectively, the active and reactive power flows, the voltages, and the currents injected by each DG in the $dq0$ reference frame. On the contrary, the three-phase power lines are modeled in the abc frame by using three-phase series RLC branches and the loads as three-phase parallel RLC Loads. The current and voltage outputs of the PI controllers as well as the PWM modules are saturated in accordance with the DGs' rated powers, resp., 380V_{ph-ph}, 32 A. All the MG's specifications are summarized in Table 11.2.

Simulations with the SymPowerSys MG model were performed by using the Runge–Kutta fixed-step solver with sampling time $T_s = 2\,\mu$s, whereas the PC and SC algorithms have been discretized using the sampling step of $\bar{T}_s = 500\,\mu$s. It is also worth to remark that both the PCs and the SCs were fed by realistic noisy measurements. Particularly, all the current and voltage measurements were converted into the $4 \div 20$ mA range with power transmission equal to 0.2 W, and then corrupted by an additive white Gaussian noise with a realistic signal-to-noise ratio of 90 dB [30].

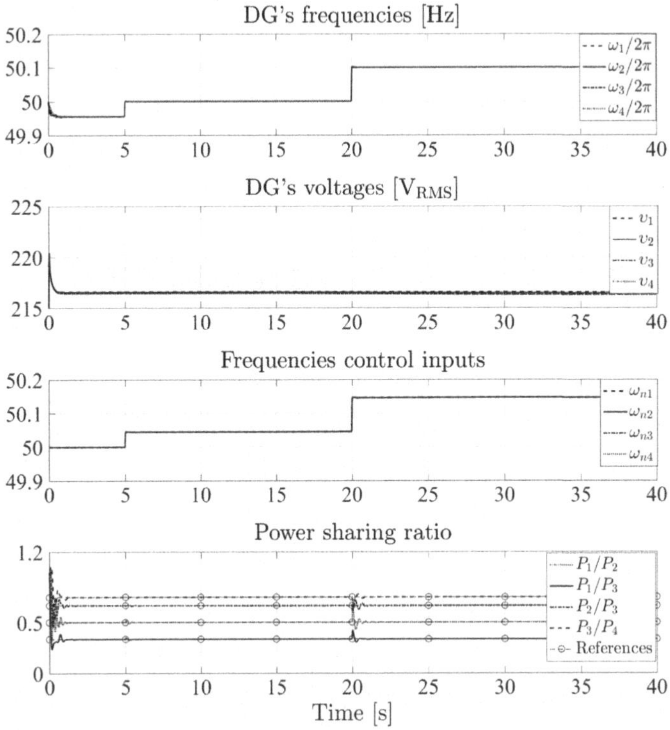

Fig. 11.4 Response of the load-flow MG model under the FTC frequency SC scheme in the Use-Case 1

Finally, to investigate the robustness of the proposed SC algorithms against sudden unplanned events, 3-ph to ground faults will be introduced on Line 2 and Line 3 of Fig. 11.1, and the consequent reconfiguration of both the physical and the SC layer are also scheduled throughout the simulations. In particular, due to the surge transient currents generated by the faults, and according to the delays of protection devices, 10 ms later, some circuit breakers located at the branch buses, respectively, ("*b*2", "*b*3") and ("*b*3","*b*4"), will properly isolate the MG in two portions in accordance with fault's location.

The control gains of the FTC in (11.20) and (11.40), (11.41) were set as

$$\varsigma_1 = 240\ , \ \varsigma_2 = 150\ , \ \alpha = 200 \tag{11.105}$$

whereas the control parameters of the A-ISM controllers (11.50)–(11.55) and (11.76)–(11.81) were set as

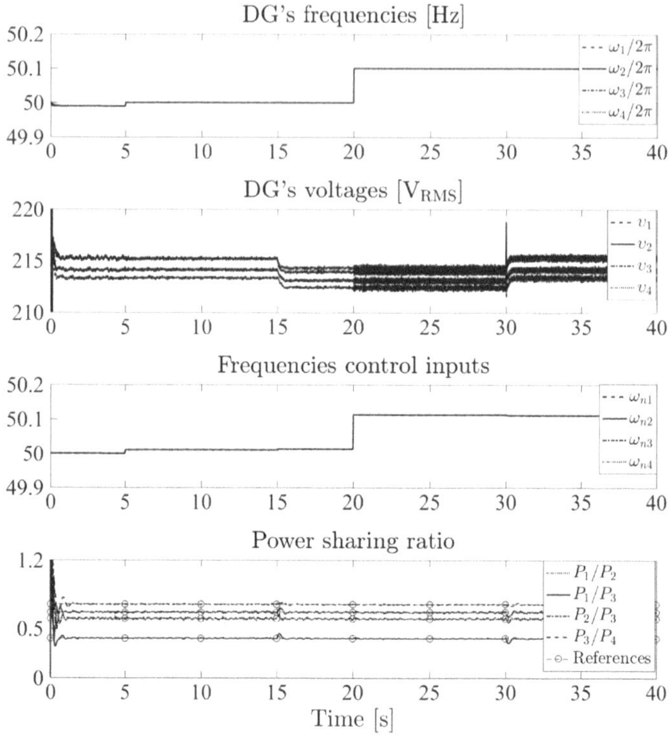

Fig. 11.5 Response of the SimPowSys MG model under the FTC frequency SC scheme in the Use-Case 1

$$\tilde{k}_{ij}(0) = [1, 1]', \quad \hat{k}_{ij,1}(0) = 4, \quad \hat{k}_{ij,2}(0) = 60, \quad \hat{\zeta}_{ij,1} = \hat{\zeta}_{ij,2} = 8 \times 10^{-3},$$
$$\tilde{\zeta}_{ij,1} = 0.5, \quad \tilde{\zeta}_{ij,2} = 0.7, \quad \hat{m}_i = 0.001, \quad \tilde{m}_i = 0.0001, \quad \forall\,(i,j) \in \mathcal{E}^c_{N+1}. \tag{11.106}$$

Notice that DG 1 is the only one that knows the SC reference setpoints of voltage v_0 and frequency ω_0, i.e., $g_1 = 1$ and $g_2 = g_3 = g_4 = 0$ [15, 19].

11.5.2 Simulations Case Studies Outline

Several use-cases are considered in order to test the performance of the proposed FTC and A-ISM control schemes. All the use-case test schedule different events throughout the simulation time. In the following subsections, for each use-case under test a short description is given.

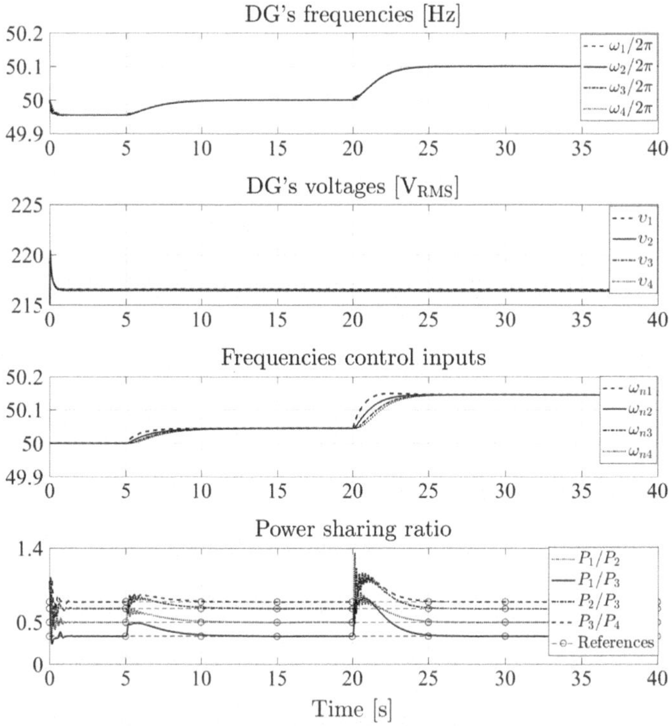

Fig. 11.6 Response of the load-flow MG model under the A-ISM frequency SC scheme in the Use-Case 1

11.5.2.1 Use-Case 1: Frequency SC Test

This use-case has the aim to test the performance of the proposed frequency restoration controllers only. Thus, the voltage restoration controller is not active, and the voltage PC operates with the constant setpoint $v_{ni} = 220\text{V}_{\text{RMS}}$. The following events are further scheduled:

- at $t = 0$ s only the PC is active with $\omega_{ni} = 2\pi 50$ Hz, $v_{ni} = 220\text{V}_{\text{RMS}}$;
- at $t = 5$ s the frequency SC is activated with setpoint $\omega_0 = 2\pi 50$ Hz;
- at $t = 15$ s load (P_{L3}, Q_{L3}) is connected to the MG;
- at $t = 20$ s the SC's frequency setpoint changes to $\omega_0 = 2\pi 50.1$ Hz;
- at $t = 30$ s load (P_{L3}, Q_{L3}) is disconnected from the MG.

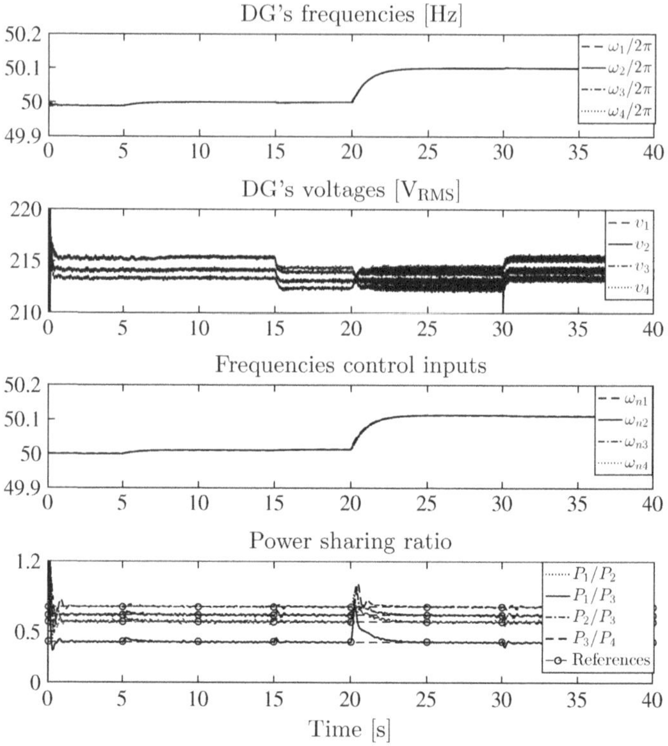

Fig. 11.7 Response of the SimPowSys MG model under the A-ISM frequency SC scheme in the Use-Case 1

11.5.2.2 Use-Case 2: Voltage SC Test

This use-case has the aim to test the performance of the proposed voltage restoration controllers only. Thus, the frequency restoration controller is not active, and the frequency PC operates with the constant setpoint $\omega_{ni} = 2\pi \cdot 50\,\text{Hz}$. The following events are further scheduled:

- at $t = 0\,\text{s}$ only the PC is active with $\omega_{ni} = 2\pi 50\,\text{Hz}$, $v_{ni} = 220\text{V}_{\text{RMS}}$;
- at $t = 10\,\text{s}$ the voltage SC is activated with setpoint $v_0 = 220\text{V}_{\text{RMS}}$;
- at $t = 15\,\text{s}$ load (P_{L3}, Q_{L3}) is connected to the MG;
- at $t = 25\,\text{s}$ the SC's voltage setpoint changes to $v_0 = 225\text{V}_{\text{RMS}}$;
- at $t = 30\,\text{s}$ load (P_{L3}, Q_{L3}) is disconnected from the MG.

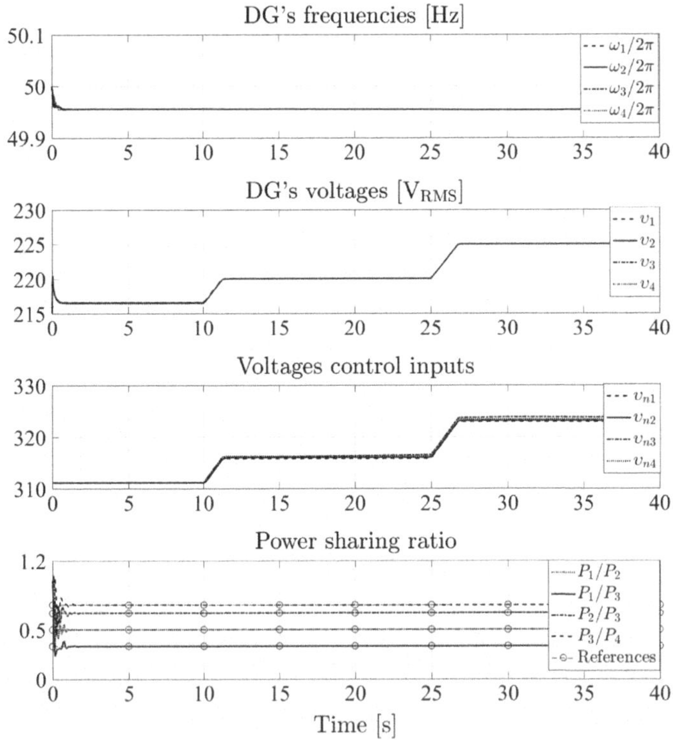

Fig. 11.8 Response of the load-flow MG model under the voltage FTC scheme in the Use-Case 2

11.5.2.3 Use-Case 3: Frequency and Voltage SC Test

This use-case involves the simultaneous action of the frequency and voltage restoration SC laws. The following events are scheduled:

- at $t = 0$ s only the PC is active with $\omega_{ni} = 2\pi 50$ Hz, $v_{ni} = 220\text{V}_{\text{RMS}}$;
- at $t = 5$ s the frequency SC is activated with setpoint $\omega_0 = 2\pi 50$ Hz;
- at $t = 10$ s the voltage SC is activated with setpoint $v_0 = 220\text{V}_{\text{RMS}}$;
- at $t = 15$ s load (P_{L3}, Q_{L3}) is connected to the MG;
- at $t = 20$ s the SC's frequency setpoint changes to $\omega_0 = 2\pi 50$ Hz;
- at $t = 25$ s the SC's voltage setpoint changes to $v_0 = 225\text{V}_{\text{RMS}}$;
- at $t = 30$ s load (P_{L3}, Q_{L3}) is disconnected from the MG.

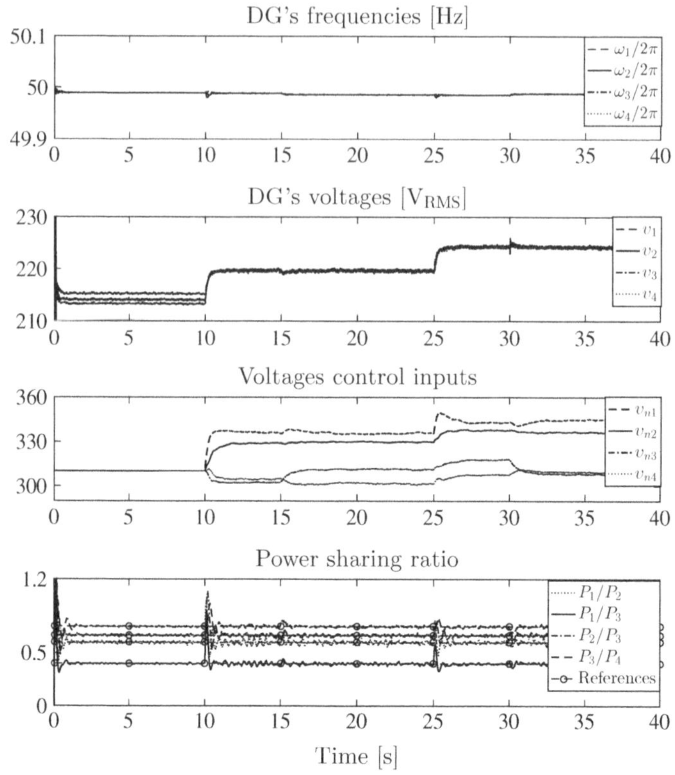

Fig. 11.9 Response of the three-phase SimPowSys MG model under the voltage FTC scheme in the Use-Case 2

11.5.2.4 Use-Case 4: SC Test with Real-Time Optimal Power Dispatch Strategies

This use-case has the aim to the test the performance of the proposed SC strategies when some modifications of the active power droop control coefficients k_{Pi} occur due to the presence of TC strategies aimed to the real-time optimization of the dispatch of power among DGs [51]. The events scheduled in this use-case are as follows:

- at $t = 0$ s only the PC is active with $\omega_{ni} = 2\pi 50$ Hz, $v_{ni} = 220\text{V}_{\text{RMS}}$;
- at $t = 5$ s the frequency SC is activated with setpoint equal to $\omega_0 = 2\pi 50$ Hz;
- at $t = 10$ s the voltage SC is activated with setpoint $v_0 = 220\text{V}_{\text{RMS}}$;
- at $t = 15$ s load (P_{L3}, Q_{L3}) is connected to the MG;

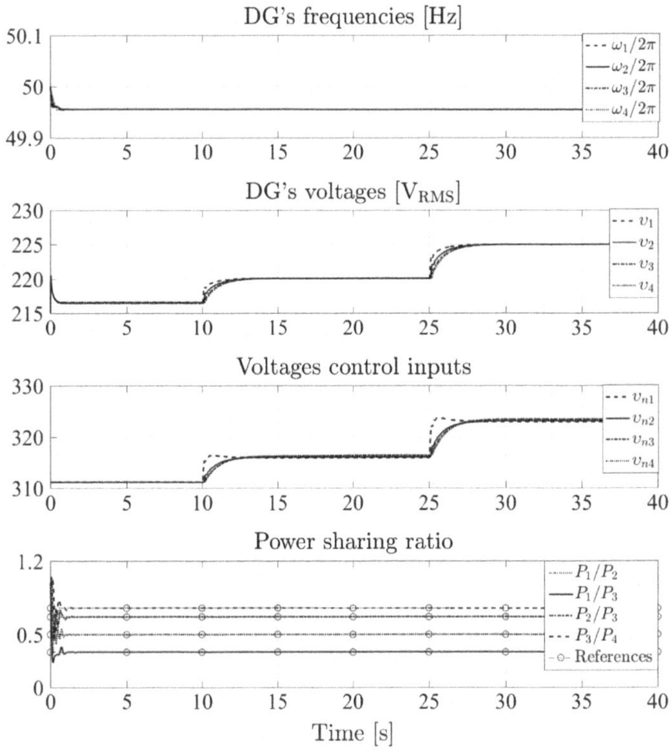

Fig. 11.10 Response of the three-phase load-flow MG model under the A-ISM voltage SC scheme in the Use-Case 2

- at $t = 25$ s the active droop gains $k_{Pi}, i = 1, 2, 3, 4$, are modified from their initial values given in Table 11.2 to the optimal values $k_{P1} = 12 \times 10^{-5}, k_{P2} = 8 \times 10^{-5}$, $k_{P3} = 6 \times 10^{-5}$, $k_{P4} = 4.8 \times 10^{-5}$;
- at $t = 30$ s load (P_{L3}, Q_{L3}) is disconnected from the MG.

For the sake of brevity, this test has been performed only using the SimPowerSys MG model [30].

11.5.2.5 Use-Case 5: Reaction of the SC Strategies to Ground Fault on Line 3

The aim of this use-case is to show how the proposed controllers react to abrupt events such as a short-circuit fault on the power lines connecting some DGs and the consequent reconfiguration of the physical electrical connection among them. In accordance with Fig. 11.3, the fault is implemented by means of a 3-ph to ground

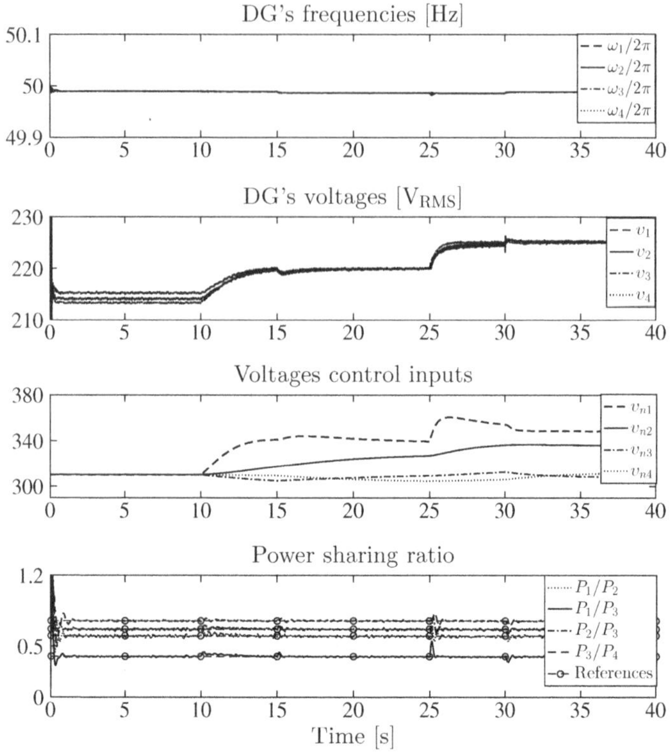

Fig. 11.11 Response of the three-phase SimPowSys MG model under the A-ISM voltage SC scheme in the Use-Case 2

fault [52] triggered at $t = 25$ s. Then, due to the surge transient current and according to the delays of protection devices, 10 ms later two circuit breakers located at the branch buses "b3" and "b4" isolate the MG into two sub-MGs, one consisting of DG 1, DG 2, and DG 3, and another composed only by DG 4 and its local load (P_{L4}, Q_{L4}). The events scheduled in this use-case are:

- at $t = 0$ s only the PC is active with $\omega_{ni} = 2\pi 50$ Hz, $v_{ni} = 220\text{V}_{\text{RMS}}$;
- at $t = 5$ s the frequency SC is activated with setpoint $\omega_0 = 2\pi 50$ Hz;
- at $t = 10$ s the voltage SC is activated with setpoint $v_0 = 220\text{V}_{\text{RMS}}$;
- at $t = 15$ s load (P_{L3}, Q_{L3}) is connected to the MG;
- at $t = 25$ s a 3-ph to ground fault occurs on the Line 2;
- at $t = 25.01$ s over-current protection devices isolate Line 3, thus DG 4 is electrically isolated by the other DGs;
- at $t = 30$ s load (P_{L3}, Q_{L3}) is disconnected from the MG.

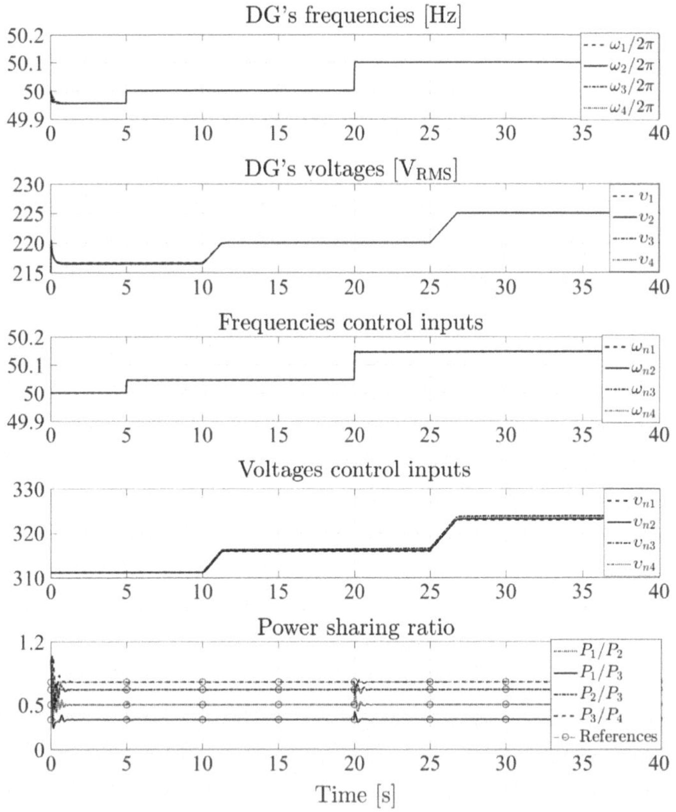

Fig. 11.12 Response of the load-flow MG model with the SC FTC architecture on Use-Case 3

11.5.3 Numerical Simulation Results

Use-Case 1. FTC controller test: We show the temporal behavior of the DG frequencies ω_i, voltages v_i, frequency control action ω_{ni}, and the power sharing ratios (11.19), respectively, for the load-flow MG model in Fig. 11.4, and for the SimPowSys MG model in Fig. 11.5 using the FTC frequency restoration scheme. It can be noted that both the MG models provides similar responses. More precisely, due to the higher degree of accuracy in modeling the MG's behavior, the SimPowSys MG models looks more sensitive to plant conditions than the simpler load-flow model as it is apparent from the voltages drawing.

In particular, in the first five seconds, both the local voltages and frequencies deviate from the corresponding setpoints. Then, just after the activation of the frequency FTC scheme the frequencies are restored to the reference value of 50 Hz.

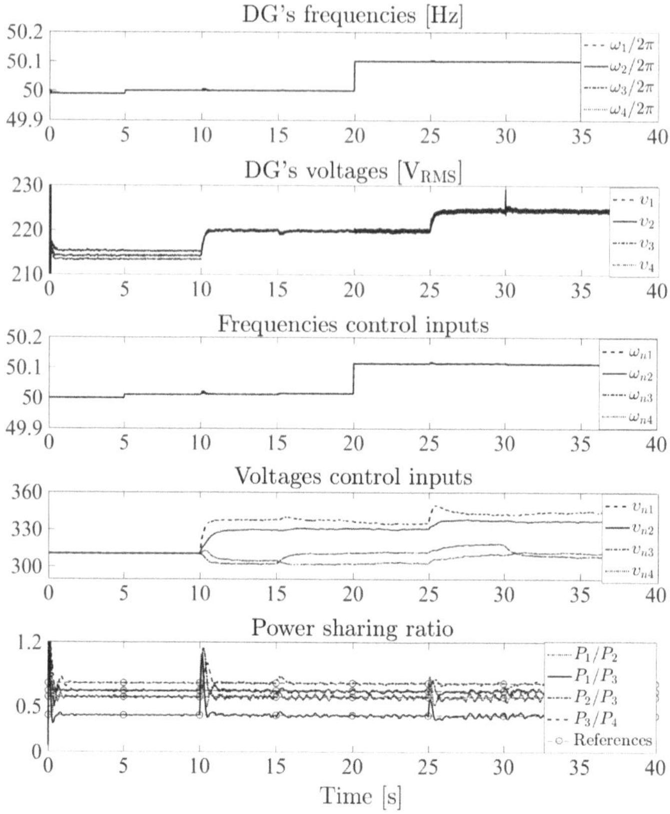

Fig. 11.13 Response of the SimPowSys MG model with the SC FTC architecture on Use-Case 3

On the other hand, since the voltage SC is not active the voltages are not restored. It is worth to remark that the proposed FTC frequency SC does not suffer from load changes, namely the connection of load (P_{L3}, Q_{L3}) at $t = 15$ s and its removal at $t = 30$ s and noisy measurements, and it preserves the constraints on the active power sharing.

Use-Case 1. A-ISM controller test: Simulation results relevant to the simplified load-flow MG model and to the more realistic SimPowSys MG model are depicted in Figs. 11.6 and 11.7, respectively. It can be seen that the considered A-ISM frequency SC scheme exhibits similar results as those provided by the FTC scheme, except for the shape of the curves during the transient which clearly corresponds to an asymptotic convergence process. This is the price to be paid in order to address delayed communications among DGs.

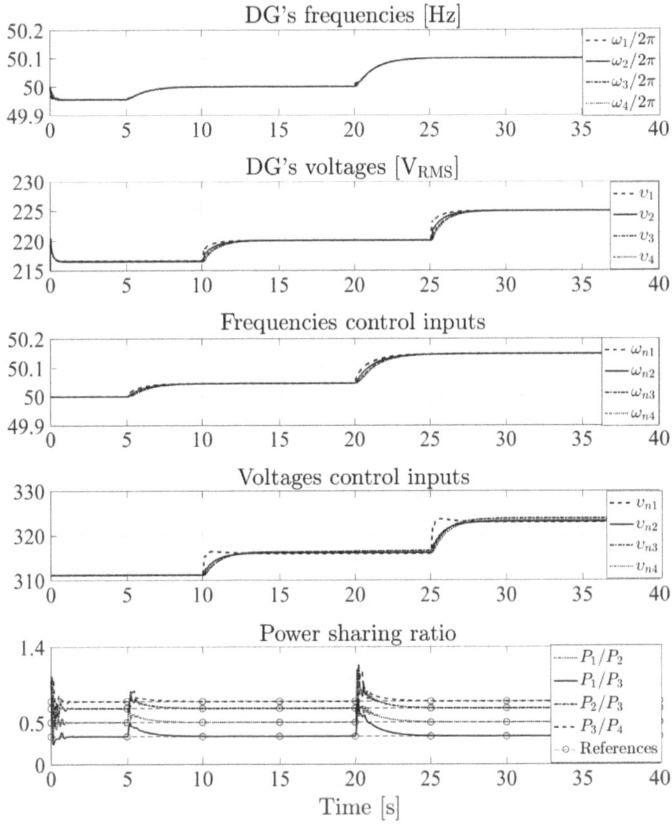

Fig. 11.14 Response of the load-flow MG model with the SC A-ISM architecture on Use-Case 3

Use-Case 2. FTC controller test: The results obtained considering the load-flow MG model are shown in Fig. 11.8, whereas those corresponding to the three-phase SimPowSys MG model are depicted in Fig. 11.9. As expected, in the first ten seconds, since only the PC loops are active, both the local voltages and frequencies deviate from the corresponding setpoints. Then, just after the activation of the voltage restoration FTC scheme, and in finite time for the load-flow MG model, the voltages are restored to reference value of $220V_{RMS}$. On the other hand, due to the presence of unmodeled dynamics in the three-phase SimPowSys model, not included in the load-flow model used for the analysis, only exponential, however fast, convergence is observed in the simulations. Proposed voltage SC proves to be robust against the load changes, i.e., the connection of load (P_{L3}, Q_{L3}), noisy measurements, and it clearly preserves the constraints on the active power sharing.

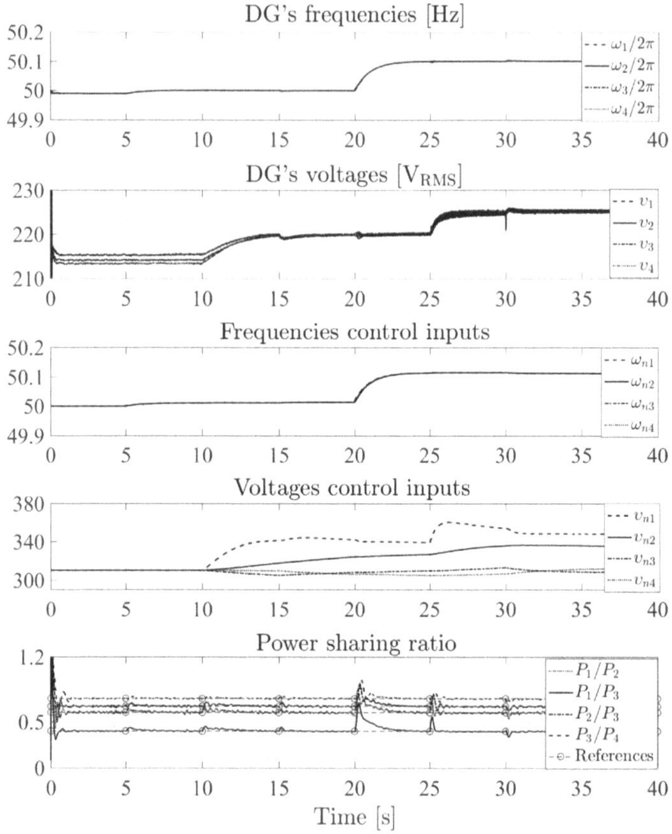

Fig. 11.15 Response of the SimPowSys MG model with the SC A-ISM architecture on Use-Case 3

Use-Case 2. A-ISM controller test: Similar to the previous paragraph, here we discuss the temporal behavior of the MG's quantities of interest under the voltage SC TD-C scheme, respectively, for the load-flow MG model in Fig. 11.10 and for the three-phase SimPowSys MG model in Fig. 11.11. It can be seen that the A-ISM voltage SC scheme exhibits similar results as those provided by the FTC scheme for the load-flow MG model, except for the transient shape which clearly reveals the occurrence of an asymptotic convergence process.

Use-Case 3. FTC controller test: Results shown, respectively, in Fig. 11.12 for the load-flow MG model, and in Fig. 11.15 for the SimPowSys model show that voltages and frequencies are restored in accordance with the reference profiles provided by DG 1, while preserving the power sharing accuracy (Fig. 11.13).

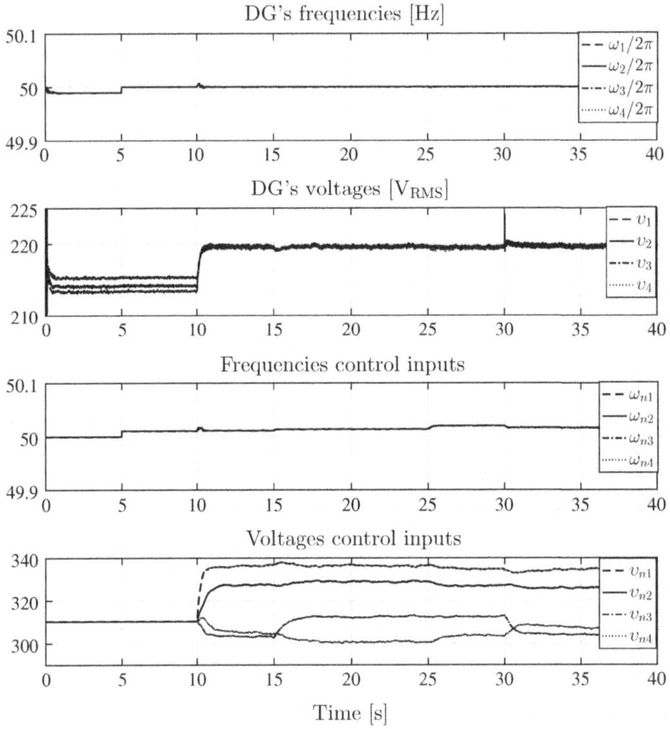

Fig. 11.16 Response of the three-phase SimPowSys MG model with the SC FTC architecture on Use-Case 4

Use-Case 3. A-ISM controller test: Similar results and considerations as those given in the previous paragraph hold also for the A-ISM architecture within the Use-Case 3. The results for this use-case are shown, respectively, in Fig. 11.14 for the load-flow MG model, and in Fig. 11.15 for the three-phase SimPowSys MG model.

Use-Case 4. FTC controller test: With reference to the Use-Case 4, in Fig. 11.16 it is shown the invariance of the proposed SC FTC architecture to the presence of strategies aimed to the real-time optimization of the dispatch of power among DGss. For the sake of brevity, this use-case has been tested only on the more realistic three-phase SimPowSys MG model. The top plot of Fig. 11.17 depicts the temporal profiles of the active droop gain coefficients k_{Pi}, whereas in the bottom plot of Fig. 11.17 it is shown the evolution of the power sharing ratios.

Use-Case 4. A-ISM controller test: Similar to the previous paragraph, in Fig. 11.18 it is shown the robustness of the proposed SC A-ISM architecture to the presence of droop coefficients variations. Only results relevant to the three-phase SimPowSys MG model are shown. The top plot of Fig. 11.19 depicts the temporal trends of the active droop gain coefficients k_{Pi} used for the simulation of Use-Case 4.

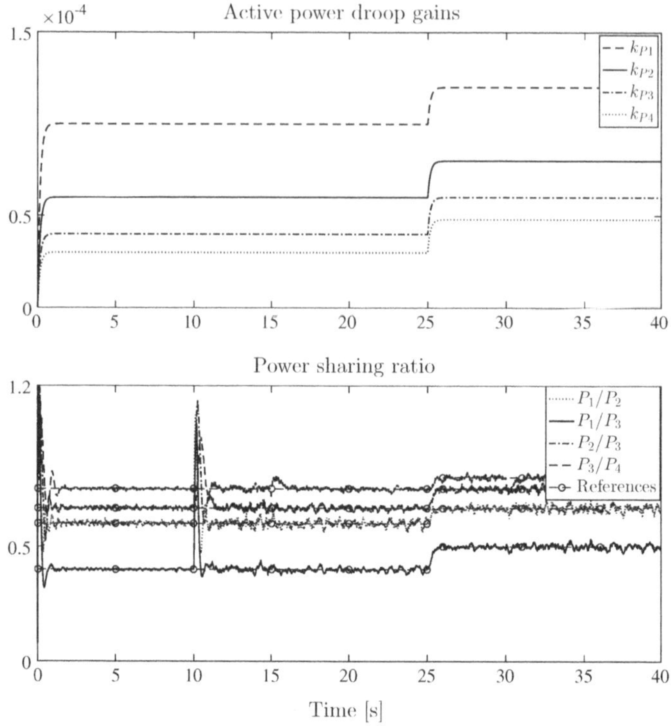

Fig. 11.17 Response of the three-phase SimPowSys MG model with the SC FTC architecture on Use-Case 4

In the bottom plot of Fig. 11.19 the actual time-varying power sharing specification are compared with the desired ones.

Use-Case 5. FTC controller test: In this paragraph, we show how the proposed SC FTC architecture reacts to abrupt unplanned events such as a short-circuit fault on the power line connecting DG 3 and DG 4, and to the consequent reconfiguration of the physical electrical connections among them, i.e., the isolation of DG 4. In particular, in Fig. 11.20 it is shown the temporal behavior of the MG's quantities of interest for the SimPowSys MG model under the SC FTC scheme. It can be noted that although the occurrence of such a critical event at $t = 25$ s, both the frequencies and voltages of all the DGs remain at the desired nominal values, respectively, $\omega_0 = 2\pi \cdot 50$ Hz and $v_0 = 220\text{V}_{\text{RMS}}$. Furthermore, the control action remains within the operating ranges without saturate. Furthermore, limited to those DGs that can share their injected power, i.e., DG 1, DG 2, and DG 3, also the power sharing constraints are still in force.

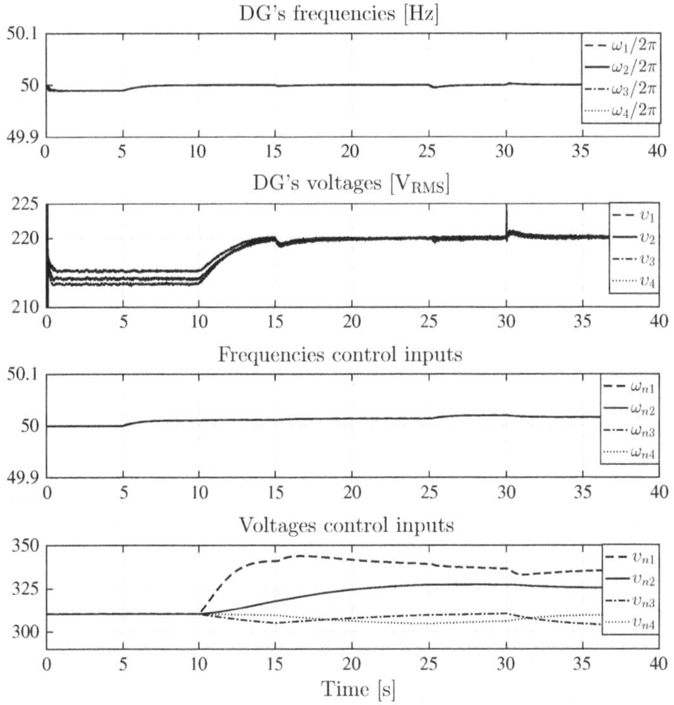

Fig. 11.18 Response of the three-phase SimPowSys MG model under the A-ISM architecture on Use-Case 4

Use-Case 5. A-ISM controller test: Similar to the previous paragraph, Fig. 11.21 shows the temporal behavior of the MG's quantities of interest for the three-phase SimPowSys MG model under the SC A-ISM scheme, but by exploiting only delayed communication among agents, both the frequencies and voltages of all the DGs remain at the desired nominal values in spite of the occurrence of a short-circuit fault on the power line connecting DG 3 and DG 4 and the consequent electrical reconfiguration of the MG. It can be further noted that the power sharing constraints for those DGs that can share their injected power such as DG 1, DG 2, and DG 3 are still preserved.

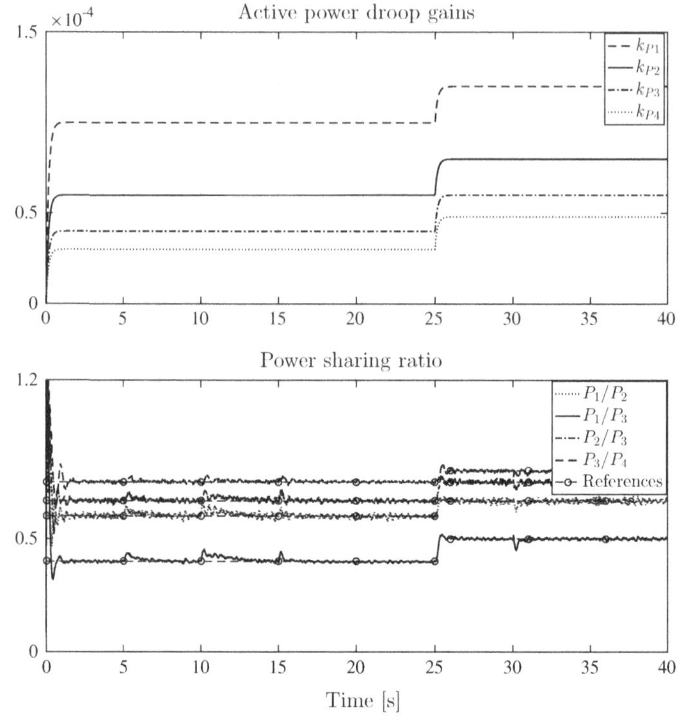

Fig. 11.19 Response of the three-phase SimPowSys MG model under the A-ISM architecture on Use-Case 4

11.6 Conclusions

In this chapter, some recent results of applications of sliding-mode control strategies for solving the AC microgrid secondary restoration control problem are discussed. To achieve the task the control problem is formulated in a distributed way in accordance with the leader–follower consensus paradigm thus avoiding centralized decision-making. The control schemes are robust in the sense that the voltage and frequency restoration of an inverter-based islanded microgrid is achieved while dispensing with the knowledge of their models, parameters, and loads. Here, two sets of continuous control strategies with discontinuous time derivatives are discussed. The first approach performs the task in finite time by exploiting instantaneous communications among distributed generators, whereas the second one attains the goal with an exponential rate by exploiting more realistic delayed communications among generators.

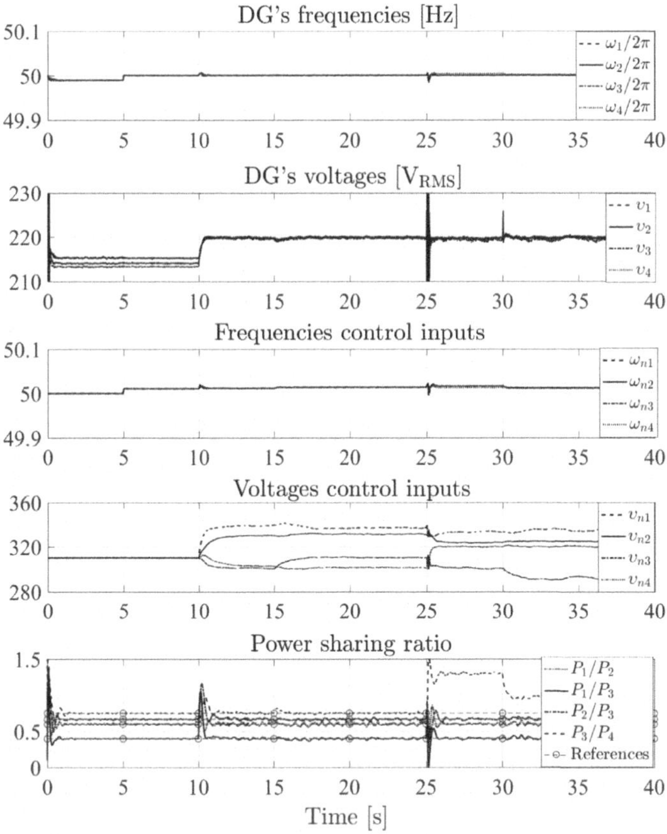

Fig. 11.20 Response of the three-phase SimPowSys MG model under the SC CT-C architecture on Use-Case 5

Next activities will be targeted on relaxing the assumed restrictions on the communication topology by covering possibly switching gossip-based asynchronous communications, as well as consider event-triggered-based distributed control strategies. Other interesting lines of investigation that represents a natural continuation of this research are, the possibility to manage active loads, or exploit seamless distributed transfer strategies for the MG to switch from islanded to grid-connected mode by means of local interactions among DGs. Worth also to mention the possibility to embed in the proposed SC architectures control techniques for managing online the power flows injected by the DGs over large-scale MGs. Experimental validations of the present research that will allow a performance assessment of the proposed techniques are also under study.

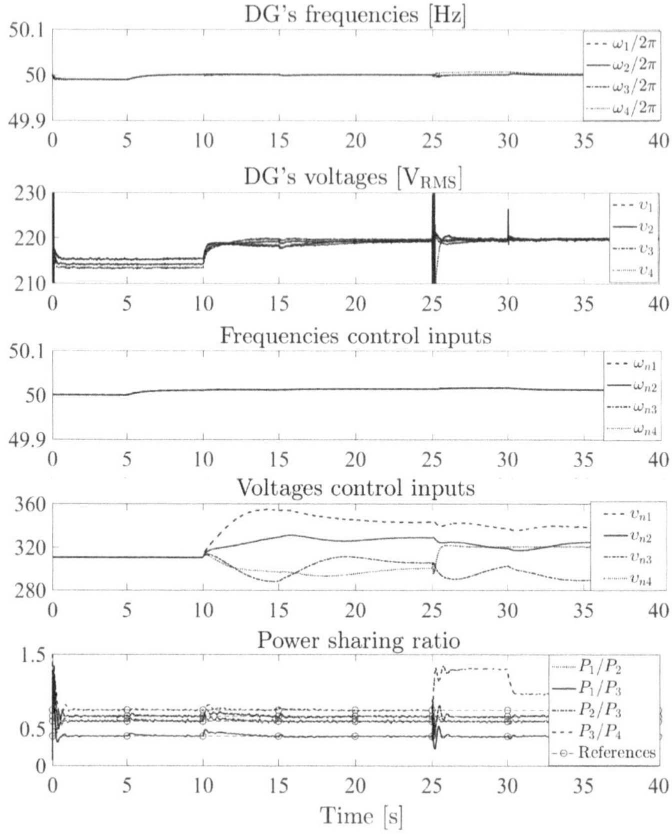

Fig. 11.21 Response of the three-phase SimPowSys MG model under the A-ISM architecture on Use-Case 5

Acknowledgements The research leading to these results has been partially supported by P.O.R. SARDEGNA F.S.E. 2014-2020 - Asse III "Istruzione e Formazione, Obiettivo Tematico: 10, Obiettivo Specifico: 10.5, Azione dell'accordo fi Partenariato:10.5.12 "Avviso di chiamata per il finanziamento di Progetti di ricerca - Anno 2017" and, by Fondazione di Sardegna under project ODIS - Optimization of DIstributed systems in the Smart-city and smart-grid settings, CUP:F72F16003170002, and by the Sardinian Regional Government, call "Cluster top-down actions (POR FESR)", project "Virtual Energy", and by project RASSR05871 MOSIMA, FSC 2014-2020, Annualita' 2017, Area Tematica 3, Linea d'Azione 3.1.

References

1. Bidram, A., Davoudi, A.: Hierarchical structure of microgrids control system. IEEE Trans. Power Syst. **3**(4), 1963–1976 (2012)
2. Guerrero, J.M., Chandorkar, M., Lee, T.-L., Loh, P.C.: Advanced control architectures for intelligent microgrids: Decentralized and hier- archical control. IEEE Trans. Ind. Electron. **60**(4), 1254–1262 (2013)

3. Lopes, J.P., Moreira, C., Madureira, A.: Defining control strategies for microgrids islanded operation. IEEE Trans. Power Syst. **21**(2), 916–924 (2006)
4. Incremona, G.P., Cucuzzella, M., Magni, L., Ferrara, A.: MPC with sliding mode control for the energy management system of microgrids. IFAC-PapersOnLine **50**(1), 7397–7402 (2017)
5. Incremona, G.P., Cucuzzella, M., Ferrara, A.: Decentralized sliding mode control of islanded AC microgrids with arbitrary topology. IEEE Trans. Ind. Electron. **64**(8), 1–8 (2017)
6. Incremona, G.P., Cucuzzella, M., Ferrara, A.: Adaptive suboptimal second-order sliding mode control for microgrids. Int. J. Control **89**(9), 1849–1867 (2016)
7. Dörfler, F., Simpson-Porco, J.W., Bullo, F.: Breaking the hierarchy: distributed control and economic optimality in microgrids. IEEE Trans. Control Netw. Syst. **3**(3), 241–253 (2016)
8. Trip, S., Cucuzzella, M., Persis, C.D., Ferrara, A., Scherpen, J.M.A.: Robust load frequency control of nonlinear power networks. Int. J. Control 1–14 (2018)
9. Andreasson, M., Dimarogonas, D.V., Sandberg, H., Johansson, K.H.: Distributed control of networked dynamical systems: static feedback, integral action and consensus. IEEE Trans. Autom. Control **59**(7), 1750–1764 (2014)
10. Qin, J., Ma, Q., Shi, Y., Wang, L.: Recent advances in consensus of multi-agent systems: a brief survey. IEEE Trans. Ind. Electron. **64**(6), 4972–4983 (2017)
11. Trip, S., Cucuzzella, M., Persis, C.D., Schaft, A.V.D., Ferrara, A.: Passivity-based design of sliding modes for optimal load frequency control. IEEE Trans. Control Syst. Technol. **99**, 1–14 (2018)
12. Guo, F., Wen, C., Mao, J., Song, Y.-D.: Distributed economic dispatch for smart grids with random wind power. IEEE Trans. Smart Grid **7**(3), 1572–1583 (2016)
13. Shafiee, Q., Guerrero, J.M., Vasquez, J.C.: Distributed secondary control for islanded microgridsa novel approach. IEEE Trans. Power Electron. **29**(1), 1018–1031 (2014)
14. Simpson-Porco, J.W., Dörfler, F., Bullo, F.: Synchronization and power sharing for droop-controlled inverters in islanded microgrids. Elsevier, Autom. **49**(9), 2603–2611 (2013)
15. Guo, F., Wen, C., Mao, J., Song, Y.-D.: Distributed secondary voltage and frequency restoration control of droop-controlled inverter-based microgrids. IEEE Trans. Ind. Electron. **62**(7), 4355–4364 (2015)
16. Simpson-Porco, J.W., Shafiee, Q., Dörfler, F., Vasquez, J.C., Guerrero, J.M., Bullo, F.: Secondary frequency and voltage control of islanded microgrids via distributed averaging. IEEE Trans. Ind. Electron. **62**(11), 7025–7038 (2015)
17. Bidram, A., Davoudi, A., Lewis, F.L., Qu, Z.: Secondary control of microgrids based on distributed cooperative control of multi-agent systems. IET Gen., Trans. Distrib. **7**(8), 822–831 (2013)
18. Bidram, A., Davoudi, A., Lewis, F.L., Guerrero, J.M.: Distributed cooperative secondary control of microgrids using feedback linearization. IEEE Trans. Power Syst. **28**(3), 3462–3470 (2013)
19. Bidram, A., Davoudi, A., Lewis, F.L., Ge, S.S.: Distributed adaptive voltage control of inverter-based microgrids. IEEE Trans. Energy Convers **29**(4), 862–872 (2014)
20. Petrillo, A., Salvi, A., Santini, S., Valente, A.S.: Adaptive synchronization of linear multi-agent systems with time-varying multiple delays. J. Franklin Inst. **354**(18), 8586–8605 (2017)
21. Jadbabaie, A., Lin, J., Morse, A.S.: Coordination of groups of mobile autonomous agents using nearest neighbor rules. IEEE Trans. Autom. Control **48**(6), 988–1001 (2003)
22. Tang, F., Guerrero, J.M., Vasquez, J.C., Wu, D., Meng, L.: Distributed active synchronization strategy for microgrid seamless reconnection to the grid under unbalance and harmonic distortion. IEEE Trans. Smart Grid **6**(6), 2757–2769 (2015)
23. Simpson-Porco, J.W., Dörfler, F., Bullo, F.: Voltage stabilization in microgrids via quadratic droop control. IEEE Trans. Autom. Control **62**(3), 1239–1253 (2017)
24. Zhong, Q.C.: Robust droop controller for accurate proportional load sharing among inverters operated in parallel. IEEE Trans. Ind. Electron. **60**(4), 1281–1290 (2013)
25. Schiffer, J., Seel, T., Raisch, J., Sezi, T.: Voltage stability and reactive power sharing in inverter-based microgrids with consensus- based distributed voltage control. IEEE Trans. Control Syst. Technol. **24**(1), 96–109 (2016)

26. Coelho, E.A.A., Wu, D., Guerrero, J.M., Vasquez, J.C., Dragičević, T., Stefanović, Č., Popovski, P.: Small-signal analysis of the microgrid secondary control considering a communication time delay. IEEE Trans. Ind. Electron. **63**(10), 6257–6269 (2016)
27. Pilloni, A., Pisano, A., Orlov, Y., Usai, E.: Consensus-based control for a network of diffusion PDEs with boundary local interaction. IEEE Trans. Autom. Control **61**(9), 2708–2713 (2016)
28. Pilloni, A., Franceschelli, M., Pisano, A., Usai, E.: Recent advances in sliding-mode based consensus strategies. In: Proceedings of the 13th IEEE International Workshop on Variable Structure Systems, pp. 1–6 (2014)
29. Mei, J., Ren, W., Ma, G.: Distributed coordinated tracking with a dynamic leader for multiple euler-lagrange systems. IEEE Trans. Autom. Control **56**(6), 1415–1421 (2011)
30. Pilloni, A., Pisano, A., Usai, E.: Robust finite-time frequency and voltage restoration of inverter-based microgrids via sliding-mode cooperative control. IEEE Trans. Ind. Electron. **65**(1), 907–917 (2018)
31. Pilloni, A., Pisano, A., Usai, E.: Voltage restoration of islanded microgrids via cooperative second-order sliding mode control. Elsevier, IFAC-PapersOnLine **50**(1), 9637–9642 (2017)
32. Bokhari, A., Alkan, A., Dogan, R., Diaz-Aguiló, M., De Leon, F., Czarkowski, D., Zabar, Z., Birenbaum, L., Noel, A., Uosef, R.E.: Experimental determination of the zip coefficients for modern residential, commercial, and industrial loads. IEEE Trans. Power Delivery **29**(3), 1372–1381 (2014)
33. Hatipoglu, K., Fidan, I., Radman, G.: Investigating effect of voltage changes on static zip load model in a microgrid environment. In: IEEE North American Power Symposium (NAPS) (2012)
34. Powell, L.: Power System Load Flow Analysis. McGraw-Hill, New York (2005)
35. Renmu, H., Ma, J., Hill, D.J.: Composite load modeling via measurement approach. IEEE Trans. Power Syst. **21**(2), 663–672 (2006)
36. Martí, J.R., Ahmadi, H., Bashualdo, L.: Linear power-flow formulation based on a voltage-dependent load model. IEEE Trans. Power Delivery **28**(3), 1682–1690 (2013)
37. Levant, A.: Higher-order sliding modes, differentiation and output-feedback control. Int. J. Control **76**(9–10), 924–941 (2003)
38. Levant, A.: Robust exact differentiation via sliding mode technique. Elsevier, Automatica **34**(3), 379–384 (1998)
39. Bartolini, G., Pisano, A., Usai, E.: First and second derivative estimation by sliding mode technique. J. Signal Process. **4**(2), 167–176 (2000)
40. Orlov, Y.: Finite time stability and robust control synthesis of uncertain switched systems. SIAM J. Control Optim. **43**(4), 1253–1271 (2004)
41. Pilloni, A., Pisano, A., Usai, E.: Voltage restoration of islanded microgrids via cooperative second-order sliding mode control. IFAC-PapersOnLine **50**(1), 9637–9642 (2017)
42. Pilloni, A., Pisano, A., Usai, E.: Finite-time consensus for a network of perturbed double integrators by second-order sliding mode technique. In: IEEE Conference on Decision and Control, pp. 2145–2150 (2013)
43. Franceschelli, M., Gasparri, A., Giua, A., Seatzu, C.: Decentralized estimation of laplacian eigenvalues in multi-agent systems. Elsevier, Automatica **49**(4), 1031–1036 (2013)
44. Cortès, J.: Finite-time convergent gradient flows with applications to network consensus. Elsevier, Automatica **42**(11), 1993–2000 (2006)
45. Utkin, V.I.: Sliding Modes in Control and Optimization. Springer Science & Business Media, Berlin (2013)
46. Xin, H., Qu, Z., Seuss, J., Maknouninejad, A.: A self-organizing strategy for power flow control of photovoltaic generators in a distribution network. IEEE Trans. Power Syst. **26**(3), 1462–1473 (2011)
47. Pilloni, A., Pisano, A., Franceschelli, M., Usai, E.: Finite-time consensus for a network of perturbed double integrators by second-order sliding mode technique. IEEE Conference on Decision and Control, pp. 2145–2150 (2012)
48. Lu, X., Yu, X., Lai, J., Wang, Y., Guerrero, J.M.: A novel distributed secondary coordination control approach for islanded microgrids. IEEE Trans. Smart Grid (2016). https://doi.org/10.1109/TSG.2016.2618120

49. Gholami, M., Pilloni, A., Pisano, A., Dashti, Z.A.S., Usai, E.: Robust consensus-based secondary voltage restoration of inverter-based islanded microgrids with delayed communications. IEEE Conf. Decis. Control **811–816** (2018)
50. Gholami, M., Pilloni, A., Pisano, A., Usai, E.: Robust consensus-based secondary frequency and voltage restoration of inverter-based islanded microgrids with delayed communications. Control Eng. Pract., Under review on Elsevier (2019)
51. Chen, G., Feng, E.: Distributed secondary control and optimal power sharing in microgrids. IEEE/CAA J. Autom. Sinica **2**(3), 304–312 (2015)
52. Mathworks, Documentation matlab R2017a: Three-phase fault (2017). https://it.mathworks.com/help/physmod/sps/powersys/ref/threephasefault.html
53. Orlov, Y.: Discontinuous Systems: Lyapunov Analysis and Robust Synthesis Under Uncertainty Conditions. Springer, Berlin (2008)
54. Pilloni, A., Pisano, A., Usai, E., Menon, P.P., Edwards, C.: Decentralized state estimation in connected systems. Elsevier. IFAC Proc. **46**(2), 421–426 (2013)
55. Yang, J., Li, S., Yu, X.: Sliding-mode control for systems with mismatched uncertainties via a disturbance observer. IEEE Trans. Ind. Electron. **60**(1), 160–169 (2013)
56. Utkin, V., Lee, H.: Chattering problem in sliding mode control systems. In: IEEE International Workshop on Variable Structure Systems, pp. 346–350 (2006)
57. Utkin, V., Shi, J.: Integral sliding mode in systems operating under uncertainty conditions. IEEE Conf. Decis. Control **4**(1), 4591–4596 (1996)
58. Levant, A.: Sliding order and sliding accuracy in sliding mode control. Taylor & Francis. Int. J. Control **58**(6), 1247–1263 (1993)
59. Khalil, H.K.: Nonlinear Systems. Prentice Hall/Pearson Educ, New Jersey (2002)
60. Edwards, C., Spurgeon, S.: Sliding Mode Control: Theory and Applications. CRC Press, Boca Raton (1998)

Chapter 12
Local and Wide-Area Sliding-Mode Observers in Power Systems

Gianmario Rinaldi, Prathyush P. Menon, Christopher Edwards and Antonella Ferrara

Abstract This chapter presents a review of sliding-mode estimation techniques recently proposed by the authors in the area of power systems. The power grid is interpreted as a large-scale system, composed of an interconnection of generator nodes and load nodes. The analysis starts at the local level by considering a single generator node for which two dynamical models are presented. The first model relies on the well-established swing equations, while the second one is a nonlinear and more accurate model accounting for the transient voltage dynamics. Dedicated sliding-mode estimators are proposed to estimate the unmeasured states of a single generator node. In addition, the assessment based on real data is discussed, which employs both the lumped generator node model for the Nordic Power System and the data relevant to the major faults which happened in 2015. Then, the attention is focused on the wide-area level by considering an interconnection of generator nodes and load nodes. A structure-preserving power network dynamical model is presented, which exhibits a differential-algebraic equations (DAE) structure. For this large-scale system, a distributed observer scheme is designed, which combines both the use of a super-twisting-like architecture and iterative algorithms for the algebraic part of the system. The distributed observer scheme is validated by using the IEEE 39 bus benchmark.

Keywords Variable structure systems · Observers · Power systems · Large-scale systems

G. Rinaldi (✉) · A. Ferrara
Department of Electrical, Computer and Biomedical Engineering, University of Pavia, Pavia, Italy
e-mail: gianmario.rinaldi01@universitadipavia.it

A. Ferrara
e-mail: antonella.ferrara@unipv.it

P. P. Menon · C. Edwards
College of Engineering, Mathematics, and Physical Sciences, University of Exeter, Exeter, UK
e-mail: P.M.Prathyush@exeter.ac.uk

C. Edwards
e-mail: C.Edwards@exeter.ac.uk

M. Steinberger et al. (eds.), *Variable-Structure Systems and Sliding-Mode Control*,
Studies in Systems, Decision and Control 271,
https://doi.org/10.1007/978-3-030-36621-6_12

12.1 Introduction

Modern power systems require timely, accurate, and robust schemes for both estimating the unmeasured state variables and for detecting and reconstructing faults and attacks causing outages in the system [1]. Moreover, there is the need to turn the existing power grids into smarter and more reliable ones, which are able to deal with the increasing issues and challenges caused by both the rapid expansion of novel heterogeneous renewable energy-based sources, and the growth of the power demands [2]. In recent years, so-called phasor measurement units (PMUs) have been implemented in practical situations to realize a near real-time measurement network in power systems [3]. However, due to economic constraints and the spatial distribution of the power systems, only a limited number of PMUs can be installed in key nodes, and only a subset of the state variables of interest can be directly measured [4]. This calls for the conception of novel dynamical state estimators capable of reconstructing in near real time the unmeasured state variables in power networks [5, 6]. Several solutions have been proposed to perform dynamical state estimation, such as extended Kalman filters [7–10], Unknown Input Luenberger observers [11–13], and sliding-mode observers [14–20]. The use of estimation schemes can also be exploited to perform timely and accurate fault detection, reconstruction, and compensation [21, 22]. This can enhance the stability and the reliability of the power network. The use of observers to estimate the state of power networks is of paramount importance in advanced control schemes, even in those relying on sliding-mode architectures [23–28]. Different strategies have been implemented, such as the use of sliding-mode observers coupled with sliding-mode controllers [29, 30] and with standard proportional–integral (PI) controllers [21]. Alternatively, Luenberger observers have been adopted both coupled with sliding-mode controllers [29, 31] and with PI controllers [32].

In this chapter, a survey of sliding-mode-based estimators recently proposed by the authors is presented. The approach starts at the local level and it considers first a single generator node for which the dynamics are discussed in Sect. 12.2. Two types of local sliding-mode-based estimators for a single generator are described and assessed. The first estimator described in Sect. 12.3 relies on the well-established swing equations modeling the generator, and it employs an original super-twisting-like sliding-mode architecture to perform dynamical reconstruction of the frequency deviation. The scheme is validated via real measurement data acquired from the Nordic Power System. The second estimator in Sect. 12.4 is proposed for a nonlinear and more accurate model of a synchronous generator accounting for the transient voltage dynamic. Then, the analysis moves to wide-area level, by introducing a large-scale system composed of generator nodes and load nodes interconnected by power transmission lines. A structure-preserving dynamical model for power grids, which exhibits a differential-algebraic equations (DAE) structure, is presented in Sect. 12.5. A network of observers suitably interconnected in a distributed fashion is employed to reconstruct frequency deviations in generation nodes, and to reconstruct voltage phase angles in load nodes.

Table 12.1 List of symbols and variables used in this chapter

Symbol and Units	Physical meaning
δ_i (rad)	Generator voltage angle
α_i (rad)	Angular position of the rotor
ω_{m_i} (rad/s)	Mechanical speed of the rotor
$\Delta\omega_i$ (p.u.) or (rad/s)	Frequency deviation
ω_i (rad/s) or (Hz)	Electrical frequency
ω_{r_i} (rad/s) or (Hz)	Electrical speed of the generator
T_{m_i},T_{e_i} (N m)	Mechanical, electrical torques
P_{e_i},P_{e_i} (p.u.)	Mechanical, electrical powers
e_{q_i} (p.u.)	q-axis transient voltage
E_{f_i} (p.u.)	Excitation voltage
V_i (p.u.)	Voltage magnitude
P_{l_i} (p.u.)	Power demand in load nodes
ϑ_i (rad)	Load voltage phase angle
p_i (-)	Pole pairs of the generator
ω_0 (Hz)	Rated value of electrical frequency
ω_{m_R} (Hz)	Rated value of mechanical speed
E_i (GJ)	Generator base energy
S_i (GVA)	Generator base power
H_i (s)	Inertia constant
J_i (kg m^2)	Rotational mass
M_i (p.u.)	Inertia
D_i (p.u.)	Droop control coefficient/damping factor
T_{d0_i} (s)	d-circuit time constant
x_{d_i},x_{q_i} (p.u.)	d- and q-axis reactance
x'_{d_i} (p.u.)	d-axis transient reactance
b_{ij} (p.u.)	Transmission line susceptance

Notation

The following (standard) notation is used in this chapter. For a given signal x, $\hat{x}$ represents its estimate. For a matrix X, X^T denotes its transpose, while X^{-1} denotes its inverse. For a continuous-time signal x and a positive real exponent p, the expression $\lceil x \rfloor^p = |x|^p \text{sign}(x)$, where $\text{sign}(\cdot)$ represents the signum function. The expression $\text{col}(x, y)$ is used to denote the column vector $[x \;\; y]^T$, where x and y are two scalars. Table 12.1 shows a list of symbols and state variables adopted in this chapter, together with their measurement units and their physical meanings. The subscript i in the form of x_i denotes that the x state variable or model parameter is associated with the ith component.

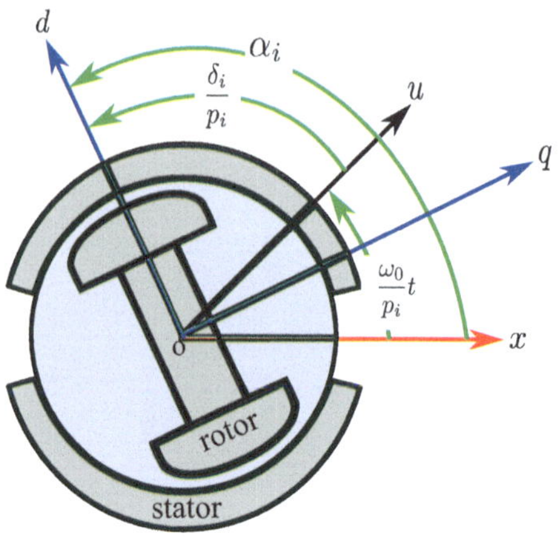

Fig. 12.1 A schematic of a synchronous generator together with the angles of interest in the swing equations

12.2 Synchronous Generator System Description

The swing equations for a synchronous generator connected to the grid are used to model the unbalance between electromagnetic and mechanical torque (see, for example, [33, 34]). This unbalance makes the mechanical speed of the rotor deviate from its synchronous value.

12.2.1 Angles Definitions

Exploiting Fig. 12.1, which shows a simplified schematic of the rotor of the ith generator, it follows that the angular position of the rotor with respect to a fixed axis x is given by the mechanical angle α_i. The reference system dqO is in-built with the rotor [33], while the axis u rotates at the constant speed ω_0/p_i, where $\omega_0 = 2\pi f_0$, f_0 is the rated value of the power grid electrical frequency, and p_i is the number of pole pairs of the generator. Note that the rotor of a synchronous generator with p_i pole pairs has a rated mechanical angular speed equal to $\omega_{m_R} = \omega_0/p_i$. It is useful to introduce the angle $\omega_0 t/p_i$ between the x and u axis. An additional angle, defined as δ_i/p_i, can be introduced in order to take into account the angular displacement between the u and d axis. From the previous development, it is apparent that the variable δ_i can be considered as the angular position of the rotor in electrical radians with respect to the synchronously rotating reference axis u. Justified by Fig. 12.1, the basic algebraic relation follows:

$$\delta_i = p_i\alpha_i - \omega_0 t. \tag{12.1}$$

Differentiating (12.1) with respect to time it yields

$$\dot{\delta}_i = p_i\dot{\alpha}_i - \omega_0 = \omega_{r_i} - \omega_0 = \Delta\omega_i, \tag{12.2}$$

where ω_{r_i} is the electrical angular speed of the rotor and $\Delta\omega_i$ is the electrical angular speed deviation from the rated value ω_0. If the electrical angular speed of the rotor is equal to its nominal values, it yields that $\dot{\delta}_i = 0$. Differentiating (12.2) yields

$$\ddot{\delta}_i = p_i\ddot{\alpha}_i = \Delta\dot{\omega}_i. \tag{12.3}$$

Equations (12.1)–(12.3) become fundamental in deriving the Swing equations.

12.2.2 Swing Equations

In order to present the dynamical model of the synchronous generator adopted in this chapter, it is necessary to recall that the kinetic energy E_i (expressed in (GJ)) stored in the rotating masses of a given generator can be defined as follows (see [33, 35]):

$$E_i = S_i H_i, \tag{12.4}$$

where S_i is the apparent power base value (expressed in (GVA)), and H_i is the inertia constant of the generator (expressed in (s)). The basic equation of motion (Newton's equation) is (see, e.g., [33, 34]):

$$J_i\dot{\omega}_{m_i} = J_i\ddot{\alpha}_i = T_{m_i} - T_{e_i}, \tag{12.5}$$

where J_i is the moment of inertia of the generator, $\dot{\omega}_{m_i} = \ddot{\alpha}_i$ is the mechanical angular acceleration, T_{m_i} is the mechanical torque and T_{e_i} is the electromagnetic torque. Using (12.3), Eq. (12.5) can be rewritten as follows:

$$J_i\frac{\ddot{\delta}_i}{p_i} = T_{m_i} - T_{e_i}. \tag{12.6}$$

By making use of (12.4), one gets

$$H_i = \frac{E_i}{S_i} = \frac{\frac{1}{2}J_i\omega_{m_R}^2}{S_i}, \tag{12.7}$$

where ω_{m_R} is the rated value of the mechanical angular speed of the generator, and S_i is the base power of the generator. From Eq. (12.7), it follows that

$$J_i = \frac{2H_iS_i}{\omega_{m_R}^2}. \tag{12.8}$$

Substituting for J_i from (12.8), Eq. (12.6) becomes

$$\frac{2H_iS_i}{\omega_{m_R}^2}\frac{\ddot{\delta}_i}{p_i} = T_{m_i} - T_{e_i}. \tag{12.9}$$

Since $\omega_{m_R} = \omega_0/p_i$, it is possible to rewrite (12.9) as follows:

$$\frac{2H_iS_i}{\frac{\omega_0 2}{(p_i)^2}}\frac{\ddot{\delta}_i}{p_i} = T_{m_i} - T_{e_i}. \tag{12.10}$$

Multiplying both the left and right sides of (12.10) by $\dot{\alpha}_i$, and using (12.2), the balance of torque (12.10) can be transformed into the balance of power as

$$\frac{2H_i}{\frac{\omega_0 2}{(p_i)^2}}\frac{\ddot{\delta}_i}{p_i}\left(\frac{\dot{\delta}_i + \omega_0}{p_i}\right) = \frac{T_{m_i} - T_{e_i}}{S_i}\dot{\alpha}_i \tag{12.11}$$

$$\frac{2H_i}{\omega_0^2}\ddot{\delta}_i\left(\dot{\delta}_i + \omega_0\right) = P_{m_i} - P_{e_i}, \tag{12.12}$$

where P_{m_i} is the mechanical input power, and P_{e_i} is the total electrical active power injected into the power grid by the ith generator. By defining the parameter $M_i := 2H_i/\omega_0$, two cases have to be considered:

$$M_i\ddot{\delta}_i = P_{m_i} - P_{e_i} \quad if \quad \dot{\delta}_i + \omega_0 \approx \omega_0, \tag{12.13}$$

$$\frac{M_i}{\omega_0}\ddot{\delta}_i = \frac{P_{m_i} - P_{e_i}}{\left(\dot{\delta}_i + \omega_0\right)} \quad \text{otherwise}. \tag{12.14}$$

Provided the mechanical angular speed remains sufficiently close to its rated value (which means $\dot{\delta}_i + \omega_0 \approx \omega_0$), Eq. (12.13) can represent the dynamics of the generator. If the mechanical angular speed is no longer close to its rated value, the equation to be considered is (12.14). It is possible to include in Eq. (12.13) the droop control action $-D_i\Delta\omega_i$ as follows:

$$M_i\ddot{\delta}_i = P_{m_i} - P_{e_i} - D_i\Delta\omega_i. \tag{12.15}$$

The resulting dynamical system is given by differential equation (12.2) together with differential equation (12.15)

$$\dot{\delta}_i = \Delta\omega_i, \tag{12.16}$$

$$M_i\Delta\dot{\omega}_i = P_{m_i} - P_{e_i} - D_i\Delta\omega_i \tag{12.17}$$

$$y_{i1} = \delta_i. \tag{12.18}$$

It is important to highlight that in Eq. (12.17), the signal P_{m_i} represents the control input of the system, while a more detailed expression of P_{e_i}, depending on power grid topology, will be introduced in the remainder of the chapter. The output in (12.18) can be measured in practical cases via PMUs [3].

12.2.3 Generator with Transient Voltage Dynamics Description

The following nonlinear model for the synchronous generator accounting for the transient voltage dynamics can be introduced [33]:

$$\left.\begin{aligned} \dot{\delta}_i &= \Delta\omega_i \\ \dot{e}_{q_i} &= \tfrac{1}{T_{d0_i}}\Big(E_{f_i} - e_{q_i} - \big(x_{d_i} - x'_{d_i}\big)\tfrac{e_{q_i} - V_i\cos(\delta_i)}{x'_{d_i}}\Big) \\ \Delta\dot{\omega}_i &= \tfrac{1}{M_i}\Big(P_{m_i} - \tfrac{V_i}{x'_{d_i}}e_{q_i}\sin(\delta_i) - \tfrac{V_i^2}{2}\big(\tfrac{1}{x_{q_i}} - \tfrac{1}{x'_{d_i}}\big)\sin\big(2\delta_i\big) - D_i\Delta\omega_i\Big) \end{aligned}\right\}. \tag{12.19}$$

The reader is referred to Table 12.1 for the physical interpretation of the introduced state variables, input signals, and model parameters. Note that (12.19) displays the same structure of (12.16)–(12.17) if e_{q_i} remains constant, and P_{e_i} is defined as

$$P_{e_i} := \frac{V_i}{x'_{d_i}}e_{q_i}\sin(\delta_i) + \frac{V_i^2}{2}\Big(\frac{1}{x_{q_i}} - \frac{1}{x'_{d_i}}\Big)\sin\big(2\delta_i\big). \tag{12.20}$$

12.3 A Super-Twisting-Like Sliding-Mode Observer for Frequency Reconstruction for Synchronous Generators

In this section, a super-twisting-like sliding-mode observer is presented. It allows one to estimate in a finite time the frequency deviation for a single generator node governed by the dynamics in (12.16)–(12.17). It is assumed the generator voltage angle as in (12.18) is measured. The underlying idea is to use the super-twisting observation algorithm [36] with the insertion of original terms depending on the output observation error, as described in the sequel.

The following assumption is now introduced.

Assumption 12.1 The signal

$$v_i := \frac{P_{m_i} - P_{e_i}}{M_i} \tag{12.21}$$

is bounded so that

$$|v_i| \leq \Psi_i, \tag{12.22}$$

where Ψ_i is a positive known constant determined from an understanding of the system.

Assumption 12.1 is always satisfied in practical cases, since the imbalance between mechanical power delivered by the turbine P_{m_i} and the alteration disturbance P_{e_i} is always bounded [37].

12.3.1 Super-Twisting-Like Sliding-Mode Observer Design

In order to estimate in finite time the frequency of the generator, the following super-twisting-like sliding-mode observer is proposed:

$$\dot{\hat{\delta}}_i = \Delta\hat{\omega}_i - a_i e_{\delta_i} - k_{1_i} \lceil e_{\delta_i} \rfloor^{1/2} \tag{12.23}$$

$$\Delta\dot{\hat{\omega}}_i = -k_{2_i} \lceil e_{\delta_i} \rfloor^{0} - a_i^2 e_{\delta_i} + a_i \Delta\hat{\omega}_i - k_{1_i} a_i \lceil e_{\delta_i} \rfloor^{1/2} \tag{12.24}$$

$$\hat{\omega}_i = \left(\Delta\hat{\omega}_i + 1\right) \omega_0, \tag{12.25}$$

where k_{1_i} and k_{2_i} are positive constants to be designed, $e_{\delta_i} := \hat{\delta}_i - \delta_i$, and $a_i := -D_i/M_i$. The estimate of the frequency in (Hz) or in (rad/s) can be obtained from its deviation measured in (p.u.) according to the algebraic equation (12.25).

Remark 12.1 Note the presence of original terms in the observer (12.23)–(12.25) with respect to the standard super-twisting sliding-mode observation architecture in [36]. More precisely, these are $-a_i e_{\delta_i}$ in (12.23) and $-a_i^2 e_{\delta_i} - k_{1_i} a_i \lceil e_{\delta_i} \rfloor^{1/2}$ in (12.24). These terms are instrumental in deriving the error dynamics in the form of the super-twisting architecture, and in having only signal v_i as matched disturbance.

By subtracting Eqs. (12.16)–(12.17) to (12.23)–(12.24), the so-called error system dynamics are obtained as

$$\dot{e}_{\delta_i} = e_{\omega_i} - a_i e_{\delta_i} - k_{1_i} \lceil e_{\delta_i} \rfloor^{1/2} \tag{12.26}$$

$$\dot{e}_{\omega_i} = -k_{2_i} \lceil e_{\delta_i} \rfloor^{0} - a_i^2 e_{\delta_i} - k_{1_i} a_i \lceil e_{\delta_i} \rfloor^{1/2} - v_i + a_i e_{\omega_i}, \tag{12.27}$$

where $e_{\omega_i} := \Delta\hat{\omega}_i - \Delta\omega_i$. By defining

$$\tilde{e}_{\omega_i} := e_{\omega_i} - a_i e_{\delta_i}, \tag{12.28}$$

after straightforward algebraic simplifications, the following error dynamics are obtained:

$$\dot{e}_{\delta_i} = -k_{1_i} \lceil e_{\delta_i} \rfloor^{1/2} + \tilde{e}_{\omega_i} \tag{12.29}$$

$$\dot{\tilde{e}}_{\omega_i} = -k_{2_i} \lceil e_{\delta_i} \rfloor^{0} - v_i. \tag{12.30}$$

Equations (12.29)–(12.30) are in the form of the super-twisting sliding-mode architecture [38]. Note that the disturbance v_i appears in the matched channel of the system in Eqs. (12.29)–(12.30). The constants k_{1_i} and k_{2_i} can be tuned according to the following equations [39, 40]:

$$k_{1_i} = 1.5\sqrt{\Psi_i} \tag{12.31}$$

$$k_{2_i} = 1.1\Psi_i. \tag{12.32}$$

As proven in [38], the conditions (12.31)–(12.32) ensure that the origin is a finite-time stable equilibrium point for the system in (12.29)–(12.30), thus getting:

$$e_{\delta_i}\left(t^*\right) = 0 \tag{12.33}$$

$$\tilde{e}_{\omega_i}\left(t^*\right) = e_{\omega_i}(t^*) - a_i e_{\delta_i}(t^*) = 0 \tag{12.34}$$

$$e_{\omega_i}\left(t^*\right) = 0 \tag{12.35}$$

in finite time t^*, and then, a correct estimate of the frequency can be achieved $\forall t \geq t^*$.

Remark 12.2 Note that the observer in the form of (12.23)–(12.24) depends on the constant $a_i = -D_i/M_i$, which is a ratio between the damping coefficient and the inertia of the generator model. In case constant a_i is not known, by defining signal

$$\tilde{v}_i = v_i - \frac{D_i}{M_i}\Delta\omega_i, \tag{12.36}$$

and by assuming (according also to Assumption 12.1) that

$$|\tilde{v}_i| \leq \tilde{\Psi}_i, \tag{12.37}$$

where $\tilde{\Psi}_i$ is a known positive constant, it is possible to introduce the following super-twisting sliding-mode observer:

$$\dot{\hat{\delta}}_i = \Delta\hat{\omega}_i - k_{1_i}\left\lceil e_{\delta_i}\right\rfloor^{1/2} \tag{12.38}$$

$$\Delta\dot{\hat{\omega}}_i = -k_{2_i}\left\lceil e_{\delta_i}\right\rfloor^{0} \tag{12.39}$$

$$\hat{\omega}_i = \left(\Delta\hat{\omega}_i + 1\right)\omega_0. \tag{12.40}$$

The finite-time convergence of the observer in the form of (12.38)–(12.39) can be easily proven as discussed above, since its error dynamics are in the form of (12.29)–(12.30) (see, e.g., [36] for further details).

12.3.2 Real Data-Based Super-Twisting-Like Observer Validation

In this section, the super-twisting-like sliding-mode observer in (12.23)–(12.25) is assessed based on real data. This approach is novel, since it is much more common to employ IEEE power network benchmarks to validate estimation and control architectures (see, e.g., [15, 27, 30]). In contrast, the idea here is to compare measurement data of the frequency available for the Nordic Power System in [41] with the estimates generated by the designed sliding-mode observer. Specifically, the lumped Nordic Power System model data presented in [35] is employed, which exhibits the same structure as in (12.16)–(12.17). A super-twisting-like sliding-mode observer is designed, which is capable of estimating the frequency. Then, the measurement data for the frequency is recovered during two major faults which took place in 2015. Technical data about these two faults can be found in [42]. The frequency measurements in [41] have been captured employing a sampling time of 0.1 s, by using a network of PMUs placed in different locations in Nordic Power System. In accordance with [42], in the present approach, it is assumed that this measurement data represents the frequency of the whole Nordic Power System. The comparison between the real frequency measurements and the estimates reveals a high level of accuracy of the observer during the first seconds after the faults. As in [35], a lumped equivalent model for the generator can be introduced in the form of (12.16)–(12.17). It is assumed that only the generator voltage angle δ_i as in (12.18) is measured. This can be easily implemented in practice by equipping the generator with an encoder tracking the angular position of the rotor [43], or via phasor measurement units (PMUs) [44]. The assessment of the observation scheme is undertaken by comparing the frequency estimates from the super-twisting-like observer with the frequency measurements coming from the real Nordic Power System.

Remark 12.3 The real Nordic Power System has complex (and unknown) dynamics which has not been considered for the design of the super-twisting-like state estimator. The unknown nonlinearities affecting the dynamics are treated as disturbances, and their impact on the performance of the proposed observer will be evaluated.

Figure 12.2 shows a schematic of the assessment to highlight the underlying ideas. A simplified and lumped dynamical model of the Nordic Power System derived in [35] is employed. This model is instrumental in designing the observer and comprises an equivalent linearized water turbine (due to the presence of hydropower generators participating in frequency regulation), and an equivalent generator. Two major faults [41], which took place in 2015 in the Nordic Power System, are chosen to assess the accuracy of the observer. Specifically:

1. On June 5, 2015, at 11:54:50 PM, there was a disconnection of a nuclear power plant from the Nordic Power System. This caused a sudden loss of generation of 878 MW. Relevant data for this fault is reported in Table 12.2;

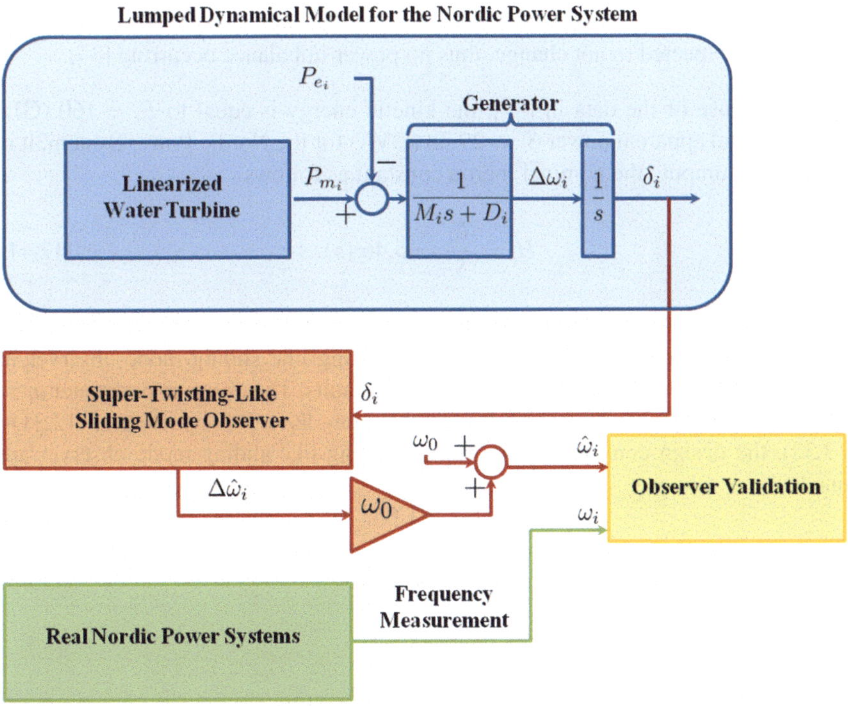

Fig. 12.2 A schematic of the lumped Nordic Power System dynamical model, together with the designed observer for the frequency estimation

Table 12.2 Relevant data of the disturbance on the June 5, 2015 (left) and on the August 8, 2015 (right)

Starting Time	11:54:00 PM	Starting Time	2:21:00 PM
Fault Time	11:54:50 PM	Fault Time	2:22:08 PM
Ending Time	11:57:00 PM	Ending Time	2:24:00 PM
Disturbance Magnitude	878 MW	Disturbance Magnitude	600 MW
Kinetic Energy	160 GJ	Kinetic Energy	160 GJ
Cause	Nuclear Power Plant	Cause	Loss of an HVDC

2. On August 8, 2015, at 2:22:08 PM, a high voltage direct current (HVDC) transmission line was lost, causing a sudden loss of consumption of 600 MW. Relevant data for this fault is reported in Table 12.2.

Remark 12.4 Note that the fault on the June 5, 2015 is characterized by a leak of power generation, while the one on the 8th August is characterized by extra-power generation. It follows that the two faults are of opposite sign. In the event that simultaneous faults with the same absolute value but opposite sign take place, the frequency deviation is expected to not change, thus no power imbalance occurring [33].

By making use of the data in [35], the kinetic energy is equal to $E_i = 160$ (GJ). Since the total apparent power $S_i = 29.28$ (GVA) for the Nordic Power System, it is possible to compute the nominal inertia constant as follows:

$$H_i = \frac{E_i}{S_i} = 5.46\ (s). \tag{12.41}$$

The nominal value of the droop coefficient is $D_i = 0.9$ (p.u.) [35]. The linearized model in Fig. 12.2 coupled with the super-twisting-like sliding-mode observer in (12.23)–(12.24) are used for the two considered faults. The observer parameter $a_i = -D_i/M_i$ is set equal to $a_i = -4.12$ (p.u.). Defining $\Psi_i = 0.5$, by using to (12.31)–(12.32), the design constants of the super-twisting-like sliding-mode observer are tuned as

$$k_{1_i} = 1.06 \tag{12.42}$$

$$k_{2_i} = 0.55. \tag{12.43}$$

By exploiting (12.21) and (12.22), one gets

$$|M_i v_i| = \left|P_{m_i} - P_{e_i}\right| \leq M_i \Psi_i. \tag{12.44}$$

By substituting the actual values of the variables, one gets

$$\left|P_{m_i} - P_{e_i}\right| \leq 0.11\ (p.u) \approx 3.20\ (\text{GW}), \tag{12.45}$$

which means that a maximum value for the power unbalance $P_{m_i} - P_{e_i}$ is equal to 3.20 (GW), which is fulfilled in practice for all the real faults described in [42]. The system as depicted in Fig. 12.2 with the proposed sliding-mode observer was implemented in MATLAB-Simulink R2017a. The fixed-step method *ode1(Euler)* was employed with an integration step size of 0.0001 s.

12.3.2.1 Fault on June 5, 2015

In the last few minutes of June 5, 2015, a nuclear power plant was suddenly disconnected from the Nordic Power System. Figure 12.3 shows both the estimates of the frequency from the super-twisting-like sliding-mode observer as a solid line and the real measured values as a dotted lines. A transient can be identified during the first seconds, due to the reaching of the sliding motion of the observer. During the pre-fault scenario, the estimated and the real values of the frequency are slightly different. Specifically, small oscillations can be noted, which are due to unmodelled dynamics. During the first seconds after the fault, the estimated and the real values are

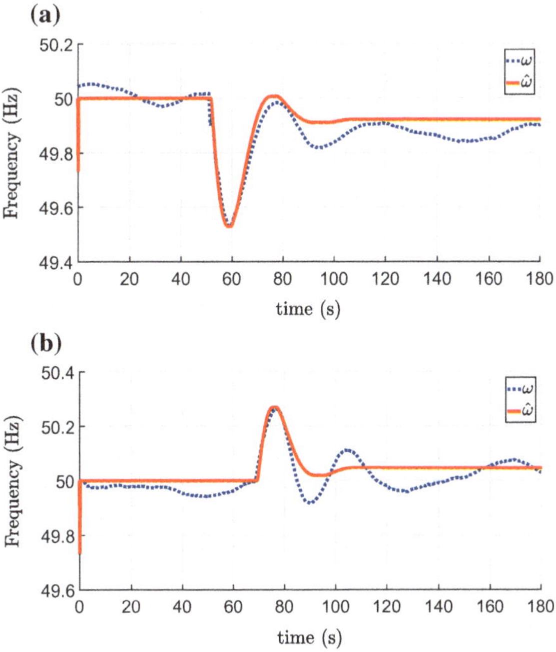

Fig. 12.3 Time evolution of the frequency measured in the Nordic Power Systems (dotted line) and the estimate value (solid line) from the proposed sliding-mode observer. **a** Fault happened on the June 5, 2015. The starting point $t = 0$ (s) in the plot corresponds to the time instant 11:54:00 PM. **b** Fault happened on the August 8, 2015. The starting point $t = 0$ (s) in the plot corresponds to the time instant 02:21:00 PM

practically the same. During the post-fault scenario, the estimated and the real values of the frequency are again slightly different. These differences can be understood by remembering that in practical cases a large class of unmodelled components, such as voltage-dependent and frequency-dependent loads, the action of electrical protection devices, and voltage oscillations are present in the power system [33]. These have not been considered in the current framework. However, the impact of these simplified assumptions for the observer design can be considered to be minimal by exploiting the difference between the real measured values and the estimated ones in Fig. 12.3a.

12.3.2.2 Fault on August 8, 2015

Following the same idea, we show now the observer-assessment for the fault which took place during the afternoon of August 8, 2015, when a high voltage direct current (HVDC) power transmission line was disconnected from the rest of the Nordic Power System. Figure 12.3b shows the results following the notation adopted. The same considerations discussed above still hold. Note that in this case, the frequency increases because there was a loss of power demand. This is in accordance with the swing equation approach presented in Sect. 12.2.2.

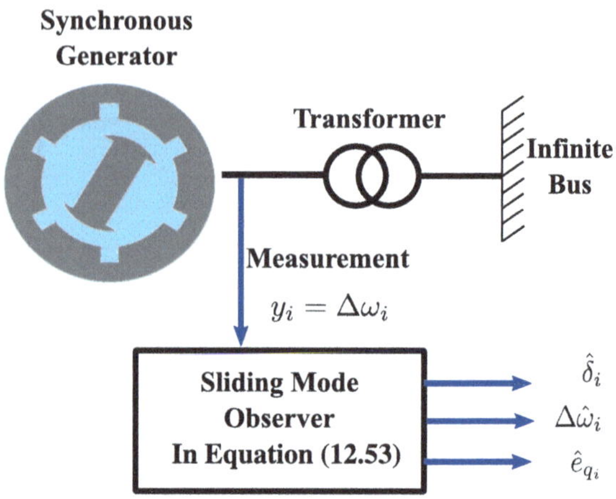

Fig. 12.4 A schematic of a synchronous generator connected to an infinite bus, together with the proposed sliding-mode observer

12.4 Sliding-Mode Observer for a Single Synchronous Generator with Transient Voltage Dynamics

In this section, a first-order sliding-mode observer is proposed to estimate at the local level both the generator voltage angle and the transient voltage of a single generator node (Fig. 12.4). It is assumed to locally measure only the frequency deviation. With respect to Sect. 12.3, the detailed dynamics description in (12.19) is employed for a single synchronous generator, which makes the approach adopted in this section more general. The synchronous generator is assumed to have nonlinear dynamics around its nominal (stable) working point, as expressed by the following assumption.

Assumption 12.2 There exists an asymptotically stable equilibrium point for the nonlinear system (12.19), given the constant input signals $\overline{T}_{m_i}$, $\overline{E}_{f_i}$, $\overline{V}_i$, which practically correspond to the nominal working point of the synchronous generator. The corresponding constant (nominal) state variables are $\left(\overline{\delta}_i, \overline{e}_{q_i}, \Delta\overline{\omega}_i = 0\right)$.

Following Assumption 12.2, it is possible to introduce time-varying perturbations ΔV_i, ΔE_{f_i}, ΔT_{m_i} of the external inputs around their nominal values as follows: $V_i = \overline{V}_i + \Delta V_i$, $E_{f_i} = \overline{E}_{f_i} + \Delta E_{f_i}$, $T_{m_i} = \overline{T}_{m_i} + \Delta T_{m_i}$. Only the perturbation ΔV_i is assumed to be known. For fixed inputs $\overline{T}_{m_i}$, $\overline{E}_{f_i}$, $\overline{V}_i$, the system in (12.19) can be represented as a combination of linear and nonlinear terms as follows:

$$\begin{aligned}
\dot{X}_{1_i} &= A_{1_i} X_{1_i} + A_{2_i} X_{2_i} + G_{1_i}(X_{1_i}, \Delta V_i) + g_{1_i}(E_{f_i}) \\
\dot{X}_{2_i} &= A_{3_i} X_{1_i} + A_{4_i} X_{2_i} + G_{2_i}(X_{1_i}, \Delta V_i) + g_{2_i}(U_i) \\
y_i &= X_{2_i} = C_i X_i,
\end{aligned} \tag{12.46}$$

where $X_i = \text{col}\left(X_{1_i},\ X_{2_i}\right)$, and the components $X_{1_i} = \text{col}\left(\delta_i,\ e_{q_i}\right)$, $X_{2_i} = \Delta\omega_i$; the vector $C_i = \left[0\ 0\ 1\right]$; $G_{1_i}(\cdot)$, $G_{2_i}(\cdot)$, $g_{1_i}(\cdot)$, and $g_{2_i}(\cdot)$ are properly defined functions, and the input $U_i = \text{col}(\Delta T_{m_i},\ \Delta E_{f_i},\ \Delta V_i)$. Only $\Delta\omega_i$ is measured by employing the PMUs [34]. The states δ_i and e_{q_i} have to be estimated for the purpose of enhancing the monitoring of the synchronous generator. Matrix A_i, is the Jacobian matrix of the nonlinear system (12.19) about an equilibrium point $\left(\overline{\delta}_i, \overline{e}_{q_i}, \Delta\overline{\omega}_i = 0\right)$, and it is given by

$$A_i = \left[\begin{array}{c|c} A_{1_i} & A_{2_i} \\ \hline A_{3_i} & A_{4_i} \end{array}\right] = \left[\begin{array}{cc|c} 0 & 0 & 1 \\ \frac{-\overline{V}_i \sin(\overline{\delta}_i)(x_{d_i} - x'_{d_i})}{T_{d0_i} x'_{d_i}} & \frac{-x_d}{T_{d0} x'_d} & 0 \\ \hline -\frac{\overline{V}_i}{M_i x'_{d_i}} \left(\frac{\overline{V}_i (x'_{d_i} - x_{q_i})}{x_{q_i}} \cos(2\overline{\delta}_i) + \overline{e}_{q_i} \cos(\overline{\delta}_i)\right) & \frac{-\overline{V}_i \sin(\overline{\delta}_i)}{M_i x'_{d_i}} & -\frac{D_i}{M_i} \end{array}\right], \tag{12.47}$$

where $A_{1_i} \in \mathbb{R}^{2\times 2}$, $A_{2_i} \in \mathbb{R}^{2\times 1}$, $A_{3_i} \in \mathbb{R}^{1\times 2}$, and $A_{4_i} \in \mathbb{R}^{1\times 1}$. Given Assumption 12.2, it follows that the Jacobian matrix A_i is Hurwitz, since the equilibrium point $\left(\overline{\delta}_i, \overline{e}_{q_i}, \Delta\overline{\omega}_i = 0\right)$ is assumed asymptotically stable. In addition, it is possible to verify by direct calculation that the pair (A_i, C_i) is detectable. The following assumption holds.

Assumption 12.3 The nonlinear functions $G_{1_i}(\cdot)$, $G_{2_i}(\cdot)$, and therefore the function $G_i(\cdot) = \mathrm{col}(G_{1_i}(\cdot), G_{2_i}(\cdot))$ are Lipschitz with respect to X_{1_i}. Let $L_{G_{1_i}}$, $L_{G_{2_i}}$, L_{G_i} be the Lipschitz constants of $G_1(\cdot)$, $G_2(\cdot)$, and $G(\cdot)$, respectively. The functions $g_{1_i}(\cdot)$, $g_{2_i}(\cdot)$, and $g_i(\cdot) = \mathrm{col}(g_{1_i}(\cdot), g_{2_i}(\cdot))$ are unknown bounded external inputs, with known positive upper bounds on their norms $\Delta_{g_{1_i}}$, $\Delta_{g_{2_i}}$, and Δ_{g_i}, respectively.

12.4.1 Observer Design

A robust sliding-mode observer is proposed to dynamically estimate the generator voltage phase angle and the transient voltage. Since (A_i, C_i) is detectable, it follows that there exists a matrix R_i such that $A_i - R_iC_i$ is stable. (Note that A_i is assumed stable, so the trivial choice $R_i = 0$ can be made.) For any $Q_i > 0$ the Lyapunov Equation [45]

$$(A_i - R_iC_i)^T P_i + P_i(A_i - R_iC_i) = -Q_i, \tag{12.48}$$

has a unique symmetric positive definite solution denoted as $P_i \succ 0$. Analogously to (12.47), the matrix P_i is partitioned as

$$P_i = \begin{bmatrix} P_{1_i} & P_{2_i} \\ P_{2_i}^T & P_{3_i} \end{bmatrix}, \tag{12.49}$$

where $P_{1_i} \in \mathbb{R}^{2\times 2}$, $P_{2_i} \in \mathbb{R}^{2\times 1}$, and $P_{3_i} \in \mathbb{R}^{1\times 1}$. A linear change of coordinates $Z_i = \mathrm{col}\left(Z_{1_i}, Z_{2_i}\right) := T_iX_i$ is now introduced for the dynamical system (12.46), where the matrix T_i is defined as

$$T_i := \begin{bmatrix} I_2 & P_{1_i}^{-1} P_{2_i} \\ 0_{12} & I_1 \end{bmatrix}. \tag{12.50}$$

Note that the change of coordinate is required to obtain a canonical form of the generator dynamics instrumental for designing the observer, in which $(A_{1_i} + P_{1_i}^{-1} P_{2_i} A_{3_i})$ is Hurwitz [46]. The matrix $\tilde{A}_i = T_i A_i T_i^{-1}$ is given by

$$\tilde{A}_i = \begin{bmatrix} \tilde{A}_{1_i} & \tilde{A}_{2_i} \\ \tilde{A}_{3_i} & \tilde{A}_{4_i} \end{bmatrix} = \left[\begin{array}{c|c} A_{1_i} + P_{1_i}^{-1} P_{2_i} A_{3_i} & \begin{array}{c} A_{2_i} - A_{1_i} P_{1_i}^{-1} P_{2_i} \\ + P_{1_i}^{-1} P_{2_i} \left(A_{4_i} - A_{3_i} P_{1_i}^{-1} P_{2_i}\right) \end{array} \\ \hline A_{3_i} & A_{4_i} - A_{3_i} P_{1_i}^{-1} P_{2_i} \end{array}\right]. \quad (12.51)$$

The system in equations in (12.46) can be written as

$$\left.\begin{aligned} \dot{Z}_{1_i} &= \tilde{A}_{1_i} Z_{1_i} + \tilde{A}_{2_i} Z_{2_i} + G_{1_i}(Z_i, \Delta V_i) + g_{1_i}(E_{f_i}) \\ &\quad + P_{1_i}^{-1} P_{2_i} \left(G_{2_i}(Z_i, \Delta V_i) + g_{2_i}(U_i)\right) \\ \dot{Z}_{2_i} &= \tilde{A}_{3_i} Z_{1_i} + \tilde{A}_{4_i} Z_{2_i} + G_{2_i}(Z_i, \Delta V_i) + g_{2_i}(U_i) \\ y_i &= Z_{2_i}. \end{aligned}\right\}. \quad (12.52)$$

Let the sliding-mode observer for (12.52) be:

$$\left.\begin{aligned} \dot{\hat{Z}}_{1_i} &= \tilde{A}_{1_i} \hat{Z}_{1_i} + \tilde{A}_{2_i} Z_{2_i} + G_{1_i}(\hat{Z}_i, \Delta V_i) + P_{1_i}^{-1} P_{2_i} \left(G_{2_i}(\hat{Z}_i, \Delta V_i)\right) \\ \dot{\hat{Z}}_{2_i} &= \tilde{A}_{3_i} \hat{Z}_{1_i} + \tilde{A}_{4_i} \hat{Z}_{2_i} + G_{2_i}(\hat{Z}_i, \Delta V_i) + \nu_i \\ \hat{y}_i &= \hat{Z}_{2_i} \\ \nu_i &= \left(\left\|\tilde{A}_{4_i}\right\| \|e_{y_i}\| + \rho_i\right) \operatorname{sign}\left(e_{y_i}\right) \end{aligned}\right\}, \quad (12.53)$$

where $\hat{Z}_{1_i}$ represents the estimate of Z_{1_i}, $\hat{Z}_{2_i}$ represents the estimate of Z_{2_i}, ρ_i is a positive constant to be designed, and ν_i is a discontinuous term depending on the output observation error $e_{y_i} := Z_{2_i} - \hat{Z}_{2_i}$. The error system dynamics is obtained by subtracting (12.53) from (12.52). By defining $e_{1_i} := Z_{1_i} - \hat{Z}_{1_i}$, it follows that

$$\dot{e}_{1_i} = \tilde{A}_{1_i} e_{1_i} + [I_2 \; P_{1_i}^{-1} P_{2_i}]\Big(G_i(Z_i, \Delta V_i) - G_i(\hat{Z}_i, \Delta V_i) + g_i(U_i)\Big) \quad (12.54a)$$

$$\dot{e}_{y_i} = \tilde{A}_{3_i} e_{1_i} + G_{2_i}(Z_i, \Delta V_i) - G_{2_i}(\hat{Z}_i, \Delta V_i) + g_{2_i}(U_i) - \nu_i. \quad (12.54b)$$

The following sliding surface is associate with (12.54a)–(12.54b):

$$\mathcal{S}_i = \left\{\left(e_{1_i}, e_{y_i}\right) | e_{y_i} = 0\right\}. \quad (12.55)$$

The following proposition is introduced.

Proposition 12.1 *Given Assumptions 12.2–12.3, the error dynamics* (12.54a)–(12.54b) *satisfy the following:*

(i) *The Euclidean norm* $\|e_{1_i}\|$ *remains bounded* $\forall t \geq 0$, *i.e.,* $\|e_{1_i}\| \leq \gamma_i$, *where* γ_i *is a positive constant.*

(ii) *If the external inputs* $g_i(U_i) = 0$ *and*

$$\lambda_{\min}\left(Q_{1_i}\right) > 2\left\|\left[P_{1_i} \;\; P_{2_i}\right]\right\| L_{G_i}, \tag{12.56}$$

then the point $e_{1_i} = 0$ *is an asymptotically stable equilibrium point for the system* (12.54a).

(iii) *System* (12.54a)–(12.54b) *is driven to the sliding surface* (12.55) *in a finite time* t_{r_i} *if the design constant* ρ_i *appearing in* ν_i *is tuned according to the following inequality:*

$$\rho_i > \left(\left\|\tilde{A}_{3_i}\right\| + L_{G_{2_i}}\right)\gamma_i + \Delta_{g_{2_i}} + \eta_i, \tag{12.57}$$

where η_i *is a positive constant.*

Proof The reader is referred to [47] for the proof of Proposition 12.1.

The following performance metric $\mathcal{P}_i$ can be introduced to evaluate the performances of the proposed observer:

$$\mathcal{P}_i := \int_{t_{r_i}}^{t} \left\|e_{y_i}(\tau)\right\| d\tau. \tag{12.58}$$

It is expected that $\mathcal{P}_i$ is almost zero. This will also be demonstrated in a simulation environment.

Remark 12.5 Suppose that a differentiable band-limited measurement noise ψ_i affects the output of the system (12.52) as $y_i = Z_{2_i} + \psi_i$. Then, the output estimation error becomes $e_{y_i} := Z_{2_i} + \psi_i - \hat{Z}_{2_i}$. By exploiting (12.54a)–(12.54b), it is possible to note that the effect of the noise can be included as part of the unknown bounded input $g_i(U_i)$, by increasing the value for Δ_{g_i} accordingly (see Assumption 12.3). As a consequence, part ***(i)*** of Proposition 12.1 still holds in this case, thus maintaining the bounded properties of $\left\|e_{1_i}\right\|$ again. As for part ***(iii)***, the sliding motion $e_{y_i} = 0$ still takes place in a finite time. However, by defining the noise-free output error as follows: $\check{e}_{y_i} := Z_{2_i} - \hat{Z}_{2_i}$, the condition $\hat{Z}_2 = Z_2 + \psi_i$ holds in a finite time, as proven in [48].

12.4.2 Simulation-Based Observer Validation

Simulation results are now provided to assess the sliding-mode observer in (12.53). A single synchronous generator connected to the grid is considered. The numerical representation of the model parameters, together with the values for both the state variables and input signals evaluated for the nominal equilibrium point, is taken from [7] and from [47]. The simulation time horizon was set equal to 10 s, and the synchronous generator was modeled in the *MATLAB-Simulink R2017b* environment by using the *Ode1-Euler* solver with an integration step size $\tau = 1 \times 10^{-4}$ s. The

(Hurwitz) Jacobian Matrix A in (12.47) evaluated for the equilibrium data together with its eigenvalues λ are

$$A = \begin{bmatrix} 0 & 0 & 377.000 \\ -20.100 & -42.830 & 0 \\ -0.180 & -0.160 & -0.005 \end{bmatrix}, \quad \lambda = \begin{bmatrix} -42.173 \\ -0.331 + 6.329j \\ -0.331 - 6.329j \end{bmatrix}. \tag{12.59}$$

For the solution of the Lyapunov Equation (12.48), matrix $Q = I_3$, and the gain matrix L is chosen equal to $R = \begin{bmatrix}376.0877 & -173.7180 & -0.0124\end{bmatrix}^T$. Then, the unique symmetric positive definite solution P of (12.48) is

$$P = \begin{bmatrix} 10.8153 & -7.2107 & -0.5481 \\ -7.2107 & 19.2845 & 3.9179 \\ -0.5481 & 3.9179 & 0.9121 \end{bmatrix}. \tag{12.60}$$

The design constant of the observer is set as $\rho = 0.5$. Matrix $\tilde{A}$ for the implementation of the observer (12.53) is obtained as follows:

$$\tilde{A} = TAT^{-1} = \begin{bmatrix} -0.020 & -0.017 & 377.000 \\ -20.125 & -42.866 & -0.001 \\ -0.180 & -0.160 & -0.005 \end{bmatrix}. \tag{12.61}$$

Two scenarios are introduced:

1. **Scenario 1**, during which the armature voltage is affected by a step variation of 0.1 (p.u.), which yields

$$E_f = 2.29 + 0.1\text{step}(t - 2). \tag{12.62}$$

   Note that this scenario has also been considered in [7].
2. **Scenario 2**, during which the mechanical input torque obeys to the following time-varying law:

$$T_m = \begin{cases} 0.8 & 0 < t \leq 2 \\ 0.8 + 0.05 \cdot (t - 2) & 2 < t \leq 4 \\ 0.9 & 4 < t \leq 10. \end{cases} \tag{12.63}$$

Note that the input disturbances E_f and T_m are unknown by the observer in the two scenario.

Three different estimation schemes are considered for the two scenarios, in order to make a comparison with existing and well-established estimation methods. More precisely, if x represents a state variables, $\hat{x}_{SM}$ denotes its estimate obtained employing the sliding-mode observer in a measurement noise-free condition, $\hat{x}_{n-SM}$ represents the estimate obtained employing a sliding-mode observer with measurement noise, while $\hat{x}_{n-EFK}$ is used for the estimate from an extended Kalman filter (EKF) with measurement noise. The same variance of the measurement noise as in [7] is chosen, which is to $\sigma^2 = 0.1$. As shown in Figs. 12.5 and 12.6, the proposed sliding-mode estimation scheme can acceptably estimate the unmeasured state variables in the noise-free case, but also in presence of measurement disturbances. It is apparent that the EKF displays a better attenuation of the measurement noise than sliding-mode observer, as revealed in Fig. 12.6. However, the EKF performance can degrade in the presence of unknown inputs affecting the system. The evolution of the performance metric $\mathcal{E}(T)$ in (12.58) as function of the variance of the measurement noise σ^2 is almost linear, as shown in Table 12.3.

12.5 Distributed Observers for State Estimation in Power Systems

In this section, a wide-area network is considered and a distributed observer scheme is formulated to estimate both the frequency deviation in each generator node and the voltage phase angle of each load node. We first introduced a description of structure-preserving power networks, which are composed of an interconnection of generator nodes and load nodes. The super-twisting-like sliding-mode observer in (12.23)–(12.24) previously adopted in this chapter for local state estimation in a single generator node, will now be used to perform distributed state estimation in all the generator nodes of a given power network. Moreover, distributed iterative algorithms are used to estimate the voltage phase angle in each load node, relying on an exchange of local information and local knowledge of the system. Finally, a simulation-based assessment is presented, which makes use of the well-established IEEE 39 bus benchmark [49]. Figure 12.7 summarizes the key ideas of the conception of a distributed observer for wide-area monitoring in power networks.

12.5.1 Power Grid System Description

A power grid can be described by a graph $G(\mathcal{B}, \mathcal{E})$ [50], with the set of nodes (commonly called also buses) $\mathcal{B} = \{1, \ldots, N\}$, and the set of edges $\mathcal{E} \subseteq \mathcal{B} \times \mathcal{B}$. The set of nodes will be partitioned into the so-called generation nodes $\mathcal{G}$ and load nodes $\mathcal{L}$, such that $\mathcal{B} = \mathcal{G} \cup \mathcal{L}$. The symbol $\mathcal{N}_i$ denotes the set of nodes directly connected to the ith node via edges. Each edge is interpreted as an unordered pair

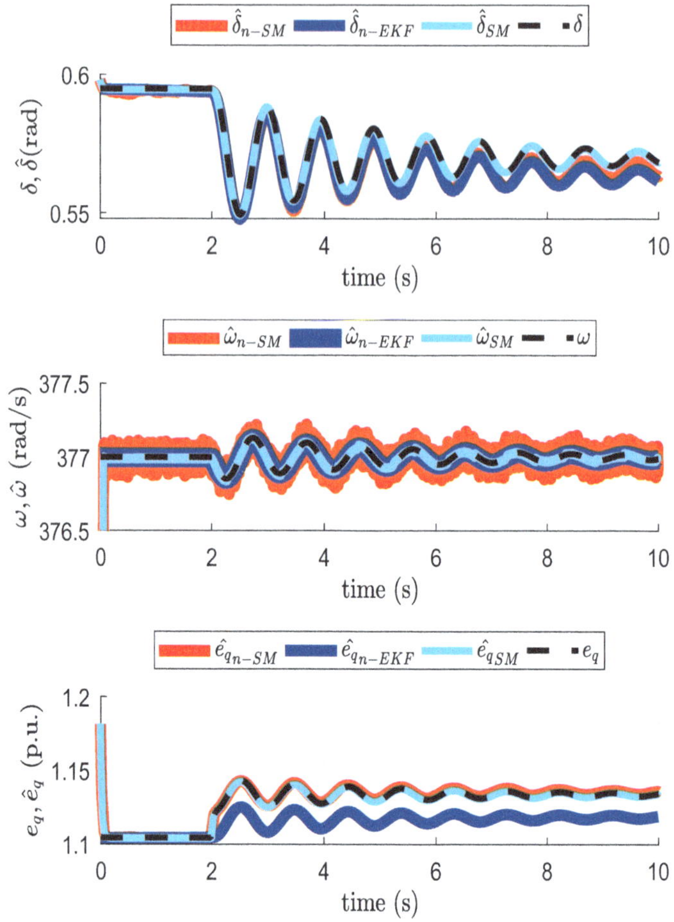

Fig. 12.5 Scenario (1) From the top: time evolution of the voltage phase angle and its estimate; time evolution of the frequency deviation and its estimates; time evolution of the transient voltage together with its estimates

of distinct nodes (i, j) and is characterized by its weight, which is the reciprocal of the corresponding power transmission line reactance, denoted as b_{ij}. It follows each edge is represented as $\{(i, j), b_{ij}\} \in \mathcal{E}$.

Definition 12.1 (*Adjacent Nodes and Neighborhood Set*) The ith node and the jth node of a power grid are said to be directly adjacent if they are linked by a power transmission line, or, in other words:

$$\exists \{(i, j), b_{ij}\} \in \mathcal{E}. \tag{12.64}$$

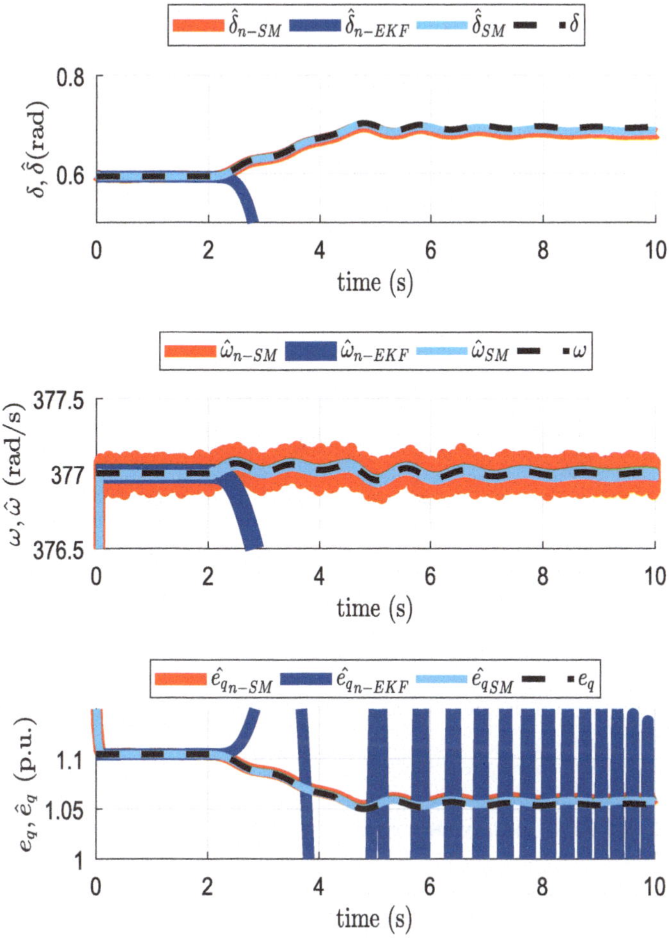

Fig. 12.6 Scenario (2) From the top: time evolution of the voltage phase angle and its estimate; time evolution of the frequency deviation and its estimates; time evolution of the transient voltage together with its estimates

The neighborhood of the ith node of a power grid is the set (denoted as $\mathcal{N}_i$) of the nodes directly adjacent to the ith node, which means

$$\mathcal{N}_i := \{k \mid \{(i, k),\ b_{ik}\} \in \mathcal{E}\}\,. \tag{12.65}$$

The set $\mathcal{N}_i$ is the union of two subsets $\mathcal{N}_i = \mathcal{M}_i \cup \mathcal{O}_i$, which are the neighboring generator node set $\mathcal{M}_i$ and the neighboring load node set $\mathcal{O}_i$.

Table 12.3 Performance metric $\mathcal{P}(T)$ defined in (12.58) as a functions of the noise variance σ^2, simulation time $T = 10$ s, $(\mu p.u.)$ =$(1e{-}6\ p.u.)$, Scenario 1

σ^2	0.00	0.02	0.04	0.06	0.08	0.10	$(p.u.)^2$
$\mathcal{P}(T)$	225	509	883	1259	1637	2026	$(\mu p.u.)$

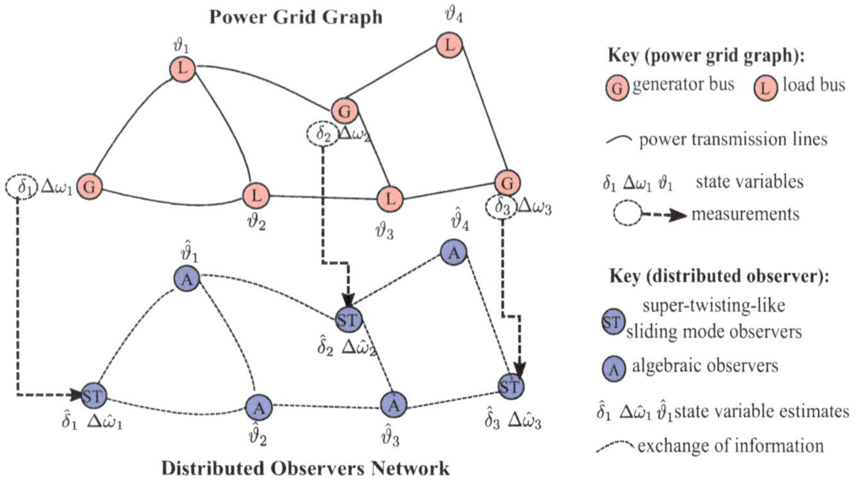

Fig. 12.7 A schematic of the proposed distributed observer scheme together with a graph of a power network

12.5.1.1 Laplacian Matrix Description

For a given power grid interpreted as a graph, let n_g be the number of generators and n_a be the number of loads. If $N := n_g + n_a$, then the power grid is said to be an N-node power grid. Suppose the enumeration of the nodes is chosen such that the first n_g nodes refer to the generators. The Laplacian matrix $L \in \mathbb{R}^{N \times N}$ captures the topology of the interconnections amongst the nodes, and its elements are defined as follows (see, e.g., [33, 34]):

$$L = \begin{cases} L_{ii} & = \sum_{k \in \mathcal{N}_i} b_{ik} \\ L_{ij} & = -b_{ij}. \end{cases} \tag{12.66}$$

The L matrix is partitioned into four submatrices as follows:

$$L = \begin{bmatrix} L^{gg} & L^{gl} \\ L^{lg} & L^{ll} \end{bmatrix}, \tag{12.67}$$

where the submatrices $L^{gg} \in \mathbb{R}^{n_g \times n_g}$, $L^{gl} \in \mathbb{R}^{n_g \times n_a}$, $L^{lg} \in \mathbb{R}^{n_a \times n_g}$, $L^{ll} \in \mathbb{R}^{n_a \times n_a}$. Furthermore, in (12.67), the matrix L^{ll} is invertible [51].

12.5.1.2 Load Dynamics

In this chapter, the static model of the loads is employed [33]. In such an approach, the electrical active power demand P_{l_i} (measured in $(p.u.)$) associated with each load node is specified and treated as a known input [52]. The following algebraic constraint holds:

$$0 = P_{l_i} - \sum_{j \in \mathcal{N}_i} P_{t_{l_j}}, \tag{12.68}$$

where $\sum_{j \in \mathcal{N}_i} P_{t_{l_j}}$ is the total electrical active power flowing from the ith load node to its neighborhood, also measured in $(p.u.)$. Equation (12.68) can be detailed employing the so-called DC power flow, which is a linear mathematical method expressing the value of the electrical active power flow through each power transmission line [52]. More precisely, consider a power transmission line (b_{ij} is the reciprocal of its reactance) linking the ith and the jth node. The DC power flow method expresses the power flowing through the transmission line, denoted as P_{ij} as

$$P_{ij} = b_{ij} \left(\vartheta_i - \vartheta_j \right), \tag{12.69}$$

where ϑ_i and ϑ_j represents the voltage phase angles associated with the ith and the jth node, respectively, and they constitute algebraic state variables. Using both the DC power flow method and the Laplacian matrix introduced earlier, Eq. (12.68) becomes

$$P_{l_i} = \overbrace{\sum_{j \in \mathcal{M}_i} L_{ij}^{lg} \delta_j + L_{ii}^{ll} \vartheta_i + \sum_{k \in \mathcal{O}_i} L_{ik}^{ll} \vartheta_k}^{\sum_{j \in \mathcal{N}_i} P_{t_{l_j}}}, \tag{12.70}$$

By considering all the n_a equations in the form of (12.70), the following algebraic relation can be obtained:

$$P_l = L^{lg} \delta + L^{ll} \vartheta, \tag{12.71}$$

where L^{ll} and L^{lg} are the submatrices in (12.67), the vector $P_l := [P_{l_1}, \dots, P_{l_{na}}]^T \in \mathbb{R}^{n_a}$, the vector $\delta := [\delta_1, \dots, \delta_{n_g}]^T \in \mathbb{R}^{n_g}$, and $\vartheta := [\vartheta_1, \dots, \vartheta_{n_a}]^T \in \mathbb{R}^{n_a}$.

12.5.1.3 Generator Dynamics

In accordance with the partition in (12.67), and using the DC power flow method again, the generator dynamics in (12.16)–(12.18) take the following form:

$$\dot{\delta}_i = \Delta\omega_i \tag{12.72}$$

$$M_i \Delta\dot{\omega}_i = -D_i \Delta\omega_i + P_{m_i} \overbrace{-\mathcal{L}_{ii}^{gg}\delta_i - \sum_{j\in\mathcal{N}_i} \mathcal{L}_{ij}^{gl}\vartheta_j}^{P_{e_i}} \tag{12.73}$$

$$y_i = \delta_i. \tag{12.74}$$

12.5.2 Algebraic Observers Design

Let the voltage phase angle of each generator be measured and the electrical active power demand of each load node be specified, using a sample period of τ seconds. Practical values for τ can be chosen according to [53]. By making use of (12.71), at the kth sampling instant, the following algebraic relation yields

$$0 = -L^{ll}\vartheta[k] - L^{lg}\delta[k] + P_l[k], \tag{12.75}$$

where the L^{ll} and L^{lg} are the submatrices in (12.67), $\vartheta[k] \in \mathbb{R}^{n_a}$, $\delta[k] \in \mathbb{R}^{n_g}$ and $P_L[k] \in \mathbb{R}^{n_a}$ are the sampled versions of ϑ, δ and P_l. By defining $b[k] := -L^{lg}\delta[k] + P_l[k]$, Eq. (12.75) takes the form

$$L^{ll}\vartheta[k] = b[k]. \tag{12.76}$$

The signal $b[k]$ is fixed within the time interval $[k\tau,\ k\tau + \tau]$, and it is possible to estimate $\vartheta[k]$ by solving, in a distributed fashion, the linear system of n_a equations

$$L^{ll}\hat{\vartheta}[k] = b[k], \tag{12.77}$$

in order to obtain an estimate $\hat{\vartheta}[k]$. This will be achieved by using an iterative scheme operating within each sampling period. Two methods are discussed for this purpose: the Jacobi method and the successive over relaxation (SOR) Method [54].

$$\hat{\vartheta}_{\text{Jac}}[k, h+1] = \left(I_{n_a} - \mathcal{D}^{-1}L^{ll}\right)\hat{\vartheta}_{\text{Jac}}[k, h] + \mathcal{D}^{-1}b[k], \tag{12.78}$$

$$\hat{\vartheta}_{\text{Sor}}[k, h+1] = \overbrace{\left(I_{n_a} - \kappa\left(\mathcal{D} - \kappa\mathcal{E}\right)^{-1}L^{ll}\right)}^{M_\kappa}\hat{\vartheta}_{\text{Sor}}[k, h] + \kappa\left(\mathcal{D} - \kappa\mathcal{E}\right)^{-1}b[k] \tag{12.79}$$

Equation (12.78) governs the Jacobi method, while Eq. (12.79) governs the SOR method. $\hat{\vartheta}_{\text{Jac}}[k, h]$ represents the hth estimate of $\vartheta[k]$ using the Jacobi method during the iteration cycle occurring in the time interval $[k\tau,\ k\tau + \tau]$, $\hat{\vartheta}_{\text{Sor}}[k, h]$ represents the hth estimate of $\vartheta[k]$ using the SOR method during the iteration cycle occurring in the time interval $[k\tau,\ k\tau + \tau]$, the matrix $\mathcal{D}$ is defined according to:

$$\mathcal{D}_{ij} = \begin{cases} L^{ll}_{ij} & \text{if } i = j, \\ 0. \end{cases} \tag{12.80}$$

As for the SOR Method, the matrix $M_\kappa := \left(I_{n_a} - \kappa\left(\mathcal{D} - \kappa\mathcal{E}\right)^{-1}\mathcal{L}^{ll}\right)$ is a function of the design weight $\kappa \in \mathbb{R}^+$, and $\mathcal{E}$ is defined according to:

$$\mathcal{E}_{ij} \begin{cases} -L^{ll}_{ij} & \text{if } i < j, \\ 0 & \text{otherwise}. \end{cases} \tag{12.81}$$

If the discrete-time linear systems in the form of (12.78)–(12.79) are asymptotically stable (a detailed stability analysis will be provided in this section), the following steady-state relations yield

$$\hat{\vartheta}_{\text{Jac}}[k,\cdot] = \left(I_{n_a} - \mathcal{D}^{-1}L^{ll}\right)\hat{\vartheta}_{\text{Jac}}[k,\cdot] + \mathcal{D}^{-1}b[k] \tag{12.82}$$

$$\hat{\vartheta}_{\text{Sor}}[k,\cdot] = \left(I_{n_a} - \kappa\left(\mathcal{D} - \kappa\mathcal{E}\right)^{-1}L^{ll}\right)\hat{\vartheta}_{\text{Sor}}[k,\cdot] + \kappa\left(\mathcal{D} - \kappa\mathcal{E}\right)^{-1}b[k], \tag{12.83}$$

where $\hat{\vartheta}_{\text{Jac}}[k,\cdot]$ and $\hat{\vartheta}_{\text{Sor}}[k,\cdot]$ are the Jacobi and SOR steady-state solutions, respectively. The relations (12.82)–(12.83) imply

$$L^{ll}\hat{\vartheta}_{\text{Jac}}[k,\cdot] = b[k] \tag{12.84}$$

$$L^{ll}\hat{\vartheta}_{\text{Sor}}[k,\cdot] = b[k]. \tag{12.85}$$

12.5.2.1 Stability Analysis of Jacobi and SOR Methods

The matrix L^{ll} is (weakly) diagonally dominant according to the Laplacian matrix definition. Then, by making use of Gershgorin's theorem [54], each eigenvalue λ_i of $\left(I_{n_a} - \mathcal{D}^{-1}L^{ll}\right)$ satisfies $|\lambda_i| \leq 1$. In addition, if all the eigenvalues of $\left(I_{n_a} - \mathcal{D}^{-1}\mathcal{L}^{ll}\right)$ lie (strictly) inside the unit disk in the complex plane, it follows (12.78) converges and $\hat{\vartheta}_{\text{Jac}}[k, h+1] \to \hat{\vartheta}_{\text{Jac}}[k, h]$ as $h \to \infty$. However, if there are eigenvalues of absolute value close to 1, the Jacobi method may converge too slowly or not converge at all. Therefore, the use of SOR method represents a valid solution. In (12.79), the weight κ has to be selected according to the convergence criteria detailed in [55] in order to ensure that the eigenvalues of M_κ lie inside the unit disk. In particular, κ has to satisfy the condition [55]:

$$\kappa\left(\mathcal{D} + k\mathcal{E}L^{llT}\right) + \kappa L^{ll}\left(\mathcal{D} + \kappa L^{llT}\right) - \kappa^2 L^{ll}L^{llT} > 0, \tag{12.86}$$

where L^{llT} denotes the transpose of the matrix L^{ll}. Moreover, the optimal value κ^* of κ satisfying Eq. (12.86), is given by

$$\kappa^* = \arg\min_{\kappa}\left(\rho(M_\kappa)\right), \tag{12.87}$$

where $\rho(M_\kappa)$ denotes the spectral radius of the matrix M_κ.

Remark 12.6 Note that the number of iterations in the algebraic schemes affects the accuracy of the estimations. More precisely, the greater the number of iterations, the greater the estimation accuracy at the expense of a prolonged sampling time. However, another aspect which influences the accuracy of the estimation is the position of the eigenvalues of the matrix M_κ in the complex plane. Specifically, if these eigenvalues are placed close to the origin (by selecting the optimal value of the weight κ according to (12.87)), a better accuracy can be achieved by using the same number of iterations. In practice, the optimization as in (12.87) can be performed off-line and only once, given a set of values for the weight $\kappa \in (0, 2)$ and the matrix $M_\kappa := \left(I_{n_a} - \kappa\,(\mathcal{D} - \kappa\mathcal{E})^{-1}\,\mathcal{L}^{ll}\right)$ is computed accordingly [55].

12.5.2.2 Initial Conditions

For both the Jacobi and the SOR method, the following initial conditions are selected:

$$\hat{\vartheta}[k, h = 0] = \begin{cases} \hat{\vartheta}[0] & \text{if } k = 1, \\ \hat{\vartheta}[k-1, \cdot] & \text{if } k > 1, \end{cases} \tag{12.88}$$

where $\hat{\vartheta}[0]$ is an arbitrarily chosen initial condition, while $\hat{\vartheta}[k-1, \cdot]$ is the steady-state estimate related to the $(k-1)$th sample period. Exploiting Eq. (12.88), for $k > 1$ the steady-state estimates $\hat{\vartheta}[k, \cdot]$ are assumed close to the previous ones $\hat{\vartheta}[k-1, \cdot]$.

12.5.2.3 RMSE Statistic

The root-mean-square error (RMSE) statistic is employed to determine the global performance of the algebraic observer schemes, defined as

$$\text{RMSE}[k, h] = \sqrt{\frac{\sum_{i=1}^{n_a}\left(\vartheta_i[k] - \hat{\vartheta}_i[k, h]\right)^2}{n_a}}, \quad \forall k. \tag{12.89}$$

According to the development in this section, the RMSE is expected to converge to zero within each time interval $[k\tau, k\tau + \tau]$.

12.5.3 *Super-Twisting-Like Sliding-Mode Observers for the Generators*

In order to estimate the frequency deviation in each generator node, it will be shown in this section that the super-twisting-like sliding-mode observer structure in (12.23)–(12.24) can still be used. The novelty here is that the observer receives the

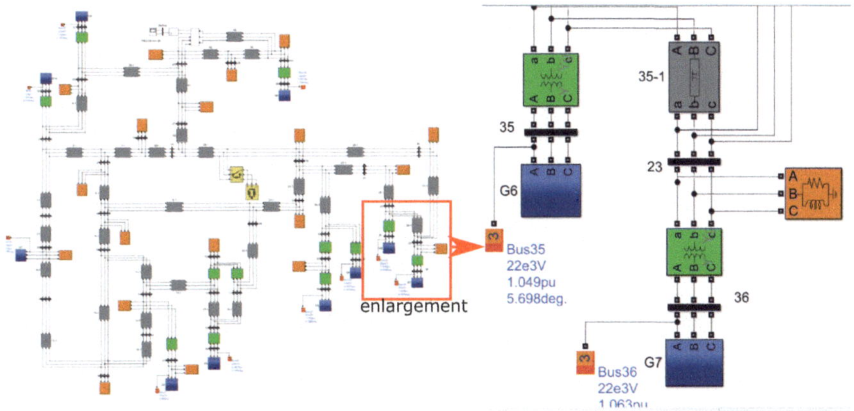

Fig. 12.8 (Left): The implementation of the IEEE 39 node benchmark in the SimPowerSystems environment. (Right): An enlarged view from which it is possible to distinguish the generators, the transformers, the loads, and the power transmission lines

estimates coming from the neighboring (load) observers, which appear in the matched channel of the error dynamics. The super-twisting-like sliding-mode observer in this contest has the following form:

$$\dot{\hat{\delta}}_i = \Delta\hat{\omega}_i - a_i e_{\delta_i} - k_{1_i} \lceil e_{\delta_i} \rfloor^{1/2} \tag{12.90}$$

$$\Delta\dot{\hat{\omega}}_i = -k_{2_i} \lceil e_{\delta_i} \rfloor^{0} - a_i^2 e_{\delta_i} + a_i \Delta\hat{\omega}_i - k_{1_i} a_i \lceil e_{\delta_i} \rfloor^{1/2} + \varphi_i^{ZOH} + \frac{P_{m_i}}{M_i}. \tag{12.91}$$

In (12.91), the ith generator super-twisting-like sliding-mode observer receives estimates from its neighboring algebraic observers. In particular:

$$\varphi_i^{ZOH} := -\frac{1}{M_i}\left(\sum_{j\in\mathcal{N}_i} L_{ij}^{gl}\hat{\vartheta}_j^{ZOH}\right), \tag{12.92}$$

where $\hat{\vartheta}_j^{ZOH}$ is the zero-order hold (ZOH) version (piecewise constant) [54] of the jth load node voltage phase angle estimate communicated by the jth neighboring algebraic observer. This exchange of local information gives the proposed observers a distributed architecture.

Remark 12.7 The ZOH implementation is required, since the algebraic observers are discrete-time systems, while the super-twisting-like ones are continuous-time systems.

The error dynamics of the super-twisting-like sliding-mode observer has the same structure of (12.26)–(12.27) and are given by

$$\dot{e}_{\delta_i} = e_{\omega_i} - a_i e_{\delta_i} - k_{1_i} \left\lceil e_{\delta_i} \right\rfloor^{1/2} \quad (12.93)$$

$$\dot{e}_{\omega_i} = -k_{2_i} \left\lceil e_{\delta_i} \right\rfloor^{0} - a_i^2 e_{\delta_i} - k_{1_i} a_i \left\lceil e_{\delta_i} \right\rfloor^{1/2} - \Phi_i + a_i e_{\omega_i}, \quad (12.94)$$

where the matched perturbation Φ_i has the following structure:

$$\Phi_i := \varphi_i^{ZOH} + \frac{1}{M_i} \left(\sum_{j \in \mathcal{N}_i} L_{ij}^{gl} \vartheta_j \right). \quad (12.95)$$

Following the same arguments presented in Sect. 12.3.1, it is possible to prove that the system in Eqs. (12.93), (12.94) converges to the origin in a finite time.

12.5.4 Numerical Test Cases

The IEEE 39 node (also called the 10-machine New-England) power system is chosen as a benchmark for the assessment and it has been implemented in a Matlab-SimPowerSystems environment. The basic data for this power system is available in the literature in [49]. Figure 12.8 shows the SimPowerSystems implementation of the IEEE 39 node benchmark. An enlarged view is also reported, in which it is possible to distinguish the generators, the power transformers, the loads, and the power transmission lines. If t is the simulation time, for $0 \le t < 5$ (s) the overall system is at steady state, which implies that there is a balance between power generation and consumption. A sudden variation in the electrical active power demand at the 16th load node takes place

$$P_{l_{16}} = 329 + 300\,\text{step}(t-5) - 300\,\text{step}(t-10) \ \ (\text{MW}). \quad (12.96)$$

This time-varying power demand makes the electrical frequency of each generator deviate from its rated value as shown in Fig. 12.11. Furthermore, when the load alteration ends (for $t = 10$ (s)), the frequency of each generator tends to reach again the rated value of 60 Hz. Two values of the sampling time τ are considered: 0.1 and 0.5 s, and within each sampling interval, a maximum of 1000 iterations can be executed. Figure 12.10 shows that the matrix $\left(I_{n_a} - \mathcal{D}^{-1}\mathcal{L}^{ll}\right)$ in the iterative scheme (12.78) has a spectral radius too close to one and the convergence speed is expected to be slow. Figure 12.10 summarizes the choice of the optimal value for the weight κ according to (12.87). The eigenvalues of the matrix M_{κ^*} are drawn in the complex plane together with the eigenvalues of $\left(I_{n_a} - \mathcal{D}^{-1}\mathcal{L}^{ll}\right)$ for the Jacobi method, and the unit disk. For $t > 5[s]$, the transient behaviors of the algebraic observers are shown in the enlargement in Fig. 12.9. This transient behavior is due to the update of the term $b[k]$ and the following iterations within each time interval $[k\tau; k\tau + \tau]$ according to Eq. (12.78). The SOR method displays a faster speed of convergence when compared to the Jacobi method, originally adopted in [56] for the two values

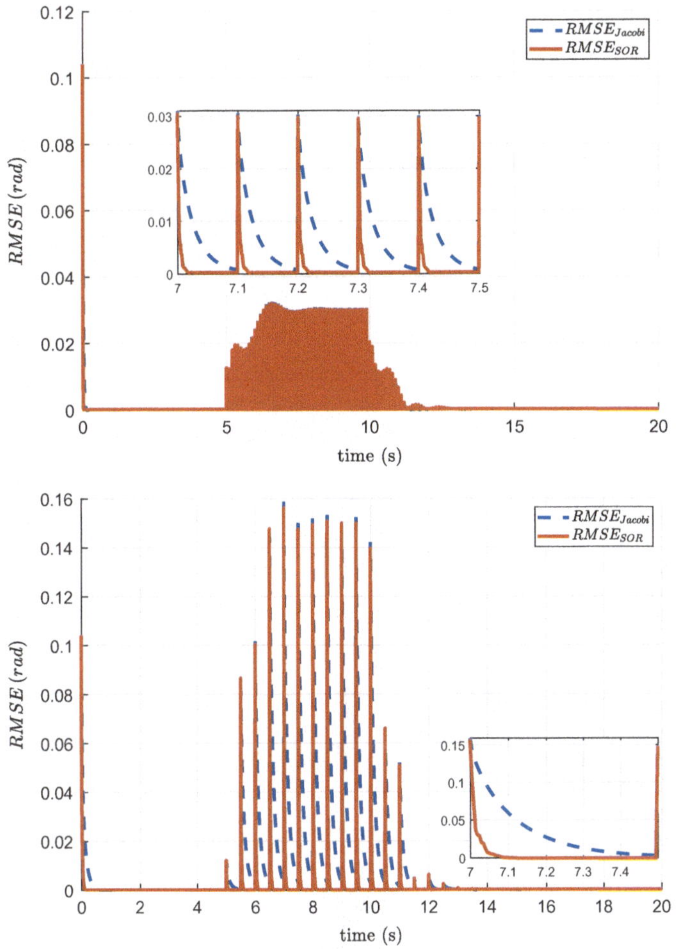

Fig. 12.9 (Top): Load node voltage phase angles RMSE error for the Jacobi and SOR method, for the value $\tau = 0.1\,(\mathrm{s})$, (Bottom): Load node voltage phase angles RMSE error for the Jacobi and SOR method, for the value $\tau = 0.5\,(\mathrm{s})$

of the sampling time considered. The design constants for the super-twisting-like sliding-mode observer are set as follows: $k_{1_i} = 5$, and $k_{2_i} = 50$. As clarified in this chapter, these estimators have continuous-time dynamics, and then, the solver *Ode1* (Euler) is select in the MATLAB-Simulink environment, with a fixed integration step size set equal to 0.1 ms. The well-established unknown input (UI) Luenberger observers [57] are compared with the proposed sliding-mode estimator, in order to demonstrate the superiority of the proposed sliding-mode estimator. Figure 12.11 shows that the accuracy of the sliding-mode observers is clearly higher than the UI Luenberger ones, as shown in Fig. 12.11.

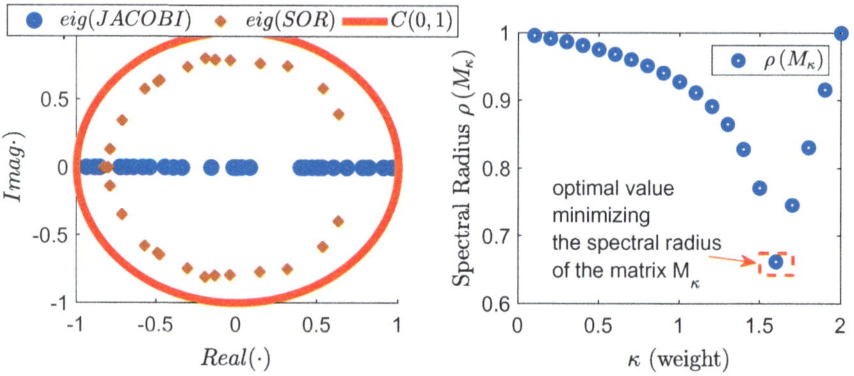

Fig. 12.10 (Left): The eigenvalues in the complex plane for the matrices of the Jacobi and SOR method. (Right): The spectral radius of the matrix M_κ and the optimal choice of the weight κ

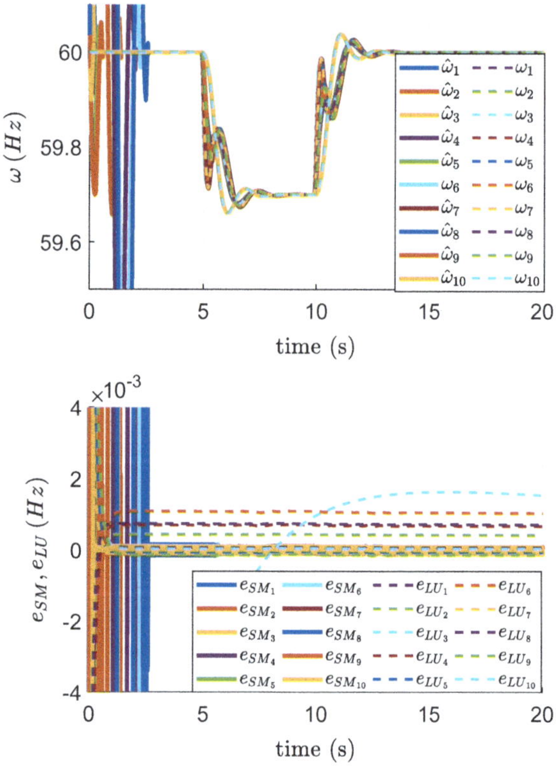

Fig. 12.11 (Top): Time evolution of the frequency of each generator together with its estimate via super-twisting-like sliding-mode observer (Bottom): Frequency estimation error for each super-twisting-like sliding-mode observer, denoted as e_{SM_i}, and for the well-established UI Luenberger observer, denoted as e_{LU_i}

12.6 Conclusions

In this chapter, local and wide-area sliding-mode estimation techniques have been discussed with application to power systems. The developed analysis started at the local level by considering a single synchronous generator. The swing equations have been derived for tutorial perspective of the manuscript. An original super-twisting-like sliding-mode observer capable of dynamically tracking the frequency deviation has been assessed via real data. This data referred to real faults taken place in the Nordic Power System. A comparison between the real measurements and the estimates has revealed the accuracy of the presented sliding-mode observer. Furthermore, the transient voltage dynamics have been considered, proposing the estimator in Sect. 12.4, with a discussion of the effect of measurement noise on the observer performances. Our attention has then moved to wide-area monitoring and a network of sliding-mode and algebraic estimators, suitable connected in a distributed fashion, has been formulated for structure-preserving DAE power networks. These have employed sliding-mode observers for the differential part of the system, and iterative algorithms for the algebraic part. The discussed numerical simulation test cases based on IEEE 39 bus benchmark have shown the effectiveness of the distributed observers.

References

1. Kayastha, N., Niyato, D., Hossain, E., Han, Z.: Smart grid sensor data collection, communication, and networking: a tutorial. Wirel. Commun. Mobile Comput. **14**(11), 1055–1087 (2014)
2. Ellabban, O., Abu-Rub, H., Blaabjerg, F.: Renewable energy resources: current status, future prospects and their enabling technology. Renew. Sustain. Energy Rev. **39**, 748–764 (2014)
3. Mohanta, D.K., Murthy, C., Sinha Roy, D.: A brief review of phasor measurement units as sensors for smart grid. Electr. Power Compon. Syst. **44**(4), 411–425 (2016)
4. Zhao, J., Zhang, G., Das, K., Korres, G.N., Manousakis, N.M., Sinha, A.K., He, Z.: Power system real-time monitoring by using pmu-based robust state estimation method. IEEE Trans. Smart Grid **7**(1), 300–309 (2016)
5. Aminifar, F., Shahidehpour, M., Fotuhi-Firuzabad, M., Kamalinia, S.: Power system dynamic state estimation with synchronized phasor measurements. IEEE Trans. Instrum. Meas. **63**(2), 352–363 (2014)
6. Kang, W., Barbot, J.P., Xu, L.: On the observability of nonlinear and switched systems. In: Emergent Problems in Nonlinear Systems and Control, pp. 199–216. Springer (2009)
7. Ghahremani, E., Kamwa, I.: Dynamic state estimation in power system by applying the extended Kalman filter with unknown inputs to phasor measurements. IEEE Trans. Power Syst. **26**(4), 2556–2566 (2011)
8. Paul, A., Kamwa, I., Joos, G.: Centralized dynamic state estimation using a federation of extended Kalman Filters with intermittent PMU data from generator terminals. IEEE Trans, Power Syst (2018)
9. Qi, J., Sun, K., Wang, J., Liu, H.: Dynamic state estimation for multi-machine power system by unscented Kalman filter with enhanced numerical stability. IEEE Trans. Smart Grid **9**(2), 1184–1196 (2018)

10. Zhao, J., Netto, M., Mili, L.: A robust iterated extended kalman filter for power system dynamic state estimation. IEEE Trans. Power Syst. **32**(4), 3205–3216 (2017)
11. Gomez-Exposito, A., Abur, A.: Power System State Estimation: Theory and Implementation. CRC Press, Boca Raton (2004)
12. Koenig, D.: Unknown input proportional multiple-integral observer design for linear descriptor systems: application to state and fault estimation. IEEE Trans. Autom. Control **50**(2), 212–217 (2005)
13. Pasqualetti, F., Dörfler, F., Bullo, F.: Attack detection and identification in cyber-physical systems. IEEE Trans. Autom. Control **58**(11), 2715–2729 (2013)
14. Cui, Y., Xu, L., Fei, M., Shen, Y.: Observer based robust integral sliding mode load frequency control for wind power systems. Control Eng. Pract. **65**, 1–10 (2017)
15. Mellucci, C., Menon, P.P., Edwards, C., Ferrara, A.: Second-order sliding mode observers for fault reconstruction in power networks. IET Control Theory Appl. **11**(16), 2772–2782 (2017)
16. Rinaldi, G., Cucuzzella, M., Ferrara, A.: Sliding mode observers for a network of thermal and hydroelectric power plants. Automatica **98**, 51–57 (2018)
17. Rinaldi, G., Ferrara, A.: Higher order sliding mode observers and nonlinear algebraic estimators for state tracking in power networks. In: Proceedings of 56th IEEE Conference on Decision and Control, Melbourne, Australia, pp. 6033–6038 (2017)
18. Rinaldi, G., Menon, P.P., Edwards, C., Ferrara, A.: Design and validation of a distributed observer-based estimation scheme for power grids. IEEE Trans. Control Syst. Technol
19. Rinaldi, G., Menon, P.P., Ferrara, A., Edwards, C.: A super-twisting-like sliding mode observer for frequency reconstruction in power systems: Discussion and real data based assessment. In: Proceedings of 15-th International Workshop on Variable Structure Systems (VSS), Graz, Austria, pp. 444–449 (2018)
20. Su, X., Liu, X., Song, Y.D.: Fault-tolerant control of multiarea power systems via a sliding-mode observer technique. IEEE/ASME Trans. Mechatron. **23**(1), 38–47 (2018)
21. Rinaldi, G., Menon, P.P., Edwards, C., Ferrara, A.: Distributed super-twisting sliding mode observers for fault reconstruction and mitigation in power networks. In: Proceedings of 57th IEEE Conference on Decision and Control, Miami, FL, USA, p. accepted (2018)
22. Tan, C.P., Edwards, C.: Sliding mode observers for robust detection and reconstruction of actuator and sensor faults. Int. J. Robust Nonlinear Control: IFAC-Affil. J. **13**(5), 443–463 (2003)
23. Cucuzzella, M., Incremona, G.P., Ferrara, A.: Design of robust higher order sliding mode control for microgrids. IEEE J. Emerg. Sel. Topics Circuits Syst. **5**(3), 393–401 (2015)
24. Loukianov, A.G., Cañedo, J.M., Utkin, V.I., Cabrera-Vázquez, J.: Discontinuous controller for power systems: sliding-mode block control approach. IEEE Trans. Ind. Electron. **51**(2), 340–353 (2004)
25. Mi, Y., Hao, X., Liu, Y., Fu, Y., Wang, C., Wang, P., Loh, P.C.: Sliding mode load frequency control for multi-area time-delay power system with wind power integration. IET Gener. Transm. Distrib. **11**(18), 4644–4653 (2017)
26. Trip, S., Cucuzzella, M., De Persis, C., van der Schaft, A., Ferrara, A.: Passivity-based design of sliding modes for optimal load frequency control. IEEE Trans. Control Syst. Technol. **99**, 1–14 (2018)
27. Trip, S., Cucuzzella, M., Ferrara, A., De Persis, C.: An energy function based design of second order sliding modes for automatic generation control. In: Proceedings of 20th IFAC World Congress, Toulouse, France, pp. 11613–11618 (2017)
28. Vrdoljak, K., Perić, N., Petrović, I.: Sliding mode based load-frequency control in power systems. Electr. Power Syst. Res. **80**(5), 514–527 (2010)
29. Li, J., Liu, X., Su, X.: Sliding mode observer-based load frequency control of multi-area power systems under delayed inputs attack. In: Proceedings of Chinese Control And Decision Conference (CCDC), Shenyang, China, pp. 3716–3720 (2018)
30. Rinaldi, G., Cucuzzella, M., Ferrara, A.: Third order sliding mode observer-based approach for distributed optimal load frequency control. IEEE Control Syst. Lett. **1**(2), 215–220 (2017)

31. Mi, Y., Fu, Y., Li, D., Wang, C., Loh, P.C., Wang, P.: The sliding mode load frequency control for hybrid power system based on disturbance observer. Int. J. Electr. Power Energy Syst. **74**, 446–452 (2016)
32. Hussein, A.A., Salih, S.S., Ghasm, Y.G.: Implementation of proportional-integral-observer techniques for load frequency control of power system. Proc. Comput. Sci. **109**, 754–762 (2017)
33. Kundur, P., Balu, N.J., Lauby, M.G.: Power System Stability and Control, vol. 7. McGraw-hill, New York (1994)
34. Sauer, P.W., Pai, M.A., Chow, J.H.: Power System Dynamics and Stability: With Synchrophasor Measurement and Power System Toolbox. Wiley (2017)
35. Ørum, E., Kuivaniemi, M., Laasonen, M., Bruseth, A.I., Jansson, E.A., Danell, A., Elkington, K., Modig, N.: Future system inertia. Technical Report, ENTSOE, Brussels (2015)
36. Davila, J., Fridman, L., Levant, A.: Second-order sliding-mode observer for mechanical systems. IEEE Trans. Autom. Control **50**(11), 1785–1789 (2005)
37. Machowski, J., Bialek, J., Bumby, J.: Power System Dynamics: Stability and Control. Wiley (2011)
38. Shtessel, Y., Edwards, C., Fridman, L., Levant, A.: Sliding Mode Control and Observation, vol. 10. Springer, Berlin (2014)
39. Fridman, L., Levant, A.: Higher order sliding modes. Sliding Mode Control Eng. **11**, 53–102 (2002)
40. Moreno, J.A., Osorio, M.: Strict Lyapunov functions for the super-twisting algorithm. IEEE Trans. Autom. Control **57**(4), 1035–1040 (2012)
41. Fingrid frequency - historical data. https://data.fingrid.fi/en/dataset/frequency-historical-data
42. Fingrid: Frequency quality analysis for year 2015. Report 2015, pp. 1–91 (2015)
43. Randall, S.P., Sudgen, D.M., Vail, W., Brown, G.T.: Rotor position encoder having features in decodeable angular positions, US Patent 5,637,972 (1997). Accessed 10 June 1997
44. Team, N.E.A.T.: Phase angle calculations: Considerations and use cases. NASPI-2016-TR-006 (2016)
45. Isidori, A.: Nonlinear Control Systems. Springer Science & Business Media, Berlin (2013)
46. Yan, X.G., Edwards, C.: Robust sliding mode observer-based actuator fault detection and isolation for a class of nonlinear systems. Int. J. Syst. Sci. **39**(4), 349–359 (2008)
47. Rinaldi, G., Menon, P.P., Edwards, C., Ferrara, A.: Sliding mode based dynamic state estimation for synchronous generators in power systems. IEEE Control Syst. Lett. **2**(4), 785–790 (2018)
48. Poznyak, A.: Stochastic output noise effects in sliding mode state estimation. Int. J. Control **76**(9–10), 986–999 (2003)
49. Hiskens, I.: IEEE PES task force on benchmark systems for stability controls. Technical Report (2013)
50. Watts, D.J., Strogatz, S.H.: Collective dynamics of 'small-world' networks. Nature **393**(6684), 440 (1998)
51. Dorfler, F., Bullo, F.: Kron reduction of graphs with applications to electrical networks. IEEE Trans. Circuits Syst. I: Regul. Pap. **60**(1), 150–163 (2013)
52. Purchala, K., Meeus, L., Van Dommelen, D., Belmans, R.: Usefulness of DC power flow for active power flow analysis. In: Proceedings of Power Engineering Society General Meeting, San Francisco, California, USA, pp. 454–459 (2005)
53. Kansal, P., Bose, A.: Bandwidth and latency requirements for smart transmission grid applications. IEEE Trans. Smart Grid **3**(3), 1344–1352 (2012)
54. Saad, Y.: Iterative Methods for Sparse Linear Systems. SIAM (2003)
55. James, K.R., Riha, W.: Convergence criteria for successive overrelaxation. SIAM J. Numer. Anal. **12**, 137–143 (1975)
56. Rinaldi, G., Menon, P.P., Edwards, C., Ferrara, A.: Distributed observers for state estimation in power grids. In: Proceedings of American Control Conference, Seattle, WA, USA, pp. 5824–5829 (2017)
57. Edwards, C., Tan, C.P.: A comparison of sliding mode and unknown input observers for fault reconstruction. Eur. J. Control **12**(3), 245–260 (2006)

Chapter 13
Sliding-Mode-Based Platooning: Theory and Applications

Astrid Rupp, Martin Steinberger and Martin Horn

Abstract This chapter discusses sliding-mode-based approaches for longitudinal control of vehicles. In the platooning applications, several vehicles are aligned in a string for economic reasons, e.g., for the sake of saving fuel. Each vehicle has an immediate impact on the followers leading to dynamic phenomena along the string. The so-called string stable controllers based on sliding-mode controllers are the focus of this work. In addition, position error overshoots have been eliminated, and thus dynamic effects leading to collisions can be avoided. The controllers have been implemented on a testbed consisting of small-scale vehicles, and experimental results are shown.

13.1 Introduction

Autonomous driving is currently extensively studied by many researchers due to its potential improvement of safety and efficiency on public roads [34]. One of the first automated multi-vehicle scenarios is platooning, where multiple vehicles are aligned in a string, following a leading vehicle at very small inter-vehicle distances in order to increase traffic throughput and decrease fuel consumption. While human drivers are not capable of maintaining small inter-vehicle distances safely, automated systems can react faster and can thus improve safety and efficiency. In order to guarantee safety, the so-called string stability has to be investigated; if distance errors, velocities, and accelerations are amplified along the string of vehicles, the platoon is said to be

A. Rupp (✉) · M. Steinberger · M. Horn
Institute of Automation and Control, Graz University of Technology, Graz, Austria
e-mail: astrid.rupp@fprimezero.com

M. Steinberger
e-mail: martin.steinberger@tugraz.at

M. Horn
e-mail: martin.horn@tugraz.at

A. Rupp
FPrimeZero GmbH, Wien, Austria

M. Steinberger et al. (eds.), *Variable-Structure Systems and Sliding-Mode Control*, Studies in Systems, Decision and Control 271,
https://doi.org/10.1007/978-3-030-36621-6_13

string unstable, otherwise string stable. Early publications have mainly focused on simple linear controllers such as PD controllers, see, e.g., [9, 35]. Already in [32], nonlinear approaches have been investigated in order to increase robustness with respect to parameter uncertainties.

Due to their robust performance, sliding-mode-based controllers have successfully been used for longitudinal control of vehicles. In [8], adaptive cruise control (ACC) is described with a constant time-headway spacing, and a first-order sliding-mode controller has been used to ensure that no position error undershoot occurs. The controller parameter is adapted such that less acceleration is applied in comfortable maneuvers, and the controller is more aggressive in safety-critical situations. The acceleration of the preceding vehicle has been assumed to be zero and has thus been neglected in the analysis. However, an undershoot in position error is not entirely excluded. Note that amplification of position errors and accelerations along a string of vehicles is not considered, since the focus lies on ACC for one vehicle only.

String stable platooning using the suboptimal controller [4] with a constant time-headway spacing for zero initial spacing errors has been proposed in [13], with extensions for slip control [3], lateral control [12], and collision avoidance [14]. The suboptimal controller for longitudinal vehicle control has also recently been published in [11]. Bidirectional control, where a vehicle receives information from both its preceding and following vehicle, has been used in [19] in order to solve the problem of nonzero initial spacing errors. An extension of this work has been presented in [16], where an integrated sliding variable is used. Nonlinear actuator dynamics have been recently discussed in [15], where parameters are adapted based on neural networks. However, the leader's velocity and acceleration are assumed to be known to all agents, and input saturation has to be considered separately, while the approaches discussed in this chapter have bounded inputs by design.

In [28], unidirectional sliding-mode-based control with an adaptive time-headway is used in order to reach and maintain very small inter-vehicle distances. This approach considers double integrator dynamics and is summarized in Sect. 13.2. Then, sliding-mode-based controllers for systems with actuator dynamics are discussed. In Sect. 13.3, the hardware and software setup of the testbed using small-scale vehicles is presented, and results of experiments are summarized. Finally, a conclusion and ideas for future work are given in Sect. 13.4.

13.2 Sliding-Mode-Based Platooning

In a first step, it is assumed that each vehicle's longitudinal dynamics is represented by a double integrator. The first vehicle, or "leader", in the string of vehicles is modeled by

$$\dot{x}_0(t) = v_0(t)\ , \quad \dot{v}_0(t) = u_\mathrm{r}(t)\ , \tag{13.1}$$

where $x_0(t) \in \mathbb{R}$, $v_0(t) \in \mathbb{R}$, and $u_\mathrm{r}(t) \in \mathbb{R}$ denote position, velocity, and acceleration of the leader, respectively. The dynamics of a leader-following vehicle, or "agent", is given by

$$\dot{x}_i(t) = v_i(t) , \quad \dot{v}_i(t) = u_i(t) , \tag{13.2}$$

with $i = 1, 2, ..., N$, and N the number of agents. $x_i(t) \in \mathbb{R}$, $v_i(t) \in \mathbb{R}$, and $u_i(t) \in \mathbb{R}$ are the position, velocity, and acceleration of agent i, respectively. In the unidirectional platooning application, agent i receives information about its preceding vehicle $i - 1$. For the sliding-mode control design published in [28], the following assumption is made.

Assumption 13.1 The acceleration u_r of the leading vehicle (13.1) is bounded by a maximum acceleration, i.e., $|u_r(t)| \leq u_{r,max}$. Moreover, the accelerations of the leader-following agents (13.2) are bounded by a maximum acceleration, i.e., $|u_i(t)| \leq u_{max}$ for all i. The leader's maximum acceleration is smaller than the maximum acceleration of the agents, i.e., $u_{r,max} < u_{max}$. The velocities v_i of the agents are positive.

Due to the physical acceleration limits of the vehicles in practice, the control inputs are bounded. The leader's acceleration being smaller than the follower's maximum acceleration is reasonable, because otherwise the vehicles may not be able to catch up with the leader, or brake in time if the leader decelerates. The assumption on positive velocities is necessary for meaningful application: on a highway, stop-and-go traffic is possible, but driving backward is not considered.

The goal of each vehicle in a platoon is then to maintain small inter-vehicle distances to the preceding vehicle. Defining the position and velocity errors for vehicle i as

$$e_{x,i} = x_{i-1} - x_i - \Delta_i , \quad e_{v,i} = v_{i-1} - v_i , \tag{13.3}$$

where Δ_i is the desired distance to the preceding vehicle, with the positions and velocities of the vehicles from (13.1), (13.2), then each vehicle steers the error to zero,

$$\lim_{t \to \infty} e_{x,i}(t) = 0 \quad \forall i \in \{1, ..., N\} . \tag{13.4}$$

For a constant desired distance Δ_i, the position error in (13.3) is also called constant distance spacing error. However, in many cases, this spacing error does not yield string stable platoons, e.g., when linear controllers are used. Hence, an additional distance is typically used to achieve string stability, and the new error definition is

$$e_{t,i} = x_{i-1} - x_i - \Delta_i - t_{h,i} v_i , \tag{13.5}$$

with the so-called time-headway $t_{h,i} > 0$, which is typically a constant parameter. This error is thus also called constant time-headway spacing error.

Compared to individual driving, special requirements for collision-free platooning have to be considered in the controller design, which are presented in Sect. 13.2.1. Then, a string stable sliding-mode controller for nonzero initial spacing errors using double integrator dynamics is discussed in Sect. 13.2.2. Actuator dynamics of first and higher order are then considered in Sects. 13.2.3 and 13.2.4, respectively. Finally, robustness with respect to lateral motion is discussed in Sect. 13.2.5.

13.2.1 String Stability

As discussed in [24], there exist various definitions for string stability, e.g., considering position error undershoot as in [10], or attenuation of position errors, velocities, and accelerations [23]. In [24], $\mathcal{L}_2$ and $\mathcal{L}_\infty$ string stability conditions have been proposed, where the $\mathcal{L}_\infty$ condition used in this chapter has been stated as follows.

Definition 13.1 ($\mathcal{L}_\infty$ *string stability*) The interconnected system (13.1) and (13.2) with unidirectional control $u_i = u_i(x_i, v_i, x_{i-1}, v_{i-1})$ is called $\mathcal{L}_\infty$ string stable if there exist class $\mathcal{K}$ functions (see [17]) γ and η such that for any initial state $\bar{e}(t_0) \in \mathbb{R}^{2N}$ at initial time t_0 and any acceleration u_r satisfying Assumption 13.1, it holds that

$$\|e_i\|_{\mathcal{L}_\infty} \leq \gamma(\|u_\mathrm{r}\|_{\mathcal{L}_\infty}) + \eta(\|\bar{e}(t_0)\|) \quad \forall i \ , \tag{13.6}$$

where the errors are defined with (13.3) by $e_i = \{e_{\mathrm{x},i}, e_{\mathrm{v},i}\}$, and the lumped error vector is denoted by $\bar{e}(t) = \begin{bmatrix} e_{\mathrm{x},1} \ e_{\mathrm{v},1} \ \ldots \ e_{\mathrm{x},N} \ e_{\mathrm{v},N} \end{bmatrix}^T$.

As discussed in [10], avoidance of position undershoot is desired in addition, i.e.,

$$e_{\mathrm{x},i}(t) \geq 0 \quad \forall t \ , \tag{13.7}$$

and the control design task is thus summarized as follows (see [28]).

Summary of Design Goals

A control law $u_i = u_i(x_i, v_i, x_{i-1}, v_{i-1})$ for the interconnected system (13.1), (13.2) has to be designed that fulfills all of the following requirements:

1. A small constant inter-vehicle distance according to (13.4) is achieved.
2. The accelerations of the vehicles do not increase along the string (and satisfy Assumption 13.1).
3. The position and velocity errors are independent of the length of the platoon N and the position in the platoon i, i.e., there is no error amplification along the platoon, fulfilling (13.6).
4. For nonzero initial spacing errors, no position error undershoots occur, i.e., (13.7) holds, and thus collisions are avoided. This condition can actually be relaxed to $e_{\mathrm{x},i}(t) > -\Delta_i$ for all times.

13.2.2 Sliding-Mode Controller Design for Nonzero Initial Spacing Errors

For the constant distance spacing (13.3), a sliding surface with relative degree two can be defined as

$$\sigma_i = e_{\mathrm{x},i} \; . \tag{13.8}$$

Using the constant time-headway spacing (13.5),

$$\sigma_{\mathrm{t},i} = e_{\mathrm{t},i} \; , \tag{13.9}$$

the sliding variable is of relative degree one. As shown in [26], both spacing policies work well in sliding, but the reaching phase is problematic regarding amplification of accelerations and position error overshoots. String stability cannot be guaranteed for nonzero initial spacing errors using the standard approaches.

Hence, an adaptive time-headway (ATH) spacing has been proposed in [28], where the time-headway $t_{\mathrm{h},i}(t)$ in the proposed sliding variable can be decreased to zero asymptotically and a constant distance formation can be achieved. The behavior is eventually approximated by the desired error dynamics

$$\dot{e}_{\mathrm{x},i} = -\frac{1}{t_{\mathrm{h}}^*} e_{\mathrm{x},i} \; , \quad t_{\mathrm{h}}^* > 0 \; , \tag{13.10}$$

where t_{h}^* is a desired convergence time constant. The main theorem of [28] is given as follows.

Assumption 13.2 The initial spacing errors are bounded and depend on the velocity of the vehicle such that $0 \leq e_{\mathrm{x},i}(t_0) \leq t_{\max,i} v_i(t_0)$, where $t_{\max,i} \geq 0$ is a parameter that can be specified. The maximum velocity error is bounded such that $|e_{\mathrm{v},i}(t_0)| < t_{\mathrm{h},i}(t_0) u_{\max}$ with

$$t_{\mathrm{h},i}(t_0) = \begin{cases} e_{\mathrm{x},i}(t_0)/v_i(t_0) & \text{if } e_{\mathrm{x},i}(t_0) > t_{\mathrm{h}}^* v_i(t_0) \text{ (Phase I)} \\ (e_{\mathrm{x},i}(t_0) + t_{\mathrm{h}}^* e_{\mathrm{v},i}(t_0))/v_{i-1}(t_0) & \text{otherwise (Phase II)} \; . \end{cases} \tag{13.11}$$

Theorem 13.1 ([28]) *Consider the interconnected system* (13.1), (13.2) *and the errors* (13.3) *with Assumptions 13.1 and 13.2. Let the control law u_i consist of two phases with a specified switching condition. In Phase I, the control law*

$$u_i = k sign(\sigma_{\mathrm{t},i}) \; , \tag{13.12}$$

with the sliding variable

$$\sigma_{\mathrm{t},i} = e_{\mathrm{x},i} - t_{\mathrm{h},i} v_i \; , \tag{13.13}$$

and an adaptive law

$$\dot{t}_{\mathrm{h},i} = -\mu_1 \frac{t_{\mathrm{h}}^* k}{v_i} , \tag{13.14}$$

$$\mu_1 = \begin{cases} 1 & -\frac{t_{\mathrm{h}}^* k}{2} \leq e_{\mathrm{v},i} < 0 \text{ and } t_{\mathrm{h},i} > t_{\mathrm{h}}^* \\ 0 & \text{otherwise} , \end{cases} \tag{13.15}$$

is applied, with t_{h}^* *defined in* (13.10). *The initial time-headway is computed as*

$$t_{\mathrm{h},i}(t_0) = \frac{e_{\mathrm{x},i}(t_0)}{v_i(t_0)} , \tag{13.16}$$

and satisfies $t_{\mathrm{h},i}(t_0) < t_{\mathrm{max},i}$ *according to Assumptions 13.1 and 13.2.*

In Phase II, control law

$$u_i = k sign(\tilde{\sigma}_i) , \tag{13.17}$$

with the combined sliding variable

$$\tilde{\sigma}_i = e_{\mathrm{x},i} - t_{\mathrm{h},i} v_i + (t_{\mathrm{h}}^* - t_{\mathrm{h},i}) e_{\mathrm{v},i} , \tag{13.18}$$

and an adaptive law

$$\dot{t}_{\mathrm{h},i} = -\mu_2 \frac{k}{v_{i-1}} t_{\mathrm{h},i} , \quad \mu_2 = \begin{cases} 1 & -\frac{t_{\mathrm{h},i} k}{2} \leq e_{\mathrm{v},i} < 0 \\ 0 & \text{otherwise} \end{cases} \tag{13.19}$$

is used. The switching time $t_{\mathrm{s},i}$ *for agent i from Phase I to Phase II is given by the time instant for which*

$$\tilde{\sigma}_i(t_{\mathrm{s},i}) = 0 , \quad \sigma_{\mathrm{t},i}(t_{\mathrm{s},i}) = 0 , \quad t_{\mathrm{h},i}(t_{\mathrm{s},i}) \leq t_{\mathrm{h}}^* \tag{13.20}$$

holds. Then, the interconnected system fulfills all design goals presented in Sect. 13.2.1.

The proof can be found in [28]. The following simulations show the performance of a platoon with seven agents, i.e., six followers and node 0 as reference node (leader). The initial positions and velocities of all agents are given by

$$\begin{aligned} x(t_0) &= \left[330\ 277\ 210\ 175\ 120\ 67\ 0\right]^T \text{ in m} , \\ v(t_0) &= \left[20\ 16.5\ 20.3\ 16.8\ 18.5\ 19.9\ 22\right]^T \text{ in m/s} , \end{aligned} \tag{13.21}$$

with the reference acceleration

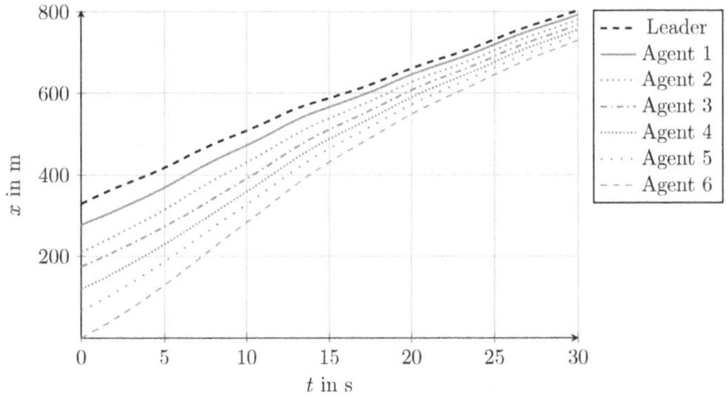

Fig. 13.1 Positions of all agents using the ATH with a time-varying reference acceleration

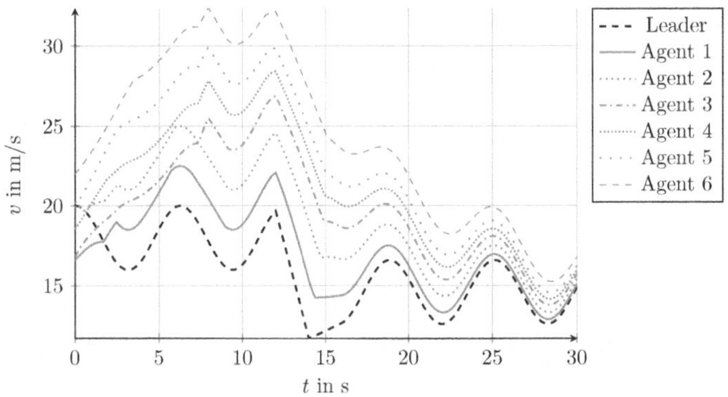

Fig. 13.2 Velocities of all agents using the ATH with a time-varying reference acceleration

$$u_r(t) = \begin{cases} -4 & t \in [12, 14) \\ 0.5 & t \in [14, 16) \\ 2\sin(t) & \text{otherwise} \end{cases} \quad \text{in m/s}^2 . \tag{13.22}$$

The desired constant distance is $\Delta_i = \Delta = 10\,\text{m}$ for followers $1, 2, \ldots, 6$. The controller parameter in (13.12), (13.17) has been chosen as $k = 5$ and the time-headway in (13.10) as $t_h^* = 1\,\text{s}$.

The positions and velocities of the agents using the proposed approach from Theorem 13.1 are shown in Figs. 13.1 and 13.2. The followers speed up in order to decrease the distance to the preceding vehicle. No position error undershoot occurs as shown in Fig. 13.3 and thus collisions can be avoided. The constant distance spacing is eventually reached in an approximated way, since the position errors and the time-headways in Fig. 13.4 approach zero. The convergence of the time-headways depends on the velocity errors displayed in Fig. 13.5; if the velocity errors are in the range given by (13.15), (13.19), the time-headways are decreased, otherwise held constant.

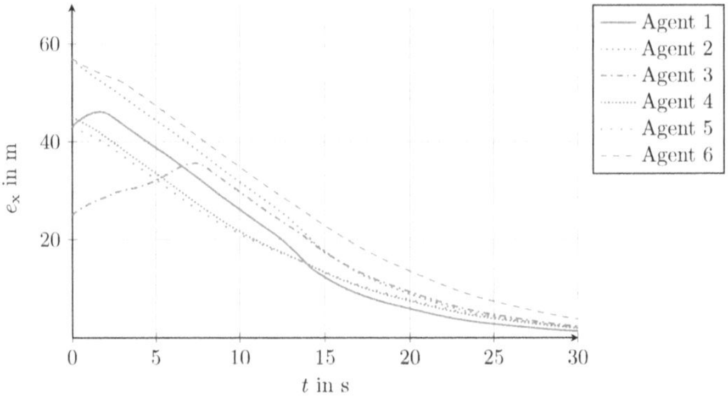

Fig. 13.3 The position errors of the agents using the ATH are nonnegative for all times and converge to zero, resulting in a constant distance spacing

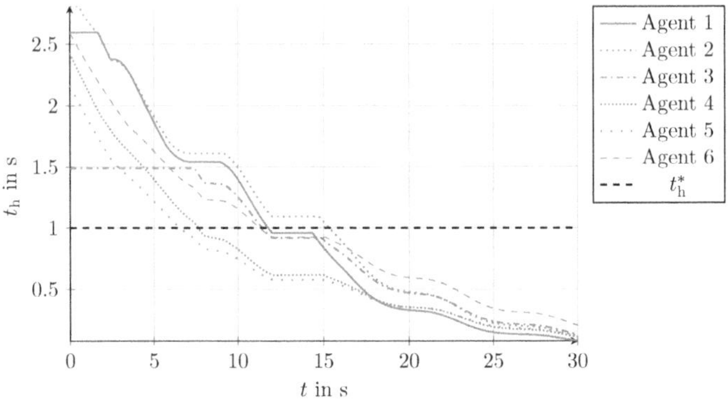

Fig. 13.4 Time-headways using the ATH with different decays in the first phase and in the second phase

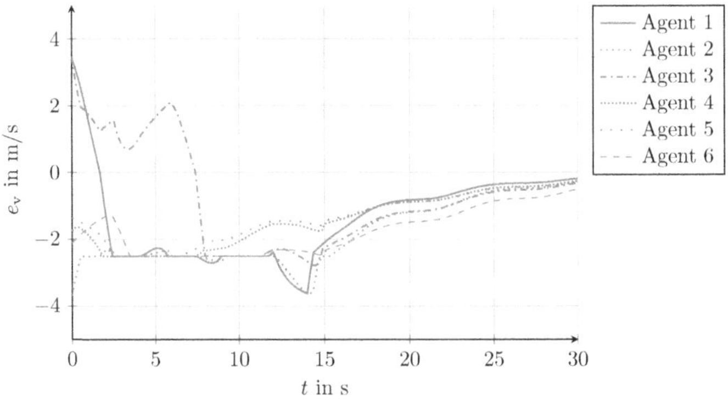

Fig. 13.5 The velocity errors of all agents using the ATH are bounded and do not depend on the position in the string i

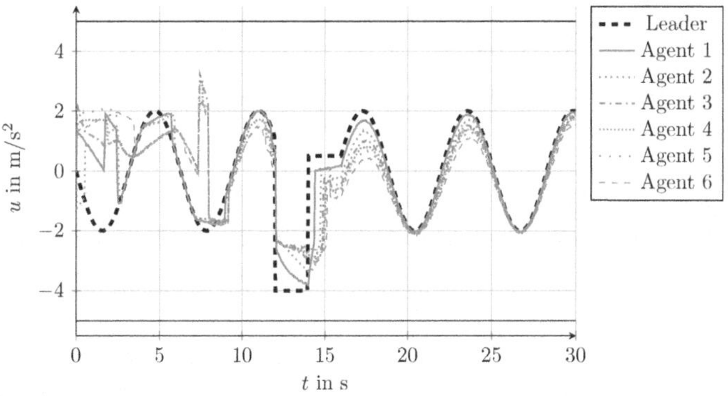

Fig. 13.6 The filtered control inputs using the ATH are not amplified along the platoon

Since the position errors, the velocity errors, and the accelerations do not increase with the position i in the string, the ATH approach is string stable. In Fig. 13.6, the filtered accelerations are displayed, where a first-order lag element with time constant $\tau = 0.01$ has been used. Note that if a predecessor applies a large deceleration (for example, starting from $t = 12$ s in (13.22)), the time-headway is held constant to avoid large velocity errors that could otherwise result in a position error undershoot.

Eventually, a constant distance spacing can be reached in an approximated way without communication of the accelerations while string stability is maintained. Note that the velocity of the preceding vehicle v_{i-1} is assumed to be known. The leader's acceleration can be arbitrary within its bounds, and the sliding-mode controllers are capable of compensating this disturbance. Thus, the proposed approach allows to reduce the time-headway to zero considering double integrator dynamics. In presence of actuator dynamics, however, a minimum time-headway has to be maintained, which is subject of the next section.

13.2.3 First-Order Actuator Dynamics

Consider the vehicle's longitudinal dynamics with first-order actuator dynamics

$$\begin{aligned} \dot{x}_i &= v_i \ , \\ \dot{v}_i &= a_i \ , \\ \tau_i \dot{a}_i &= -a_i + u_i \ , \end{aligned} \tag{13.23}$$

where u_i is the control input, a_i is the actual acceleration of the vehicle, and τ_i is the time constant of the actuator dynamics. In [30], the constant distance spacing and constant time-headway spacing have been analyzed using first-order sliding-mode control and the suboptimal controller. The results are briefly summarized in this section.

A first-order sliding-mode-based controller (FOSMC) is given by

$$u_i = k\mathrm{sign}(\sigma_{\mathrm{d},i}) \,, \tag{13.24}$$

using the constant distance spacing sliding variable of relative degree one

$$\sigma_{\mathrm{d},i} = e_{\mathrm{x},i} + ce_{\mathrm{v},i} \,. \tag{13.25}$$

The first derivative of (13.25) considering the dynamics (13.23) is governed by

$$\dot{\sigma}_{\mathrm{d},i} = e_{\mathrm{v},i} + ce_{\mathrm{a},i} \,, \tag{13.26}$$

with $e_{x,i} = x_{i-1} - x_i$, $e_{v,i} = v_{i-1} - v_i$, and $e_{a,i} = a_{i-1} - a_i$. The second derivative of the sliding variable reads as

$$\ddot{\sigma}_{\mathrm{d},i} = e_{\mathrm{a},i}\left(1 - \frac{c}{\tau}\right) + \frac{c}{\tau}u_{i-1} - \frac{c}{\tau}k\mathrm{sign}(\sigma_{\mathrm{d},i}) \,. \tag{13.27}$$

As discussed in [30], oscillations in the sliding variable may be excited, which are then propagated along the string and can thus result in string unstable behavior.

Using the constant time-headway spacing (13.9) as sliding variable, the control input becomes

$$u_i = k\mathrm{sign}(\sigma_{\mathrm{t},i}) \,, \tag{13.28}$$

and the dynamics of the sliding variable is given by

$$\ddot{\sigma}_{\mathrm{t},i} = a_{i-1} + a_i\left(\frac{t_{\mathrm{h}}}{\tau} - 1\right) - \frac{t_{\mathrm{h}}}{\tau}k\mathrm{sign}(\sigma_{\mathrm{t},i}) \,. \tag{13.29}$$

With $t_{\mathrm{h}} = \tau$, the sliding dynamics is governed by

$$\ddot{\sigma}_{\mathrm{t},i} = -k\mathrm{sign}(\sigma_{\mathrm{t},i}) + a_{i-1} \,. \tag{13.30}$$

Note that acceleration of the preceding vehicle acts as a disturbance bounded by $|a_{i-1}| < k$. However, it cannot be guaranteed that sliding is maintained for any $a_{i-1} \neq 0$, and oscillations occur in the case of any deviation from the origin, which are then propagated along the string of vehicles, yielding string unstable behavior. Thus, as soon as the sliding variable is deflected from the surface using the FOSMC, string unstable behavior is encountered and the approaches (13.24) and (13.28) for dynamics (13.23) are not robust.

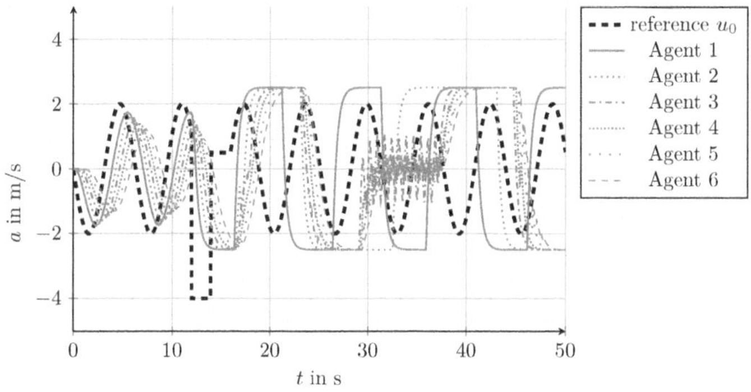

Fig. 13.7 Accelerations using the first-order sliding-mode controller

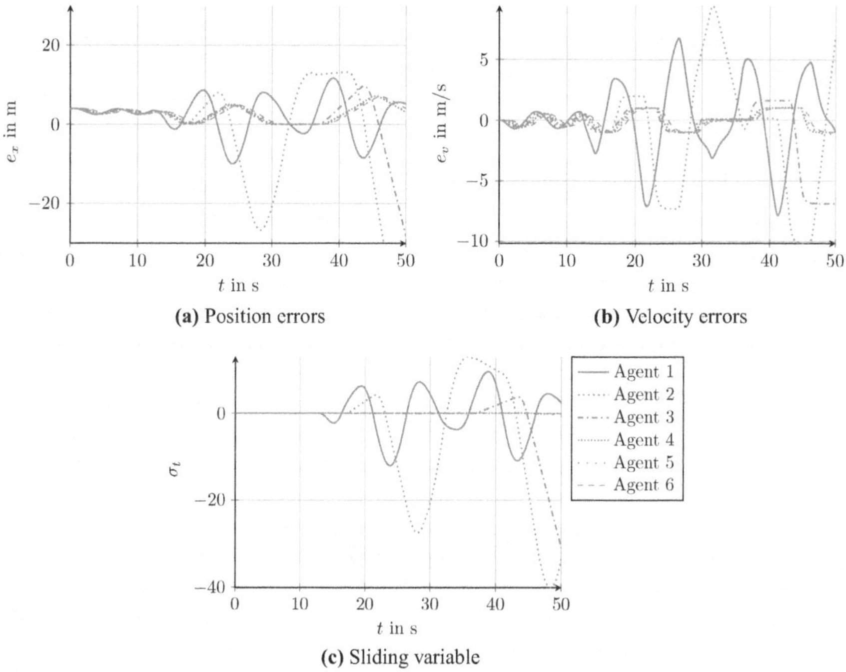

Fig. 13.8 Disturbances are amplified along the platoon and collisions occur using the first-order sliding-mode controller

Simulation results of the FOSMC using the constant time-headway spacing are shown in Figs. 13.7 and 13.8 when starting in sliding. The accelerations are shown in Fig. 13.7, and the errors are given in Fig. 13.8. Since the approach is not robust, collisions occur as soon as $e_{x,i} < -\Delta_i$, and string stability cannot be guaranteed.

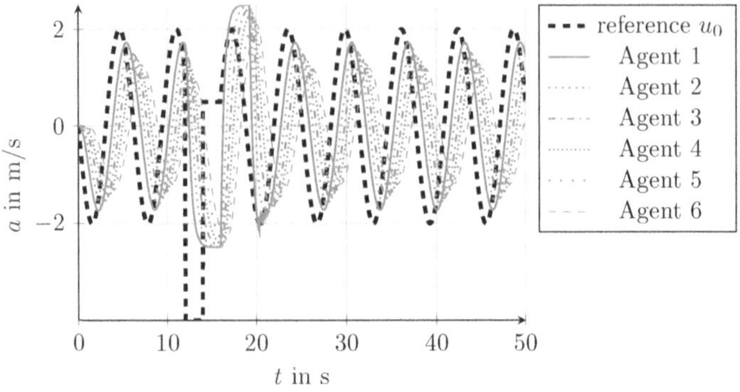

Fig. 13.9 Accelerations using the SOC

In the second example, the suboptimal controller (SOC)

$$u_i = \alpha_i(t) k_{\mathrm{SO}} \mathrm{sign}(\sigma_{\mathrm{t},i} - \beta\sigma_{\mathrm{M},i}) , \tag{13.31}$$

with parameter k_{SO} and adaptive parameter $\alpha_i(t)$ has been used that can steer the sliding variable to $\sigma_{t,i} = 0$ (for details on the suboptimal controller, see [4]). The platoon is thus string stable in a robust sense as shown in Figs. 13.9 and 13.10: the position errors and accelerations are attenuated along the string of vehicles, and no collisions occur. Note that the peaks in the sliding variable of Agent 1 have no effect on the other vehicles, i.e., robustness is given when the SOC is used.

Remark 13.1 The closed-loop system using the SOC is robust with respect to these small deviations. However, unmodeled higher-order dynamics may result in oscillations. Thus, increasing the order of any sliding-mode controller indefinitely is not reasonable.

Instead, the parameter t_{h} can be exploited to attenuate oscillations of unmodeled dynamics, which is focus of the following section.

13.2.4 Higher-Order Actuator Dynamics

The leader's longitudinal dynamics are now given by

$$\dot{x}_0(t) = v_0(t) , \quad \dot{v}_0(t) = a_0(t) , \tag{13.32}$$

with position $x_0(t) \in \mathbb{R}$, velocity $v_0(t) \in \mathbb{R}$, and acceleration $a_0(t) \in \mathbb{R}$. The reference acceleration a_0 is assumed to be unknown to the other agents, and thus, there is no communication between the vehicles. The vehicle-following agent's longitudinal dynamics are in the same fashion governed by

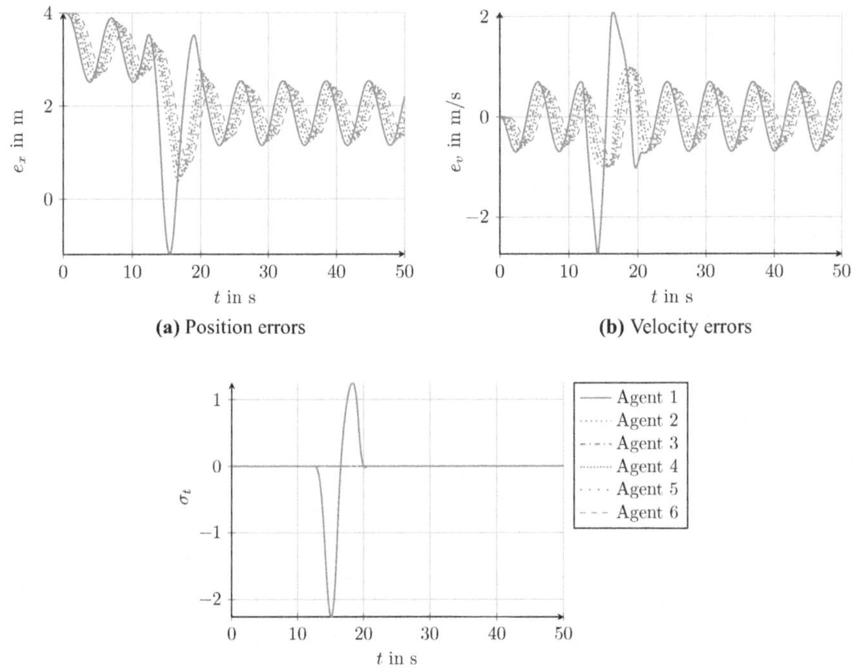

(a) Position errors

(b) Velocity errors

(c) Sliding variable

Fig. 13.10 No oscillations occur due to the robust performance of the SOC

$$\dot{x}_i(t) = v_i(t) , \quad \dot{v}_i(t) = a_i(t) , \tag{13.33}$$

for $i = 1, 2, ..., N$, where N is the number of agents, and $x_i(t) \in \mathbb{R}, v_i(t) \in \mathbb{R}, a_i(t) \in \mathbb{R}$ are the position, velocity, and the acceleration of agent i, respectively. It is assumed that each vehicle is subject to unmodeled higher-order actuator dynamics of the form

$$A_j(s) = W(s)U_j(s) , \quad j = 0, \ldots, N , \tag{13.34}$$

where $A_j(s) = \mathscr{L}\{a_j(t)\}$, and $U_j(s) = \mathscr{L}\{u_j(t)\}$. $W(s)$ is the transfer function of an asymptotically stable linear system that is used for all agents.

When sliding-mode-based control is used, these dynamics yield periodic motions of the sliding variable, as discussed, e.g., in [33]. Frequency domain analysis has been used for sliding-mode-based controllers for single systems without external disturbances in [6], and the SOC with external signals has been investigated in [7]. These considerations are adopted in this chapter in order to investigate string stability in frequency domain. In the case of linear controllers, string stability is obtained if

$$\|T_t\|_\infty = \max_\omega(|T_t(j\omega)|) \leq 1 , \tag{13.35}$$

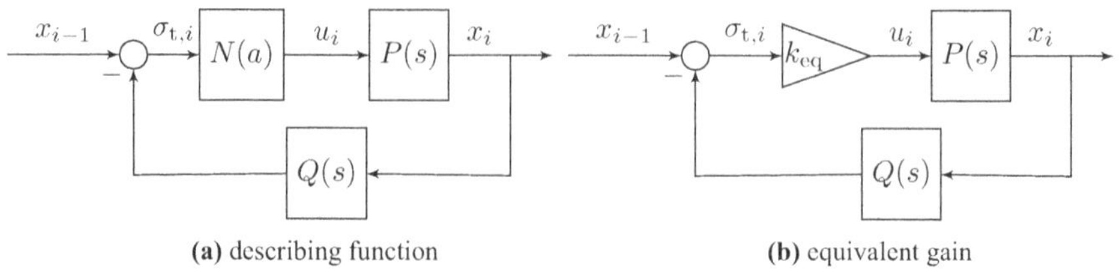

Fig. 13.11 Schematic of the closed-loop system for describing function analysis of agent i

with

$$T_t(s) = \frac{E_{t,i}(s)}{E_{t,i-1}(s)} = \frac{X_i(s)}{X_{i-1}(s)} = \frac{V_i(s)}{V_{i-1}(s)} \tag{13.36}$$

with $X_i(s) = \mathscr{L}\{x_i(t)\}$, $V_i(s) = \mathscr{L}\{v_i(t)\}$, $E_{t,i}(s) = \mathscr{L}\{e_{t,i}(t)\}$, which holds in homogeneous platoons [36]. This condition has also been used for sliding-mode-based approaches on the sliding surface enforcing linear dynamics as in [13]. Therein, string stability has been guaranteed in sliding using a constant time-headway spacing, i.e., if the error $e_{t,i}$ is zero for all times. In presence of unmodeled dynamics and thus oscillations in the sliding variable, string stability cannot be concluded. Hence, the effects of oscillations in the sliding variable arising from unmodeled actuator dynamics can be investigated via frequency domain analysis, which is focus of this section.

The agent's dynamics in frequency domain reads as

$$P(s) = \frac{X_i(s)}{U_i(s)} = \frac{1}{s^2} W(s) . \tag{13.37}$$

The closed-loop transfer function with a linear controller $R(s)$ and the constant time-headway spacing (13.5) is then given by

$$T_t(s) = \frac{R(s)P(s)}{1 + R(s)P(s)Q(s)} , \tag{13.38}$$

where the time-headway spacing exhibits a filtering effect as

$$Q(s) = t_h s + 1 , \tag{13.39}$$

see [31]. In the linear case, it has been shown that the time-headway can be chosen based on the plant and the controller parameters such that (13.35) holds, see, e.g., [10].

For sliding-mode-based analysis, consider the schematic in Fig. 13.11, where the sliding-mode controller has been replaced by the describing function $N(a)$, which is a function of the oscillation's amplitude a (and may also be a function of the oscillation's frequency, see [5]). For the first-order sliding-mode controller with constant

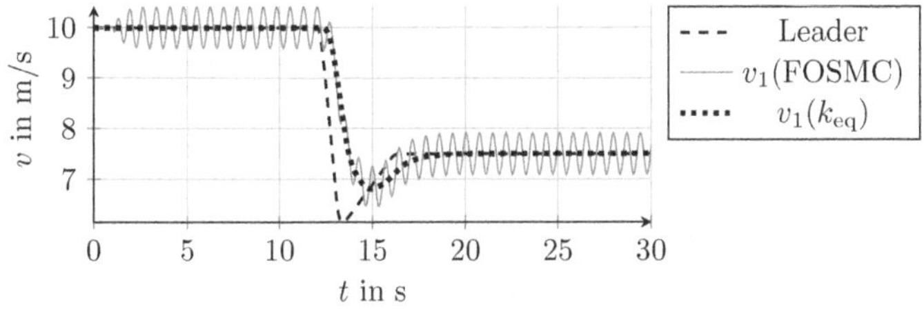

Fig. 13.12 The equivalent gain is used to approximate the mean values of the signals in the closed-loop system

time-headway spacing

$$u_i = k\text{sign}(\sigma_{t,i}) \,, \tag{13.40}$$

with parameter $k > 0$, the describing function is a function of the amplitude a only and given by

$$N(a) = \frac{4k}{\pi a} \,. \tag{13.41}$$

The equivalent gain describes the mean value of the system response to slow time-varying reference signals as depicted in Fig. 13.11b. For the FOSMC (13.40) it is given by[1]

$$k_{\text{eq}} = \frac{2k}{\pi a_{\text{c}}} \,. \tag{13.42}$$

In Fig. 13.12, the velocity of one agent is shown using the FOSMC and its equivalent gain. The amplitude a_{c} can be found either analytically, or numerically: it is computed according to the harmonic balance equation by solving

$$N(a)P(j\omega)Q(j\omega) = -1 \,, \tag{13.43}$$

i.e., by the intersection

$$P(j\omega_{\text{c}})Q(j\omega_{\text{c}}) = -\frac{1}{N(a_{\text{c}})} \,. \tag{13.44}$$

For example, if the transfer function in (13.34) is governed by

[1]Note that the locus of perturbed relay systems (LPRS) approach can be used to achieve an exact value for the equivalent gain [7]; however, since the approximations of the describing function analysis are not as involving as the LPRS method and match the simulations and experiments, the latter approach is used.

$$W(s) = \frac{1}{(s\tau + 1)^2}, \tag{13.45}$$

the amplitude a_c using (13.41) in (13.44) can be computed analytically by

$$a_c = \frac{4kt_h^2\tau}{2\pi(t_h - 2\tau)}. \tag{13.46}$$

The same approach can be used, e.g., for the suboptimal controller, where the describing function is given by

$$N_S(a) = \frac{4k}{\pi a}(\sqrt{1-\beta^2} + j\beta), \tag{13.47}$$

and the equivalent gain is

$$k_{eq} = \frac{2k}{\pi a_c}(1+\beta). \tag{13.48}$$

Then, the transfer function of the closed-loop system using sliding-mode-based controllers with respect to slow motions can be approximated by

$$T_{eq}(s) = \frac{k_{eq}P(s)}{1 + k_{eq}P(s)Q(s)}. \tag{13.49}$$

As proposed in [27], one can then select the time-headway t_h so that (13.35) holds using (13.49), and consequently, a string stable platoon is achieved. Note that for any equivalent gain $k_{eq} > 0$, the approximated linear system with constant distance spacing (13.3) cannot achieve string stability, which has been discussed in detail for linear systems, e.g., in [10]. Hence, in case of periodic oscillations, the constant time-headway spacing has to be used, and Algorithm 1 as published in [27] can be performed to arrive at a proper choice of the time-headway, resulting in a string stable platoon for any number of agents. Note that a small time-headway is desired for small inter-vehicle spacings and thus high traffic throughput, but it has to be larger than a critical value in order to guarantee string stability, as discussed for linear systems in [18].

Remark 13.2 The amplitudes of the oscillations have to be smaller than the minimum distance Δ_i in order to achieve collision-free platooning; note that string stability does not imply collision-free platooning.

Remark 13.3 Since Algorithm 1 uses approximations of the closed-loop system, a proper time-headway can be found with respect to a certain resolution, where $\Delta t_h = 0.1$ has been chosen.

Simulations show that the fast periodic motions, which cannot be avoided for one agent, do not affect a following agent, if the equivalent closed-loop system (13.49) is

Algorithm 1 Proper choice of time-headway [27]

Initialization
choose sliding-mode controller and parameters
choose initial $t_{\mathrm{h}}(t = 0) = t_{\mathrm{h},0}$

Step 1: Find amplitude of the periodic motion of $\sigma_{\mathrm{t},i}$
compute $\tilde{P} = P(s)Q(s)$
obtain amplitude a_{c} and frequency ω_{c} via (13.44)

Step 2: approximate slow motions of signals of agent i
compute k_{eq} according to (13.42) or (13.48), and $T_{\mathrm{eq}}(s)$ according to (13.49)

Step 3: check for string stability
check if (13.35) holds for $T_{\mathrm{eq}}(s)$
if yes **then**
  reduce t_{h} (e.g., bisection method), go to Step 1
else
  increase t_{h} or use value of last iteration and stop

Table 13.1 Simulation results

Control	t_{h}	a_{sim}	a_{nyq}	String stable	
				Bode	Sim.
FOSMC	1.5	0.571	0.551	Yes	Yes
	1	0.413	0.398	Yes	Yes
	0.5	0.279	0.265	No	No
SOC	1.5	0.166	0.154	Yes	Yes
	1	0.118	0.110	Yes	Yes
	0.5	0.072	0.066	Yes	Yes

string stable according to (13.35), i.e., if the time-headway is chosen properly. For the following simulation results, a platoon of $N = 6$ following agents and one leader has been simulated. The FOSMC and SOC have both been implemented with controller parameter $k = 5$ for all agents, with a fixed-step sampling time $T_{\mathrm{s}} = 1 \times 10^{-4}$ s. All vehicles start in formation with $e_{\mathrm{t},i}(t_0) = 0$, $e_{\mathrm{v},i}(t_0) = 0$ in (13.3), (13.5) for $i = 1, 2, \cdots, N$. The leader's reference control input has been chosen as

$$u_0 = \begin{cases} -4 & t \in [12\,\mathrm{s}, 14\,\mathrm{s}] \\ 2\sin(t + \pi) & \text{otherwise} \end{cases} . \tag{13.50}$$

The actuator dynamics have been chosen according to (13.45) with $\tau = 0.1$. The Nyquist plot of $P(j\omega)Q(j\omega)$ exhibiting intersections with both FOSMC and SOC is shown in Fig. 13.13, with the resulting amplitudes given in Table 13.1. The amplitudes found in simulation are denoted by a_{sim}, and the amplitudes obtained from the Nyquist plot a_{nyq}.

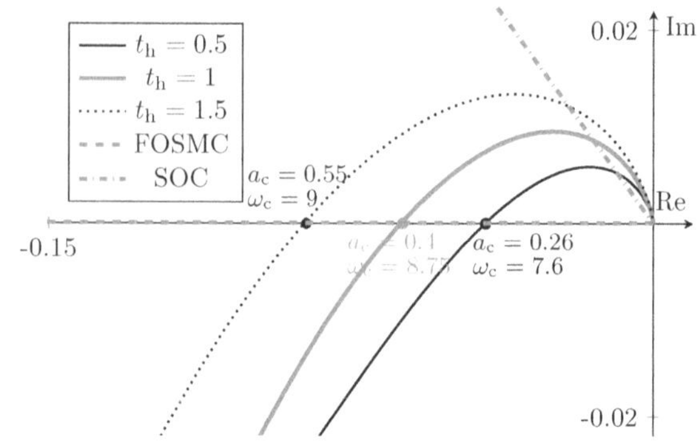

Fig. 13.13 Nyquist plot of $P(j\omega)Q(j\omega)$ with different time-headways

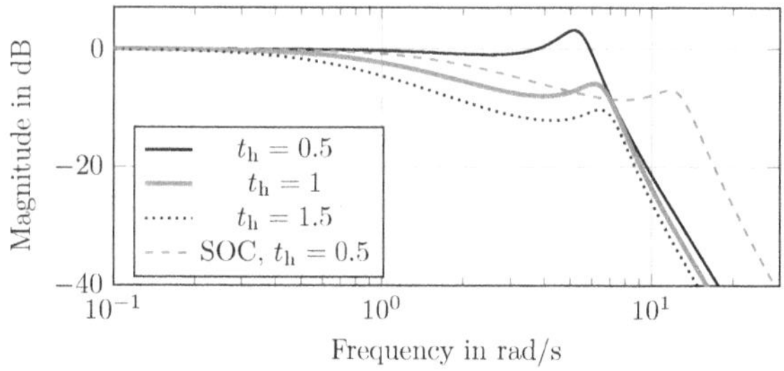

Fig. 13.14 Magnitude plots of the closed-loop system with equivalent gain and different time-headways

When the equivalent gain of the sliding-mode controllers have been obtained, string stability can be assessed from the bode plots given in Fig. 13.14: The platoon with FOSMC is string unstable for $t_h = 0.5$ and string stable for $t_h \geq 1$, while the SOC achieves string stability with $t_h = 0.5$. In Fig. 13.15, simulation results using the FOSMC with $t_h = 0.5$ are shown, where the oscillations are amplified along the string, i.e., the amplitudes are increased for increasing position i in the string. The results using the SOC with $t_h = 0.5$ are shown in Fig. 13.16, where the amplitudes are smaller and the oscillations are faster than with the FOSMC. The resulting closed-loop system is string stable and the oscillations are not amplified along the string.

Remark 13.4 In order to reach the minimum time-headway for each vehicle, one can first compute the minimum time-headway and then apply the ATH in order to decrease the actual time-headway to this minimum.

13.2.5 Robustness with Respect to Lateral Deviation

The robustness of sliding-mode-based controllers can also be exploited for rejection of disturbances that arise from lateral motion. For example, the longitudinal distance shall be maintained during a lane change, or if the vehicles are on adjacent lanes

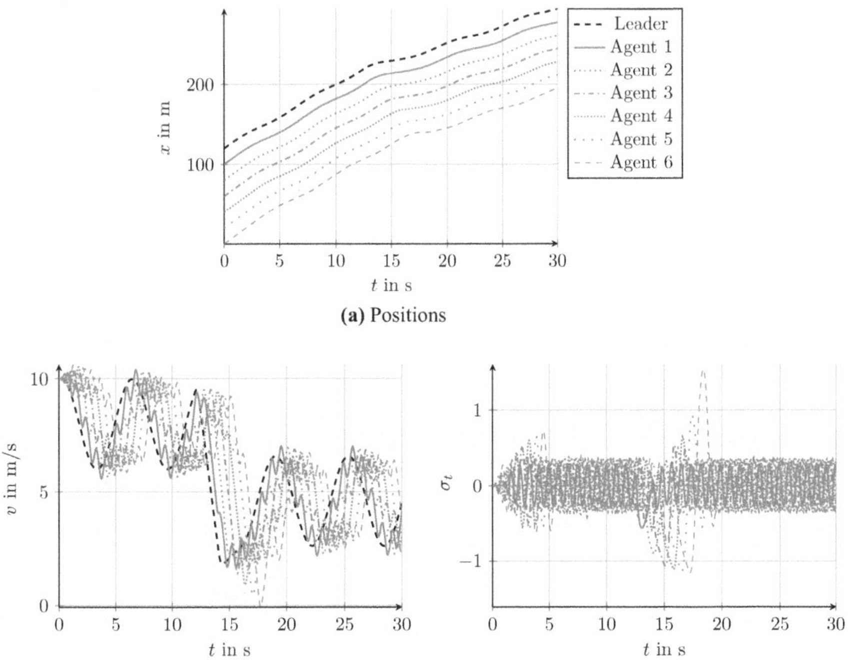

(b) Velocities

(c) Sliding variables

Fig. 13.15 String unstable platooning with FOSMC and $t_h = 0.5$

in a road bend. This section shows that these disturbances are matched and can be rejected by sliding-mode-based controllers.

13.2.5.1 Formation Control on Adjacent Lanes

Consider a road segment depicted in Fig. 13.17 with constant curvature $\kappa_1 = 1/R_1$, i.e., a segment of a circle with radius R_1, where the radius is defined with respect to the center of the inner lane. The curvature on the second lane is then defined by $\kappa_2 = 1/R_2$, with $R_2 = R_1 + w_{\text{lane}}$ and width w_{lane}. If the relative position between the vehicles should be maintained, the vehicle on the outer lane has to drive at a faster velocity than the vehicle on the inner lane, due to a larger waylength s_i according to

$$s_i = R_i \phi \,, \tag{13.51}$$

with angle ϕ. Then for constant curvatures, the requirement for a constant angular velocity results in

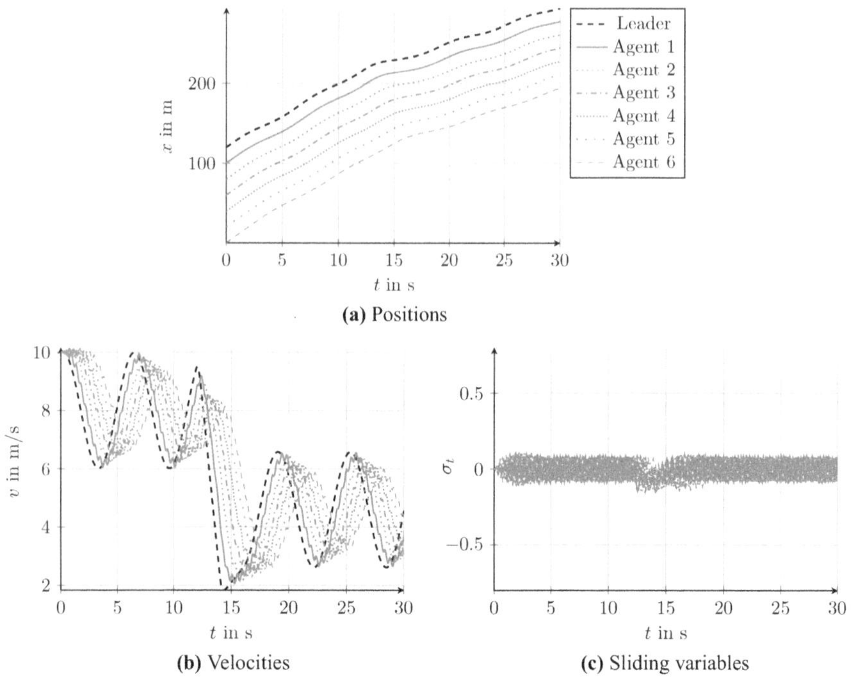

Fig. 13.16 String stable platooning with SOC and $t_h = 0.5$

Fig. 13.17 In road bends, the vehicle on the outer lane (blue) has to drive faster than the vehicle on the inner lane (red) in order to keep a constant angular velocity

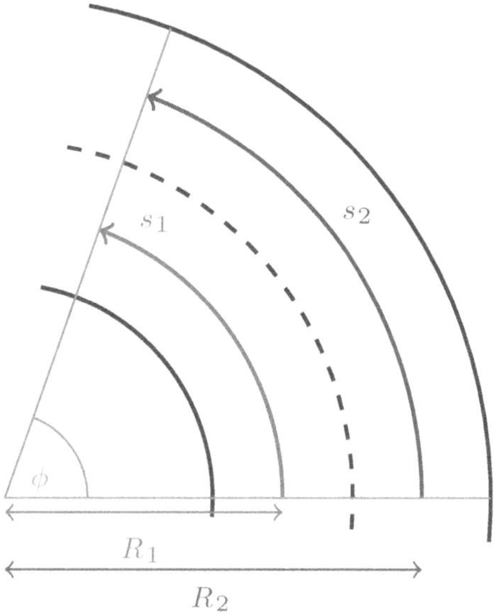

$$\dot{\phi} = \frac{\dot{s}_1}{R_1} = \frac{\dot{s}_2}{R_2} , \tag{13.52}$$

and thus, the velocities $\dot{s}_1$, $\dot{s}_2$ are related by

$$\dot{s}_1 = \dot{s}_2 \frac{R_1}{R_2} . \tag{13.53}$$

This means that the vehicle on the outer lane has to maintain a faster speed than the vehicle on the inner lane in order to maintain the same angular velocity. Then, the longitudinal distance between the vehicles can be maintained. Hence, the curvature can be interpreted as a disturbance δ_i on the longitudinal velocity and is modeled as an additive disturbance subsequently. The agent's longitudinal dynamics are then given by

$$\begin{aligned} \dot{x}_i &= v_i + \delta_i , \\ \dot{v}_i &= u_i , \end{aligned} \tag{13.54}$$

where x_i, v_i, u_i denote position, velocity, and acceleration of vehicle i, respectively. The curvature typically increases linearly on highway roads [21], i.e., they are typically designed as clothoid segments as depicted in Fig. 13.18a. A point (x, y) on the clothoid is given by

$$\begin{aligned} x(l) &= A\sqrt{\pi} \int_0^l \cos\left(\frac{\pi t^2}{2}\right) dt , \\ y(l) &= A\sqrt{\pi} \int_0^l \sin\left(\frac{\pi t^2}{2}\right) dt , \end{aligned} \tag{13.55}$$

with parameter l. The length of the path is given by $L = Al\sqrt{\pi}$ and the curvature is $\kappa(l) = \frac{\sqrt{\pi}}{A} l$ for a constant parameter $A > 0$.

For a road consisting of such clothoid segments, the derivative of the curvature is bounded, and thus the vehicle's longitudinal dynamics are affected by a bounded matched disturbance $\tilde{\delta}_i$,

$$\begin{aligned} \dot{x}_i &= v_i , \\ \dot{v}_i &= u_i + \dot{\delta}_i = u_i + \tilde{\delta}_i , \end{aligned} \tag{13.56}$$

which is bounded by $|\tilde{\delta}_i| < \Gamma_i$. If the acceleration of the preceding vehicle is small such that $|u_{\text{eq},i}(t) + \tilde{\delta}_i| \leq u_{\max}$ holds, then the formation can be maintained even if the vehicles are on different lanes.

13.2.5.2 Lane Changes

Similarly, a lane change can also be considered as a disturbance that arises from different curvatures. During a lane change, the lateral position of the vehicle on the road y_i is changed in a smooth way due to a bounded lateral acceleration.

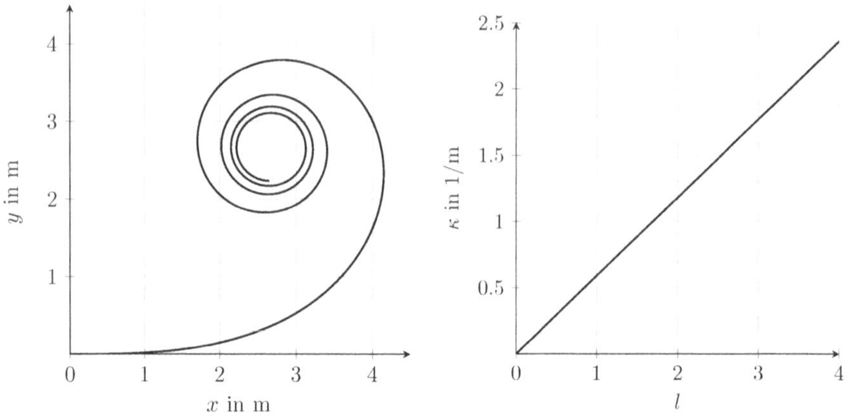

(a) Clothoid **(b)** Curvature versus l

Fig. 13.18 Clothoids and corresponding curvature as a function of the parameter l, which are typically used for design of road segments

Fig. 13.19 Road coordinate system with corresponding velocities

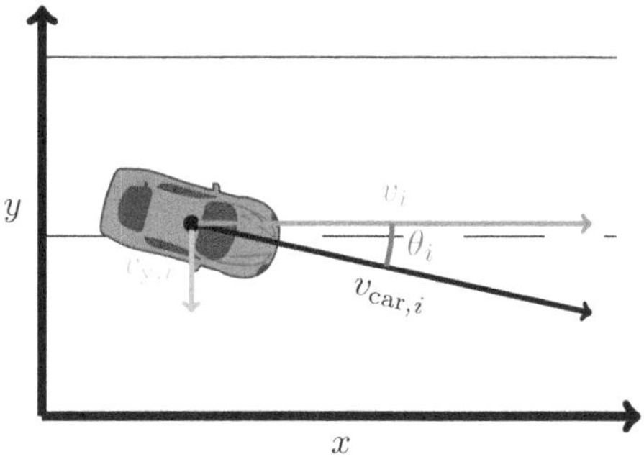

Note that the states x_i, v_i in (13.56) are considered with respect to the longitudinal road coordinates as depicted in Fig. 13.19. For the longitudinal control in (13.56), it is assumed that the relative distance to other objects $\tilde{e}_{x,ij} = x_j - x_i$ and velocities v_i, v_{i-1} are available with respect to the road coordinate system.

The lateral velocity of the vehicle $v_{y,i}$ as shown in Fig. 13.19 is larger than zero during a lane change, and the velocity of the vehicle $v_{car,i}$ is not equal to the road velocity v_i in (13.56); specifically,

$$v_{car,i} = \sqrt{v_i^2 + v_{y,i}^2}\ . \tag{13.57}$$

Assuming a point mass, the vehicle can be described in the x, y coordinate system by

$$
\begin{aligned}
v_i &= \dot{x}_i = v_{\mathrm{car},i}\cos(\theta_i) \ , \\
v_{\mathrm{y},i} &= \dot{y}_i = v_{\mathrm{car},i}\sin(\theta_i) \ , \\
\dot{\theta}_i &= \omega_i \ ,
\end{aligned}
\tag{13.58}
$$

where θ_i is the orientation with respect to the x-axis as depicted in Fig. 13.19 and ω_i is the angular velocity. The acceleration of the car $\dot{v}_{\mathrm{car},i} = u_{\mathrm{car},i}$ and the acceleration u_i in (13.56) differ as well. By differentiation of (13.58), the dynamics along the x-axis are

$$
\begin{aligned}
\dot{v}_i &= \dot{v}_{\mathrm{car},i}\cos(\theta_i) - v_{\mathrm{car},i}\sin(\theta_i)\dot{\theta}_i \\
&= u_{\mathrm{car},i}\cos(\theta_i) - \dot{y}_i\dot{\theta}_i \\
&= f(v_i)u_{\mathrm{car},i} - d_i \ .
\end{aligned}
\tag{13.59}
$$

The orientation θ_i is assumed to be small, which is practicable on highways, thus $\cos(\theta_i) \approx 1, \sin(\theta_i) \approx 0$. Then, $0 < f_{\min} \leq |f(v_i)| \leq 1$. Assuming a bounded ω_i, the matched disturbance d_i is bounded, i.e., there exists a D_i such that $|d_i| < D_i$. Since u_i is bounded, one can simplify these disturbances to $\dot{v}_i = u_i + \tilde{d}_i$ and $|\tilde{d}_i| \leq \tilde{D}$, which yields the dynamics

$$
\begin{aligned}
\dot{x}_i &= v_i \ , \\
\dot{v}_i &= u_i + \tilde{d}_i \ .
\end{aligned}
\tag{13.60}
$$

This means that the longitudinal formation can be maintained while driving on a curved road or while merging onto one lane, if the disturbances $\tilde{\delta}_i$ in (13.56) and/or $\tilde{d}_i$ in (13.60) can be rejected by the control input u_i. In the formation with constant distance spacing and a sliding-mode-based controller, the control input is given by $u_{\mathrm{eq},i} = u_{\mathrm{r}}$. Then, if

$$
u_{\mathrm{r}} + \Gamma_i + \tilde{D}_i \leq u_{\max} \tag{13.61}
$$

holds, both disturbances can be rejected at the same time, if an appropriately designed controller is used, and the formation can be maintained.

Remark 13.5 The lateral acceleration of the vehicle during a maneuver consists of the lateral acceleration from the curvature of the road that is bounded by Γ_i, and the lateral acceleration of the lane change that is bounded by $\tilde{D}_i$. Typically, a lane change trajectory is computed considering the road curvature in such a way that a certain overall lateral acceleration is not exceeded (see [29]).

A robust longitudinal controller has to be used in order to maintain the formation during a lane change or in bends. The disturbance $\hat{d}_i = \tilde{d}_i + \tilde{\delta}_i$ with (13.56) and (13.60) can be handled by the ATH in certain scenarios; the sliding dynamics using the ATH of Sect. 13.2.2 with the disturbed dynamics of one agent (13.60) reads as

$$
\dot{\tilde{\sigma}}_i = e_{\mathrm{v},i} + (t_{\mathrm{h}}^* - t_{\mathrm{h},i})u_{i-1} + t_{\mathrm{h}}^*(k\,\mathrm{sign}(\tilde{\sigma}_i) + \hat{d}_i) + \mu_2 k t_{\mathrm{h},i} \ . \tag{13.62}
$$

Then, if condition

$$|e_{v,i} + (t_h^* - t_{h,i})u_{i-1} + t_h^* \hat{d}_i + \mu_2 k t_{h,i}| < t_h^* k \tag{13.63}$$

is fulfilled, sliding can be maintained. Note that (13.63) cannot be guaranteed in general for all times. However, the disturbance compensation is only important for small position errors. Then, the time-headways $t_{h,i}$ and the velocity errors are small. In the case that the longitudinal velocity is much larger than the lateral velocity, the maximum disturbance D_i is very small, especially if compared to the other variables in (13.63). If the vehicles' control is $u_i \approx u_r$ and (13.61) holds, then the disturbance can be compensated. Note that if the disturbance cannot be compensated, the position error to the preceding vehicle increases during the lane change due to a larger waylength, which is not safety-critical. Further analysis of the ATH in presence of disturbances is subject to future work.

In order to test the robustness of the algorithm with respect to lane changes or different curvatures, an example of a lane change is considered. As a sampling time, $T_s = 0.001$ s is used. Note that the path of the lane change in Fig. 13.20a is chosen to start rather abruptly on purpose in order to investigate the effects of different lateral accelerations.

The velocity of the car as shown in Fig. 13.19 is larger during the lane change in order to maintain a constant velocity along the road as shown in Fig. 13.20b. The position errors of the two followers are presented in Fig. 13.20c, where no effects caused by the lane change can be detected. Note that the accelerations in Fig. 13.20e have been filtered with a first-order lag element with time constant $\tau = 0.01$ s. The disturbances are bounded and small as shown in Fig. 13.20e. They can thus be rejected, and sliding can be maintained as shown in Fig. 13.20d.

13.3 Experiments

In this section, the testbed as presented in [29] is first described in detail, and results on longitudinal sliding-mode control are presented in the second part. The testbed can be used for various advanced driver assistance systems, e.g., lane keeping assist, adaptive cruise control (ACC), motorway chauffeur, or platooning.[2]

13.3.1 Description of the Testbed

The requirements for the testbed have been stated in [29] as follows:

- The focus lies on control engineering, i.e., controller design, trajectory generation, and estimation techniques. Hence, the effort spent on localization should be minimal to allow fast prototyping.

[2]Videos of the testbed can be found at [1].

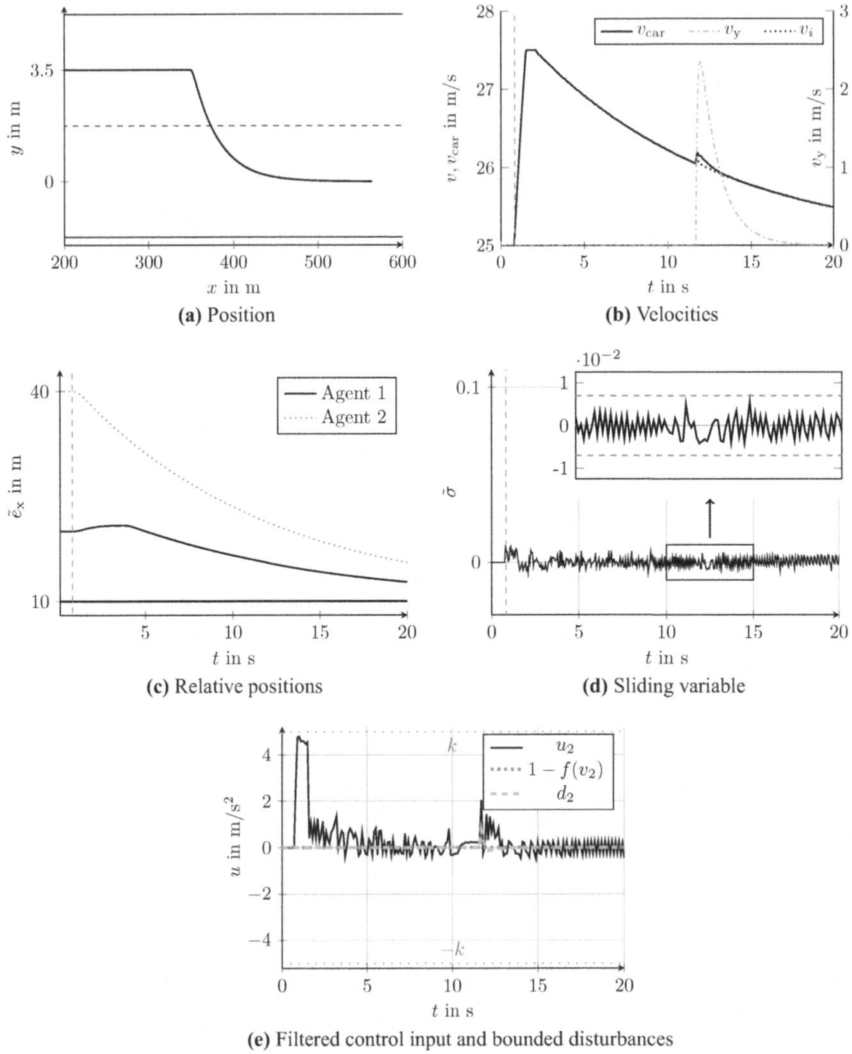

Fig. 13.20 Results of disturbance compensation during a lane change scenario

- Safety must be guaranteed at all times.
- Hardware should be replaceable easily if damaged, and extendable.
- The testbed should be scalable so that many different vehicles can be used.
- Distributed computation is desired and the software runs on embedded systems on each vehicle.
- Real-time capability must be ensured.
- Access to the testbed should be possible at any time.

(a) Trucks

(b) Passenger cars

Fig. 13.21 Small-scale vehicles equipped with BeagleBone Black boards

Real testbeds currently do not fulfill these requirements, since the installation of sensors is very complex, test drivers have to be trained appropriately, hardware is expensive, and access to test tracks is not easily obtained. Hence, the small-scale automated driving testbed presented in [29] has been designed, which allows fast and easy prototyping and thus an evaluation of different control algorithms. It is similar to the one at KTH Stockholm [2], but uses a significantly cheaper motion capture system and is dedicated to autonomous ground vehicles only. The hardware and software components of the setup are described subsequently.

13.3.1.1 Hardware

Two different kinds of small-scale vehicles have been acquired: trucks for scenarios such as parking with trailers or platooning, and passenger cars to test standard maneuvers such as ACC or overtaking, which are described subsequently.

Trucks

Three trucks[3] in scale 1:14 are available on the testbed, where two of them are shown in Fig. 13.21a. The vehicles have been equipped with a BeagleBone Black,[4] which is a low-cost single-board computer. A WiFi module[5] has been installed in order to receive the actual position in the room or to communicate with other vehicles. The trucks are actuated by a motor and servos for gear changing and steering via PWM signals. The steering angle is within the range $|\delta_{\text{truck}}| \leq 17°$ and the maximum allowable speed of the trucks has been set to $1.5\,\text{m/s}$ due to indoor testing and thus limited space for the testbed, but speeds may also be larger in different settings.

[3]http://www.tamiya.de/de/produkte/rcmodelltrucks.htm.

[4]http://beagleboard.org/black.

[5]http://www.tp-link.com/at/products/details/cat-9_TL-WR802N.html.

Passenger Cars

Three cars[6] in scale 1:10 as shown in Fig. 13.21b have been built up, which are also equipped with a BeagleBone Black and a WiFi module. However, the actuation upon acquisition differed from the trucks: the speed of the included motors strongly depended on the state of charge of the battery and did not yield accurate results. Hence, brushless DC (BLDC) motors with Hall sensors[7] and motor speed controllers[8] have been used to achieve better performance at low speeds. In addition, the transmission gears have been replaced, which also improves driving at low speeds. The steering of the cars is more accurate than the trucks' steering, but the maximum steering angle is in a similar range with $|\delta_{\text{car}}| \leq 20°$. Currently, three passenger cars are available for experiments on the proposed testbed.

Sensors

Webcams[9] on the ceiling and AprilTags [22] on the vehicles are used for data acquisition for the platooning scenarios: This setup permits GPS emulation, which is described in Sect. 13.3.1.1. Other sensors, e.g., velocity sensors, can be added in future in order to improve the performance of the vehicles. The use of ultrasonic sensors on the vehicles is briefly described in [29].

Actuation

In order to actuate the motor of the vehicles via PWM (pulse-width modulated) signals, a lookup table was created by measuring the velocity of the vehicle with a camera[10] for various PWM values. For the steering actuation, no Lookup table was required due to a linear dependency between steering angle and PWM signal. The slope and the offset were calculated by measuring the radius of the circle driven by the vehicle for different steering angles.

Projection of Virtual Components

In order to visualize virtually defined components such as lanes, virtual obstacles, or traffic signs, two ultrashort throw projectors[11] are used. In the current status of the testbed, the road and static obstacles can be projected onto the floor in order to demonstrate that, e.g., lane keeping assists work properly. In future work, different dynamic environments including virtual traffic participants will be projected.

Position Tracking

It is necessary to obtain position information of the controlled vehicle and surrounding obstacles for platooning and other automated driving functionalities. For this

[6]http://www.conrad.at/de/reely-strassenmodel-audi-rs6-brushed-110-rc-modellauto-elektro-strassenmodell-allradantrieb-rtr-24-ghz-238002.html.

[7]http://www.amainhobbies.com/lrp-vector-x20-brushless-motor-6.5-lrp50674/p226984.

[8]http://vedder.se/2015/01/vesc-open-source-esc.

[9]http://www.logitech.com/de-at/product/c930e-webcam.

[10]http://gopro.com *all links accessed: 2018-04-09.*

[11]http://www.optoma.de/projectorproduct/x320ust.

purpose, the positions of moving objects are estimated via webcams and AprilTags, which are mounted on the vehicle as shown in Fig. 13.21. The position tracking code is based on the AprilTags C++ Library,[12] which provides fast and robust 3D position estimation and the tracking algorithm is insensitive to bad light conditions. The camera intrinsics have been calibrated using AprilCal [25], which is necessary only once for each webcam.

To cover the testbed area used for the presented experiments, four webcams mounted on the ceiling have been used for this position estimation. Note that for driving on a planar environment, it is sufficient to estimate a 2D position; hence, for the calibration of the camera extrinsics, a planar calibration target based on AprilTags has been used. Note that the calibration of the world coordinate system has to be executed only if the position of a camera changes and can be performed automatically using configuration files; thus it is straightforward to build up the testbed in a different environment (e.g., for presentation purposes), and the testbed is not limited to one room, which is a major benefit.

The tag detection algorithm is executed on a dedicated Linux workstation, called "Position Tracking Computer", and the webcams are connected to this Position Tracking Computer via USB as depicted in Fig. 13.23. The AprilTag detection software then delivers the AprilTags' positions and orientations with a rate of 10 Hz and broadcasts the information via WiFi using UDP.

High-Definition Map

The road is shown in Fig. 13.22a and has been defined virtually, i.e., there are no lane markings on the floor. The road is stored on the vehicles, The (x, y) coordinates with orientation θ, the waylength s and the curvature κ of the inner lane are stored with a precision of 1 cm on each vehicle, emulating a high-definition road map for self-driving vehicles. The field of view using four webcams is shown in Fig. 13.22b, where the road has been plotted in software on the images. Note that the road is not marked on the floor in order to allow various scenarios and road definitions, e.g., switching from a highway to an intersection, a parking lot, etc., but is projected onto the floor to improve the demonstrations.

13.3.1.2 Overall Setup

In Fig. 13.23, the overall setup of the testbed is shown. On the "ADAS" computer, code is generated from an easy-to-use environment in MATLAB/Simulink,[13] as described in Sect. 13.3.1.3, which is then made available on an FTP server. Via SSH connection, the BeagleBone Black on the vehicle has access to the code. Then, the code is compiled and the program is executed on the BeagleBone Black, and relevant data can be logged on each vehicle separately. During an experiment, the positions of the vehicles are tracked by the AprilTag detection program on the Posi-

[12] http://people.csail.mit.edu/kaess/apriltags.

[13] http://de.mathworks.com/products/simulink.html.

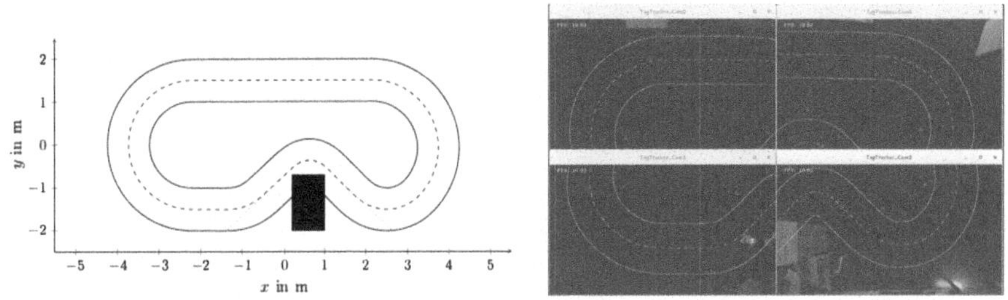

(a) Definition of the road

(b) Field of view of the webcams and the virtually defined road

Fig. 13.22 Road coordinates defined for testing in the foyer of the institute

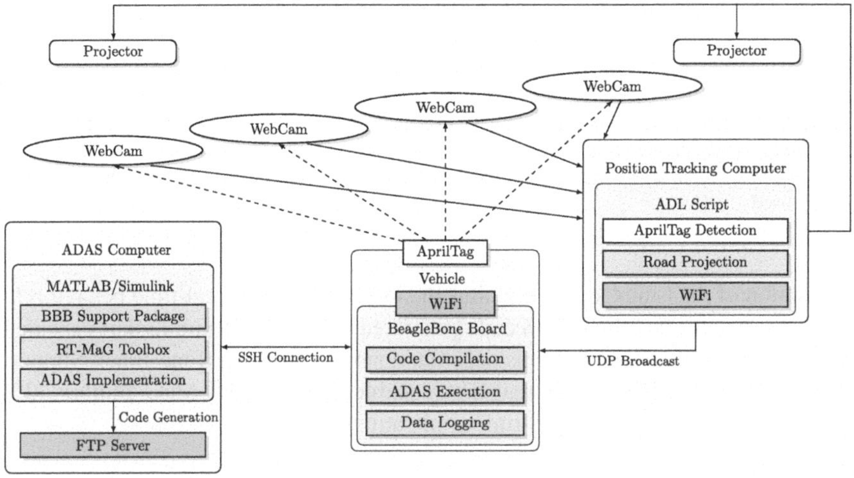

Fig. 13.23 Setup of the small-scale testbed for control engineering purposes

tion Tracking Computer using the webcams as only sensors. The information is then sent via UDP to the vehicles.

13.3.1.3 Software

Each vehicle should be capable of executing ADAS functionalities on the BeagleBone Black in real-time. For this purpose, the software has been set up as explained subsequently.

Real-Time Operating System

On the BeagleBone Black, the programs are executed on a Debian Linux-based operating system (OS) with RT PREEMPT patch (Version 4.4),[14] which enables full

[14]http://elinux.org/BeagleBoardDebian#Mainline_.284.4.x_lts.29.

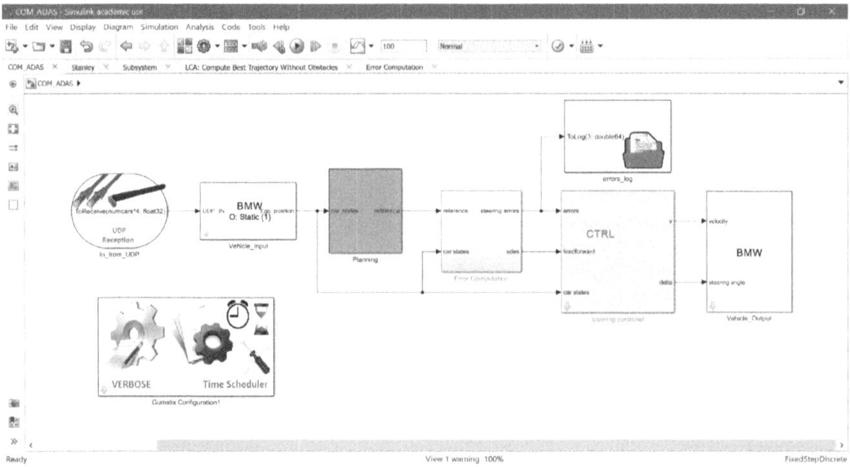

Fig. 13.24 Simulink block diagram with inputs and outputs (white blocks), the planning level (dark gray), error computation (gray with gradient), and the tracking controllers (gray). The RT-MaG toolbox configuration block is shown in the bottom left corner, and a data logging block (folder icon) is used

preemption of the Linux kernel and thus the OS exhibits hard real-time behavior. In order to guarantee that the ADAS code generated in MATLAB/Simulink is executed in real time, additional software is necessary, since the BeagleBone Black Support Package from MATLAB is not real-time capable. Thus, the RT-MaG Toolbox [20] has been used for multi-rate real-time applications.

MATLAB/Simulink

All ADAS functionalities are implemented in MATLAB/Simulink as shown in Fig. 13.24. Inputs to the Simulink models are available at the UDP input blocks, and the outputs are generated with the BeagleBone Black Support package for Simulink.

Since there are typically different people working on the testbed, a Simulink library has been created. Longitudinal and lateral controllers, different planning algorithms, computations for input or actuation, and also kinematic or bicycle vehicle models for simulation studies are included and can be extended by all users.

Data Logging

Data is logged locally on each BeagleBone Black using the RT-MaG toolbox for MATLAB/Simulink and can be easily transferred to the ADAS computer via SSH connection. Since it is also possible to log the CPU execution times of the different tasks, the real-time capability of the generated code on the BeagleBone Black can be analyzed easily.

Car-to-Car Communication

The RT-MaG Toolbox originally allows to use only one UDP input and one UDP output block, where the same IP address for both blocks has to be used. Thus, necessary adaptations in the MATLAB code of the toolbox that generates the C code have been made. With this adapted version, several input blocks and different IP addresses can be used, and it is possible to receive both the position data from the Position Tracking Computer and information of the preceding vehicle. Each vehicle can finally send its desired acceleration to the following vehicle.

13.3.2 Results

The results of sliding-mode-based longitudinal controllers are presented in this section. First, the results of the ATH with first-order sliding-mode control are shown. Then, the resulting oscillations are analyzed and the time-headway is chosen so that string stable behavior can be observed.

13.3.2.1 Platooning with Nonzero Initial Spacing Errors

Figure 13.25 shows the results using the ATH with a FOSMC. Since the sliding variable in Fig. 13.25a exceeds a specified tolerance of the boundary layer several times, the time-headway in Fig. 13.25b is reset several times.

The velocity and the computed acceleration of the vehicle are shown in Fig. 13.26. Due to measurement noise and packet dropouts, the velocities of both agents vary strongly and hence, the system has to deal with large perturbations and uncertainties. However, the agent is capable of following the leader while avoiding collisions, which can be concluded from the phase plane of the errors in Fig. 13.25c. The ATH is real-time capable with a sampling time of $T_{\text{ctrl}} = 10$ ms as depicted in Fig. 13.25d. Note, however, that the time-headway is reset, if the sliding variable exceeds a threshold of $\sigma_T = 0.5$. Otherwise, oscillations are amplified and collisions occur in the experiments due to higher-order dynamics. Note that this happens only if the time-headway is below a certain value; hence, setting a minimum time-headway is reasonable and will remove these effects.

13.3.2.2 Platooning Considering Actuator Dynamics

Experimental results of a platoon have been presented in [30] and are shown in Fig. 13.27 for two following agents and a leader with constant velocity using a constant distance spacing with $\Delta_i = 1$ m using a FOSMC with $k = 0.2$. Oscillations in the sliding variable occur due to unmodeled actuator dynamics and additional effects on the testbed, e.g., discretization effects, time delays, and measurement

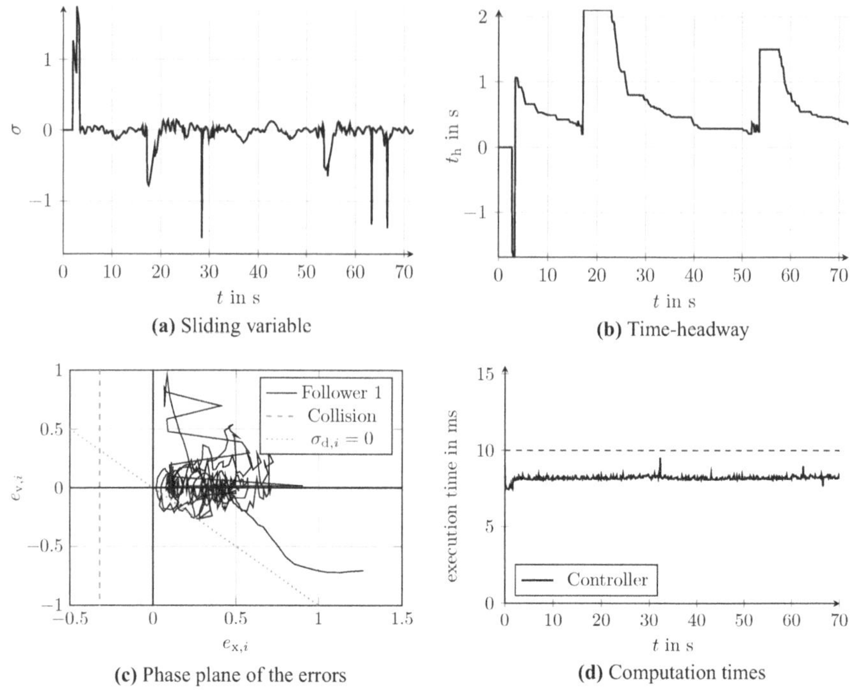

Fig. 13.25 Results of one agent approaching the leader using the ATH with a FOSMC

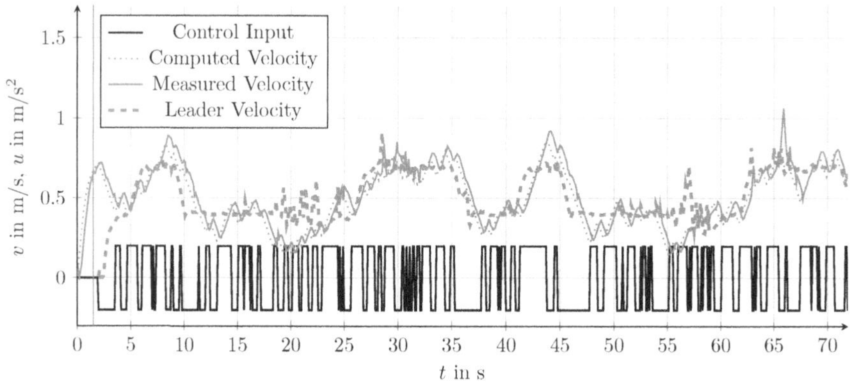

Fig. 13.26 Computed velocity and control input in the experiment of the ATH with a FOSMC

noise. These oscillations can be observed in the velocities and the distances between the vehicles, and the amplitudes of the oscillations of the second follower are much larger than the amplitudes of the first vehicle's oscillations. Although no collisions occur for two following agents, these amplifications will result in collisions for a larger number of vehicles, and the platoon is string unstable.

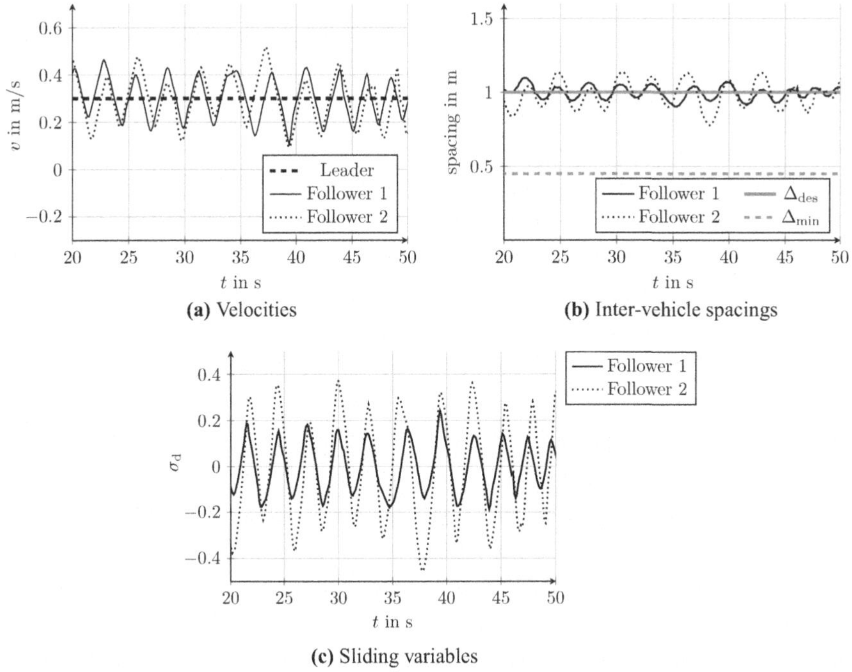

Fig. 13.27 Results of the platooning tests with a FOSMC and the constant distance spacing: amplification of oscillations along the string cannot be avoided

The results of the constant time-headway spacing with three following agents and a constant leader's velocity using a FOSMC are shown in Fig. 13.28. Again, oscillations arise and string stability cannot be concluded in the experiments.

Higher-Order Actuator Dynamics

The bode plots of the closed-loop system with equivalent gain are displayed in Fig. 13.30. Therein, both controllers yield string stable systems for $t_h \geq 2$. For $t_h = 1$, the FOSMC results in string unstable behavior, while the SOC yields string stable behavior. These results are summarized in Table 13.2, with a_{nyq} the amplitude found in the Nyquist plot, a_{sim} the amplitude of the sliding variable in simulation and a_{exp} the amplitude from the sliding variable plots obtained from experiments (Fig. 13.29). It can be seen that string stability can be checked in presence of oscillations. Experimental results of the FOSMC with $t_h = 1$ are shown in Fig. 13.31, where collisions occur, since the spacings and velocities are amplified along the string. Increasing the time-headway to $t_h = 2$ yields the string stable performance shown in Fig. 13.32. The errors are not propagated along the string for a time-varying leader's reference. Finally, the results of the SOC are shown in Fig. 13.33, and string stable behavior can be achieved even for $t_h = 1$.

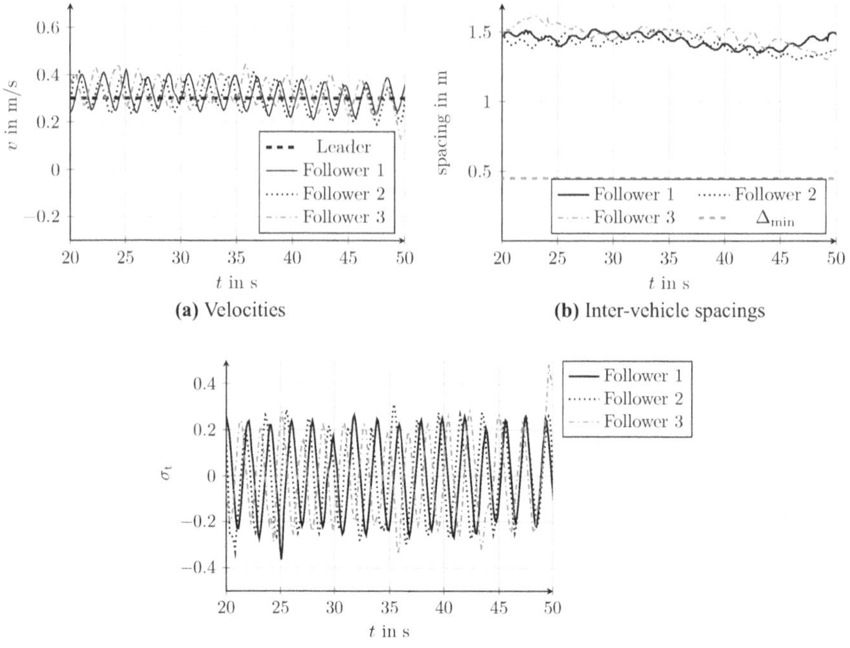

(a) Velocities

(b) Inter-vehicle spacings

(c) Sliding variables

Fig. 13.28 Results of the platooning tests with three following agents using a FOSMC and the constant time-headway spacing with $t_\mathrm{h} = 3\,\mathrm{s}$: oscillations occur due to unmodeled dynamics, but are not amplified along the string

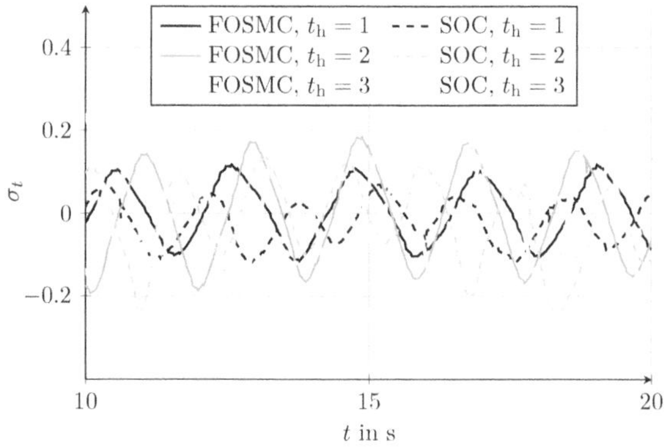

Fig. 13.29 Sliding variables of agent 1 in the experiments

13.3.2.3 Formation Control

Robustness to both different lanes and lane changes have been tested in the experiments with $k = 0.6$ and $t_\mathrm{h} = 3\,\mathrm{s}$ with a minimum distance of $\Delta_i = 0.5\,\mathrm{m}$, where the current lane of each vehicle is shown in Fig. 13.34 and the results are shown

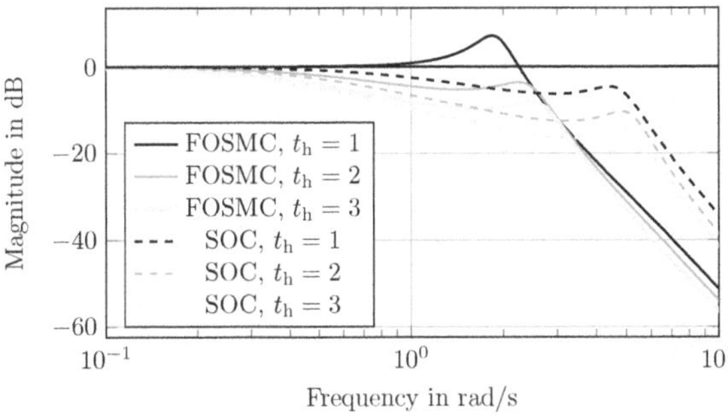

Fig. 13.30 Bode plots of closed-loop system with equivalent gain with different time-headway spacings in the experiment

Table 13.2 Oscillations in the experiments

Control	t_h	a_{nyq}	a_{sim}	a_{exp}	String stable		
					Bode	Sim.	Exp.
FOSMC	3	0.115	0.118	0.22	Yes	Yes	Yes
	2	0.085	0.09	0.15	Yes	Yes	Yes
	1	0.064	0.065	0.1	No	No	No
SOC	3	0.032	0.036	0.15	Yes	Yes	Yes
	2	0.023	0.025	0.1	Yes	Yes	Yes
	1	0.0148	0.015	0.05	Yes	Yes	Yes

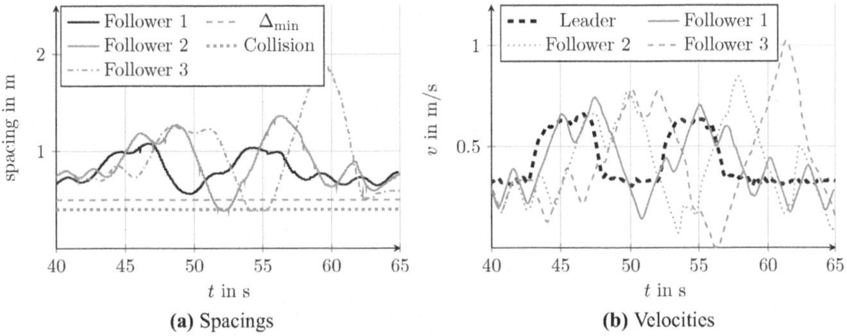

Fig. 13.31 String unstable platooning with FOSMC and $t_h = 1$

in Fig. 13.35. Note that Followers 1 and 3 have to increase the velocities due to a larger waylength on the outer lanes and during lane changes. No additional effects in the sliding variable occur, as displayed in Fig. 13.36. Note that the amplitude of the oscillations is larger due to a larger controller parameter k.

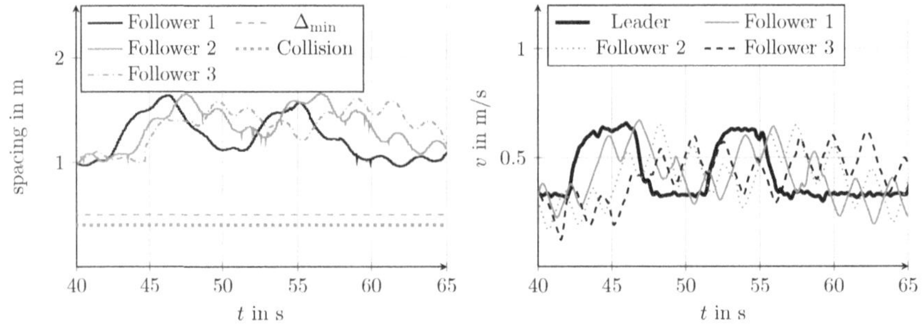

(a) Spacings (b) Velocities

Fig. 13.32 String stable platooning with FOSMC and $t_h = 2$

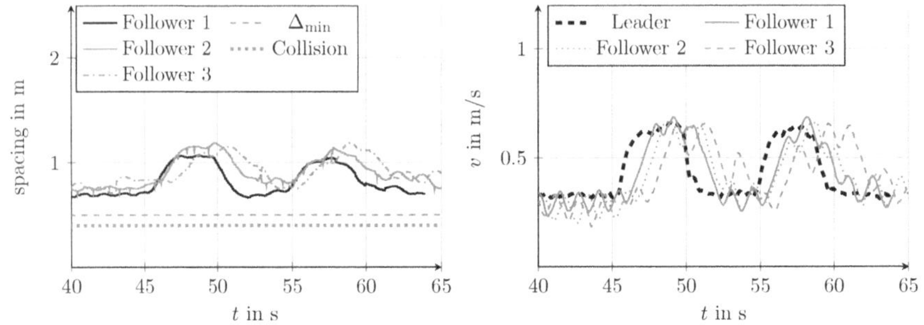

(a) Spacings (b) Velocities

Fig. 13.33 String stable platooning with SOC and $t_h = 1$

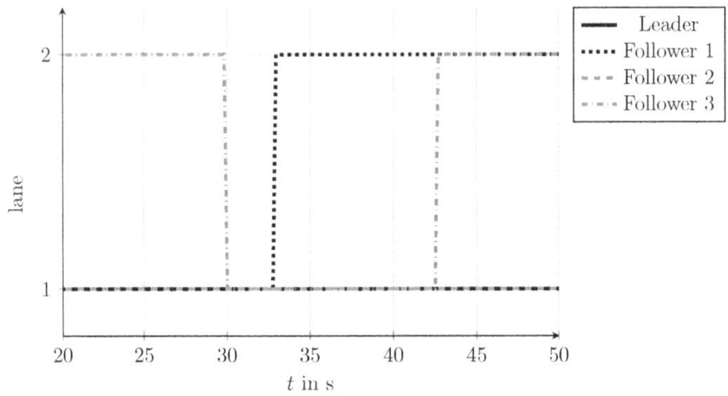

Fig. 13.34 Lanes of the vehicles, where lane 2 has a larger waylength

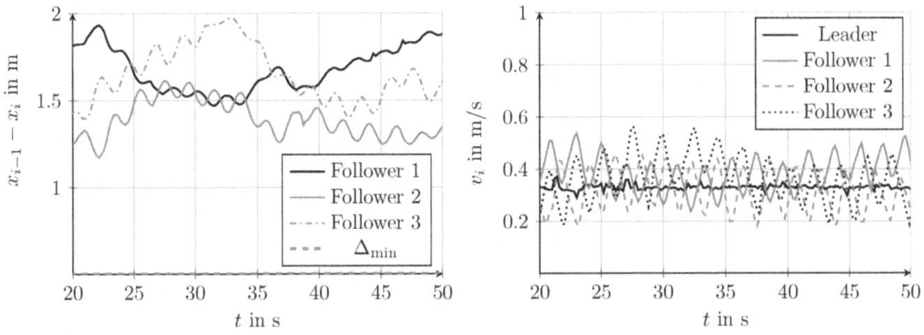

(a) Spacings **(b)** Velocities

Fig. 13.35 Results of the lane change experiment

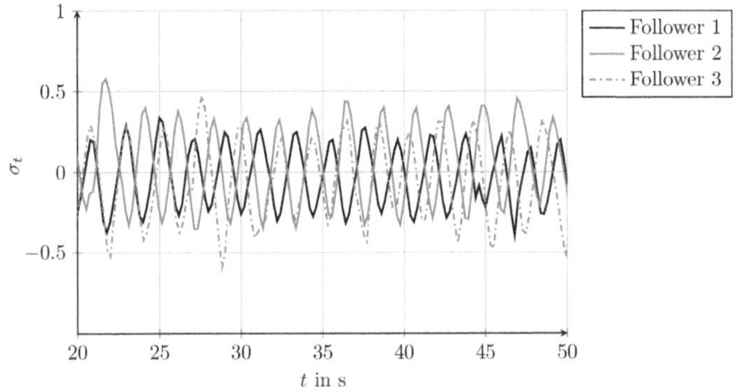

Fig. 13.36 Sliding variables of the following vehicles during lane changes

Large oscillations occur when sliding-mode-based approaches are used on the small-scale vehicles due to slow actuator dynamics and additional unmodeled effects, such as time delays. These effects are subject to future work.

13.4 Conclusion and Future Work

In this work, sliding-mode-based controllers have been outlined and tested on small-scale vehicles. While the constant distance spacing may achieve string stable behavior for double integrator dynamics in simulations, vehicles in practice have unmodeled actuator dynamics that require a minimum time-headway for string stability. Experiments show that a minimum time-headway can be found for the small-scale vehicles, even if the sliding variable oscillates. These oscillations have to be handled in future work. Moreover, cooperative driving, e.g., platoon merging maneuvers, will be tested on the small-scale testbed.

Acknowledgements This work was accomplished in cooperation with the VIRTUAL VEHICLE Research Center in Graz, Austria, supported by the industrial partners AVL List GmbH and MAGNA STEYR Engineering AG & Co KG. The authors would like to acknowledge the financial support of the Austrian COMET K2 - Competence Centers for Excellent Technologies Programme of the Austrian Federal Ministry for Transport, Innovation and Technology (bmvit), the Austrian Federal Ministry of Science, Research and Economy (bmwfw), the Austrian Research Promotion Agency (FFG), the Province of Styria, and the Styrian Business Promotion Agency (SFG).

ADOLABO
Graz University of Technology

References

1. Automated driving lab, institute of automation and control. https://www.tugraz.at/en/institutes/irt/automated-driving-lab/videos/ (2018). Accessed 24 Jan 2018
2. KTH Smart Mobility Lab. https://www.kth.se/en/ees/omskolan/organisation/avdelningar/ac/research/control-of-transport/smart-mobility-lab/smart-mobility-lab-1.441539 (2017). Accessed 19 Oct 2017
3. Amodeo, M., Ferrara, A., Terzaghi, R., Vecchio, C.: Slip control for vehicles platooning via second order sliding modes. In: 2007 IEEE Intelligent Vehicles Symposium, pp. 761–766 (2007)
4. Bartolini, G., Ferrara, A., Usai, E.: Output tracking control of uncertain nonlinear second-order systems. Automatica **33**(12), 2203–2212 (1997)
5. Boiko, I.: Discontinuous Control Systems: Frequency-Domain Analysis and Design. Springer, Berlin (2008)
6. Boiko, I., Fridman, L., Pisano, A., Usai, E.: Analysis of chattering in systems with second-order sliding modes. IEEE Trans. Autom. Control **52**(11), 2085–2102 (2007)
7. Boiko, I., Fridman, L., Pisano, A., Usai, E.: On the transfer properties of the "generalized sub-optimal" second-order sliding mode control algorithm. IEEE Trans. Autom. Control **54**(2), 399–403 (2009)
8. Eigel, T.: Integrierte Längs-und Querführung von Personenkraftwagen mittels Sliding-Mode-Regelung. Ph.D. thesis, Technische Universität Braunschweig (2009)
9. Eyre, J., Yanakiev, D., Kanellakopoulos, I.: String stability properties of AHS longitudinal vehicle controllers. In: Transportation Systems, pp. 71–76 (1997)
10. Eyre, J., Yanakiev, D., Kanellakopoulos, I.: A simplified framework for string stability analysis of automated vehicles. Vehicle Syst. Dyn. **30**(5), 375–405 (1998)
11. Ferrara, A., Incremona, G.P.: Sliding modes control in vehicle longitudinal dynamics control. In: Advances in Variable Structure Systems and Sliding Mode Control - Theory and Applications, pp. 357–383. Springer (2018)
12. Ferrara, A., Librino, R., Massola, A., Miglietta, M., Vecchio, C.: Sliding mode control for urban vehicles platooning. In: 2008 IEEE Intelligent Vehicles Symposium, pp. 877–882 (2008)
13. Ferrara, A., Vecchio, C.: Controlling a platoon of vehicles via a second order sliding mode approach. IFAC Proc. **39**(12), 439–444 (2006)
14. Ferrara, A., Vecchio, C.: Second order sliding mode control of vehicles with distributed collision avoidance capabilities. Mechatronics **19**(4), 471–477 (2009). Robotics and Factory of the Future, New Trends and Challenges in Mechatronics
15. Guo, X., Wang, J., Liao, F., Teo, R.S.H.: CNN-based distributed adaptive control for vehicle-following platoon with input saturation. IEEE Trans. Intell. Transp. Syst. **19**(10), 3121–3132 (2018)

16. Guo, X., Wang, J., Liao, F., Teo, R.S.H.: Distributed adaptive integrated-sliding-mode controller synthesis for string stability of vehicle platoons. IEEE Trans. Intell. Transp. Syst. **17**(9), 2419–2429 (2016)
17. Khalil, H.K.: Nonlinear Systems. Prentice-Hall, Upper Saddle River (2002)
18. Klinge, S., Middleton, R.H.: Time headway requirements for string stability of homogeneous linear unidirectionally connected systems. In: Proceedings of the 48th IEEE Conference on Decision and Control, 2009 held jointly with the 2009 28th Chinese Control Conference. CDC/CCC 2009, pp. 1992–1997. IEEE (2009)
19. Kwon, J., Chwa, D.: Adaptive bidirectional platoon control using a coupled sliding mode control method. IEEE Trans. Intell. Transp. Syst. **15**(5), 2040–2048 (2014)
20. Manecy, A., Marchand, N., Viollet, S.: RT-MaG: an open-source SIMULINK toolbox for linux-based real-time robotic applications. In: 2014 IEEE International Conference on Robotics and Biomimetics (ROBIO 2014). Institute of Electrical and Electronics Engineers (IEEE) (2014)
21. Marzbani, H., Jazar, R., Fard, M.: Better road design using clothoids. In: Denbratt, I., Subic, A., Wellnitz, J. (eds.) Sustainable Automotive Technologies 2014. Lecture Notes in Mobility, Chapter 3, pp. 25–40. Springer, Berlin (2015)
22. Olson, E.: AprilTag: a robust and flexible visual fiducial system. In: 2011 IEEE International Conference on Robotics and Automation. Institute of Electrical and Electronics Engineers (IEEE) (2011)
23. Öncü, S., van de Wouw, N., Heemels, W.P. M.H., Nijmeijer, H.: String stability of interconnected vehicles under communication constraints. In: 2012 IEEE 51st IEEE Conference on Decision and Control (CDC), pp. 2459–2464. IEEE (2012)
24. Ploeg, J., Shukla, D.P., van de Wouw, N., Nijmeijer, H.: Controller synthesis for string stability of vehicle platoons. IEEE Trans. Intell. Transp. Syst. **15**(2), 854–865 (2014)
25. Richardson, A., Strom, J., Olson, E.: AprilCal: assisted and repeatable camera calibration. In: 2013 IEEE/RSJ International Conference on Intelligent Robots and Systems. Institute of Electrical and Electronics Engineers (IEEE) (2013)
26. Rupp, A.: Trajectory Planning and Formation Control for Automated Driving on Highways. Ph.D. thesis, Graz University of Technology (2018)
27. Rupp, A., Reichhartinger, M., Horn, M.: String stability analysis for sliding mode controllers in platoons with unmodeled actuator dynamics: a frequency domain approach. In: 2019 European Control Conference (ECC) (2019)
28. Rupp, A., Steinberger, M., Horn, M.: Sliding mode based platooning with non-zero initial spacing errors. IEEE Control Syst. Lett. **1**(2), 274–279 (2017)
29. Rupp, A., Tranninger, M., Wallner, M., Zubaca, J., Steinberger, M., Horn, M.: Fast and low-cost testing of advanced driver assistance systems using small-scale vehicles. In: Proceedings of the 9th IFAC Symposium on Advances in Automotive Control (AAC 2019) (2019)
30. Rupp, A., Wallner, R., Koch, R., Reichhartinger, M., Horn, M.: Sliding mode based platooning with actuator dynamics. In: 2018 15th International Workshop on Variable Structure Systems (VSS) (2018)
31. Seiler, P., Pant, A., Hedrick, K.: Disturbance propagation in vehicle strings. IEEE Trans. Autom. Control **49**(10), 1835–1842 (2004)
32. Swaroop, D., Hedrick, J.K.: String stability of interconnected systems. IEEE Trans. Autom. Control **41**(3), 349–357 (1996)
33. Utkin, V., Lee, H.: Chattering problem in sliding mode control systems. In: International Workshop on Variable Structure Systems, 2006. VSS'06., pp. 346–350 (2006)
34. Watzenig, D., Horn, M. (eds.): Automated Driving 2016: Safer and More Efficient Future Driving. Springer, Berlin (2017)
35. Yanakiev, D., Kanellakopoulos, I.: A simplified framework for string stability analysis in AHS. In: Proceedings of the 13th IFAC World Congress, vol. 182, pp. 177–182 (1996)
36. Zhou, J., Peng, H.: String stability conditions of adaptive cruise control algorithms. IFAC Proc. Vol. **37**(22), 649–654 (2004). IFAC Symposium on Advances in Automotive Control 2004, Salerno, Italy, 19–23 April 2004

Chapter 14
Single-Loop Integrated Guidance and Control Using High-Order Sliding-Mode Control

Michael A. Cross and Yuri B. Shtessel

Abstract With threats becoming more sophisticated, improvements at the sensor and subsystem level may not be enough for missile interceptors. In this chapter, a single-loop integrated guidance and control (G&C) intercept strategy is introduced that establishes a direct link between the sensor feedback and the fin actuators. By only requiring the normal relative velocity for feedback, target acceleration limits, and the interceptor's input–output relative degree controller design is simple. Robustness is also evident from the black-box treatment of the interceptor. The single-loop structure reduces system complexity by requiring less feedback sensors. This enables operation in sensor-denied areas and improves vehicle maneuverability from the weight reduction. An added benefit is its ability to avoid instability from spectral separation between the guidance and autopilot subsystem control loops. These capabilities stem from the high-order sliding-mode controller. Simulations compare it to a baseline G&C system, and then its practical implementation is tested through unmodeled actuator dynamics of various speeds.

14.1 Introduction

In this chapter, a single-loop guidance and control (G&C) structure is introduced for a missile interceptor. Despite its unconventional topology, it efficiently minimizes miss distances using a traditional intercept strategy and a sophisticated high-order sliding-mode controller. The result is a high-performing system with a simple structure. It is

M. A. Cross (✉) · Y. B. Shtessel
Department of Electrical Engineering, University of Alabama in Huntsville, Huntsville, AL 35763, USA
e-mail: michael.cross@dynetics.com

Y. B. Shtessel
e-mail: shtessel@eng.uah.edu

M. Steinberger et al. (eds.), *Variable-Structure Systems and Sliding-Mode Control*, Studies in Systems, Decision and Control 271,
https://doi.org/10.1007/978-3-030-36621-6_14

easy to design with a near black-box consideration toward the interceptor and target only requiring minimal state feedback.

This chapter consists of five sections. The introduction notes critical developments in guidance and control, integrated guidance and control pertaining to missile interceptors, and single-loop G&C. The problem formulation introduces the relevant interceptor equations of motion and formalizes the problem. The methodology examines how sliding-mode control is used to aid in solving the problem. The simulation results show both the formal application of the methodology and compare this approach to a traditional G&C intercept strategy. Unmodeled dynamics and other effects are also considered in this section. Finally, the conclusion examines the simulation results and identifies potential improvements for this technology.

14.1.1 Traditional Guidance and Control

Guidance and control (G&C) has traditionally been treated as two separate processes. The Lark program beginning shortly after World War II solidified this concept. Lark was the first *homing* missile system using onboard active guidance. It was developed by the U.S. Navy in 1944. What makes the program special is that nearly every interceptor developed over the next six decades used the proportional navigation guidance principles first introduced in the program. It was also the first application of the three-loop guidance and control design structure [1].

Multi-loop structures have many names including successive loop and cascaded loop. Within the aerospace community, the block diagram in Fig. 14.1 is commonly referred to as the *three-loop autopilot*. In it, the guidance is synthesized using engagement kinematics while the autopilot stabilizes the airframe dynamics and tracks acceleration commands provided by the guidance. This creates a successive control loop structure where the inner-loop handles the high bandwidth (autopilot) processes and the outer-loop handles the lower bandwidth (guidance) processes. Designs are developed separately and then assembled. When the overall system performance is inadequate, the subsystems are separated and redesigned. The process is iterative because engineer's design is rooted by assumptions from their counterpart. This design process is costly, time-consuming, and does not ensure satisfactory results. The large number of sensory feedback required to achieve success is another drawback to this structure.

Despite these topology issues, most of the advances made in G&C since Lark have happened at the subsystem level. Advances in homing guidance include proportional navigation (PN), commanded line-of-sight (CLOS), pure proportional navigation (PPN), true proportional navigation (TPN), augmented proportional navigation (APN), beam-rider, zero-effort miss (ZEM), and GENEX just to name a few [2–7]. Some of the more advanced guidance algorithms listed not only achieve intercept but also control the impact angle of the missile interceptor at impact. Despite their complexities, all of these algorithms are rooted in the notion of the *collision triangle*, in which minimizing the line-of-sight change between the interceptor and target

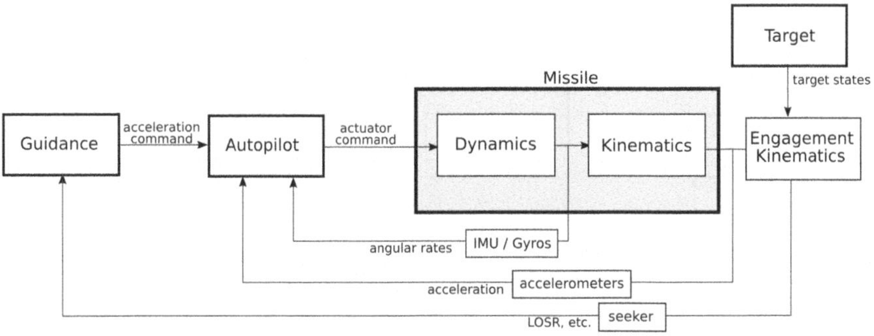

Fig. 14.1 Three-loop autopilot block diagram

leads to a collision. With the advent of the digital computer, many of the sensors and modern control strategies theorized for autopilot design became possible including model predictive control. Though the subsystem technology advances were made for guidance and control, little was done concerning the original structure.

14.1.2 Integrated Guidance and Control

In contrast to the traditional three-loop autopilot structure, integrated guidance and control (IGC) is a unified framework where guidance and control is considered together instead of independently. A block diagram of an IGC is shown in Fig. 14.2. An advantage of IGCs are the following:

- Synergy—capitalize on interactions between the guidance and control subsystems by exploiting synergies between their processes.
- Direct Optimization—lumping the G&C into a single framework improves the optimization potential while reducing the amount of design iterations.
- Flexibility—the structure allows more or less sensor feedback paths to be established. Very beneficial when considering things like GPS denied environments.
- Cost & Time Savings—the potential for fewer design iterations.
- Improved Performance—improve stability margin for spectrally separated subcomponents.

An example of guidance and control *synergy* can be found by considering the block diagrams of the conventional missile guidance and control system and IGC in Fig. 14.2. In the conventional approach, the guidance law has no knowledge of the rotational rates or the sensed acceleration of the missile. Instead, the guidance only knows the relative position and velocity of the engagement. As the range to target decreases, small changes in geometry lead to large acceleration commands that can exceed the autopilot performance limits. If integral feedback loops are used in the autopilot for increased steady-state tracking, instability can occur. Furthermore,

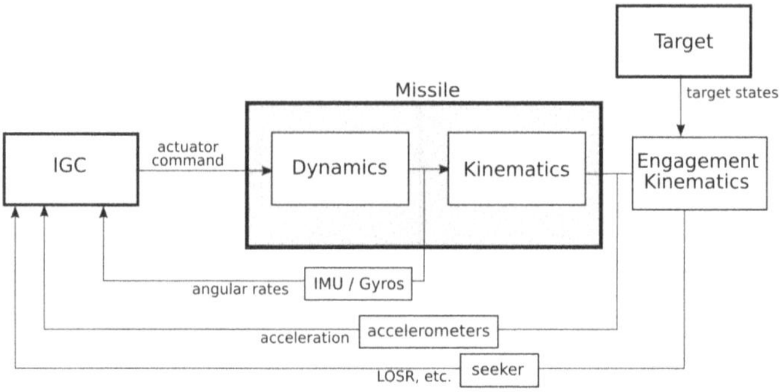

Fig. 14.2 Integrated guidance and control (IGC) block diagram

autopilots can not adjust themselves based on relative engagement kinematics. As a result, conventional G&C systems have relied on driving the autopilot time constant as small as possible to improve stability. The autopilot's time constant dictates the miss distance in conventionally designed G&C systems [1, 8]. This instability is caused by the spectral separation between the two subsystems.

Advances in IGC have come through a variety of methods including the game-theoretic approach [9, 10], state-dependent Riccati equation (SDRE) [11–13], back-stepping [13, 14], and sliding-mode control (SMC) [15, 16]. These have seen varying levels of success, but most require strong knowledge of the interceptor's dynamic properties for model inversion, more complete knowledge of the target, or a bevy of sensor feedback. The algorithms also have a tendency of being complex.

14.1.3 Single-Loop Guidance and Control

Single-loop guidance and control is a subset of IGC. It offers the same advantages of a standard IGC systems. Additionally, it offers the opportunity to control the vehicle using a single feedback path. This path makes the vehicle less reliant on standard sensors, less expensive to manufacturer, smaller, lighter, and easier to design. These wonderful features come at a steep price—a high input–output relative degree. Sliding-mode control is one of the few control schemes capable of handling this issue. An example of single-loop block diagram is shown in Fig. 14.3.

Shima was the first to use SMC in a *single-loop* IGC structure. Unlike Shtessel and many other authors, in [17, 18], Shima chose the zero-effort miss distance (ZEM) as the sliding variable. He favored it for its meaningful measure of miss distance in a two-sided differential game problem and it reduces an n-dimensional guidance problem to a scalar one. By estimating time-to-go and other measurements, the ZEM sliding surface is able to reduce to a relative degree of one with respect to the fin actuator. Shima was able to demonstrate postponement of instability near terminal

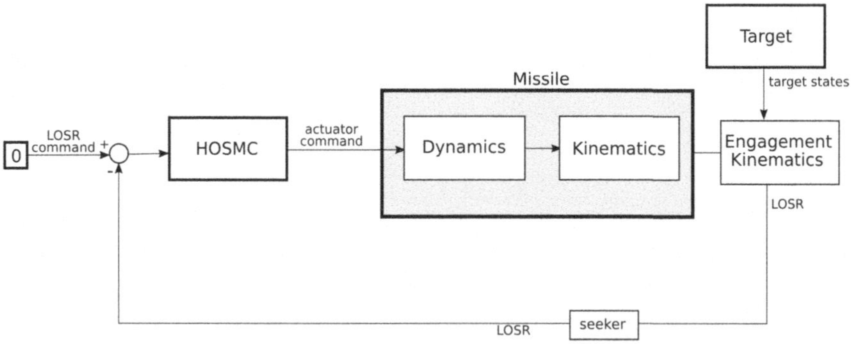

Fig. 14.3 Single-loop autopilot block diagram

time. It especially showed promise against highly maneuverable targets. Idan, the coauthor of Shima's papers, extended this concept to on–off actuators in [19].

A second approach toward IGC using SMC was presented in [20] by Harl and Balakrishnan using terminal second-order sliding-mode (TSM) control. This control law was first developed by the authors in [21]. Contrary to high-order sliding-mode (HOSM) control, TSM relies on the finite-time reaching phase to meet the surface at final time, rather than relying on the sliding phase to hold the sliding variable to the sliding surface. The advantage of TSM is in its ability to shape the sliding variable's trajectory to the surface. This is a significantly different approach than previous sliding-mode control techniques. The TSM algorithm guides toward a predicted impact point (PIP). Two heading errors are defined to guide toward the PIP. The system has a relative degree of three, with a system output of heading error and an input of fin deflection. The TSM control law is generated using an estimate of time-to-go and accounts for the high relative degree. The study showed the method's feasibility, but did not compare it to other methods. Later, in [22] the authors expanded this method to include terminal impact angle as a desired end-game parameter.

Shima and Harl have shown that sliding-mode control is a capable method for handling single-loop guidance and control. There is much more to be explored in this area, especially when considering the benefits of higher-order sliding-mode control techniques which neither author has used. Thus, it is the purpose of this chapter to introduce an integrated guidance and control strategy that has a simple, single-loop structure, that is easy to design, and greatly improves miss distances over conventional methods. All, while requiring minimal state feedback including, and a near black-box consideration of the interceptor and target. This problem will be developed, and high-order sliding-mode control strategies will be covered before implementation and testing.

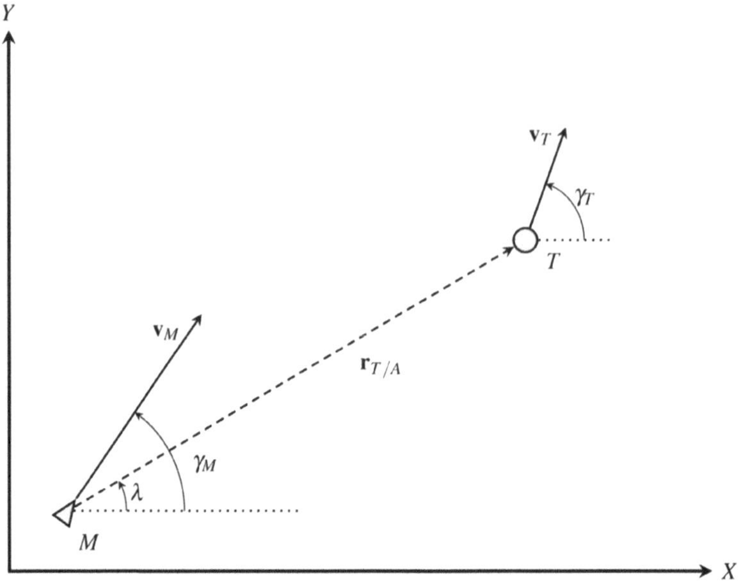

Fig. 14.4 Missile engagement geometry during homing phase of flight

14.2 Problem Formulation

The missile-target engagement scenario consists of a missile M attempting to intercept a target T via heading changes. During homing guidance, onboard sensors are used to steer the vehicle until collision occurs. It is assumed midcourse guidance is successful and the velocities of the missile v_M and target v_T are on a near-collision course. These become the initial conditions to this problem. Figure 14.4 shows how this is arranged in Euclidean space. Since the target maneuvers and the initial heading is never completely correct, course corrections are required to achieve intercept. This section formulates the missile-target engagement problem including all subject matter required for accurate modeling including the engagement kinematics, missile dynamics, integrated kinematic-dynamic state-space model, intercept strategy, and the input–output relative degree.

14.2.1 Engagement Kinematics

Kinematics is the branch of mechanics that studies the motion of objects without considering the forces that cause the motion. It emphasizes the geometric properties of the problem, beginning with the positioning of the objects and ending with expressions for their velocities and accelerations.

Consider the engagement geometry where the missile M and target T, are specified by the position vectors $\boldsymbol{r}_M$ and $\boldsymbol{r}_T$ Fig. 14.4. These vectors are measured from the fixed $\boldsymbol{X}, \boldsymbol{Y}, \boldsymbol{Z}$ coordinate system. The line-of-sight (LOS) vector, $\boldsymbol{r}_{T/M}$, specifies the target position relative to the missile. If it is measured from the fixed coordinate system, then the three vectors can simply be related through vector addition by the equation

$$\boldsymbol{r}_T = \boldsymbol{r}_M + \boldsymbol{r}_{T/M}. \tag{14.1}$$

Since all three vectors are measured with respect to an inertial coordinate system, differentiation yields absolute velocities $\boldsymbol{v}_T = \boldsymbol{v}_M + \boldsymbol{v}_{T/M}$. Similarly, the accelerations become $\boldsymbol{a}_T = \boldsymbol{a}_M + \boldsymbol{a}_{T/M}$. Unfortunately, many of the states that are tracked in practical systems are measured with respect to moving vehicles. Therefore, a rotating coordinate system $\boldsymbol{r}, \boldsymbol{n}, \boldsymbol{\omega}$ is required to capture these kinematics. This coordinate system is fixed to the missile and assumed to always have its primary axis pointing along the line-of-sight. Meaning, it translates and rotates with respect to the fixed inertial $\boldsymbol{X}, \boldsymbol{Y}, \boldsymbol{Z}$ system.

Differentiating Eq. (14.1) twice yields the relative velocity of

$$\boldsymbol{v}_T - \boldsymbol{v}_M = \dot{r}\hat{\boldsymbol{r}} + r\Omega_\omega\hat{\boldsymbol{n}} \tag{14.2}$$

and the relative acceleration of

$$\boldsymbol{a}_T - \boldsymbol{a}_M = (\ddot{r} - r\Omega_\omega^2)\hat{\boldsymbol{r}} + (r\dot{\Omega}_\omega + 2\dot{r}\Omega_\omega)\hat{\boldsymbol{n}} + (r\Omega_r\Omega_\omega)\hat{\boldsymbol{\omega}}, \tag{14.3}$$

where r is the scalar range between the missile and target, and $\Omega_\omega = \dot{\lambda}$ is the line-of-sight rate. When these two equations are written as components for $\boldsymbol{r}$ and $\boldsymbol{n}$, four equations emerge

$$(\boldsymbol{v}_T - \boldsymbol{v}_M)\cdot\hat{\boldsymbol{r}} = \dot{r}\hat{\boldsymbol{r}} \tag{14.4a}$$

$$(\boldsymbol{a}_T - \boldsymbol{a}_M)\cdot\hat{\boldsymbol{r}} = \ddot{r} - r\Omega_\omega^2\hat{\boldsymbol{r}} \tag{14.4b}$$

$$(\boldsymbol{v}_T - \boldsymbol{v}_M)\cdot\hat{\boldsymbol{n}} = r\Omega_\omega\hat{\boldsymbol{n}} \tag{14.4c}$$

$$(\boldsymbol{a}_T - \boldsymbol{a}_M)\cdot\hat{\boldsymbol{n}} = r\dot{\Omega}_\omega + 2\dot{r}\Omega_\omega\hat{\boldsymbol{n}}. \tag{14.4d}$$

These form the basis for the kinematic state-space representation for the missile engagement. The velocity of the target relative to the missile within the inertial frame of reference can be shortened to $(\boldsymbol{v}_T - \boldsymbol{v}_M) = \boldsymbol{v}$, and extended to their respective coordinates with subscripting, $\boldsymbol{v}\cdot\hat{\boldsymbol{r}} = v_r$ and $\boldsymbol{v}\cdot\hat{\boldsymbol{n}} = v_n$. These values are very important for the intercept strategy. Furthermore, the derivative for the line-of-sight rate can be expressed as

$$\dot{\Omega}_\omega = \frac{\dot{v}_n}{r} - \frac{v_n\dot{r}}{r^2}. \tag{14.5}$$

Substituting the aforementioned terms into Eq. (14.4a), (14.4b), (14.4c), (14.4d) and arranging to create the state vector $x = [r, v_r, \lambda, v_n]^\top$ produces the kinematic state-space representation of

$$\begin{pmatrix} \dot{r} \\ \dot{v}_r \\ \dot{\lambda} \\ \dot{v}_n \end{pmatrix} = \begin{bmatrix} v_r \\ \frac{v_n^2}{r} \\ \frac{v_n}{r} \\ \frac{-v_r v_n}{r} \end{bmatrix} + \begin{bmatrix} 0 & 0 \\ -1 & 0 \\ 0 & 0 \\ 0 & -1 \end{bmatrix} \begin{pmatrix} a_{M_r} \\ a_{M_n} \end{pmatrix} + \begin{bmatrix} 0 \\ a_{T_r} \\ 0 \\ a_{T_n} \end{bmatrix} \tag{14.6}$$

This representation has the form $\dot{\boldsymbol{x}} = \boldsymbol{a}(\boldsymbol{x}) + \boldsymbol{B}\boldsymbol{u} + \boldsymbol{\rho}$, where $\boldsymbol{x}$ is the state vector, $\boldsymbol{u}$ is the control vector, and $\boldsymbol{\rho}$ is the disturbance. Though this accurately describes the kinematics of the engagement, it neglects the dynamics of the missile and creates an impractical control vector.

14.2.2 Missile Dynamics

As was shown in the engagement kinematics, a missile must accelerate (a_{M_n}) to intercept a maneuvering target. For an endoatmospheric missile, body rotation is the most efficient means of generating the necessary forces to accelerate the vehicle and counteract the target's maneuvering. Therefore, the rotational missile dynamics must be modeled accurately.

As was the case for the engagement kinematics, motion is constrained to the longitudinal plane. Consider the rotated missile body in Fig. 14.5, where $\boldsymbol{x}$, $\boldsymbol{z}$ is the coordinate system attached to the body of the vehicle, $\boldsymbol{U}$ is the primary axis attached to the wind frame, and $\boldsymbol{X}$ is the primary axis attached to the same inertial frame of reference used in the engagement kinematics. The angle of attack α, flight path angle γ, and pitch angle θ, track the orientation of these coordinate systems relative to each other. In this scenario, the angle of attack α leads to an aerodynamic force $\boldsymbol{F}$ on the body at the vehicle's center of pressure. This force is then represented as a set of forces ($\boldsymbol{F}_x$ and $\boldsymbol{F}_z$) and a moment M at the center of mass. The forces ($\boldsymbol{F}_x$ and $\boldsymbol{F}_z$) arise from the aerodynamic force F at the center of pressure, being projected onto the body coordinate system axes.

Deriving a set of dynamic equations from the forces and moments previously described is well understood. It begins as a ninth-order nonlinear system before it is decoupled into a longitudinal and lateral set of equations. Further approximations are made to linearize the systems and separate them into modes. These modes are characterized by their grouping of open-loop poles. Due to the quick reactions required in the engagement scenario, the high-frequency open-loop poles of the short-period mode are most important. The short-period approximation

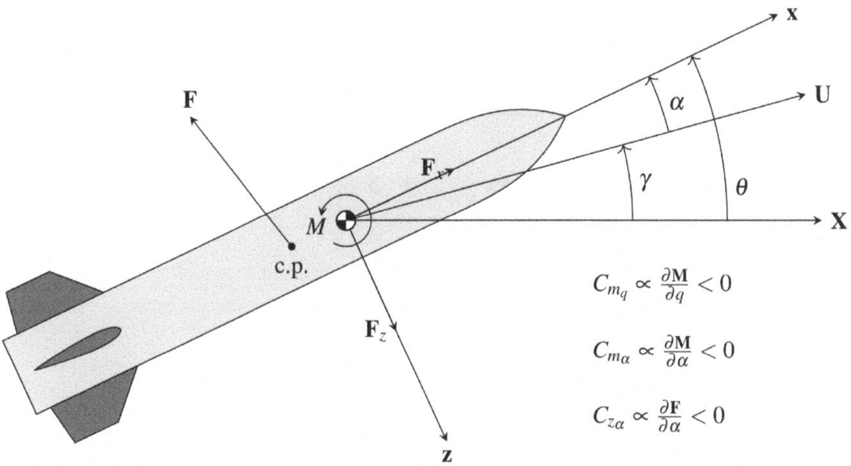

Fig. 14.5 A free body diagram of the aerodynamic forces on a maneuvering missile. The diagram includes all terms necessary for analysis of the free response of the short-period approximation

$$\begin{pmatrix} \dot{\alpha} \\ \dot{q} \\ \dot{\theta} \end{pmatrix} = \begin{bmatrix} \frac{\bar{q}S}{mU_0} C_{z_\alpha} & 1 & 0 \\ \frac{\bar{q}Sl}{I_{yy}} C_{m_\alpha} & \frac{\bar{q}Sl^2}{I_{yy}2U_0} C_{m_q} & 0 \\ 0 & 1 & 0 \end{bmatrix} \begin{pmatrix} \alpha \\ q \\ \theta \end{pmatrix} + \begin{bmatrix} \frac{\bar{q}S}{m} C_{z_\delta} \\ \frac{\bar{q}Sl}{I_{yy}} C_{m_\delta} \\ 0 \end{bmatrix} \delta \tag{14.7}$$

is a modified version of Pamadi's [23], where the stability derivative terms are C_{z_α}, C_{m_α} and C_{m_q}, and the control derivative terms are C_{z_δ} and C_{m_δ}. This system takes the form $\dot{\boldsymbol{x}} = A\boldsymbol{x} + Bu$, a linear state-space representation, in which the state vector is $\boldsymbol{x} = [\alpha, q, \theta]^\top$ and the control vector is $u = \delta$.

The free response of the system, that is, $\dot{\boldsymbol{x}} = A\boldsymbol{x}$ such that $u = 0$, is dependent on the stability derivative terms. Stability and frequency of the short period is dictated by the value of C_{m_α}. A negative value indicates a stable mode, and is confirmed by the center of gravity's forward position with respect to the vehicle's center or pressure The mode's damping is reliant on the magnitude of C_{m_q}, and the vehicle's acceleration effectiveness is determined by C_{z_α}. All of the terms required for free response analysis are conceptualized in Fig. 14.5.

The forced response of the system ($u \neq 0$) requires the stability derivative terms from Fig. 14.5 and the control derivatives from Fig. 14.6. It is important to think of the fins as a *moment* generators and not force generators. The control derivative term C_{m_δ} is responsible for producing that moment. It is a product of the force generated at the fin and the moment arm from fin center of pressure to the vehicle's center of gravity. Provided there is a fixed moment arm length, increasing the surface areas of the actuation fins will lead to greater rotational authority. However, this comes at a cost. Increasing the force also increases the C_{z_δ} term. This term, in tail fin controlled missiles, is critical in analyzing the closed-loop acceleration control,

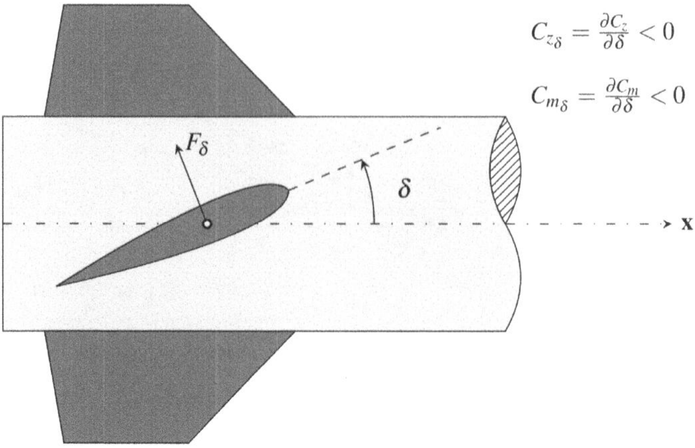

Fig. 14.6 A free body diagram of the aerodynamic forces on the control actuation system of a missile. These terms and diagram are used for analysis of the forced response of the short-period approximation

because it leads to non-minimum phase. In this study, the fin size is considered small enough to approximate $C_{z_\delta} \approx 0$ and a moment arm large enough for $C_{m_\delta} > 0$

Ultimately, the engagement kinematics require information on the missile's acceleration vector. Thus, the state output matrix C from $\boldsymbol{y} = C\boldsymbol{x} + D\boldsymbol{u}$ must be formulated as

$$\begin{pmatrix} \alpha \\ q \\ \theta \\ a_z \end{pmatrix} = \begin{bmatrix} 1 & 0 & 0 \\ 0 & 1 & 0 \\ 0 & 0 & 1 \\ \frac{\bar{q}S}{m} C_{z_\alpha} & 0 & 0 \end{bmatrix} \begin{pmatrix} \alpha \\ q \\ \theta \end{pmatrix} + \begin{bmatrix} 0 \\ 0 \\ 0 \\ \frac{\bar{q}S}{m} C_{z_\delta} \end{bmatrix} \delta, \tag{14.8}$$

where the states are included at the output.

14.2.3 Integrated State-Space Model

The integrated state-space model having the form $\dot{\boldsymbol{x}} = \boldsymbol{a}(\boldsymbol{x}, t) + \boldsymbol{B}\boldsymbol{u}(t) + \boldsymbol{\rho}$ is a combination of Eqs. (14.6) and (14.8). The control variable in from the kinematics $\boldsymbol{a}_T$ is expressed in terms of the new control variable δ. This transformation requires a LOS-to-body coordinate transformation and aerodynamic parameters [23]. Extending the state-space formulation to include the missile dynamics and engagement kinematics results in

$$\begin{pmatrix} \dot{r} \\ \dot{v_r} \\ \dot{\lambda} \\ \dot{v_n} \\ \dot{\alpha} \\ \dot{q} \\ \dot{\theta} \end{pmatrix} = \begin{pmatrix} v_r \\ \frac{v_n^2}{r} + \frac{\bar{q}S}{m} C_{z_\alpha} \alpha \sin(\lambda - \theta) \\ \frac{v_n}{r} \\ \frac{-v_r v_\lambda}{r} + \frac{\bar{q}S}{m} C_{z_\alpha} \alpha \cos(\lambda - \theta) \\ \frac{\bar{q}S}{mU_0} C_{z_\alpha} \alpha + q \\ \frac{\bar{q}Sl}{I_{yy}} C_{m_\alpha} \alpha + \frac{\bar{q}Sl^2}{I_{yy} 2U_0} C_{m_q} q \\ q \end{pmatrix} + \begin{pmatrix} 0 \\ 0 \\ 0 \\ 0 \\ \frac{\bar{q}S}{m} C_{z_\delta} \\ \frac{\bar{q}Sl}{I_{yy}} C_{m_\delta} \\ 0 \end{pmatrix} \delta + \begin{pmatrix} 0 \\ a_{T_r} \\ 0 \\ a_{T_n} \\ 0 \\ 0 \\ 0 \end{pmatrix}. \tag{14.9}$$

This is used extensively in the case studies and simulations that follow. In addition, relative degree of the system and control strategy are based on this. The ρ term identifies uncertain disturbances on the system. A high-order sliding-mode controller is able to compensate for these uncertainties if they are bounded and continuously differentiable. Also, the actuator dynamics are neglected in the controller's design, but their effects are studied more closely in the first case study.

14.2.4 Intercept Strategy

The strategy behind intercept is not a new one. Sailors for hundreds of years have used constant bearing decreasing range (CBDR) as a method for *avoiding* collisions with other ships at sea. The engagement strategy employed in this section uses the same concept but to *intercept* instead of avoid. Missile-target interception occurs when the range-to-target r decreases below the hit range threshold r_{hit}, making $(||r|| < r_{hit})$ a necessary and sufficient condition for interception. In practice, this makes sense because the target has some volume and a warhead has some amount of effectiveness. However, it also important to recognize that as the range-to-target r approaches zero, many of the states from Eq. (14.9) approaches a singularity point from r being in the denominator. Once the range satisfies the threshold, the analysis stops thereby avoiding the singularities.

The range-to-target's time rate of change, $\dot{r}(t)$, is specified by Eq. (14.4a)

$$\dot{r}(t) = v_r(t).$$

A more common term for this value is the *closing velocity*, v_c, where $(v_c \equiv -v_r)$. If the interceptor is initially closing on the target $(v_c(0) > 0)$, and it can maintain a positive value, then the interceptor will inevitably hit the target. Thus, $(v_c > 0, \quad \forall t > 0)$ is a sufficient condition for interception. The closing velocity kinematics

$$\dot{v}_c = \boldsymbol{a}_M \cos(\lambda - \theta)_M - \boldsymbol{a}_T \cdot \hat{\boldsymbol{r}} - \frac{v_n^2}{r}$$

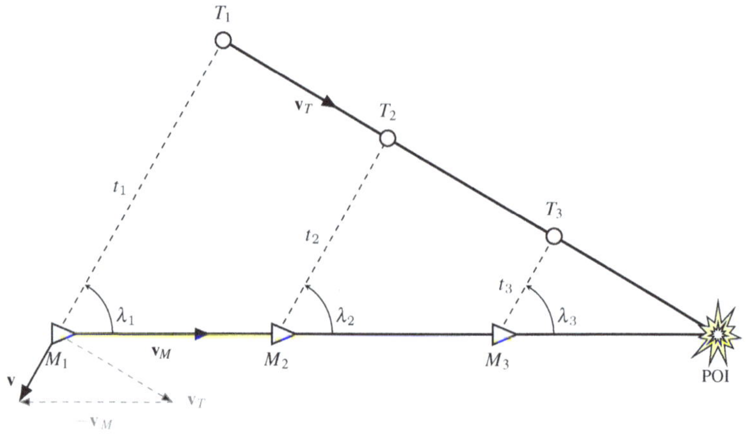

Fig. 14.7 Constant bearing collision course triangle

provide the answer to maintaining a positive rate of closure. The $(\lambda - \theta)$ term refers to the missile's acceleration angle with respect to the LOS. Assuming a non-maneuvering target and the missile acceleration can only act normal to the LOS, then the kinematics simplify to

$$\dot{v}_c = -\frac{v_n^2}{r}.$$

Due to the squared normal relative velocity term, any magnitude leads to a decay in the closing velocity. Geometrically, this means the missile's velocity should be angled such that the vector summation of $\boldsymbol{v} = \boldsymbol{v}_T - \boldsymbol{v}_M$ should be along the LOS but opposite the target's direction. This establishes a collision course, or collision triangle from Fig. 14.7 for a non-maneuvering target. Minimizing the normal relative velocity or line-of-sight rate, its counterpart, is the foundation for nearly all forms of proportional navigation.

Remark 14.2.1 Given an interceptor closing on a target $(v_c(0) > 0)$ with negligible acceleration capability along the line-of-sight $(\boldsymbol{a}_T \cdot \hat{\boldsymbol{r}} = 0)$, if the relative normal velocity v_n can be minimized such that

$$\int_{t_0}^{t_f} \frac{v_n^2(t)}{r(t)} dt < v_c(0)$$

then closing velocity stays positive and interception occurs at t_f.

Thus, one way to achieve intercept for the integrated state-space model from Eq. (14.9) is to control δ such that

$$\delta : v_n \to 0. \tag{14.10}$$

This is the intercept strategy used throughout the duration of this research.

14.2.5 Guidance Strategies

There are many different types of guidance schemes, but none more popular than proportional navigation. It and its many variants all stem from the intercept strategy in Eq. 14.10. This relationship between these is shown by considering the relative normal velocity dynamics

$$\dot{v_n} = -\frac{v_r v_n}{r} - a_{M_n} + a_{T_n}. \tag{14.11}$$

Its Laplace transform, assuming the relative velocity along the line-of-sight v_r and the range-to-target r are constant for small periods of time, and that the missile acceleration is of the form $a_{M_n} = \Lambda\, v_n$, results in the transfer function

$$\frac{v_n(s)}{a_{T_n}(s)} = \frac{1}{s + (\frac{v_r}{r} + \Lambda)}, \tag{14.12}$$

where a $\Lambda > N\frac{V_c}{r}$ for $N > 1$ stabilizes the first-order system. Substituting this stability criteria back into the generalized missile acceleration expression results in

$$a_{M_n} = N\frac{v_c}{r}v_n = N v_c \dot{\lambda}$$

the very definition for proportional navigation. Using a similar process, many of the guidance algorithms can be related back to the simple intercept strategy of Sect. 14.2.4.

14.2.6 The Difficulty of Single-Loop G&C

When designing a controller, the relative degree of the plant is of great importance. Relative degree, simply stated, is the difference in the polynomial degree between the poles and zeros of a transfer function. Systems with high relative degree are harder to control than those with lower relative degree. The PID controller is the most commonly used controller in the world. However, one downside to the controller is that it's only capable of handling systems up to two relative degrees. This makes it incapable of directly implementing the guidance laws to the actuator in a single-loop G&C structure. Instead, additional parameters are required for feedback to ensure the plant does not exceed two relative degrees. Arranging these in a successive loops enables the modelers to break the process into multiple pieces. This inevitably led to the three-loop autopilot structure with a guidance and control loop.

Unlike many of the other techniques, sliding-mode control excels at handling plants with higher relative degree. This enables designers to directly control the actuator through guidance. As stated previously there is great potential for this technique.

14.2.7 Problem Statement

The purpose of this chapter is to design a control δ from Eq. 14.9 using a single-loop structure that forces the relative velocity between the missile and target $\boldsymbol{v}$ in the direction perpendicular to the line-of-sight ($v_n = \boldsymbol{v} \cdot \hat{\boldsymbol{n}}$) to zero in finite time while in the presence of unknown but bounded target maneuvers. ($\boldsymbol{a}_T = [a_{T_r}, a_{T_n}]^\top$).

This approach, unlike proportional navigation, is designed to minimize the magnitude of acceleration at end-game. By forcing v_n to zero in finite time, the missile is only required to mimic the target's motion requiring only a slight acceleration advantage, unlike proportional navigation that requires an acceleration advantage of three to four times the target.

14.3 Methodology

The purpose of this section is to provide information and techniques for making solving the problem statement stated in Sect. 14.2.7. The fundamentals of sliding-mode control, higher-order sliding-mode control, and higher-order sliding-mode differentiators are all covered in this section.

14.3.1 Fundamentals of Sliding-Mode Control

Sliding-mode control (SMC) is a control method [24, 25] that originated from Russia during the 1960s. It is essentially a two-phase controller. The first phase, the reaching phase, occurs as the state trajectory, beginning at its initial condition, moves toward a predefined hyperplane in the state space. The second phase, the sliding phase, occurs after the reaching phase as the state trajectory moves along the hyperplane toward an equilibrium point. The hyperplane, known as the sliding surface, is where this control methodology gets its name. The sliding surface can be designed in a multitude of ways to provide a favorable state trajectory during the sliding phase. Consider the single-input single-output (SISO) system,

$$\dot{\boldsymbol{x}} = a(t, \boldsymbol{x}) + b(t, \boldsymbol{x})u \tag{14.13}$$

$$\sigma = \sigma(t, \boldsymbol{x}), \tag{14.14}$$

where $\boldsymbol{x} \in \boldsymbol{R}^n$ is a state vector, $\sigma \in \boldsymbol{R}^1$ is the sliding variable, $u \in \boldsymbol{R}^1$ is the control input, $a(t, \boldsymbol{x}), b(t, \boldsymbol{x}) \in \boldsymbol{R}^n$ are Lipschitz vector fields, and $\sigma(t, \boldsymbol{x}) \in \boldsymbol{R}^1$ is sufficiently smooth function and understood in the Filippov sense.

The system in Eq. 14.13 is classified by its input–output relative degree r_d. The relative degree is determined by successively differentiating the output σ with respect to time until the control u appears for the first time

$$\sigma^{(r_d)} = h(t, x) + g(t, x)u, \tag{14.15}$$

where $h(t, x) = L_{\tilde{a}}^{r_d}\sigma$ and $g(t, x) = L_{\tilde{b}} L_{\tilde{a}}^{r_d - 1}\sigma \neq 0$. $L\{\cdot\}$ is the Lie derivative operation. Also note that $g(t, x)$ is the multiplicative disturbance and $h(t, x)$ the nonlinear system states that are bounded by the Lipschitz constant L.

The purpose of control u is to finite-time stabilize the sliding set

$$\sigma, \dot{\sigma}, \ldots, \sigma^{(r_d - 1)} = 0 \tag{14.16}$$

for the uncertain system Eq. 14.13. Maintaining this sliding set is essential to constraining the sliding variable to the sliding surface. This concept is fundamental to sliding-mode control. Any continuous control

$$u = \varphi(\sigma, \dot{\sigma}, \ldots, \sigma^{(r_d - 1)}) \tag{14.17}$$

forcing $\sigma = 0$ would first need to satisfy $u = \varphi(0, 0, \ldots, 0) = -h(t, x)/g(t, x)$ from Eq. 14.15. Consider the two systems having the same form as Eq. 14.15

$$\sigma^{(r_d)} = C + K_m u \qquad \text{and} \qquad \sigma^{(r_d)} = -C + K_m u$$

such that $0 < K_m \leq g(t, x) \leq K_M$ and $|h(t, x)| \leq C$. It is clear that there is no continuous control from Eq. 14.17 that will simultaneously satisfy both of these systems. Thus, the control defined in Eq. 14.17 must be *discontinuous* to constrain a state trajectory to the sliding surface [26–28].

14.3.2 High-Order Sliding-Mode Control

Sliding modes are classified by their smoothness degree of the constraint function along the system trajectory. Their classification comes from the sliding-mode $\sigma \equiv 0$ and is defined by the first total time derivative $\sigma^{(r_d)}$ where the discontinuity in the control u (i.e., $\text{sign}(\sigma)$) occurs. The number r is the sliding order of the controller. The controllers are called "finite-time-convergent r-sliding-mode controllers"; however, they are shortened to "r-sliding controller". First, second, and r-order sliding-mode controllers are listed as: 1-sliding, 2-sliding, and r-sliding [27]. The last family represents the arbitrary-order sliding-mode controllers, which is of considerable interest for the single-loop IGC application.

High-order sliding-mode control (HOSMC) refers to sliding-mode controllers that are r-sliding mode greater than two. The two most common are the nested sliding controller and the quasi-continuous sliding controller. The quasi-continuous sliding-mode controller is investigated due to the high frequency leaping characteristic of the nested sliding controller. The method is *quasi-continuous* sliding-mode control (QCSMC) because it is continuous everywhere except on the r-sliding set given by Eq. (14.16). A version of a quasi-continuous controllers is

$$u = -\tilde{\alpha}\Phi_{r-1,r}(\sigma, \dot{\sigma}/\lambda, \ldots, \sigma^{(r-1)/\lambda^{r-1}}).$$

Controllers for $r \leq 4$ are defined below. The $\tilde{\alpha}$ term is based on the magnitude of $h(t, x)$. When $\tilde{\alpha}$ exceeds its bounded limit it is capable of compensating for the disturbance and maintains the sliding mode through high-frequency switching. Also, in this version of QCSMC, the scalar value λ controls the rate of convergence during the reaching phase. In this sense, the λ term can be thought of as the tuning parameter for the reaching phase, while $\tilde{\alpha}$ is the tuning parameter for the sliding phase [27].

$$
\begin{aligned}
&1.\quad u = -\tilde{\alpha}\text{sign}(\sigma)\\
&2.\quad u = -\tilde{\alpha}\frac{\dot{\sigma} + \lambda|\sigma|^{1/2}\text{sign}(\sigma)}{|\dot{\sigma}| + \lambda|\sigma|^{1/2}}\\
&3.\quad u = -\tilde{\alpha}\frac{\ddot{\sigma} + 2\lambda^{3/2}\left(|\dot{\sigma}| + \lambda|\sigma|^{2/3}\right)^{-1/2}\left(\dot{\sigma} + \lambda|\sigma|^{2/3}\text{sign}(\sigma)\right)}{|\ddot{\sigma}| + 2\lambda^{3/2}\left(|\dot{\sigma}| + \lambda|\sigma|^{2/3}\right)^{1/2}}\\
&4.\quad u = -\tilde{\alpha}\frac{\varphi_{3,4}}{N_{3,4}} \quad \text{where}\\
&\quad \varphi_{3,4} = \dddot{\sigma} + 3\lambda^2\left[\ddot{\sigma} + \lambda^{4/3}\left(|\dot{\sigma}| + 0.5\lambda|\sigma|^{3/4}\right)^{-1/3}\left(\dot{\sigma} + 0.5\lambda|\sigma|^{3/4}\text{sign}(\sigma)\right)\right]\\
&\quad\quad \left[|\ddot{\sigma}| + \lambda^{4/3}\left(|\dot{\sigma}| + 0.5\lambda|\sigma|^{3/4}\right)^{2/3}\right]^{1/2}\\
&\quad N_{3,4} = |\dddot{\sigma}| + 3\lambda^2\left[|\ddot{\sigma}| + \lambda^{4/3}\left(|\dot{\sigma}| + 0.5\lambda|\sigma|^{3/4}\right)^{2/3}\right]^{1/2}.
\end{aligned}
\tag{14.18}
$$

14.3.3 High-Order Sliding-Mode Differentiators

Implementing the HOSM controller given by Eq. 14.17 requires observation or real-time differentiation of the $\dot{\sigma}, \ldots, \sigma^{(r-1)}$ or real-time differentiation. A proven methodology for observing these values is a high-order sliding-mode differentiator. It is described from the following [27]:

$$
\begin{aligned}
\dot{z}_0 &= v_0, \qquad v_0 = -\lambda L^{\frac{1}{k+1}} |z_0 - \sigma(t)|^{\frac{k}{k+1}} \operatorname{sign}(z_0 - \sigma(t)) + z_1 \\
\dot{z}_1 &= v_1, \qquad v_1 = -\lambda_{k-1} L^{\frac{1}{k}} |z_1 - v_0|^{\frac{k-1}{k}} \operatorname{sign}(z_1 - v_0) + z_2 \\
&\ \ \vdots \\
\dot{z}_{k-1} &= v_{k-1}, \qquad v_{k-1} = -\lambda_1 L^{\frac{1}{2}} |z_{k-1} - v_{k-2}|^{\frac{1}{2}} \operatorname{sign}(z_{k-1} - v_{k-2}) + z_k \\
\dot{z}_k &= \lambda_0 L \operatorname{sign}(z_k - v_{k-1}),
\end{aligned}
\tag{14.19}
$$

where $\lambda_1, \lambda_2, \ldots, \lambda_k > 0$ and $|\sigma^{k+1}(t)| \leq L$.

14.4 Simulation and Results

This section shows the application of the problem statement and methodologies. Then three case studies are considered. The first examines the impact of unmodeled dynamics on the simulated system while the last two studies examine the higher-order sliding-mode controller versus a traditional G&C structure for short- and long-range shots.

14.4.1 Integrated Single-Loop G&C Using HOSMC

This section integrates the methodologies introduced in Sect. 14.3 with the problem formulation from Sect. 14.2. As was shown previously, the IGC is attempting to guide and control the missile interceptor using a single control loop. Recall that the fully integrated kinematic-dynamic state-space representation from Eq. 14.9 had the form $\dot{\boldsymbol{x}} = \boldsymbol{a}(\boldsymbol{x}, t) + \boldsymbol{B}\boldsymbol{u}(t) + \boldsymbol{\rho}$. Note that no output was specified for this system.

As mentioned previously standard three-loop autopilots have several feedback states. As a control designer, more feedback states provide more options for defining the sliding variable. In previous works, Harl in [20] selected a sliding variable for their terminal second-order sliding-mode control law to guarantee a zero heading error at the time of impact using the predicted impact point (PIP), while Shima [17], uses zero-effort miss (ZEM) for his sliding variable. However, in this scenario, extreme limitations are going to placed on the system. A seeker measuring the relative transversal velocity is considered the *only* feedback state. The sliding variable is simply defined as

$$
\sigma = v_n. \tag{14.20}
$$

This is the most direct choice and is based explicitly on the intercept strategy from Eq. (14.10). Recall the block diagram illustrating the single-loop structure is shown in Fig. 14.3. With the sliding variable selected, three Lie derivatives are taken to determine the input/output relative degree r_d. This results in

$$\sigma^{(3)} = h(t, x) + g(t, x)\delta, \tag{14.21}$$

where

$$g(t, x) = -a_{M_z} \sin(\lambda - \theta)\frac{\bar{q}Sl}{I_{yy}}C_{m_\delta}.$$

Based on the work of [28], a third-order quasi-continuous sliding-mode controller

$$u = -\tilde{\alpha}\frac{\ddot{\sigma} + 2\lambda^{3/2}(|\dot{\sigma}| + \lambda|\sigma|^{2/3})^{-1/2}(\dot{\sigma} + \lambda|\sigma|^{2/3}\text{sign}(\sigma))}{|\ddot{\sigma}| + 2\lambda^{3/2}(|\dot{\sigma}| + \lambda|\sigma|^{2/3})^{1/2}s} \tag{14.22}$$

is selected to force v_n to the sliding surface [27]. Though the control is continuous through the reaching phase, it is still discontinuous on the sliding set from Eq. 14.16. In order to mitigate the high-frequency switching, a variation to the sliding variable can be made. Instead of designing about $u = \delta$, the control can be taken as $u = \dot{\delta}$. This improves the smoothness of the end effector such that $\delta = \int udt$, but it also increases the relative degree of the system ($r_d = 4$). This can be changed by transforming the sliding variable σ into a new sliding variable S such that

$$S = \dot{\sigma} + c\sigma. \tag{14.23}$$

This sliding variable also has the added benefit of partial dynamical collapse. If designed correctly this can effectively serve as a first-order filter for the state variables. It is apparent that this system is back to a 3-sliding-mode controller ($r = 3$) and can still use the same quasi-continuous sliding-mode controller developed in Eq. 14.22, but with S in place of σ. This results in the control effort

$$u = -\tilde{\alpha}\frac{\ddot{S} + 2\lambda^{3/2}(|\dot{S}| + \lambda|S|^{2/3})^{-1/2}(\dot{S} + \lambda|S|^{2/3}\text{sign}(S))}{|\ddot{S}| + 2\lambda^{3/2}(|\dot{S}| + \lambda|S|^{2/3})^{1/2}s}. \tag{14.24}$$

This changes the control surface so that v_n and $\dot{v}_n$ are both driven to zero in finite time. Though the system is of 3-sliding mode, it now requires three exact differentiations of the output σ. This is accomplished with a high-order sliding-mode differentiator developed in Sect. 14.3.3. Recall, that for a HOSMC, the acceleration of the target (a_{T_n}) is treated as a disturbance on the system. As long as it is bounded and smooth, the HOSMC will be able to handle it with finite-time convergence.

14.4.2 Case Studies

In this section, the quasi-continuous sliding-mode controller (QCSMC) is evaluated as a control technique for missile interception through numerical simulation. The evaluation consists of two simulations. The first simulation analyzes the effects of actuator dynamics, because they are not considered in the development of the controller. The second simulation illustrates the advantages and disadvantages of the controller over a more traditional G&C setup.

14.4.2.1 Simulation Setup

An outbound maneuvering target is pursued by a missile interceptor. At the onset of the simulation, the missile exits its midcourse homing phase and begins its terminal G&C scheme. It is traveling at a speed of 500 m/s. The missile is out of propellant and considered constant mass. Drag, however, continues to slow the vehicle. Furthermore, the vehicle is limited to less than ± 10 g of accelerations and the actuators to $\pm 20°$. A summary of the interceptor's parameters and initial conditions are provided in Tables 14.1 and 14.4, respectively.

At one second into the simulation, the target, traveling at 100 m/s, begins its evasive maneuvering. It weaves in a ± 6 g sinusoid trying to fool the interceptor. The parameters and initial conditions for the target are summarized in Table 14.2. The target does not have mass or moments of inertia because it is following a prescribed weaving motion.

The simulation is run using Euler numerical integration at a time step of 0.0001 s. This value is required for precise differentiation of state feedback. The quasi-continuous sliding-mode controller values used throughout all three experiments are summarized in Table 14.3. The baseline proportional navigation PID (PNPID) G&C values are summarized in Table 14.5. These are only used in Experiments I and II. These tables provide all the necessary information to repeat each of the simulations.

14.4.2.2 Simulation Test I: The Effects of Actuator Dynamics and Sigmoid on HOSMC

In the first test, the effects of unmodeled dynamics of varying speed are considered, and an approximation function is introduced to help. Unmodeled dynamics are dynamical elements included in a real-world or simulated system that are not considered in the development of the controller. In this study, first-order actuator dynamics are the unmodeled dynamics, as they are not considered in the development of the controller. For every degree of unmodeled dynamics, a degree is added to the input–output relative degree; assuming the unmodeled dynamics occur within the input–output definition. Not accounting for the increase in relative degree can adversely affect the controller's performance.

There are two ways of dealing with the unmodeled dynamics. The first method is to compensate for the dynamics by increasing the degree of the HOSM controller [29]. The second is to use an approximating the sign$(\cdot)$ function. In this study, the HOSM controller's degree is not increased, because doing so requires an additional estimate of the sliding variable. The controller introduced in Eq. 14.23 already has an additional time derivative of the original sliding surface ($\dot{\sigma}$).

The approximating function chosen in this study was the sigmoid$(\cdot)$ function

$$\text{sigmoid}(S) = \frac{S}{|S| + \epsilon}. \quad (14.25)$$

Remark 14.4.1 The tunable parameter ϵ is for smoothing the control effort. It is tuned through simulation observations. The goal is to make $\epsilon > 0$ as small as possible while also achieving control without chattering. The replacement of sign(S) function by the sigmoid operator in Eq. 14.25 provokes a boundary layer about the sliding mode $S = 0$ that contains the motion of the sliding variable. This is a common attempt at achieving a reasonable trade-off between a sliding mode in $S = 0$, with its insensitivity to the matched perturbation and elimination of control chattering, while losing the insensitivity to the matched perturbation. Note that there is no guarantee that control chattering will be eliminated, but rather transformed into "smoother" oscillations with a lower frequency. However, usually, it is possible to find (via simulations) a reasonably small $\epsilon > 0$ that achieves this trade-off.

This study considers the effects of the unmodeled dynamics on different speed actuators and examines the impact of using the sigmoid(S) function.

The first-order actuators modeled include slew rate limits and angular limits. A slow, fast, and instantaneous actuator are considered. The slow actuator has a time constant of 0.1 s, the fast actuator has a time constant of 0.01, and actuator dynamics are neglected in the case of the instantaneous actuator. The slew rate for the slow and fast actuator is limited to 600° per second, and the actuator stroke for all three is limited to ±20°. These values are based on commercial off-the-shelf parts representative of what is used on current missile interceptors. Besides the actuator speeds, all other parameters are identical. More detail can be found in Tables 14.1, 14.2, 14.3 and 14.4 in the appendix.

A target at a range of 4 km undergoing 6 g sinusoidal maneuvers beginning one second into the time of flight was simulated for all three actuator speeds. Figure 14.8 shows the actuator motion for all three actuator speeds. Despite the significant difference in actuator motion, all three achieved miss distances under 9 cm. It is clear that the oscillation amplitudes exhibited in Fig. 14.8 increase as the actuator speed is lowered. This is due to the slower actuator having more difficulty tracking the discontinuities brought on by sign(S) near the sliding surface leading to overshoot. The faster actuator also experiences overshoot, but at a much smaller amplitude. Though the actuator speeds render little difference in the performance of the system (e.g., miss distance, sliding variable trajectory, etc.) the oscillatory nature makes it impractical on a real vehicle. Next, the sigmoid$(\cdot)$ function is introduced ($\epsilon = 3$) with identical target parameters. The vehicles were simulated and the actuator positions were compared. The actuator motion in Fig. 14.9 is clearly reduced for the high-speed actuator and slightly for the slower actuator. The faster actuator benefits more from the sigmoid$(\cdot)$ function because of the time it dwells near the sliding surface $S = 0$. This is clear when considering Fig. 14.8. This effectively eliminates the oscillations from the fast actuator, making it a practical solution. The sliding variable for all three cases is shown in Fig. 14.10. It is clear from this that the beginning

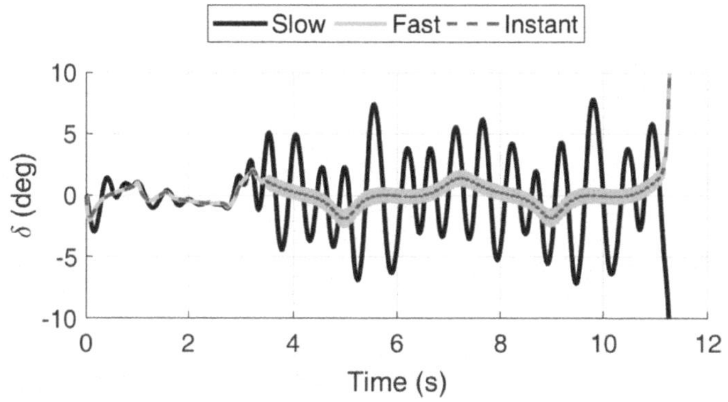

Fig. 14.8 The fin position time history for the slow, fast, and instantaneous actuator cases during a missile interception at end-game using the sign function

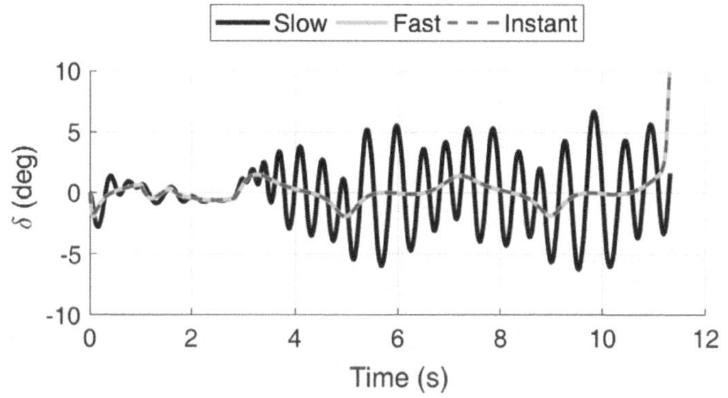

Fig. 14.9 The fin position time history for the slow, fast, and instantaneous actuator cases during a missile interception at end-game using the sigmoid function ($\epsilon = 3$)

of the sinusoidal maneuvers began at one second. Also, the faster actuator has no problems constraining the sliding variable to the surface despite the 6 g maneuvers of the target. The slower actuator is not able to keep up. However, both actuators managed to keep the normal velocity component near zero (Fig. 14.11) resulting in miss distances under 10 cm for all three cases. All of the subsequent tests use the sigmoid approximation and use an $\epsilon = 3$.

14.4.2.3 Simulation Test II: PNPID Versus QCSMC at a 3 km Range

In this test, the QCSMC is compared to an established G&C setup. This setup consists of a proportional navigation (PN) guidance system and a PID autopilot controller. For the sake of brevity, this system is abbreviated PNPID. From a topological standpoint,

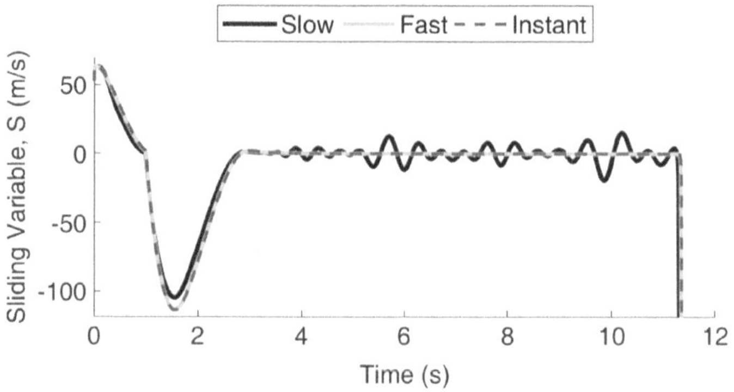

Fig. 14.10 The sliding variable time history for the slow, fast, and instantaneous actuator cases during a missile interception at end-game using the sigmoid function ($\epsilon = 3$)

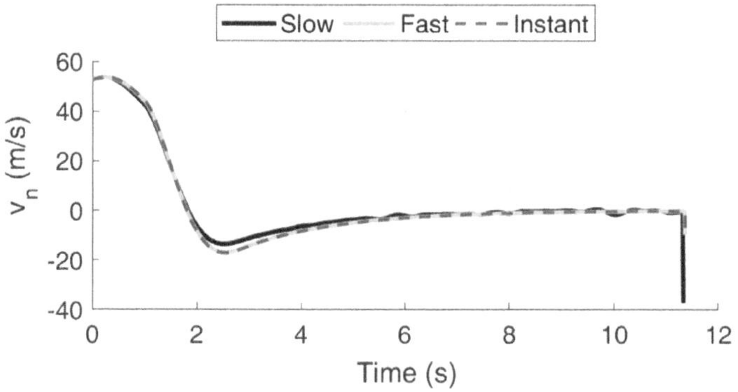

Fig. 14.11 The normal velocity relative to the target for the slow, fast, and instantaneous actuator cases during a missile interception at end-game using the sigmoid function ($\epsilon = 3$)

the two systems are quite different. The PNPID requires state feedback from the line-of-sight rate, closing velocity, body acceleration, and body angular rates (Fig. 14.1), while the QCSMC only requires the normal velocity relative target (Fig. 14.3). This lack of information immediately puts the QCSMC at a disadvantage.

The actuator dynamics used in both cases is the same as the *fast* case from simulation test I. Also, the same controller settings for the QCSMC from simulation test I are used. The proportional navigation guidance uses $N = 4$ and the PID controller uses proportional values of $K_o = -1/15$ on the outer loop and $K_i = -100$ on the inner loop.

At a distance of 3 km, the two controllers are simulated. The QCSMC outperforms the PNPID with a miss distance of 9.7 cm compared to the PNPID's 1117 cm. The PNPID is considered a miss because it is over the 1-meter hit threshold. The trajectory for each is shown in Fig. 14.12. An explanation for this difference begins by

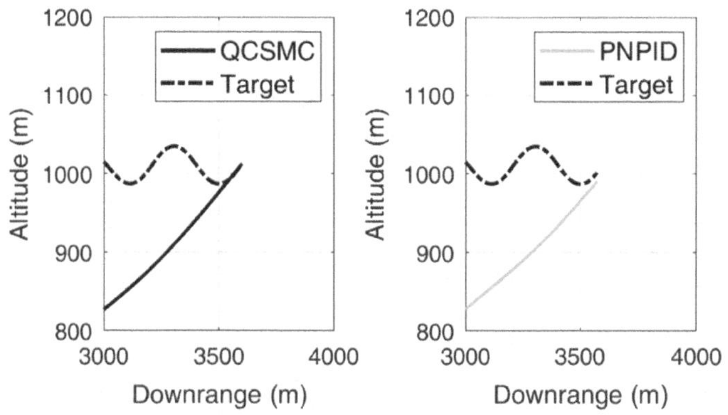

Fig. 14.12 The normal velocity relative to the target for the QCSMC and PNPID controllers during the 3 km scenario at end-game

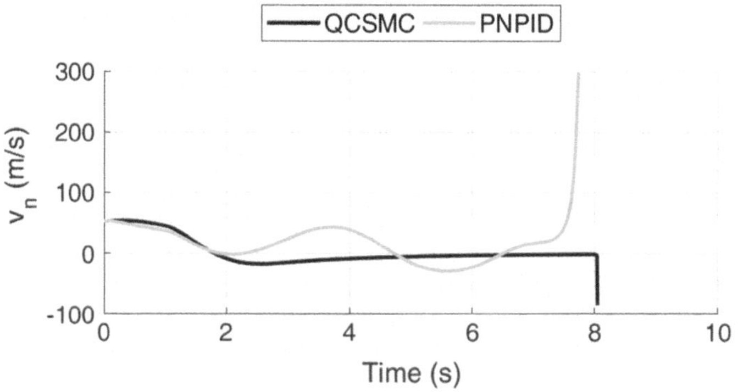

Fig. 14.13 The normal velocity relative to the target for the QCSMC and PNPID controllers during the 3 km scenario at end-game

comparing the normal velocity with respect to the target for each case (Fig. 14.13). The PNPID controller does not react to the 6 g maneuvering as early or aggressively as the QCSMC. This allows the relative normal velocity between the missile and target to wander. As the missile gets closer the target, the PNPID attempts to negate the normal velocity, but at this point, it requires a significant amount of acceleration to achieve this gap. This is shown clearly in Fig. 14.14 where the PNPID reaches the 10 g limit of the aircraft. Evidence to the intensity of the maneuver can also be seen in the fin deflections from Fig. 14.15. Even if the PNPID controller was not limited by acceleration as shown by the dashed line in Fig. 14.14, the vehicle would still have a miss distance of 825 cm.

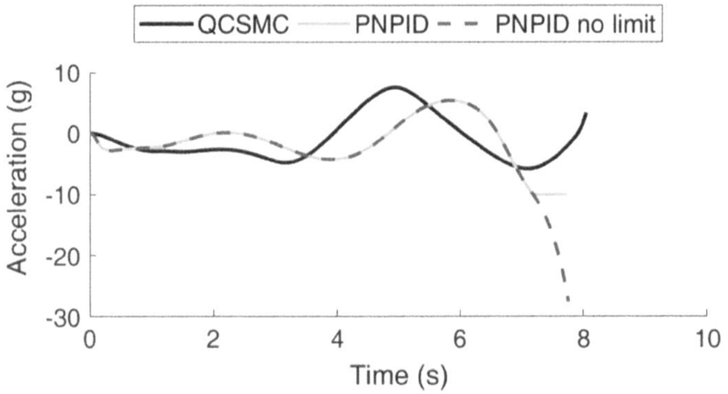

Fig. 14.14 The body acceleration of the missile for the QCSMC and PNPID controllers during the 3 km scenario at end-game

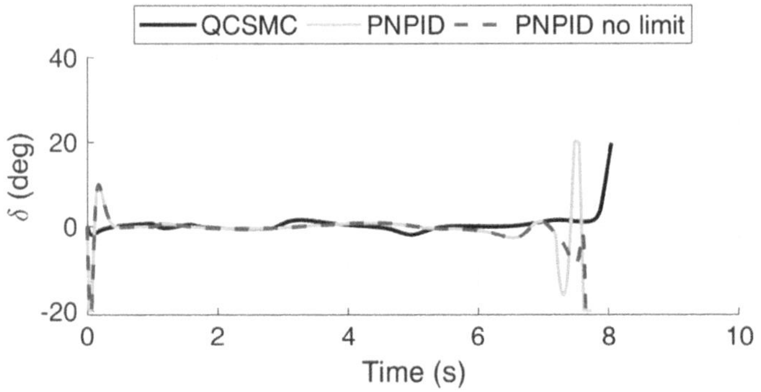

Fig. 14.15 The fin position time history for the QCSMC and PNPID controllers during the 3 km scenario at end-game

The large body rotations required at end-game are not ideal for ordinance effectiveness nor its kinematic capability. Separation in the aerodynamic flow can lead to reduced effectiveness of the missile. Furthermore, the large maneuver reduces the velocity of the missile (Fig. 14.16) resulting in a lower dynamic pressure and less capability.

14.4.2.4 Simulation Test III: PNPID Versus QCSMC at a 6 km Range

In this test, the range to target is increased from 3 to 6 km while all other variables are kept the same as simulation test II. At this distance, the missile has further to travel and is closer to midcourse than end-game. The trajectories for both cases are shown in Fig. 14.17. Comparing these trajectories, it is clear that the PNPID gets within the

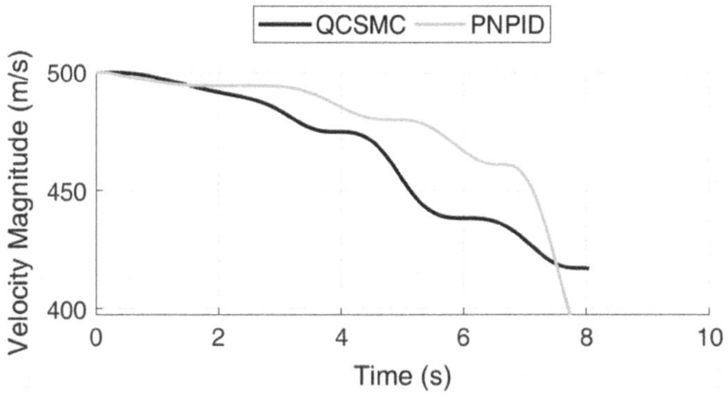

Fig. 14.16 The velocity magnitude of the missile for the QCSMC and PNPID controllers during the 3 km scenario at end-game

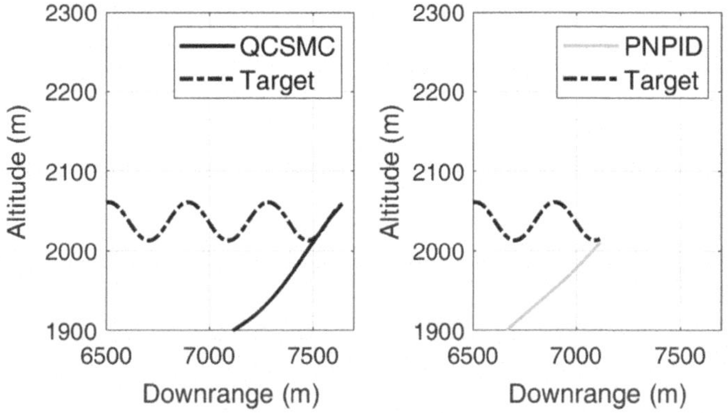

Fig. 14.17 The trajectory of the target and missile interceptor to the target for the QCSMC and PNPID controllers during the 6 km scenario at end-game

vicinity of the target earlier than the QCSMC. The PNPID arrives at 15.2 s while the QCSMC arrives 5.5 s later at 20.7 s.

In the case of the PNPID, it does not attempt aggressively react to the target's sinusoidal oscillations at long range. This is clear by the growth of the acceleration profile shown in Fig. 14.19. During this time, the relative normal velocity in Fig. 14.18 is allowed to wander about zero at the frequency equal to the target's maneuver. By the time the proportional navigation's acceleration command ramps up to negate the relative normal velocity—it is too late. The acceleration required greatly exceeds the vehicle's 10 g capability. This results in a miss distance of 482 cm. Even when the 10 g limitation of the vehicle is ignored, the miss distance only improves to 374 cm.

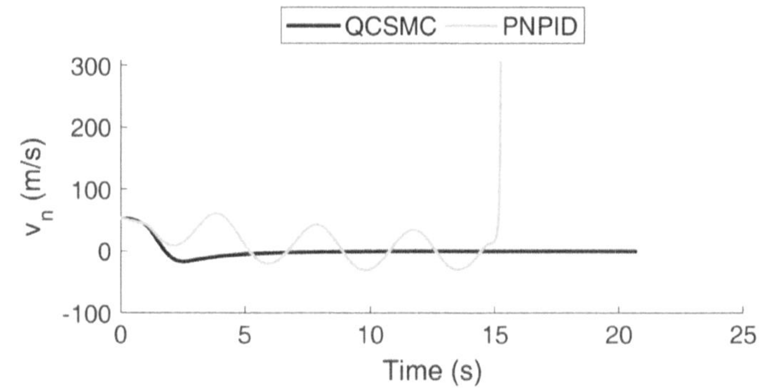

Fig. 14.18 The normal velocity relative to the target for the QCSMC and PNPID controllers during the 6 km scenario at end-game

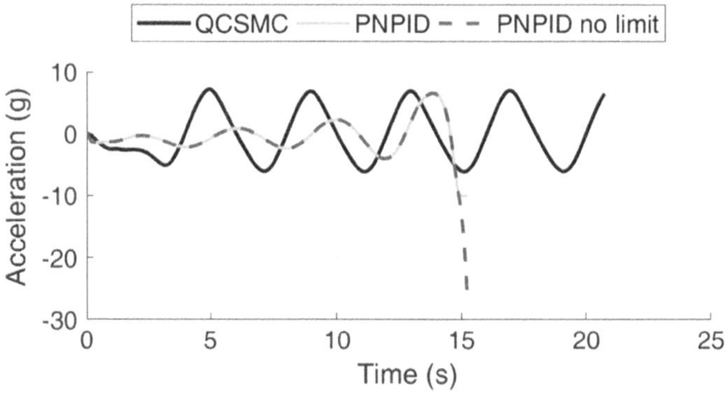

Fig. 14.19 The body acceleration of the missile for the QCSMC and PNPID controllers during the 6 km scenario at end-game

Conversely, the QCSMC consistently attempts to negate the relative normal velocity in Fig. 14.18, caused by the 6 g sinusoidal maneuver of the target. This produces consistently high peak accelerations early in flight (Fig. 14.19). Though, by negating the relative normal velocity earlier, the acceleration requirements do not grow out of control at closer geometries. This results in a miss distance of 9.3 cm, but delays the intercept by 5.5 s.

It is important to note that the QCSMC in this form should not be used as a long-range guidance strategy solution. The velocity magnitude for this engagement shown in Fig. 14.20 illustrates how much velocity is sapped by negating the relative normal velocity early in flight. Each hump in the curve corresponds to a crest in the target maneuver. If the vehicles had been further apart, the missile may not have had enough dynamic pressure to ensure a hit at end-game. Though the PNPID controller better conserves its kinetic energy, there is still a rapid depletion of it at end-game. In a more advanced aerodynamic simulation, the large rotations required can cause major problems.

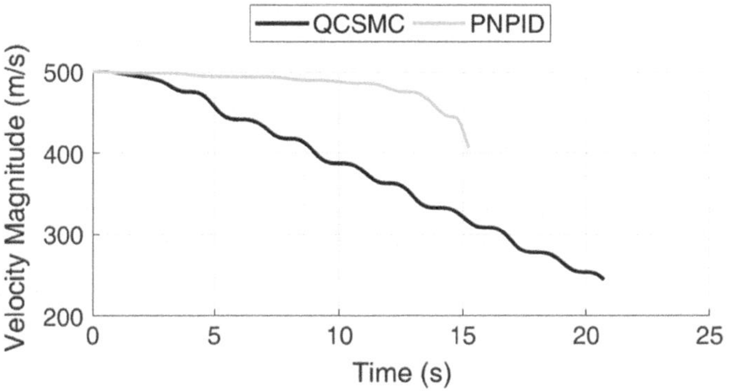

Fig. 14.20 The velocity magnitude of the missile for the QCSMC and PNPID controllers during the 6 km scenario at end-game

14.4.3 Conclusion

In this chapter, a single-loop high-order quasi-continuous sliding mode controller was developed using only the relative degree of the system and a bounded disturbance value for the target's maximum acceleration. With only this information and state feedback for the normal relative velocity, a controller was designed and then tested against a nonlinear model. Despite its lack of knowledge of the vehicle, target, and feedback states, it was able produce smaller miss distances than a traditional PNPID system. Furthermore, it was also shown how this strategy could be adapted to work on a practical vehicle with the introduction of the sigmoid function without significant loss in performance. As such, it is a good solution for developing a guidance and control technique quickly with minimal information.

Although the control strategy was successful it was not optimal. At distances far from the target, the interceptor attempted to negate the relative normal velocity between the missile and target due to the sinusoidal target maneuvers. At long range, this can unnecessarily saps too much of its kinetic energy, limiting its performance. For future work, augmenting the sliding variable may provide flexibility in solving this issue. Another technique that could be fruitful is the use of adaptive continuous high-order sliding-mode controller. These techniques should be investigated for this application in the future.

Appendix

Tables provided in this section allow for the recreation of Experiments I, II, and III.

Table 14.1 Missile and kinematic parameters

Parameter	Symbol	Value	Units
Actuator angular limit	$\delta_{\max}$	±20	deg
Actuator slew rate limit	$\dot{\delta}_{lim}$	±600	deg/s
Fast actuator time constant	τ_{fast}	0.01	
Slow actuator time constant[a]	τ_{slow}	0.01	
Body acceleration limit	$a_{M,\text{limit}}$	±10	g
Moment of inertia	I_{yy}	1000	$\frac{\text{kg m}^2}{\text{s}}$
Empty mass	m	100	kg
Aerodynamic normal force with respect to A.o.A	C_{N_α}	0.11	
Aerodynamic pitch stability	C_{M_α}	−0.01	
Aerodynamic pitch control derivative	C_{M_δ}	−0.015	
Aerodynamic pitch damping	C_{M_q}	−0.001	
Aerodynamic surface reference	S	0.7854	m^2
Time step	Δt	0.0001	s

[a]Only used in Experiment I

Table 14.2 Target parameters and initial conditions

Parameter	Symbol	Value	Units
Flight path angle	γ_T	0	deg
Velocity	V_T	100	$\frac{\text{m}}{\text{s}}$
Maneuver amplitude	a_T	±6	g
Maneuver period	T_T	4	s
Start maneuver	t_{start}	1	s

Table 14.3 Quasi-continuous sliding-mode controller (QCSMC) parameters

Parameter	Symbol	Value
Magnitude parameter	$\tilde{\alpha}$	100
Convergence parameter	λ	8
Dynamic collapsing term	C	2
Sigmoid smoothing parameter	ϵ	3

Table 14.4 Missile initial conditions

Parameter	Symbol	Value	Units
Initial range to target	r_0	4[a], 3[b], 6[c]	km
Initial angle of attack	α_0	0	deg
Initial line-of-sight	λ_0	20	deg
Initial flight path angle	γ_0	10	deg
Initial velocity	V_0	500	$\frac{\text{m}}{\text{s}}$

[a]Used in Experiment I
[b]Used in Experiment II
[c]Used in Experiment III

Table 14.5 Proportional navigation PID (PNPID) controller parameters

Parameter	Symbol	Value
Outer-loop proportional gain	$K_{o,p}$	0.0667
Inner-loop proportional gain	$K_{i,p}$	−25
Guidance constant	N	4

References

1. Zarchan, P.: Tactical and Strategic Missile Guidance. Progress in Astronautics and Aeronautics, vol. 199. American Institute of Aeronautics and Astronautics, Reston (2002)
2. Janus, J.P.: Homing Guidance (A Tutorial Report). Technical report, DTIC Document (1964)
3. Murtaugh, S.A., Criel, H.E.: Fundamentals of proportional navigation. IEEE Spectr. **3**(12), 75–85 (1966)
4. Guelman, M.: A qualitative study of proportional navigation. IEEE Trans. Aerosp. Electron. Syst. **AES-7**(4), 637–643 (1971)
5. Shukla, U.S., Mahapatra, P.R.: The proportional navigation dilemma-pure or true? IEEE Trans. Aerosp. Electron. Syst. **26**(2), 382–392 (1990)
6. Ohlmeyer, E.J., Phillips, C.A.: Generalized vector explicit guidance. J. Guid. Control Dyn. **29**(2), 261–268 (2006)
7. Palumbo, N.F., Blauwkamp, R.A., Lloyd, J.M.: Basic principles of homing guidance. Johns Hopkins APL Tech. Dig. (Appl. Phys. Lab.) **29**(1), 25–41 (2010)
8. Ohlmeyer, E.J.: Root-mean-square miss distance of proportional navigation missile against sinusoidal target. J. Guid. Control Dyn. **19**(3), 563–568 (1996)
9. Lin, C.F., Wang, Q., Spayer, J.L., Evers, J.H., Cloutier, J.R.: Integrated estimation, guidance, and control system design using game theoretic approach. In: American Control Conference, pp. 3220–3224. IEEE (1992)
10. Menon, P.K., Vaddi, S., Ohlmeyer, E.: Finite-horizon robust integrated guidance-control of a moving-mass actuated kinetic warhead. In: AIAA Guidance, Navigation, and Control Conference and Exhibit, p. 6787 (2006)
11. Menon, P.K., Ohlmeyer, E.J.: Integrated design of agile missile guidance and autopilot systems. Control Eng. Pract. **9**(10), 1095–1106 (1999)
12. Palumbo, N.F., Jackson, T.D.: Integrated missile guidance and control: a state dependent Riccati differential equation approach. In: Proceedings of the 1999 IEEE International Conference on Control Applications, vol. 1, pp. 243–248. IEEE (1999)
13. Menon, P.K., Sweriduk, G., Ohlmeyer, E.J., Malyevac, S.: Integrated guidance and control of moving-mass actuated kinetic warheads. J. Guid. Control Dyn. **27**(1), 118–126 (2004)
14. Hwang, T,W., Tahk, M.J.: Integrated backstepping design of missile guidance and control with robust disturbance observer. In: International Joint Conference on SICE-ICASE, pp. 4911–4915. IEEE (2006)
15. Shkolnikov, I.A., Shtessel, Y.B., Lianos, D.: Integrated guidance-control system of a homing interceptor- Sliding mode approach. In: AIAA Guidance, Navigation, and Control Conference and Exhibit, (August) (2001)
16. Shtessel, Y.B., Shkolnikov, I.A.: Integrated guidance and control of advanced interceptors using second order sliding modes. In: Proceedings. 42nd IEEE Conference on Decision and Control, vol. 5, pp. 4587–4592. IEEE (2003)
17. Shima, T., Idan, M., Golan, O.M.: Sliding-mode control for integrated missile autopilot guidance. J. Guid. Control Dyn. **29**(2), 250–260 (2006)
18. Idan, M., Shima, T., Golan, O.M.: Integrated sliding mode autopilot-guidance for dual-control missiles. J. Guid. Control. Dyn. **30**(4), 1081–1089 (2007)

19. Koren, A., Idan, M., Israel, T.: Integrated sliding mode guidance and control for a missile with on – off actuators. J. Guid. Control. Dyn. **31**(1) (2008)
20. Harl, N., Balakrishnan, S.N., Phillips, C.: Sliding mode integrated missile guidance and control. In: AIAA Guidance, Navigation, and Control Conference, number August, pp. 7741 (2010)
21. Harl, N., Balakrishnan, S.N.: Reentry terminal guidance through sliding mode control. J. Guid. Control Dyn. **33**(1), 186–199(2010)
22. Harl, N., Balakrishnan, S.N.: Impact time and angle guidance with sliding mode control. IEEE Trans. Control Syst. Technol. **20**(6), 1436–1449 (2012)
23. Pamadi, B.N.: Performance, Stability, Dynamics, and Control of Airplanes. AIAA, Reston (2004)
24. Utkin, V.: Sliding Modes in Optimization and Control Problems. Springer, Berlin (1992)
25. Utkin, V., Guldner, J., Shi, J.: Sliding Mode Control in Electro-Mechanical Systems. CRC Press, Boca Raton (2009)
26. Shtessel, Y.B., Edwards, C., Fridman, L., Levant, A.: Sliding Mode Control and Observation. Birkhäuser Basel, Boston (2014)
27. Levant, Arie: Higher-order sliding modes, differentiation and output-feedback control. Int. J. Control **76**(9–10), 924–941 (2003)
28. Levant, A.: Quasi-continuous high-order sliding-mode controllers. In: Proceedings of the 42nd IEEE Conference on Decision and Control, vol. 5, pp. 4605–4610. IEEE (2003)
29. Panathula, C.B., Rosales, A., Shtessel, Y.B., Fridman, L.M.: Closing gaps for aircraft attitude higher order sliding mode control certification via practical stability margins identification. IEEE Trans. Control Syst. Technol. (2017)

The manufacturer's authorised representative in the EU is Springer Nature Customer Service Centre GmbH, Europaplatz 3, 69115 Heidelberg, Germany. If you have any concerns regarding our products, please contact ProductSafety@springernature.com

Printed and bound by CPI Group (UK) Ltd, Croydon, CR0 4YY
30/09/2026
02225602-0016